More Gene Manipulations in Fungi

Contributors

Margaret Alic
Flora Banuett
Saloman Bartnicki-Garcia
J. W. Bennett
Allen D. Budde
John H. Burnett
Lorna A. Casselton
Meryl A. Davis
Luc Giasson
Gerhard Gierz
N. Louise Glass
Michael H. Gold
Michael J. Hynes
Anne R. Kubelik
Myra B. Kurtz
Linda L. Lasure
Landry Le Chevanton
Sally A. Leong
Ian A. J. Lorimer
Charles P. Novotny
Berl R. Oakley
Stephen A. Osmani
John F. Peberdy
David D. Perkins
Marguerite Picard
Patricia J. Pukkila
Peter J. Punt
J. Antoni Rafalski
Jean-Luc Rossignol
Stewart Scherer
Eric U. Selker
Paul L. Skatrud
Daniel Z. Skinner
Charles A. Specht
Mary M. Stankis
William E. Timberlake
Scott V. Tingey
Robert C. Ullrich
Robert F. M. Van Gorcom
Cees A. M. J. J. van den Hondel
John G. K. Williams
Huiling Yang
Denise Zickler

More Gene Manipulations in Fungi

Edited by

J. W. Bennett

Department of Biology
Tulane University
New Orleans, Louisiana

Linda L. Lasure

Panlabs, Incorporated
Bothell, Washington

1991

ACADEMIC PRESS, INC.
HARCOURT BRACE JOVANOVICH, PUBLISHERS
San Diego New York Boston
London Sydney Tokyo Toronto

This book is printed on acid-free paper. ∞

Academic Press, Inc.
San Diego, California 92101

United Kingdom Edition published by
Academic Press Limited
24–28 Oval Road, London NW1 7DX

Library of Congress Cataloging-in-Publication Data

More gene manipulations in fungi / edited by J.W. Bennett, Linda L. Lasure.
p. cm.
Includes index.
ISBN 0-12-088642-1
1. Fungi--Genetic engineering. 2. Fungi--Genetics. 3. Fungal molecular biology. I. Bennett, J. W. II. Lasure, Linda L.
QK602.M67 1991
589.2'0415--dc20 91-11605
CIP

PRINTED IN THE UNITED STATES OF AMERICA
91 92 93 94 9 8 7 6 5 4 3 2 1

Contents

2 Predicting the Molecular Basis of Mycelial-Yeast Dimorphism with a New Mathematical Model of Fungal Morphogenesis

Saloman Bartnicki-Garcia and Gerhard Gierz

II. Tools of the Trade

3 Cloning and Analysis of Fungal Genes

William E. Timberlake

4 Molecular Karyotype Analysis of Fungi

Daniel Z. Skinner, Allen D. Budde, and Sally A. Leong

III. Molecular Morphogenesis

IV. Meiosis, Mating Types, and Other Mysteries of Fungal Sex

11 Repeat-Induced Point Mutation and DNA Methylation

Eric U. Selker

12 *Ascobolus immersus* and *Podospora anserina:* Sex, Recombination, Silencing, and Death

Jean-Luc Rossignol and Marguerite Picard

13 *Sordaria macrospora:* The Transition to the Age of Gene Manipulation

Landry Le Chevanton and Denise Zickler

V. Mycotechnology

14 Fungal Protoplasts

John F. Peberdy

15 Genetics and Molecular Biology of the Lignin-Degrading Basidiomycete *Phanerochaete chrysosporium*

Margaret Alic and Michael H. Gold

16 Molecular Genetics of Human Fungal Pathogens

Myra B. Kurtz and Stewart Scherer

17 Molecular Biology of the Beta-Lactam-Producing Fungi

Paul L. Skatrud

18 Heterologous Gene Expression in Filamentous Fungi

Cees A. M. J. J. van den Hondel, Peter J. Punt, and Robert F. M. Van Gorcom

VI. Appendices

Appendix A Genetic Analysis with RAPD Markers

John G. K. Williams, Anne R. Kubelik, J. Antoni Rafalski, and Scott V. Tingey

Appendix B Growth Media

J. W. Bennett and Linda L. Lasure

Appendix C Genetic Coding Conventions

Linda L. Lasure and J. W. Bennett

Contributors

Numbers in parentheses indicate the pages on which the authors' contributions begin.

Margaret Alic (319), Department of Chemical and Biological Sciences, Oregon Graduate Institute of Science and Technology, Beaverton, Oregon 97006

Flora Banuett (217), Department of Biochemistry and Biophysics, University of California, San Francisco, California 94143

Saloman Bartnicki-Garcia (27), Department of Plant Pathology, University of California, Riverside, California 92521

J. W. Bennett (441, 459), Department of Cell and Molecular Biology, Tulane University, New Orleans, Louisiana 70118

Allen D. Budde (87), USDA-ARS Plant Disease Resistance Research Unit, Madison, Wisconsin 53706

John H. Burnett, Oxford University, Oxford OX1 3PF, England

Lorna A. Casselton (126), School of Biological Sciences, Queen Mary and Westfield College, University of London, London E1 4NS, England

Meryl A. Davis (151), Department of Genetics, The University of Melbourne, Parkville, Victoria 3052, Australia

Luc Giasson (234), Biotechnology Laboratory, University of British Columbia, Vancouver, British Columbia, Canada V6T 1W5

Gerhard Gierz (27), Department of Mathematics and Computer Science, University of California, Riverside, California 92521

N. Louise Glass (193), Department of Botany and Biotechnology Laboratory, University of British Columbia, Vancouver, British Columbia, Canada V6T 1W5

Michael H. Gold (319), Department of Chemical and Biological Sciences, Oregon Graduate Institute of Science and Technology, Beaverton, Oregon 97006

Michael J. Hynes (151), Department of Genetics, The University of Melbourne, Parkville, Victoria 3052, Australia

Anne R. Kubelik (432), Central Research and Development Department, E.I. du Pont de Nemours and Company, Inc., Wilmington, Delaware 19880

Myra B. Kurtz (342), Infectious Disease Research, Merck Sharp & Dohme Research Laboratories, Rahway, New Jersey 07065

Linda L. Lasure (441, 459), Panlabs, Incorporated, Bothell, Washington 98011-8805

Landry Le Chevanton (291), Institut de Génétique et Microbiologie, Université Paris-Sud, 91405 Orsay, Cedex, France

Sally A. Leong (87), USDA-ARS Plant Disease Resistance Research Unit, and Department of Plant Pathology, University of Wisconsin, Madison, Wisconsin 53706

Ian A. J. Lorimer (193), Department of Biochemistry and Biotechnology Laboratory, University of British Columbia, Vancouver, British Columbia, Canada, V6T 1W5

Charles P. Novotny (234), Department of Microbiology and Molecular Genetics, University of Vermont, Burlington, Vermont 05405

Berl R. Oakley (107), Department of Molecular Genetics, Ohio State University, Columbus, Ohio 43210

Stephen A. Osmani (107), Department of Cell Biology, Baylor College of Medicine, Houston, Texas 77030

John F. Peberdy (307), Department of Botany, University of Nottingham, Nottingham NG7 2RD, England

David D. Perkins (3), Department of Biological Sciences, Stanford University, Stanford, California 94305

Marguerite Picard (266), Institut de Génétique et Microbiologie, Université Paris-Sud, 91405 Orsay, Cedex, France

Patricia J. Pukkila (126), Department of Biology and Curriculum in Genetics, University of North Carolina, Chapel Hill, North Carolina 27599

Peter J. Punt (396), TNO Medical Biology Laboratory, 2280 AA Rijswijk, The Netherlands

J. Antoni Rafalski (431), Agricultural Products Department, E.I. du Pont de Nemours and Company, Inc., Wilmington, Delaware 19880

Jean-Luc Rossignol (266), Institut de Génétique et Microbiologie, Université Paris-Sud, 91405 Orsay, Cedex, France

Stewart Scherer (342), Department of Microbiology, University of Minnesota Medical School, Minneapolis, Minnesota 55455

Eric U. Selker (258), Institute of Molecular Biology, University of Oregon, Eugene, Oregon 97403

Paul L. Skatrud (364), Department of Molecular Genetics, Lilly Research Laboratories, Indianapolis, Indiana 46285

Daniel Z. Skinner (87), USDA-ARS and Department of Agronomy, Kansas State University, Manhattan, Kansas 66506

Charles A. Specht (234), Department of Botany, University of Vermont, Burlington, Vermont 05405

Mary M. Stankis (234), Department of Microbiology and Molecular Genetics, University of Vermont, Burlington, Vermont 05405

William E. Timberlake (51), Departments of Genetics and Plant Pathology, University of Georgia, Athens, Georgia 30602

Scott V. Tingey (431), Agricultural Products Department, E.I. du Pont de Nemours and Company, Inc., Wilmington, Delaware 19880

Robert C. Ullrich (234), Department of Botany, University of Vermont, Burlington, Vermont 05405

Robert F. M. Van Gorcom (396), TNO Medical Biology Laboratory, 2280 AA Rijswijk, The Netherlands

Cees A. M. J. J. van den Hondel (396), TNO Medical Biology Laboratory, 2280 AA Rijswijk, The Netherlands

John G. K. Williams (431), Central Research and Development Department, E.I. du Pont de Nemours and Company, Inc., Wilmington, Delaware 19880

Huiling Yang (234), Department of Botany, University of Vermont, Burlington, Vermont 05405

Denise Zickler (291), Institut de Génétique et Microbiologie, Université Paris-Sud, 91405 Orsay, Cedex, France

George Beadle and Edwin Tatum, Stockholm, 1958.

Foreword

Some books are to be tasted, others to be swallowed, and some few to be chewed and digested.

Francis Bacon (1625)

One duty of a historian is said to be to record 'significant' change.

G. C. Ainsworth (1976)

I have no hesitation whatsoever in commending *More Gene Manipulations in Fungi,* which admirably complements its predecessor, *Gene Manipulations in Fungi,* as marking a watershed in the development of mycology, and I am equally certain that it is a book to be studied "with diligence and attention."

Science advances in two ways: through the rare quantum leaps in understanding of original minds, and through the application of techniques that those discoveries suggest or make possible. Since the publication of Watson and Crick's (1953) seminal proposals for the structure of DNA and its implications for genetics, there has been a spectacular yet steady development of techniques for their study and application to biological problems.

In one sense, the 1980s marked the beginning of the repayment of the debt incurred by genetics and biochemistry to the classic studies of Lindegren (1931), Winge (1942), and Beadle and Tatum (1941). Those pioneers demonstrated, unequivocally, both the utility and advantages of fungi for these disciplines and, in part, directed their development into the analysis of cellular function.

In the ensuing half century, the one-to-one relationship between genes and proteins and the colinearity of the genetic material, first demonstrated for eukaryotes in a fungus, coupled with the understanding of the structure of DNA, have led to the chemical circumscription of genes and the ability to isolate them from one cell type and then transfer them either to a similar, or to a quite different cell type, wherein they can function. That knowledge has enabled living matter to be manipulated at the most fundamental level in ways that were unimaginable fifteen years ago, let alone fifty. Now these methods have begun to be applied to fungi.

The eighteen contributions in this volume demonstrate both the range of problems now capable of investigation through the use of these new genetic manipulative methods, as well as the nature of the techniques themselves and their effectiveness. The editors have chosen their topics and authors to illustrate how these techniques can be used to dissect the cellular biology of the fungi and so to reformulate a range of fundamental problems in the integration and development of both the mycelium and of the basis of reproductive behavior. Some of the techniques available and some of the biotechnological applications that they have rendered possible will surely play ever increasing and novel roles in the longstanding relationship between the life of man and molds.

Like all good scientific accounts, the topics in this volume record only partial solutions, but each is a stimulant to new thinking. As is the way of scientific progress, once one issue is resolved, other, more intriguing, unexpected problems are revealed. But the contributions in this book are not simply probings, increasingly narrow in scope, into topics of inverse interest to the depth of the analysis. Many of them raise, in a new form, some of the most fundamental problems of the fungal cell and its attributes. In some cases the interest is almost exclusively mycological, for example, the application of an ingenious mathematical model to the molecular basis of the widespread phenomenon of mycelial–yeast dimorphism.

At last there is some further progress with the seventy-year-old puzzle of the fundamental nature of basidiomycete mating types: a problem that bedeviled those concerned with understanding them in the 1950s and 1960s because of their intractability to analysis by the methods then available. Yet, although their structure is becoming clearer and there is hope of elucidating their mode of action, their origin and evolution has become even more difficult to imagine. Equally surprising is the apparent simplicity of the dimictic mating system of the Ascomycetes, which has now been shown to mask an astonishing and, at present, somewhat inscrutable complexity. But there is the possibility of exciting biological generalizations ahead. There is still a long way to go, but the presence of DNA-binding regions in the mating type factors of Ascomycetes, a smut fungus, and other Basidiomycetes suggests a basis for a commonality of function in the mating process. The excitement of detecting such apparently universal features suggested by these findings is reinforced more explicitly in those contributions where structures, such as tubulin and microtubules, or functions, such as regulatory circuits, all of which are common to all living organisms, are examined.

Other contributions assess the present "state of the art," for example, the prospect for cloning and analysis of fungal genes, still not free from technical problems; the new horizons that the application of molecular methods has already brought, or are bringing to fungal genetics; and that least studied and exploited of subjects, the fungal karyotype. New ways and opportunities are described with protoplasts, lignin degradation, pathogens, antibiotics, and synthesis in novel cellular environments.

Inevitably, in dealing with such a diversity of new applications, the fungi of choice have been those already fairly well known genetically, but there is no doubt that it is now only a matter of time before it will be possible to apply the new techniques to other fungi: that is one of their great attractions. Not surprisingly, therefore, one of the most significant contributions to this volume is a plea to widen the horizons of study beyond the bounds of the genetically known fungi, a plea all the more significant, coming as it does from one whose career has been devoted almost wholly to *Neurospora!* But that is only right and when it comes about, as assuredly it will, this admirable volume will continue to be both a guide and stimulant for the new mycologists of the 21st century.

Sir John H. Burnett
Oxford, England

Preface

Fungi are eukaryotic, heterotrophic organisms with an absorptive mode of nutrition. Most fungi are both multicellular and multinucleate, with rigid chitinous cell walls, and usually exhibit mycelial or yeast-like growth habits. In recent years, in recognition of the unique nutritional and morphological attributes of fungi, taxonomists have elevated their taxonomic ranking to a separate kingdom, "The Fifth Kingdom," on equal footing with plants and animals.

The economic importance of fungi is a reflection of their diverse life forms and metabolic virtuosity. Yeast is the most widely used industrial microorganism; in addition, it is the premier model system for eukaryotic molecular biology. Filamentous fungi are major agents of plant pathogenesis and biodegradation. Fungi also cause diseases in animals. Their role as human pathogens is increasing, a reflection of the spread of AIDS and the widespread use of immunosuppressive drugs. In traditional biotechnology, fungi are major producers of enzymes, amino acids, and pharmacologically active secondary metabolites; in the new biotechnology, both yeasts and molds are efficient hosts for heterologous gene expression. The diverse life cycles and reproductive strategies found within the Fifth Kingdom have long fascinated biologists, and several species have been developed as model systems for the analysis of recombination.

Gene Manipulations in Fungi was published in 1985. The book was written at a time when yeast molecular biology had just come of age, while the molecular study of filamentous fungi was in its infancy. It was aimed at scientists who were otherwise "separated by discipline, organism, geography, and objective" who, nevertheless, shared "common goals in applying modern genetic techniques to fungi." The book appeared at an interface between the decline of pure classical fungal genetics and the ascent of modern genetic engineering, and reflected the excitement of a young and promising discipline.

Since 1985, "molecular mycology" has come of age. As has often been the case in molecular biology, progress has been even faster than predicted.

Saccharomyces cerevisiae remains a formidable model system, but the technical gap betwen yeasts and molds is not now as wide as it once was. Transformation systems and genomic libraries are available for a number of fungal species. Gene cloning is now commonplace, and the genes for many unique fungal products have been sequenced, expressed, and engineered. Both yeasts and molds have been developed as expression hosts for industrial fermentations. Important plant and animal pathogens that had proven intractable to classical genetic analysis are succumbing to molecular techniques. Fungal mating type genes continue to pique the curiosity of scientists and expand our theories about the evolution of sexual recombination.

It is clearly time for another book, despite the knowledge that any volume on fungal molecular biology will be somewhat out-of-date from the moment it appears. We have asked our authors to review state-of-the-art research, with an intent to inform the community of scientists about what is possible when fungal systems are analyzed using the tools of molecular biology. *More Gene Manipulations in Fungi* is a current reference, with the texture of a "work in progress."

We are particularly pleased that our book is appearing in 1991, the fiftieth anniversary of the publication of Beadle and Tatum's "one gene–one enzyme" theory. Beadle and Tatum had started their studies on the biochemical effects of the gene by using eye color mutants in *Drosophila.* Their decision to work with the bread mold *Neurospora crassa* introduced many of the basic techniques of microbial genetics, which later became commonplace in bacteria during the golden age of *Escherichia coli.* Their experiments with nutritionally deficient mutants of *N. crassa* became the cornerstone of modern biochemical genetics. Some historians of science date the birth of molecular biology to the publication of the one gene–one enzyme hypothesis. Others (e.g., Gunther Stent) view it as a culmination of classical theory that merely rendered "valuable midwife service" to the birth of molecular biology. However categorized, Beadle and Tatum's research represented a turning point in the experimental analysis of gene function. In 1958 they shared the Nobel Prize in Medicine or Physiology with Joshua Lederberg. Our frontispiece photograph shows them standing side by side at the Nobel Award ceremonies in Stockholm. *More Gene Manipulations in Fungi* is a testament to the continuing strength of fungi as experimental tools in modern biology.

J. W. Bennett
Linda L. Lasure

Acknowledgments

Our thanks to Jean Thomson-Black and Phyllis Moses at Academic Press for encouraging us to do a new book along the lines of our 1985 monograph *Gene Manipulations in Fungi*. We also thank David Perkins, Bob Metzenberg, and Bill Timberlake for their suggestions as to content, format, and philosophy.

A number of our peers reviewed chapters for us. Special thanks go to John Fincham, Nancy Keller, Paul Lehmann, Bob Metzenberg, Ron Morris, and Eric Selker for their help.

Patricia Lawler Crickenger provided help in many dimensions. Her love of language, her impeccable spelling and typing, and her ready humor were all important during the various stages of putting together this book. She died unexpectedly on December 11, 1990, before she could see the finished product to which she had contributed so much.

Lori Velimesis also gave cheerful and competent help along the way.

Finally, J. W. B. thanks Shubha Kale and Toby Feibelman for keeping things going in the lab, and John, Dan, and Mark Bennett for being supportive at home.

I

The Fifth Kingdom: Diversity and Dimorphism

1

In Praise of Diversity

DAVID D. PERKINS
Department of Biological Sciences
Stanford University
Stanford, California

If the Lord Almighty had consulted me before embarking upon the creation, I should have recommended something simpler.

—Alfonso X of Castile, 13th century (Attributed)

Haldane, dining at Oxford with a group of theologians, is said to have been asked what can be concluded about the nature of the Creator from a study of his creation. His legendary response, "An inordinate fondness for beetles," reflected the fact that more species of beetles have been described than of any other order of organism. Other creatures have since been suggested as rivals to the Coleoptera for the Deity's favors, based not only on species diversity but on other criteria such as biomass (May, 1989). To my knowledge, no one has yet come forward to propose the fungi. Yet any mycologist in Haldane's position would respond that the fungi must be among God's favorite organisms. Fungi cannot compete on a numerical

MORE GENE MANIPULATIONS
IN FUNGI

basis: only 100,000 named species compared with over 300,000 beetles. What is the secret of their appeal?

It could be argued that the Deity's fondness for fungi is reflected in the many beautiful and intricate adaptations that have intrigued both amateurs and professionals: a harpoon that discharges to penetrate and inoculate an impinging rotifer (Fig. 1); a morphogenetic response that directs spore discharge precisely toward a point of light (see Ingold, 1971); a launching platform (Fig. 2); an intricate cup optimally designed to ensure splash dispersal by raindrops (Fig. 3); the reprogramming of host-plant development by which *Ustilago violacea* transforms female flowers of *Silene* into male and replaces pollen with teliospores in the anthers (see Day and Garber, 1988).

Before Darwin, adaptations such as these were rationalized as revealing the beauty of creation and the wisdom of the creator. For a modern biologist, they must be explained in terms of evolution, without which "nothing in biology makes sense" (Dobzhansky, 1942; Simpson, 1947). This is a more exacting task than that of the creationist. The modern biologist's view, that every living thing is the product of 2000 million years of experimentation, is no less awe inspiring than the world view of pre-Darwinian theology. During all that vast time, organisms have been engaged in genetic manipulations that put to shame anything we can think of doing in the laboratory.

Every organism now alive, without exception, can boast an equally remarkable lineage: not one of its myriad ancestors since the origin of life has ever died before reproducing! We eukaryotes all trace back to a common ancestor who invented meiosis. All our pedigrees are equally ancient. The adaptations that were essential for the survival and reproductive success of our progenitors in each generation were the product not of design but of "evolutionary tinkering," as Jacob (1977) has aptly called it. Opportunities for tinkering have been no fewer for fungi than for insects, angiosperms, or vertebrates. Every trait, be it molecular, morphological, or behavioral, has a historical basis (see Delbrück, 1949). The divergent results of all this tinkering in the various lineages are what makes biology so challenging and so endlessly fascinating.

The appeal of fungi for biologists may have come initially not only from their adaptations but also from their strangeness. They seem to be so different from us and from the plants and animals with which we are immediately familiar that they might be thought to have come from another planet. The basic plan of filamentous fungi is a microscopic cylinder that grows at the tip and becomes multicellular by septation. The linear hyphal growth confers an important advantage for life in or on solid substrates. Unlike unicellular forms, the organism can actively invade fresh territory rather than remain stranded as a colony on unfavorable or ex-

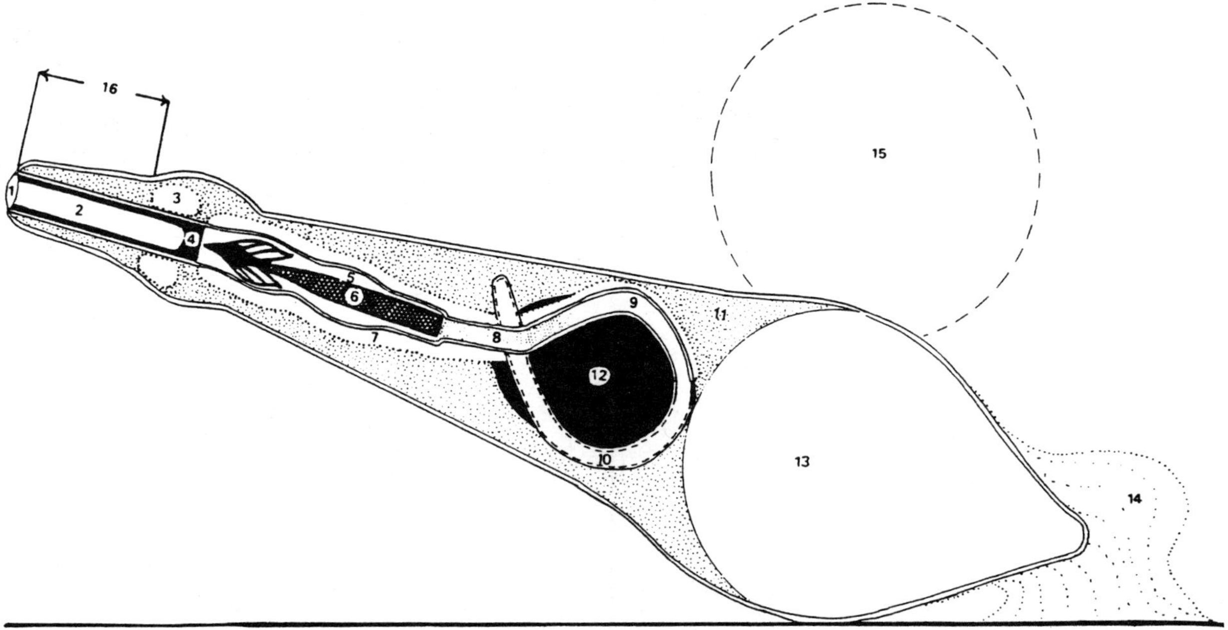

Fig. 1. Diagram of an undischarged gun cell of *Haptoglossa mirabilis.* 1. Muzzle. 2. Bore with extra wall layer. 3. Apical vacuolar system. 4. Plug. 5. Projectile chamber. 6. Projectile. 7. Lateral vacuolar system. 8–10. Lower flexuous tube. 11. Protoplasm. 12. Nucleus. 13. Basal vacuole. 14. Adhesive pad. 15. Empty zoospore cyst. 16. Apical cone. Reproduced, with permission, from Barron (1987).

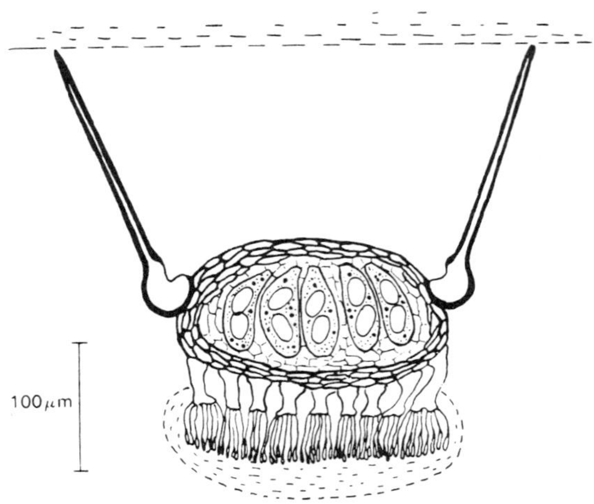

Fig. 2. Vertical section of the cleistothecium of the powdery mildew *Phyllactinia guttata (corylea)* showing two of the five or more "leg" appendages. Each leg is a unicellular thorn-like structure with a bulb at the proximal end. The inner half of the bulb has a thin wall so that, on drying, the appendages change shape, freeing the cleistothecium from surrounding mycelium by levering it from its original prone position on the undersurface of a hazel leaf into the position shown. The appendages then function as vanes to orient the cleistothecium as it falls. Mucilage secreted by special cells sticks the cleistothecium to a target surface, with appendages outward. After overwintering, the cleistothecium fractures at the equator and hinges open in orientation for discharge of the two-spored asci from the inverted hemisphere (Cullum and Webster, 1977; Webster, 1979; see Fig. 150C of Webster, 1980). Reproduced, with permission, from Ingold (1971).

hausted substrates (see Metzenberg, 1979). Formation of complex multicellular structures entails septation, aggregation, and cohesion of hyphae rather than the repeated cell division seen in plants and animals (see, for example, Read and Beckett, 1985). Morphogenesis may involve further septation and differentiation of the aggregates. Fungi succeed in using this seemingly primitive mode of development to form precise and intricate macroscopic structures. At the microscopic level, differentiated fungal cells can rival in complexity the specialized cells of plants, animals, and protists.

In contrast to the overt strangeness of fungi, we now know that the fungal genetic apparatus—nucleus, chromosomes, and genes—is typically eukaryotic, and that fungal meiosis, including recombination, is essentially identical to that of plants and animals. Despite the many surface differences, the fungi are undoubtedly our kin. Their genetic hardware is basically the same as ours. Their developmental programming is refreshingly different.

In contrast to mycologists and students of natural history, most geneticists probably have been drawn to the fungi for utilitarian rather than

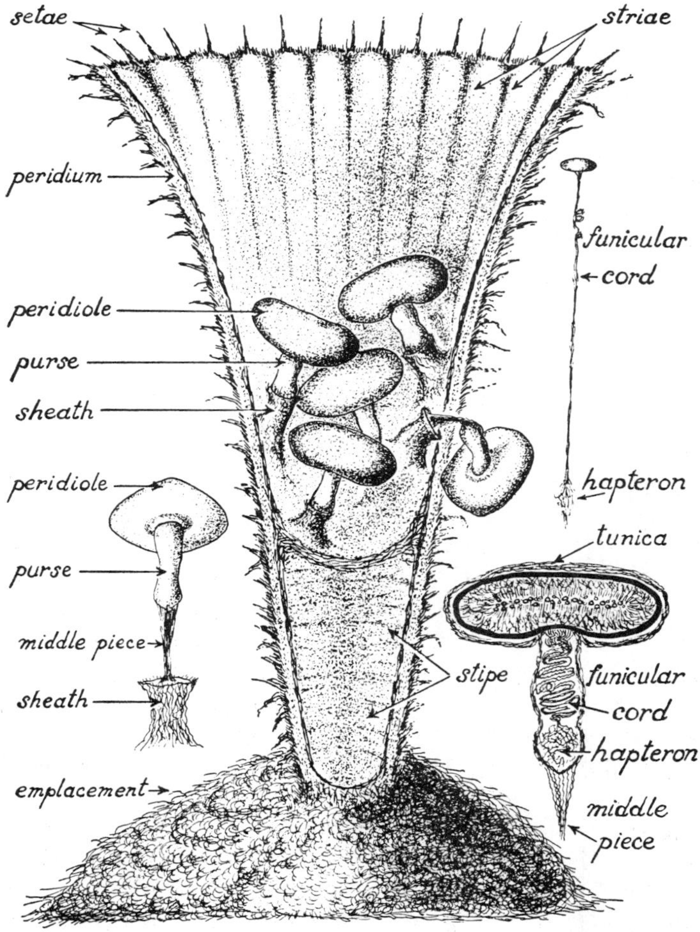

Fig. 3. Section of the fruit body of the bird's nest fungus *Cyathus striatus*. Raindrops cause the spore-bearing peridioles to be thrown as much as 1.2 m. The peridiole is splashed out of the cup, releasing the sticky hapteron. When the hapteron adheres to a plant in the line of flight it causes the cord to be wound around a stem, securing the peridiole where it can be eaten by an herbivore. Central figure magnified ~ 7.5×. Reproduced, with permission, from Brodie (1951).

aesthetic reasons. Many fungi are blessed with well-known advantages for genetic research. The usual litany of favorable features would cite haploidy, low DNA content, short generation times, large numbers of sexually derived progeny [10^{12} basidiospores for a large puffball, by Buller's (1909) estimate], ease of cultivation and preservation, simplicity of nutritional requirements, production of propagules suitable for plating, rapid growth (5 mm/hr for *Neurospora*), ability to grow as heterokaryons, and recoverability of all four products of individual meioses. In some fungi, autonomously expressed differences in ascospore color or morphology permit visible detection of rare recombinational events or signal the presence of chromosome rearrangements. It would be difficult for a student of recombination to design an experimental system more perfect than an eight-spored ascus in which each ascospore contains the information that was present earlier in one of the eight polynucleotide strands (half-chromatids) at the consummation of meiotic recombination.

Fungal genetics has been largely laboratory bound since its inception. Almost all the variants used in genetic, biochemical, and molecular studies have been obtained as single-gene mutants that were induced and selected in the laboratory. Rarely have phenotypically scorable genetic markers been obtained as variants from nature, and only now are restriction fragment length polymorphisms (RFLPs) beginning to be widely used. There have been exceptions, of course. Studies of mating type, vegetative incompatibility, pathogenicity, and meiotic drive have mostly employed naturally occurring variants (for other examples, see Day and Garber, 1988; Perkins and Turner, 1988). However, genetic manipulations with auxotrophs and other clearly scored induced markers made possible the landmark discoveries of molecular genetics reviewed by Fincham (1985) in his introductory chapter to the previous volume of *Gene Manipulations in Fungi.* Induced mutations have proved equally indispensable for the genetic dissection of complex traits such as circadian rhythms (see Dunlap, 1990).

Past neglect of natural variation and of evolutionary problems cannot be attributed entirely to the fact that most practitioners of fungal genetics were not trained as mycologists or ecologists. Many evolutionary problems require that a base of genetic information be available before the species concerned can be investigated effectively. Establishing that basic information requires a long period of laboratory research. Also, new developments in evolutionary and molecular genetics have made it possible to formulate many questions more effectively. The essential empirical and theoretical base for fungal evolutionary genetics now seems to be in place.

The spectacular successes of laboratory-based genetics for cellular and molecular biology have no doubt diverted attention from the organisms themselves and from the naturally occurring genetic manipulations that were involved in their evolution. This chapter concerns some of these neglected aspects. The discussion will focus on ascomycetes and basidiomycetes disproportionately, because they are the groups for which genetic knowledge is most advanced.

I. GENETICS OF NATURALLY OCCURRING DIVERSITY

Evolution has produced both adaptations and species. Adaptations have been forged by natural selection acting on genetic differences within species. Over and above the incorporation of adaptations, genetically discrete populations have differentiated into separate species.

Genetic analysis of the diversity has not been easy. Intraspecific variation is predominantly cryptic or polygenic in nature. Although crosses can readily be made, genetic analysis is not simple because most of the variant genes are not clearly distinguishable phenotypically. On the other hand, interspecific diversity, while it may be more obvious, is usually inaccessible by conventional genetic analysis because crosses cannot be made.

II. INTRASPECIFIC DIVERSITY

What we know of natural populations of outbreeding fungal species suggests that they are fundamentally similar in make-up to populations of diploid plants and animals, in which gene pools are characterized by balanced gene complexes of regulatory and modifier genes whose individual effects are small and by many loci showing genetic polymorphism. [Genetic polymorphism (Ford, 1940) is defined as the occurrence in a sexually reproducing population of two or more genotypes in frequencies such that the rarest cannot be accounted for merely by recurrent mutation.] Because most intraspecific variability is hidden, genetic polymorphisms long went unrecognized by geneticists, while taxonomists defined species in terms of type specimens. Naturally occurring allelic variants, so-called isoalleles, were occasionally identified in well-studied organisms but they remained a curiosity until Lewontin and Hubby (1966) and Harris (1966) used protein polymorphisms to show dramatically that a large fraction of loci are polymorphic in *Drosophila* and in humans. A similar high level of genetic polymorphism was subsequently found in *Neurospora* (Spieth, 1975), demonstrating enormous stores of genetic variability that previously had gone unsuspected.

The discovery that polymorphisms are maintained in a haploid organism such as *Neurospora* raised questions about their adaptive significance. A haploid vegetative phase makes it difficult, but not impossible, to evoke the hypothesis of heterozygote advantage. Several other hypotheses have been proposed. Genetic polymorphisms might permit adjustment to heterogeneous environments (the hypothesis favored by Spieth) and they might buffer the population against onslaughts by parasites (see, for example, Bremerman, 1980; Hamilton *et al.,* 1990).

Most of our information on genetic polymorphisms in fungi has come from isozyme differences. Although no one appears to have made a systematic study of RFLPs among individuals within the same local population, thousands of nucleotide differences are known to exist among *Neurospora crassa* strains from different localities (Metzenberg *et al.,* 1984). Most of these cryptic intraspecific differences either are adaptively neutral or have individual effects that are small and difficult to demonstrate. In contrast, a major effect may be exerted by the genetically polymorphic vegetative incompatibility genes, which effectively preclude heterokaryon formation in natural populations of *N. crassa* (Mylyk, 1976) and restrict the transfer of cytoplasmic factors, affecting pathogenicity in *Cryphonectria parasitica* (Anagnostakis, 1988) and *Ophiostoma ulmi* (Brasier, 1984).

In diploid organisms, deleterious recessive genes are expected to be present in sheltered condition in genetically polymorphic outbreeding populations. The counterparts of these detrimental recessives in predominantly haploid fungi are phase-specific genes that are expressed in the dikaryotic phase of the life cycle or the brief diplophase. Wild strains of *N. crassa* carry a sizable load of recessive alleles that have serious deleterious effects when they are homozygous in a cross (Leslie and Raju, 1985). Inbreeding in a species containing deleterious recessives, even those with minor effects, would be expected to result in reduced fecundity; this indeed has been shown to occur in *N. crassa* (Raju *et al.,* 1987).

III. SPECIES AND SPECIES DIVERSITY

No single species, no matter how generalized and polymorphic, can be adapted for efficient exploitation of more than a small range of the diverse niches available on earth. For example, no single species could exploit the environmental resources used by a hummingbird and those used by a pelican. If individuals adapted for the two niches were interfertile and shared the same gene pool, their recombinant progeny would be grotesque and unable to live in either world. Reproductive discontinuity is essential. Dobzhansky (1950) has defined a species as the largest, most inclusive population of sexual and cross-fertilizing individuals that share in a common gene pool. This concept seems as valid for fungi as for animals and plants. From this perspective, each species is a unique evolutionary experiment. Once speciation has occurred, a new species is free to go its own evolutionary way. Thus the diversity between discrete species can eventually exceed by far any variation that sets off one individual from another within the same species.

With the advent of molecular biology, it has become possible to identify in detail many previously hidden ways in which closely related species differ from one another (see, for example, Aquadro, 1990). Tests for homology and function can readily be made across species, but knowledge of the genetic differences that exist between related species does not reveal the process of speciation, because present differences may have arisen subsequent to speciation or may have been incidental rather than intrinsic to the process.

IV. SPECIATION IN FUNGI

Speciation in fungi has been considered from different points of view by Burnett (1983), Kemp (1985), and Brasier (1987). The discussion that follows presents a somewhat different perspective. I shall begin by outlining a conceptual framework that has been influenced by recent advances in the genetics of speciation in plants and animals, and especially by the ideas of Carson (1982, 1985, 1987, 1989). Fungi that have adopted different breeding systems will then be examined. Pseudohomothallic species will receive special consideration because pseudohomothallism depends on heterokaryosis and is therefore unique to fungi. An epilogue will consider the role of species used as model organisms. Practical aspects of species identification will be largely ignored, not out of any disrespect for the importance of taxonomy but to avoid possible confusion between problems of evolutionary mechanism and problems of cataloguing.

A. Origin of Species

Carson (1982, 1985) has proposed a unified theory of species formation for both plants and animals. His scenario appears to apply equally well to fungi. Carson's theory proposes that gene-pool reorganization is the fundamental feature of speciation, whereas reproductive isolation is a secondary development. The theory concerns primarily the origin of new cross-fertilizing species from parental populations that are likewise cross-fertilizing. This is because evidence from both animals and plants indicates that speciation occurred most abundantly in lineages of outbreeding biparental species rather than in lines that had abandoned cross-fertilization. Carson's main points are summarized in the following paragraphs.

Each well-established cross-fertilizing species possesses a unique gene pool that has evolved over many generations into a complex, highly organized, polygenic system that involves many regulatory and modifier genes. The components of the system are balanced and held in the gene pool by stabilizing selection.

Formation of new species occurs by subdivision and branching rather than by gradual transformation of the entire population as has often been assumed by paleontologists. Two successive events must occur in order for a new species to originate from an older one, in Carson's view—first disorganization of the gene pool, then reorganization. In the first phase, the original balanced gene pool or a subsegment of it is disorganized stochastically, by chance events. Drastic reduction in population size is required; this usually occurs when a small daughter population has become spatially separated from the original population. Disorganization of the small gene pool may involve founder effects, shifting balance, or hybridization between discordant genomes. [Rapid disorganization can result from activation of transposable elements (McClintock, 1978).] The resultant gene pool is poorly integrated. The disorganization phase, resulting in a nascent new species, can occur rapidly, in relatively few generations.

Reorganization, the second phase of species formation, is slow, encompassing many generations during which adaptations are gradually formed and integrated by mutation, selection, and recombination. The adaptations are both internal, to the changed genetic environment (Mayr, 1954), and external, to the ambient environment. The first 100 to 1000 generations are thought to be crucial. Eventually the new gene pool comprises a coadapted system that is molded to a new niche.

Carson's organization theory of speciation owes much to Sewall Wright's shifting balance theory of evolutionary change based on his adaptive-landscape model (Wright, 1932; see Dobzhansky, 1970, pp. 24–28). In Wright's theory, gene combinations that enable organisms to inhabit certain environments are represented by adaptive peaks of unequal height. Unfavorable gene combinations that render their possessors unfit are symbolized by adaptive valleys. The number of adaptive peaks is great and they are discontinuous. Each species occupies a different adaptive peak. Speciation occurs when a subpopulation deviates sufficiently in genetic make-up that it starts moving up the slope to a new peak.

B. Significance of Reproductive Isolation

Zoologists and botanists have tended to differ in their thinking about speciation. In animals, reproductive isolation has usually been emphasized as an essential feature of species formation. In plants, reproductive isolation has been less emphasized because many good species are capable of hybridizing in nature but nevertheless maintain their integrity as separate species. In fungi, as in animals, reproductive isolation commonly has been viewed as a prerequisite for speciation (see, for example, Burnett, 1983; Perkins and Raju, 1986; Brasier, 1987; Perkins and Turner, 1988). It may be appropriate now to reconsider assumptions about reproductive isolation in the context of Carson's theory, which proposes that reproductive isolation is not a necessary precondition of species formation but rather an incidental by-product of the essential genomic reorganization.

Fertility barriers are not an essential prerequisite for speciation if, as is thought, species originate allopatrically, in geographically isolated subpopulations. For many new species, reproductive isolation may indeed be one of the initial results of reorganization. For other new species, however, reorganization may not have resulted in reproductive isolation before the new species comes in contact again with its progenitor species. Carson points out that when interspecies crosses occur in such a situation, they are not a threat to the integrity of well-established species because introgressive gene flow in zones of contact does not result in disorganization of large integrated gene pools. Anderson (1949) has provided many examples of plant species between which introgression has occurred without disrupting species integrity.

Even when reorganization has not initially resulted in reproductive isolation, mechanisms that impede gene flow between species are expected eventually to accumulate. Reproductive isolation thus remains a generally useful criterion—a sufficient though not necessary condition for designating populations as separate species.

C. Species in Fungi with Different Breeding Systems

Species concepts rooted in population genetics have been developed almost exclusively for plants and animals. It will be of interest to know whether new insights into species formation may be provided by fungi as a result of their unique characteristics and their very different mode of life.

Consideration of the species concept would be greatly simplified if all fungi were heterothallic. However, fungi display a diverse array of breeding systems. We are confronted not only with heterothallic species but also with homothallic and pseudohomothallic species and with a myriad of fungi that appear to have abandoned sexual reproduction.

The last category is genetically trivial. The deuteromycetes ("Fungi Imperfecti") are vegetatively propagated asexual clones that have been described as taxonomic species on the basis of morphology. Some are no doubt single-mating-type strains of heterothallic species (our knowledge is imperfect). Many are probably truly asexual (the life cycle of the organism is imperfect). The origin of asexual derivatives from fully sexual ancestors can be visualized readily. Mutations impairing meiosis or other stages of the sexual phase are extremely common. Because vegetative growth in most fungi is potentially immortal and because vegetative propagules are produced, mutational loss of sexuality can produce a new clonally reproducing lineage that is at no immediate disadvantage. The asexual fungus may thrive in its given environment. Acquisition of simple morphological differences could then qualify it for taxonomic description as a new asexual (anamorphic) species. Such lineages, consisting of individuals that have abandoned sexual reproduction and no longer pool their genes, are thought to be phylogenetic dead ends (Wagner, 1970; Carson, 1982) in the sense that phyletic change within them is limited to the gradual accumulation of mutations since they are denied the opportunity to generate new species by the quantum jumps that result when speciation occurs by genomic reorganization in biparental sexual organisms (see Stanley, 1975).

Much confusion might be avoided if the term species were reserved for sexually reproducing populations and asexual lineages were given another name.

Homothallic fungi, in contrast to the deuteromycetes, are still sexual, retaining karyogamy and meiosis although sacrificing the advantages of genetic heterogeneity and polymorphism that accompany biparental reproduction and outbreeding. The abundance of homothallic fungal species testifies to a positive selective value of sexual reproduction. Without selection, sexuality would quickly be lost because surely hundreds of genes are needed in concert for the realization of karyogamy and meiosis, and each of these genes is subject to mutational impairment. The ability of homothallic strains to complete the sexual cycle rapidly, without seeking a mate, cannot itself explain the retention of sex, because rapid propagation can also be accomplished vegetatively. Why, then, is the sexual cycle retained?

An important immediate advantage of sex is thought to be its ability to purge the genome of deleterious mutations. Sexual reproduction recombines and exposes different combinations of genes to natural selection. Mutation at one or another locus has a high probability of occurring during each generation, and most new mutations that are not neutral are deleterious. If mutations at different loci act synergistically in their adverse effects on fitness, retention of the sexual stage can confer a substantial selective advantage because meiotic assortment and recombination prevent accumulation of the deleterious mutations (Kondrashov, 1988; Charlesworth, 1990). This would favor retaining sexual reproduction in homothallic fungi, provided that genetically unlike nuclei are able to come together in meiosis. In fact, crosses involving unlike nuclei have long been known to occur in homothallic species such as *Aspergillus nidulans, Sordaria fimicola, Sordaria macrospora,* and *Neurospora africana,* in which the segregation of markers from crossed perithecia has been used for genetic analysis in the laboratory. An additional advantage has been realized by species in which ascospores or basidiospores serve as long-lived resistant propagules.

Studies of mating type genes have shown that the genetic and molecular basis of homothallism in *Neurospora* and its relatives is quite different from that in the facultatively homothallic yeasts *Saccharomyces cerevisiae* and *Schizosaccharomyces pombe* (Glass and Lorimer, 1991).

D. Pseudohomothallism as an Adaptation Unique to Fungi

In many genera, some species are heterothallic and have opted for outbreeding, while others are homothallic and highly, if not completely, inbred. A third breeding type is also found repeatedly, both in ascomycetes and in basidiomycetes. This is pseudohomothallism ("secondary homothallism"), which requires and maintains heterokaryon formation.

Pseudohomothallic species are basically heterothallic. They possess mating types homologous to those of their heterothallic relatives, from which they appear to have been derived. Spore formation is reprogrammed, however. The number of ascospores or basidiospores is halved and spores are formed in such a way that each spore contains nuclei representing two different meiotic products. The included nuclei are of opposite mating type in a majority of spores, and most of the germlings are therefore self-fertile dikaryons capable of completing the sexual life cycle without seeking a mate.

The pseudohomothallic condition has apparently arisen independently in each genus in which it occurs. If asci or basidia in the heterothallic species of a genus are eight-spored, those in the pseudohomothallic species are four-spored, as in *Podospora anserina* and *Neurospora tetrasperma;* if heterothallics are four-spored, the pseudohomothallics are two-spored, as in *Agaricus brunnescens* (see, for example, Summerbell *et al.*, 1989). If meiosis produces two *A* and two *a* products, and if the four nuclei are distributed at random into two binucleate spores, two-thirds of the spores are expected to receive one *A* and one *a* nucleus (Langton and Elliot, 1980). This has been demonstrated experimentally in a two-spored *Coprinus* species (Elliot and Challen, 1983).

In some species, the fraction of ascospores that receives nuclei of opposite mating type has been increased from the random distribution expectation of 67% to nearly 100%. This is accomplished in the four-spored species *N. tetrasperma* by a precisely regulated series of events in the ascus (Fig. 4A). The mating type alleles *A* and *a* are located near the centromere of chromosome 1 in *N. tetrasperma,* so that *A* and *a* almost always go with the centromeres to opposite poles of the first division spindle. The second division spindles then overlap instead of occurring in tandem as in the eight-spored *N. crassa.* A third division brings *A* and *a* nuclei close together in four pairs, and walls are formed to enclose each pair in a single large ascospore of nuclear constitution *A* + *a* (Dodge, 1927).

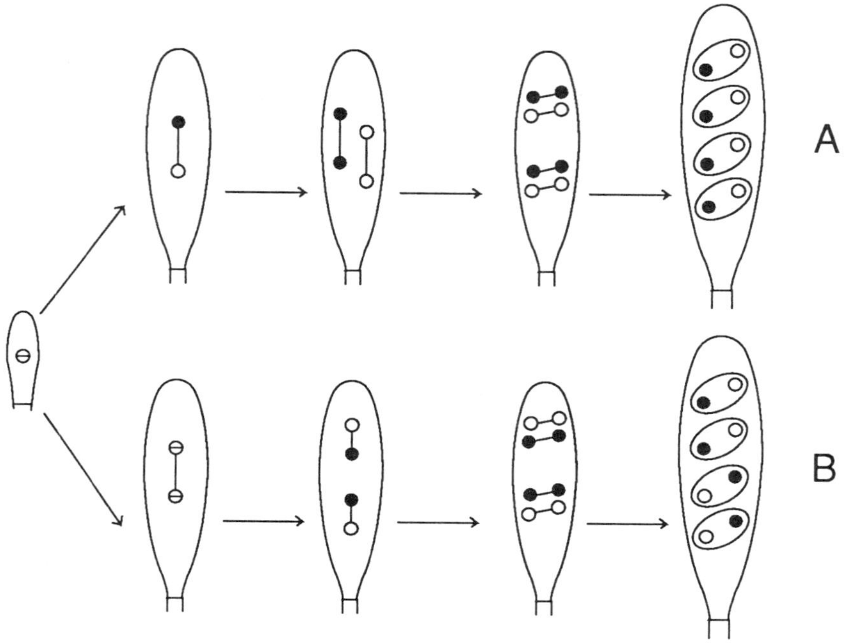

Fig. 4. Schematic diagram showing the origin of heterokaryotic ascospores in *Neurospora tetrasperma* (A) and *Podospora anserina* (B). The mating type genes (● and ○) segregate at the first meiotic division in *N. tetrasperma,* but at the second division in *P. anserina.* Spindles overlap at the second division in *N. tetrasperma,* but not in *P. anserina.* In both species, nuclei of opposite mating type are brought into close proximity at the third division. Adapted from Burnett (1975), with modifications.

The pseudohomothallic species *P. anserina,* a close relative of *Neurospora,* resembles *N. tetrasperma* in producing asci in which each of the four large spores contains nuclei of opposite mating type. This result is arrived at by a quite different route in *Podospora,* however (Rizet and Engelmann, 1949; Fig. 4B). The mating type alleles + and − in chromosome 1 are separated from the centromere by an interval in which a single reciprocal crossover occurs in 99% of meioses. Second division spindles do not overlap. Nuclei of opposite mating type are then brought into proximity at the third division and the unlike nuclei are enclosed pairwise in four large heterokaryotic ascospores. The end result in indistinguishable from that in *N. tetrasperma.*

The genetic and cytological basis of developmental programming that underlies pseudohomothallism is an intriguing problem, but understanding development would not address the problem of adaptational significance. Both of these four-spored species are widespread and highly successful in nature in completely different ecological situations. (*Neurospora* grows on burned vegetation, *Podospora* on dung.) Perhaps pseudohomothallism is maintained despite the different circumstances because it combines advantageous features of both homothallism and heterothallism. As in homothallic species, individual germlings do not require a compatible mate. Pseudohomothallic species can therefore complete the sexual cycle and produce the next generation of ascospores more rapidly than their heterothallic counterparts. (In the laboratory, a heterokaryotic *N. tetrasperma* culture produces abundant ascospores several days earlier than mixed *A* and *a* strains of *N. crassa.*) The selective advantage conferred by a shortened generation time is not trivial (see Gould's 1977 essay). As for homothallic species, timely production of ascospores and basidiospores may also be advantageous because their dormancy, longevity, and resistance to adverse conditions tide the organism over unfavorable periods.

Unlike true homothallic species, pseudohomothallic species retain the ready option of cross-fertilization. They remain facultatively heterothallic. This is realized in several ways. For example, in *N. tetrasperma,* (1) a small fraction of large ascospores is homokaryotic for mating type as a result of crossing over in the interval proximal to mating type; (2) some asci produce two or more small homokaryotic ascospores; (3) one nuclear component may not survive in some ascospores that were initially heterokaryotic; and (4) homokaryotic conidia are produced in large numbers during vegetative growth of self-fertile single-ascospore cultures. N. B. Raju (personal communication) has shown that at least 20% of conidia from a self-fertile strain give rise to self-sterile *A* or *a* colonies. Homokaryotic cultures function as though they belong to a heterothallic species. As a result, *N. tetrasperma* is potentially cross-fertilizing like its heterothallic relatives.

Despite the potentiality for outcrossing in pseudohomothallic species, inbreeding would be expected to prevail in the absence of countermeasures. When the heterokaryotic progeny from sexual spores of a pseudohomothallic species undergo self-fertilization to produce the next sexual generation, the partners in each mating are nuclei that originated from two products of the same meiosis. This is incest carried to an extreme, and it might be expected to result in a precipitous loss of genetic variability. Contrary to intuition, however, intra-ascus matings result in a significantly slower decline of heterozygosity than do matings between ordinary siblings that originate from different meioses (Zakharov, 1968; Kirby, 1984). In fact, if matings were exclusively between members of the same ascus in a species such as *N. tetrasperma,* and if mating types always segregated at the first division, allelic differences initially present at the centromeres of all seven chromosomes would be permanently retained in heterokaryotic or heterozygous condition, generation after generation (Sansome, 1946; Zakharov, 1986). This situation is approximated in the two-spored pseudohomothallic mushroom *A. brunnescens,* in which 96% of binucleate basidiospores produced heterokaryotic germlings that were heteroallelic at all of four randomly selected and presumably unlinked RFLP-marked loci (Summerbell *et al.,* 1989). *A. brunnescens,* like *N. tetrasperma* and *P. anserina,* must have developed some mechanism for increasing the fraction of heterokaryotic spores above the random expectation of two-thirds. Because it is unlikely that four randomly selected loci are all close to centromeres, the high frequency of ascospores heterokaryotic for all four markers indicates that crossing over is repressed in *A. brunnescens.* Crossing over is also repressed or regionally restricted in *N. tetrasperma* (Howe and Haysman, 1966) and *P. anserina* (Marcou *et al.,* 1979).

The results with *Agaricus* suggest that a pseudohomothallic species can maintain heterozygosity and continue indefinitely as a complex balanced heterokaryon, contrary to what would be expected with simple inbreeding. Numerous examples of permanent balanced heterozygosity have been described in plants and animals (Carson, 1967; Luykx, 1987). *Oenothera,* with its translocation complexes, is best known. Apparently, pseudohomothallic fungi have used a quite different genetic manipulation to achieve the same result.

The dual-mating-type spores of pseudohomothallic fungi could have a special significance for colonization and speciation. It seems likely that airborne ascospores are agents of long-distance dispersal (Perkins and Turner, 1988). Colonization by a single ascospore of a heterothallic species such as *N. crassa* would result in a single clone that is unable to reproduce sexually for lack of a mate. In contrast, an ascospore of *N. tetrasperma* could act as founder for a new sexually reproducing population.

V. MODEL ORGANISMS AND THEIR ROLE

Most genetic investigations depend directly or indirectly on information from crosses, and crossing can usually be accomplished only between members of the same species. Experimental geneticists are usually wed, therefore, to a particular research organism, and many geneticists are devoted to one or another of a relatively short list of favorite species headed by *Drosophila melanogaster, Zea mays, Mus musculus, Escherichia coli* and its phages, and more recently, *Saccharomyces cerevisiae* and *Caenorhabditis elegans.* Following closely in the eukaryotic list are several plants, including *Lycopersicon esculentum* (tomato), and the microorganisms *N. crassa, Aspergillus nidulans,* and *Chlamydomonas reinhardii.* Each of these organisms has features that initially attracted geneticists. The longer the species has been used, the more powerful and attractive it has become, not only for genetic manipulations but also as a model for other organisms. As Fink (1988) has pointed out, encyclopedic knowledge about sophisticated genetic models affects the nature of questions that can be asked about less well-studied species:

> The answer to the question "Is this like *E. coli*?" has profound meaning because the quality and quantity of work on this model organism elevates the criteria for comparison. If the answer is "Yes", then the question has been answered. . . . If the answer is "No", then a new phenomenon has been uncovered.

Because of biological diversity, no single species can serve as a universal model. Each species has its peculiar strengths and limitations. Having established bacteriophage as models for molecular genetics, Delbrück adopted a very different subject, *Phycomyces,* as a potential model organism for sensory physiology. In selecting *Caenorhabditis,* Brenner obtained an organism superbly fit for developmental genetics; at the same time he provided a needed model for the class Nematoda. *Chlamydomonas* likewise met needs for which previous model species were inadequate.

Since the beginnings of genetics as a research field, the choice of organisms has been made by those doing the research, and concentration of effort on particular species has evolved naturally, by trial, error, and selection (see Sturtevant, 1971). The spectacular advances of the past 40 years have been driven by the initiative and judgment of individual scientists, not by programmatic research dictated from above.

Tension inevitably exists between the desire to investigate a few organisms in depth and the desire to carry out broad studies on many organisms. Palca (1989) quotes a prominent biologist who asserts that progress in plant molecular biology has been slowed by the multitude of different plants being studied and goes on to state: "It was Max Delbrück who started the concept that you can't do that. You can't work on a whole bunch of different organisms. You have to work on one, and only one." (The out-of-context quotation may distort what was intended; in fact the biologist who was quoted uses several quite different organisms in his own research.) The stark quotation is disturbing not so much for its overstatement and its implications, which can readily be countered, as for the context in which it appears. The article reports that rival granting agencies in the United States are competing for roles in genome-cloning projects, and that one agency proposes to adopt a particular plant species as a prospective model organism to which it would commit major resources.

These developments have generated controversy regarding the extent to which scarce funds should be allocated to large centrally initiated projects focused on a few designated species. Only a small fraction of research proposals from individual investigators are now being funded, and there is wide perception that awards from granting agencies discriminate in favor of biologists working on a few favorite model species. The situation may also affect academic hiring if use of an organism on the short list of popular models becomes a criterion for employment.

In response to these problems, Fulton (1989) has cited examples of major discoveries made with species that are not among the most popular. He concludes:

> For certain goals it is wise, even essential in the case of megaprojects such as the genome games, to focus on certain research subjects. Yet to structure the overall support system to restrict biologists to a chosen few organisms, or even to excessively focus on them, would create a world where understanding is locked on yeast, fruit flies, and mice, and now, perhaps a roundworm and a miniweed. Such focus would also miss the marvelous opportunities for fundamental discoveries still offered by the evolutionary diversity of organisms. A proper balance between emphasis on a few organisms in depth and a broader use of other organisms that are favorable for particular problems is crucial. Biologists studying fundamental problems should work on a suitable organism or organisms for good reasons, but they should not necessarily work only on an organism that is in vogue this week. Let us continue to creatively pursue interesting biological problems and choose organisms suitable to these pursuits, not just suitable genomes to sequence.

It is appropriate to remember that several of the currently favorite organisms experienced a long lag phase before achieving recognition. The first yeast conference in 1961 was attended by only 11 persons (von Borstel, 1963). Dougherty and Calhoun's (1948) early attempts to arouse interest in *Caenorhabditis* ended in failure. In America, research with *Arabidopsis* was limited for many years to a single laboratory (Redei, 1975).

Diverse organisms need to be studied to avoid overgeneralization. As long as *Drosophila* remained the primary genetic model for animals, it was assumed incorrectly that sex determination in other organisms is also mediated by the ratio of sex chromosomes to autosomes. It took studies with mice to reveal that maleness in mammals was determined by the Y-chromosome. Without *Podospora,* the incorrect assumption would have been made from *N. tetrasperma* that mating type genes must be located close to the centromere in other pseudohomothallic species. Without *Neurospora,* the incorrect assumption would have been made from *Podospora* that senescence in other fungi is the result of excision of plasmid sequences from the mitochondrial chromosome, rather than insertion. If only *Saccharomyces* were studied, it might be assumed incorrectly that homothallism is universally based on extra copies of mating type genes carried in silent cassettes, contrary to what has been found in *Neurospora.*

In fact, we need many model organisms, not just a few. A new model species should show promise for solving new problems or better ways for solving old ones, or it should represent a phylogenetic group in which no model has been available. Molecular karyotyping (Skinner *et al.,* 1991) and molecular mapping of DNA polymorphisms (e.g., Metzenberg and Grotelueschen, 1990; Williams *et al.,* 1990, 1991) can greatly speed the characterization of novel genomes. Nevertheless, developing essential genetic information for a previously uninvestigated species remains a daunting task on which few can afford to embark as a major undertaking. Economy favors the use of organisms that have already been domesticated genetically but are not yet recognized as stars or superstars.

Where need for a new model is apparent, a realistic practical strategy for established investigators could be to continue ongoing work with an already well-developed system, but at the same time to undertake exploratory work on a promising new species as a secondary effort. Although most geneticists are understandably wed to a single species, bigamy (Fink, 1988) and polygamy are to be encouraged. Prior research experience with a genetically sophisticated organism is an inestimable advantage for anyone wishing to court a virgin or neglected species. Witness Beadle, taking up *Neurospora* after experience with maize and *Drosophila,* Pontecorvo going to *Aspergillus* after *Drosophila,* Roman and Fogel going to yeast after maize, Brenner to *Caenorhabditis* after phage and *E. coli,* Sager to *Chlamydomonas* after maize, or Herskowitz *fils* to yeast and *Ustilago* after phage and *E. coli.*

Plant pathology provides an example of the need for new genetic models (Valent, 1990). Nonpathogenic models may be used by default (see, for example, Raper, 1988), but they cannot solve the central problems of pathogenicity. The ideal would be a combination of host and parasite in which both are genetically tractable and highly developed. Prospective candidates may be *Ustilago maydis* (Banuett and Herskowitz, 1988; Banuett, 1991) and *Cochliobolus heterostrophus* (Yoder, 1988), both pathogens of maize, and *Magnaporthe grisea* (Valent, 1990), a pathogen of rice. *Ustilago violacea* (Day and Garber, 1988) has the advantage that wild populations of its host plant, *Silene,* are available. It may thus be possible to examine natural variation and coevolution critically using both host and parasite (Alexander and Antinovics, 1988; Stratton, 1990). Other potentially useful systems are described by Sidhu (1988).

A number of filamentous fungi stand out as more general genetic models or prospective models. Best known are the ascomycetes *Neurospora crassa, Aspergillus nidulans, Podospora anserina, Ascobolus immersus, Sordaria macrospora, S. fimicola,* and *S. brevicollis;* the basidiomycetes *Schizophyl-*

lum commune, Coprinus cinereus, Ustilago maydis, and *U. violacea;* and the zygomycete *Phycomyces blakesleeanus.* Progress has been impressive, but even the best developed of these organisms has recently been overshadowed by the magnificent blossoming of yeast molecular biology and genetics.

Saccharomyces has rightly become the eukaryotic counterpart of *E. coli,* with unprecedented power for many molecular, cellular, and genetic problems. Even features of yeast that originally seemed unfavorable have been put to use. Chromosomes so small that examination by light microscopy is precluded were turned to advantage for electrophoretic separation of intact chromosomal DNA. Although mitosis and meiosis cannot be observed by traditional methods, they are being dissected genetically and molecularly with a precision and detail never before achieved.

But yeast, like *E. coli,* has its limitations. A universe of problems remains that cannot be solved with yeast but that can be attacked effectively using filamentous fungi. For example, *Saccharomyces* is unsuitable for studying multicellular differentiation and pattern formation, pathogenicity, or interactions in heterokaryons. The fact that filamentous fungi are not as quick or convenient as yeast or *E. coli* for certain molecular manipulations does not negate their value. Diversity within the fungi may well be as great as that in the animal kingdom, where sponges and coelenterates would hardly be considered adequate as models for insects and mammals. Study is needed of fungi representing different phylogenetic groups and different ecological situations, sampling the diverse ways in which different organisms have solved the problems of existence. Research is needed on different breeding systems—inbreeding and outbreeding, homothallic and heterothallic, bipolar and tetrapolar—to provide information on population structure and evolutionary dynamics. Differences among sibling species need to be examined to see what changes have occurred during speciation.

Knowledge of a flute or a kettledrum is not sufficient to understand all the other instruments in a symphony orchestra or to predict their characteristics. Nor is knowledge of a single species, however complete, adequate for understanding diverse species. Diversity of research organisms in the laboratory must at least dimly reflect the diversity of species in nature if the scope and beauty of evolutionary improvisations are to be appreciated and the genetic manipulations that brought them about are to be understood.

REFERENCES

Alexander, H. M., and Antinovics, J. (1988). Disease spread and population dynamics of anther-smut infection of *Silene alba* caused by the fungus *Ustilago violacea. J. Ecol.* **76,** 91–104.

Anagnostakis, S. L. (1988). *Cryphonectria parasitica,* cause of chestnut blight. *Adv. Plant Path.* **6,** 123–136.

Anderson, E. (1949). "Introgressive Hybridization." John Wiley & Sons, New York.

Aquadro, C. F. (1990). Contrasting levels of DNA variation in *Drosophila* species revealed by "six-cutter" restriction map surveys. *In* "Molecular Evolution" (M. Clegg and S. O'Brien, eds.), pp. 179–189. Liss, New York.

Banuett, F., and Herskowitz, I. (1988). *Ustilago maydis,* smut of maize. *Adv. Plant Path.* **6,** 457–482.

Banuett, F. (1991). Life cycle determinants of the plant pathogen *Ustilago maydis. In* "More Gene Manipulations in Fungi" (J. W. Bennett and L. L. Lasure, eds.) pp. 217–233. Academic Press, San Diego.

Barron, G. L. (1987). The gun cell of *Haptoglossa mirabilis. Mycologia* **79,** 877–883.

Brasier, C. M. (1984). Intermycelial recognition systems in *Ceratocystis ulmi:* Their physiological properties and ecological importance. *In* "The Ecology and Physiology of the Fungal Mycelium" (D. H. Jennings and A. D. M. Raynor, eds.), pp. 451–497. Cambridge University Press, Cambridge.

Brasier, C. M. (1987). The dynamics of fungal speciation. *In* "Evolutionary Biology of Fungi" (A. D. M. Rayner, C. M. Brasier, and D. Moore, eds.), pp. 231–260. Cambridge University Press, Cambridge.

Bremerman, H. J. (1980). Sex and polymorphism as strategies in host–pathogen interactions. *J. Theor. Biol.* **87,** 671–702.

Brodie, H. J. (1951). The splash-cup dispersal mechanism in plants. *Can. J. Bot.* **29,** 224–234.

Buller, A. H. R. (1909). "Researches on Fungi," Vol. 1. Longmans, Green, & Co., London.

Burnett, J. H. (1975). "Mycogenetics." John Wiley & Sons, London.

Burnett, J. H. (1983). Speciation in fungi. *Trans. Br. Mycol. Soc.* **81,** 1–14.

Carson, H. L. (1967). Permanent heterozygosity. *Evol. Biol.* **1,** 143–168.

Carson, H. L. (1982). Speciation as a major reorganization of polygenic balances. *In* "Mechanics of Speciation" (C. Barigozzi, ed.), pp. 411–433. Liss, New York.

Carson, H. L. (1985). Unification of speciation theory in plants and animals. *Systematic Bot.* **10,** 380–390.

Carson, H. L. (1987). The genetic system, the deme, and the origin of the species. *Ann. Rev. Genet.* **21,** 405–423.

Carson, H. L. (1989). Genetic imbalance, realigned selection, and the origin of the species. *In* "Genetics, Speciation and the Founder Principle" (L. V. Giddings, K. V. Kaneshiro, and W. W. Anderson, eds.), pp. 345–362. Oxford University Press, New York.

Charlesworth, B. (1990). Mutation-selection balance and the evolutionary advantage of sex and recombination. *Genet. Res.* **55,** 199–221.

Cullum, F. J., and Webster, J. (1977). Cleistocarp dehiscence in *Phyllactinia. Trans. Brit. Mycol. Soc.* **68,** 316–320.

Day, A. W., and Garber, E. D. (1988). *Ustilago violacea,* anther smut of the Caryophyllaceae. *Adv. Plant Path.* **6,** 457–482.

Delbrück, M. (1949). A physicist looks at biology. *Trans. Connecticut Acad. Arts Sci.* **38,** 173–190. [Reprinted in "Phage and the Origins of Molecular Biology" (J. Cairns, G. S. Stent, and J. D. Watson, eds.), pp. 9–22. Cold Spring Harbor Laboratory Press, Cold Spring Harbor, New York (1966).]

Dobzhansky, T. (1942). Biological adaptation. *Scientific Monthly* **15,** 391–402.

Dobzhansky, T. (1950). Mendelian populations and their evolution. *Amer. Nat.* **84,** 401–418.

Dobzhansky, T. (1970). "Genetics of the Evolutionary Process." Columbia University Press, New York.

Dodge, B. O. (1927). Nuclear phenomena associated with heterothallism and homothallism in the ascomycete *Neurospora. J. Agr. Res.* **35,** 289–305.

Dougherty, E. C., and Calhoun, H. G. (1948). Possible significance of free-living nematodes in genetic research. *Nature (London)* **161,** 29.

Dunlap, J. C. (1990). Closely watched clocks: Molecular analysis of circadian rhythms in *Neurospora* and *Drosophila. Trends Genet.* **6,** 159–165.

Elliot, T. J., and Challen, M. P. (1983). Genetic ratios in secondarily homothallic basidiomycetes. *Exp. Mycol.* **7,** 170–174.

Fincham, J. R. S. (1985). From auxotrophic mutants to DNA sequences. *In* "Gene Manipulations in Fungi" (J. W. Bennett and L. L. Lasure, eds.), pp. 3–34. Academic Press, New York.

Fink, G. R. (1988). Notes of a bigamous biologist. *Genetics* **118,** 549–550.

Ford, E. B. (1940). Polymorphism and taxonomy. *In* "The New Systematics" (J. S. Huxley, ed.), pp. 493–513. Clarendon Press, Oxford.

Fulton, C. (1989). Favored organisms. *Science* **245,** 1311.

Glass, N. L., and Lorimer, I.A. J. (1991). Ascomycete mating types. *In* "More Gene Manipulations in Fungi" (J. W. Bennett and L. L. Lasure, eds.), pp. 194–215. Academic Press, San Diego.

Gould, S. J. (1977). Organic wisdom, or why should a fly eat its mother from inside? "Ever since Darwin" (S. J. Gould), pp. 91–96. Norton, New York.

Hamilton, W. D., Axelrod, R., and Tanese, R. (1990). Sexual reproduction as an adaptation to resist parasites (a review). *Proc. Nat. Acad. Sci. U.S.A.* **87,** 3566–3573.

Harris, H. (1966). Enzyme polymorphisms in man. *Proc. Roy. Soc. Ser. B* **164,** 298–310.

Howe, H. B., and Haysman, P., (1966). Linkage group establishment in *Neurospora tetrasperma* by interspecific hybridization with *N. crassa. Genetics* **54,** 293–302.

Ingold, C. T. (1971). "Fungal Spores: Their Liberation and Dispersal." Clarendon Press, Oxford.

Jacob, F. (1977). Evolution and tinkering. *Science* **196,** 1161–1166.

Kemp, R. F. O. (1985). Do fungal species really exist? A study of basidiomycete species with special reference to those in *Coprinus* section *lanatuli. Bull. Brit. Mycol. Soc.* **19,** 34–39.

Kirby, G. C. (1984). Breeding systems and heterozygosity in populations of tetrad-forming fungi. *Heredity* **52,** 35–41.

Kondrashov, A. S. (1988). Deleterious mutations and the evolution of sexual reproduction. *Nature (London)* **336,** 435–440.

Langton, F. A., and Elliot, T. J. (1980). Genetics of secondarily homothallic basidiomycetes. *Heredity* **45,** 99–106.

Leslie, J. F., and Raju, N. B. (1985). Recessive mutants from natural populations of *Neurospora crassa* that are expressed in the sexual diplophase. *Genetics* **111,** 759–777.

Lewontin, R. C., and Hubby, J. L. (1966). A molecular approach to the study of gene heterozygosity in natural populations. II. Amount of variation and degree of heterozygosity in natural populations of *Drosophila pseudoobscura. Genetics* **54,** 595–609.

Luykx, P. (1987). Variation in sex-linked interchange heterozygosity in the termite *Incisitermes schwarzi* Banks (Insecta: Isoptera) on the island of Jamaica. *Genome* **29,** 319–325.

McClintock, B. (1978). Mechanisms that rapidly reorganize the genome. *Stadler Genetics Symp.* **10,** 25–47.

Marcou, D., Masson, A., Simonet, J.-M., and Piquepaille, G. (1979). Evidence for non-random spatial distribution of meiotic exchanges in *Podospora anserina:* Comparison between linkage groups 1 and 6. *Mol. Gen. Genet.* **176,** 67–79.

May, R. M. (1989). An inordinate fondness for ants. *Nature (London),* **341,** 386–387.

Mayr, E. (1954). Change of genetic environment and evolution. *In* "Evolution as a Process" (J. Huxley, A. C. Hardy, and E. B. Ford, eds.), pp. 157–180. George Allen and Unwin, Ltd. London.

Metzenberg, R. L. (1979). Implications of some genetic control mechanisms in *Neurospora. Microbiol. Rev.* **43,** 361–383.

Metzenberg, R. L., and Grotelueschen, J. (1990). *Neurospora crassa* restriction polymorphism map. *In* "Genetic Maps" (S. J. O'Brien, ed.), 5th Ed., pp. 3.22–3.29. Cold Spring Harbor Laboratory Press, Cold Spring Harbor, New York.

Metzenberg, R. L., Stevens, J. N., Selker, E. U., and Morzycka-Wroblewska, E. (1984). A method for finding the genetic map position of cloned DNA fragments. *Neurospora Newsl.* **31,** 35–39.

Mylyk, O. M. (1976). Heteromorphism for heterokaryon incompatibility genes in natural populations of *Neurospora crassa. Genetics* **83,** 275–284.

Palca, J. (1989). Genome projects are growing like weeds. *Science* **245,** 131.

Perkins, D. D., and Raju, N. B. (1986). *Neurospora discreta,* a new heterothallic species defined by its crossing behavior. *Exp. Mycol.* **10,** 323–338.

Perkins, D. D., and Turner, B. C. (1988). *Neurospora* from natural populations: Toward the population biology of a haploid eukaryote. *Exp. Mycol.* **12,** 91–131.

Raju, N. B., Perkins, D. D., and Newmeyer, D. (1987). Genetically determined nonselective abortion of entire asci in *Neurospora crassa. Can. J. Bot.* **65,** 1539–1549.

Raper, C. S. (1988). *Schizophyllum commune,* a model for genetic studies of the Basidiomycotina. *Adv. Plant Path.* **6,** 511–522.

Read, N. D., and Beckett, A. (1985). The anatomy of the mature perithecium in *Sordaria humana* and its significance for fungal multicellular development. *Can. J. Bot.* **63,** 281–296.

Redei, G. P. (1975). *Arabidopsis* as a genetic tool. *Ann. Rev. Genet.* **9,** 111–127.

Rizet, G., and Engelmann, C. (1949). Contribution a l'étude génétique d'un ascomycète tétrasporé: *Podospora anserina* (Ces.) Rehm. *Rev. Cytol. Biol. Vég.* **11,** 201–304. (English summary, pp. 296–300.)

Sansome, E. R. (1946). Maintenance of heterozygosity in a homothallic species of the *Neurospora tetrasperma* type. *Nature (London)* **157,** 484.

Sidhu, G. S. (ed.) (1988). "Genetics of Plant Pathogenic Fungi" *Advances in Plant Pathology,* Vol. 6. Academic Press, San Diego.

Simpson, G. G. (1947). The problem of plan and purpose in nature. *Scientific Monthly* **64,** 481–495.

Skinner, D. Z., Budde, A. D., and Leong, S. A. (1991). Molecular karyotype analysis of fungi. *In* "More Gene Manipulations in Fungi" (J. W. Bennett and L. L. Lasure, eds.), pp. 86–103. Academic Press, San Diego.

Spieth, P. T. (1975). Population genetics of allozyme variation in *Neurospora intermedia. Genetics* **80,** 785–805.

Stanley, S. M. (1975). A theory of evolution above the species level. *Proc. Nat. Acad. Sci. U.S.A.* **72,** 646–650.

Stern, C. (1962). In praise of diversity. *Am. Zoologist* **2,** 575–579.

Stratton, D. A. (1990). Genetics and population structure of *Ustilago violacea. Mycol. Soc. Am. Newsl.* **41,** 39 (Abstr.).

Sturtevant, A. H. (1971). On the choice of materials for genetic studies. *Stadler Genet. Symp.* **1, 2,** 51–57.

Summerbell, R. C., Castle, A. J., Horgen, P. A., and Anderson, J. B. (1989). Inheritance of restriction fragment length polymorphisms in *Agaricus brunnescens. Genetics* **123,** 293–300.

Valent, B. (1990). Rice blast as a model system for plant pathology. *Phytopathology* **80,** 33–36.

von Borstel, R. C., ed. (1963). Carbondale yeast genetics conference (November 16–18, 1961). *Microbial Genet. Bull.* **19** (Suppl.), 1–21.

Wagner, W. H., Jr. (1970). Biosystematics and evolutionary noise. *Taxon.* **19,** 146–151.

Webster, J. (1979). Cleistocarps of *Phyllactinia* as shuttlecocks. *Trans. Brit. Mycol. Soc.* **72,** 489–490.

Webster, J. (1980). "Introduction to Fungi," 2d Ed. Cambridge University Press, Cambridge.
Williams, J. G. K., Kubelik, A. R., Livak, K. J., Rafalski, J. A., and Tingey, S. V. (1990). DNA polymorphisms amplified by arbitrary primers are useful as genetic markers. *Nucleic Acids Res.* **18,** 6531–6535.
Williams, J. G. K., Kubelik, A. R., Rafalski, J. A., and Tingey, S. V. (1991). Genetic analysis with RAPD markers. *In* "More Gene Manipulations in Fungi" (J. W. Bennett and L. L. Lasure, eds.), pp. 431–439. Academic Press, San Diego.
Wright, S. (1932). The roles of mutation, inbreeding, crossbreeding, and selection in evolution. *Proc. 6th Intern. Congr. Genet.* **1,** 356–366.
Yoder, O. C. (1988). *Cochliobolus heterostrophus,* cause of Southern corn leaf blight. *Adv. Plant Path.* **6,** 93–112.
Zakharov, I. A. (1968). Homozygosity in intratetrad and intraoctad fertilization in fungi. *Genetika* **4,** 98–105 (In Russian). English translation in *Soviet Genetics* **4,** 636–642.
Zakharov, I. A. (1986). Some principles of the gene localization in eukaryotic chromosomes. Formulation of the problem and analysis of nonrandom localization of the mating-type loci in some fungi. *Genetika* **22,** 2620–2624 (In Russian). English translation in *Soviet Genetics* **22,** 1415–1419.

NOTE ADDED IN PROOF

After this chapter was completed I discovered that the same title has been used by Curt Stern (1962) for a presidential address to the American Society of Zoologists. Stern stressed the importance of studying not just the shared general properties of organisms but also the diverse and unique ways in which those general properties are represented in different organisms and groups.

2

Predicting the Molecular Basis of Mycelial–Yeast Dimorphism with a New Mathematical Model of Fungal Morphogenesis

SALOMON BARTNICKI-GARCIA
Department of Plant Pathology
University of California
Riverside, California

GERHARD GIERZ
Department of Mathematics and Computer Science
University of California
Riverside, California

MORE GENE MANIPULATIONS
IN FUNGI

I. INTRODUCTION

Mycelial–yeast (M–Y) dimorphism, the ability of fungi to produce two basically different types of somatic (vegetative) morphology, one mycelial, the other yeast-like, has been an intriguing topic since its discovery about the middle of the last century (for a historical review, see Bartnicki-Garcia, 1963). Because of its relative simplicity and ease of experimental manipulation, M–Y dimorphism has often been regarded as a good model system to study the basis of fungal morphogenesis. There is already a wealth of information on morphological, physiological, biochemical, and genetic studies on this topic. A volume edited by Szaniszlo and Harris (1985) covers the most outstanding examples of M–Y dimorphism in great detail (see also reviews by San-Blas and San-Blas, 1984; Cihlar, 1985; Shepherd *et al.,* 1985; Maresca and Kobayashi, 1989). This veritable mountain of information, however, provides no decisive insight into the molecular basis of dimorphism. Various macromolecular features have been explored in a number of dimorphic fungi, including studies on DNA, RNA, proteins, and cell-wall polysaccharides. In addition, there have been numerous investigations comparing levels of various intermediary metabolites in mycelial and yeast forms of dimorphic fungi. All these findings have augmented our knowledge of the biochemistry of these fungi but have not given us a definitive or a unified view of any particular biochemical event(s) as the cause of mycelium or yeast development. Substantial differences have been recorded between yeast and mycelial cells of a given species but these differences reflect metabolic changes elicited by the drastically different cultural conditions used to obtain the two phenotypes (for examples, see Soll, 1985).

II. TOWARD A UNIFIED THEORY OF MYCELIAL–YEAST DIMORPHISM

This chapter is an attempt to develop a unified theory of M–Y dimorphism in fungi based on the following premises.

1. The molecular and cellular bases of dimorphism should be essentially the same for all fungi, probably differing in specifics but obeying common principles.
2. Since the shape of the fungal cell is determined by the shape of its wall, dimorphism is basically the ability of a fungus to construct cell walls with two different morphologies.
3. The shape of the wall is not primarily determined by the chemical composition of the wall. Hyphae of different fungi can have totally different chemical composition (Bartnicki-Garcia, 1970) but identical shape. Conversely, cells with totally different morphology (hyphal and yeast forms) may share closely similar composition (see Reid and Bartnicki-Garcia, 1976).
4. The shape of the cell wall is established by the manner in which the wall is built.
5. Cell walls of fungi are made from the materials (enzymes and preformed polymers) discharged by cytoplasmic vesicles.
6. The pattern of exocytotic events around the cell surface, that is, the spatial distribution of vesicle discharges, is the final determinant of the shape of the cell wall.
7. The secret of dimorphism, as well as of much of fungal morphogenesis, resides in the cellular events that govern the pattern of spatial distribution of wall-building vesicles.

The purpose of this chapter is to show that a surprisingly simple mechanism can account for the morphological duality we call dimorphism.

III. DIMORPHISM: TWO PATTERNS OF WALL GROWTH

With few exceptions, for example, zoospores, the shape of a fungal cell is dictated by its wall. There is autoradiographic evidence (Bartnicki-Garcia and Lippman, 1969, 1977; Gooday, 1971; Biely *et al.,* 1973) to show that fungal shape is directly related to the pattern of wall growth. Accordingly, dimorphism can be viewed as the ability of a fungus to express two distinctly different patterns of wall growth: (1) a polarized pattern would be responsible for the formation of cylindrical cells, for example, hyphae or germ tubes; and (2) a nonpolarized diffuse pattern of wall growth would give rise to spherical yeast-like cells (Bartnicki-Garcia and Lippman, 1969; Fig. 1). Intermediate morphologies would be produced by controlling the timing and extent of polarization of wall growth.

In either type of development, the fungal cell multiplies by the formation of new centers of wall growth, responsible for sprouting cells with the same

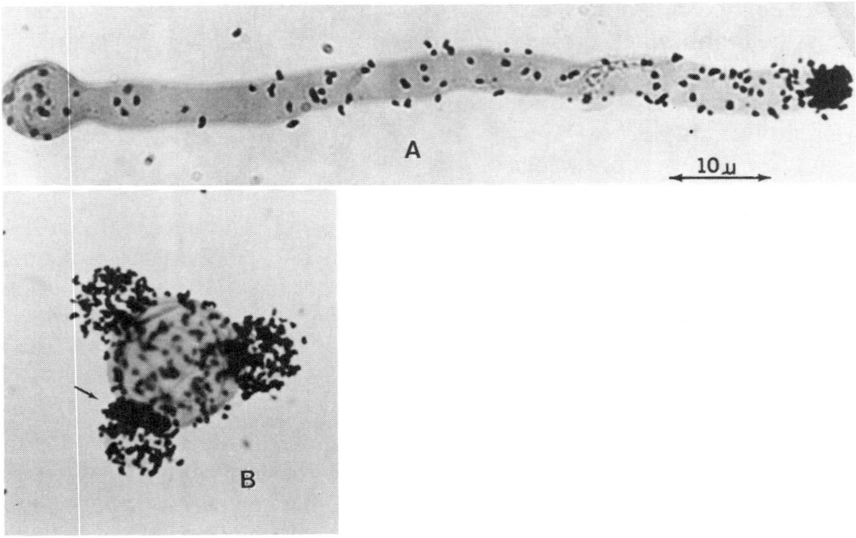

Fig. 1. Dimorphism in *Mucor rouxii:* patterns of cell-wall construction. Autoradiographs show the two drastically different patterns of cell wall deposition (chitin and chitosan) in hyphal and yeast cells. A. Hypha showing a highly localized apical pattern of wall deposition. B. Yeast cell with three buds, all showing a uniformly scattered pattern of wall deposition. [In one bud (arrow) the heavy deposit of wall at the base of the bud represents septum wall synthesis.] Reproduced, with permission, from Bartnicki-Garcia and Lippman (1969).

morphogenetic program. In hyphae, the sprouts become branches; in yeast cells, they grow as spheroidal/ellipsoidal buds.

IV. MOLECULAR AND CELLULAR BASIS OF FUNGAL SHAPE

Basically, we lack an understanding of the key molecular events responsible for fungal morphogenesis, particularly events related to the genesis of a cell wall. We do have mounting information on the biosynthesis of individual polymers, namely chitin, β-1,3-glucan, and α-mannan, and others (see reviews by Ballou, 1976; Gooday and Trinci, 1980; Wessels and Sietsma, 1981; Cabib and Roberts, 1982; Wessels, 1986), but we lack an understanding of how these polymers become assembled orderly into a cell wall (Seilitrennikoff, 1983). So far, chitin has been studied the most, and considerable knowledge exists of the biochemistry, cytology, and genetics of chitin synthesis (see reviews by Gooday and Trinci, 1980; Bartnicki-Garcia and Bracker, 1984; Ruiz-Herrera, 1984; Cabib, 1987; see also, Bulawa *et al.,* 1986; Orlean, 1987; Leal-Morales *et al.,* 1988; Au-Young and Robbins, 1990; Flores-Martinez *et al.,* 1990), yet we have little knowledge of the regulation of chitin synthesis.

Although we know very little about the specific sequence of macromolecular events that gives rise to a cell wall, we expect it to follow the conventional path:

$$\text{DNA} \longrightarrow \text{mRNA} \longrightarrow \text{enzymes} \longrightarrow \text{cell wall polymers}$$

However, no matter how precisely we define this sequence, including the regulatory factors that govern its operation, such a unidimensional view of molecular events can account for the substance but *not* for the morphology of the cell wall. To generate shape, we must invoke an additional mechanism responsible for giving spatial orientation to these processes. Present knowledge of fungal biology indicates that this morphogenetic role belongs to the system of secretory cytoplasmic vesicles and the largely unknown mechanism(s) that regulates their movement. These mobile vesicles deliver to the cell surface the ingredients needed for wall construction. The migration pattern of these vesicles plays a decisive role in determining the shape of the cell wall. Therefore, any serious discourse on the origin of fungal shape must be centered on the factors that guide and govern the movement of cytoplasmic vesicles.

A. Vesicular Concept of Wall Growth

The notion that cytoplasmic vesicles play a key role in fungal cell-wall growth is well established. Vesicles bring to the cell surface all polymeric materials (preformed proteins and polysaccharides, as well as enzymes for *in situ* polymer synthesis) needed to construct the cell wall. Circumstantial proof for the correlation between vesicles and wall growth is most readily evident in hyphal tips of fungi (McClure *et al.,* 1968; Girbardt, 1969; Grove and Bracker, 1970; Grove, *et al.,* 1970; Heath *et al.,* 1971; Howard, 1981). The vesicle concept provides an attractive foundation for the notion that cell-wall formation is a discontinuous process; in other words, the growth of a cell wall should not be regarded as a diffuse continuous process but rather as the sum of numerous separate submicroscopic growth events or "units of wall growth" (Bartnicki-Garcia, 1973, 1990). Each discrete growth event results from the materials discharged by an individual vesicle.

B. Two Secretory Pathways in Wall Biogenesis

Electron-microscopic studies on hyphal-tip cytology disclosed the existence of two vesicle populations with substantially different sizes: macrovesicles and microvesicles. Current knowledge indicates a conspicuous division of labor since these vesicle types perform entirely different functions (Bartnicki-Garcia, 1990). The polymers and enzymes that constitute the amorphous phase of the wall are secreted in macrovesicles, whereas the enzyme that makes the microfibrillar chitin skeleton of the walls of most fungi—chitin synthetase—is transported separately in microvesicles called *chitosomes* (Bracker *et al.,* 1976; Bartnicki-Garcia *et al.,* 1978, 1984; Hanseler *et al.,* 1983; Leal-Morales *et al.,* 1988).

Given the complexity of the wall and manifested need for more than one type of vesicle, the concept was advanced that the growth of the cell wall required the coordinated supply of more than one kind of vesicle responsible for delivering ingredients for various biochemical functions: enzymes for making microfibrils; plasticizing enzymes; and preformed nonfibrillar matrix polymers. Accordingly, a unit of cell wall growth was defined as the amount of growth produced by the minimum combination of vesicles required to deliver the ingredients and catalysts necessary to perform these functions (Bartnicki-Garcia, 1973).

C. Vesicles and Morphogenesis

A vesicle-mediated process of wall growth provides a good basis for understanding how the spatial regulation of wall synthesis could take place and, hence, it constitutes the foundation for explaining the origin of cell shape (Bartnicki-Garcia, 1973). A recent exercise in computer simulation of fungal morphogenesis (Bartnicki-Garcia *et al.,* 1989, 1990) has established a simple mathematical equation to define the shape of fungal cells and a novel and much simpler insight into the probable mechanism that governs the spatial distribution of wall-destined vesicles and, hence, morphogenesis.

Briefly, we have advanced the hypothesis that cell morphogenesis results from the displacement of the source of wall-destined vesicles. In the mathematical model that supports this concept, vesicles are assumed to arise from an idealized point source named the VSC or *vesicle supply center;* in real cells, the VSC may be viewed as the geometric center of a complex system of the vesicle-producing or vesicle-releasing[1] structures present in a fungal cell. This could be the geometric center of the various vesicle-producing cisternae (Golgi or equivalent) or a distribution center for vesicles collected from distant sources. The overall function of the VSC is to release

[1] A distinction between production and release of vesicles needs to be expressed here. There is good evidence, particularly in hyphal cells, that vesicles need not move directly to the cell surface after being produced but can be transported to the apical region over a relatively long distance. If the places of formation and release are different, the VSC refers to the place where vesicles are released on their final journey toward the cell surface.

vesicles in all directions. The model obviates the need for a more elaborate mechanism that would target individual vesicles to predetermined sites on the cell surface (e.g. the apical pole). Vesicles need only be endowed with the ability to move toward the cell surface in any random direction. The morphogenetically critical factor is the net movement of the VSC. The observed or predicted changes in wall properties in the regions of wall growth, for example, elasticity/rigidification (Robertson, 1965; Saunders and Trinci, 1979), plasticity (Bartnicki-Garcia, 1973), or polymer cross-linking (Vermeulen and Wessels, 1984; Wessels, 1986), are probably not the cause of morphogenesis but a reflection of the pattern of vesicle discharge, which generates a graded distribution of biochemical and biophysical activities on the cell surface. Likewise, the specific ingredients of the wall should not make much difference, as long as they meet the architecture requirements of a eukaryotic cell wall (Bartnicki-Garcia, 1984), namely, a skeleton of microfibrils embedded in amorphous cement. Hyphal walls with the same shapes are made with totally different polymers by fungi belonging to different phylogenetic lines (Bartnicki-Garcia, 1970).

V. A VESICLE-BASED COMPUTER SIMULATION OF FUNGAL MORPHOGENESIS

A two-dimensional computer simulation of fungal morphogenesis was created by imagining that a cell is a container under pressure and is bombarded continuously from *within* by a myriad of tiny vesicles. Upon impact, each vesicle becomes inserted into the wall of the container; since wall thickness remains constant, vesicle insertion increases the surface area of the cell by one unit. The model was formulated in two dimensions but the conclusions can be extrapolated to the corresponding three-dimensional solid of revolution (Bartnicki-Garcia *et al.*, 1989, 1990).

A. The VSC Concept

The model assumes that vesicles are released at random in all directions from an idealized point source inside the cell, the VSC. The vesicles travel in the direction in which they were initially emitted until they reach the cell boundary, where they become incorporated and thus increase cell area by one unit. For the sake of simplicity, the model considers that one type of vesicle delivers all necessary materials to produce a unit of growth, including ingredients needed to give the wall a localized measure of transient plasticity. The movement of the VSC is the key to cell morphogenesis. Thus, if the VSC remains stationary while vesicles are released randomly in all directions, the resulting shape generated by the computer simulation would be a growing circle, that is, a two-dimensional simulation of a growing spherical cell (Fig. 2, a–h). If the VSC is displaced while continuing to release vesicles, different forms would be generated; the exact shapes produced would depend on the direction and relative velocity of displacement of the VSC.

B. Computer Simulation of Hyphal Growth

In this vesicle-based simulation of fungal morphogenesis, an elementary maneuver—making the VSC advance linearly while releasing vesicles continuously—was all that was needed to generate a tip-growing tubular shape identical to that of a hypha (Fig. 2, i–r). This finding led us to conclude that a simpler mechanism than was previously anticipated could explain the apical growth of fungal hyphae; by merely advancing an existing vesicle-generating or vesicle-distributing apparatus in a continuous linear fashion, a fungus could automatically establish a polarized apical pattern of surface expansion that would give rise to a hypha.

C. Mathematics of Hyphal Growth: The Hyphoid Equation

The vesicle-based simulation of hyphal growth can be described by a surprisingly simple geometric function:

$$y = x \cot (xV/N)$$

which defines the shape and size of a fungal hypha by two *physiological* parameters. N is the number of wall-destined vesicles released from the VSC per unit time; V is the rate of linear displacement of the VSC (see Bartnicki-Garcia *et al.,* 1989, 1990). This equation, called the *hyphoid,* describes the ideal shape of a hypha in longitudinal median cross-section (Fig. 3). Remarkably, the shape of the hyphoid curve is nearly identical to the actual shape of many fungal hyphae shown in published images photographed through either optical or electron microscopes. One virtue of this equation is that it describes the profile of a hypha in its entirety, from the apex to the base of the tube. An excellent correspondence between shape predicted by the model and the shape of real hyphae has been previously illustrated for diverse fungi (Bartnicki-Garcia *et al.,* 1989; 1990).

The mathematical model shows that, in theory, a single source of vesicles, that is, a single Golgi apparatus, could generate a hypha, provided that during the course of cell growth it advanced linearly as it released its vesicles. A single source of vesicles could not possibly account for the prodigious rate of hyphal extension so common in fungi (Grove and Bracker, 1970; Gooday and Trinci, 1980). Instead, fungi seem to have

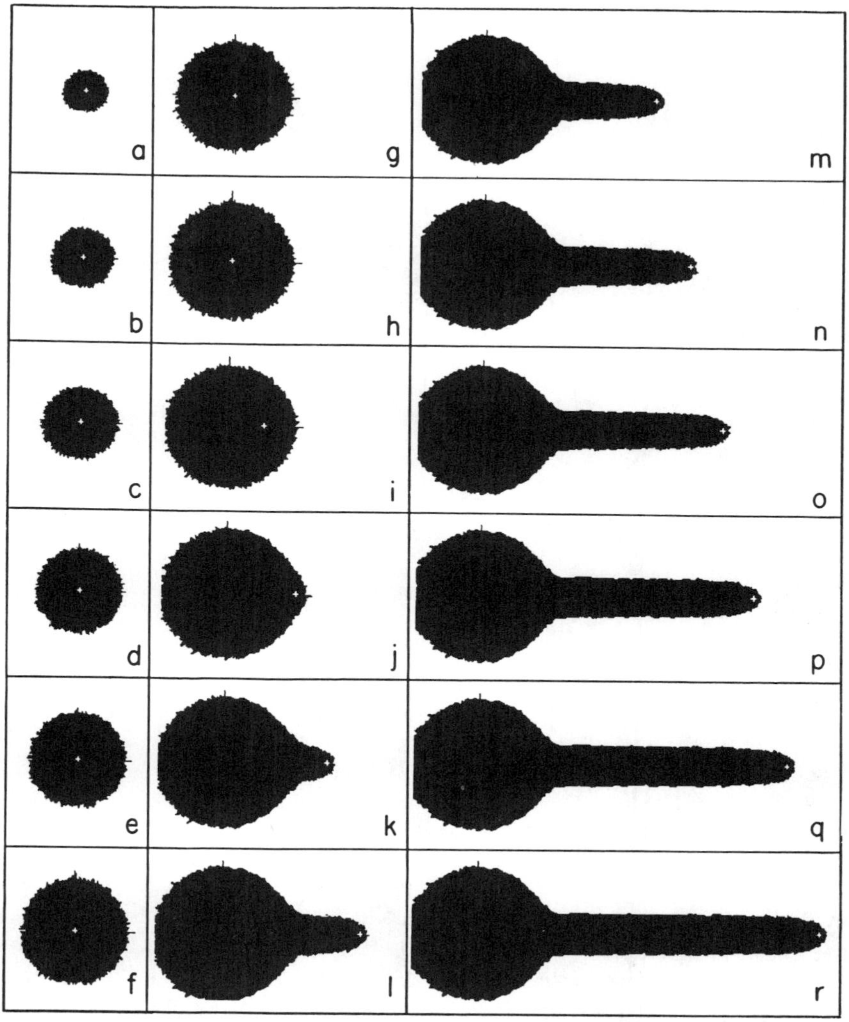

Fig. 2. Computer simulation of fungal morphogenesis: a hypha arising from a spherical cell. This series emulates either the two-stage sequence of morphological development during fungal spore germination (Bartnicki-Garcia, 1981) or the transition from spherical yeast cell to a long tubular hypha. In each frame, the same number of vesicles was released from the VSC (+). Frames a–h simulate the development of a spherical cell by the random discharge of vesicles from a stationary VSC. Frames i–k simulate the emergence of a germ tube, that is, initiation of hyphal morphogenesis. This was achieved by displacing the VSC toward the periphery while maintaining the same rate of vesicle discharge. Frames l–r show that a continued linear displacement of the VSC produces a distinct hyphal shape. Reproduced, with permission, from Bartnicki-Garcia *et al.* (1990).

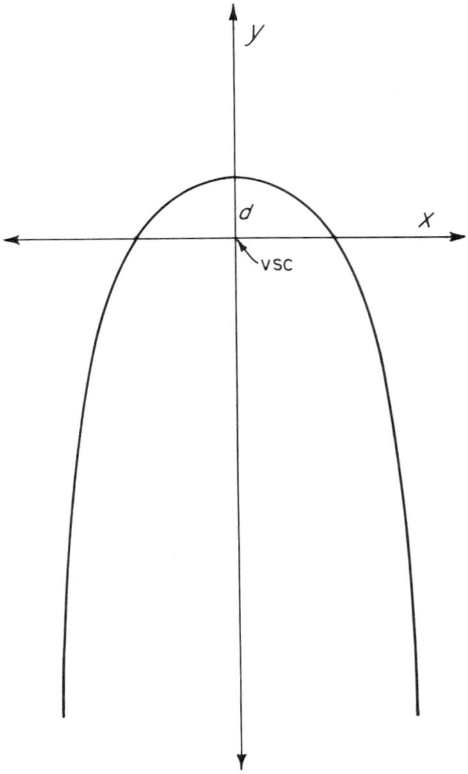

Fig. 3. A hyphoid curve plotted on an arbitrary scale from equation $y = x \cot (xV/N)$. Reproduced, with permission, from Bartnicki-Garcia *et al.* (1989).

evolved an efficient mechanism to collect vesicles from the subapical cytoplasm (Grove and Bracker, 1970; Heath *et al.,* 1971; Barstow and Lovett, 1974; Collinge and Trinci, 1974; Howard and Aist, 1979) and translocate them to the tip. The model predicts that the collected vesicles would be first delivered to a distribution center (VSC) from which they would be free to migrate in any random direction toward the cell surface. Accordingly, the VSC represents a terminal collection point for vesicles that may be traveling along cytoskeletal tracks to the tip. Since the apex appears to be a preferred nucleation site for cytoplasmic microtubules in fungal hyphae (Hoch and Staples, 1985), we speculated that the VSC of a hypha might be a microtubule organizing center or a structure intimately associated with it (Bartnicki-Garcia *et al.,* 1989, 1990).

The availability of a mathematical equation to describe hyphal morphology has opened the door to quantitating the morphological plasticity of fungi. For example, the diameter of the hypha would be determined by the V/N ratio, that is, the number of vesicles released per unit time relative to the rate of displacement of the VSC. Also, displacement of the VSC away from the longitudinal axis would produce hyphal shapes with different degrees of curvature typical of many fungi (e.g., the loops of nematophagous fungi or basidiomycete clamp connections) (S. Bartnicki-Garcia and G. Gierz, unpublished results).

D. The VSC and the Spitzenkörper

A remarkable correlation emerged from the computer model (Bartnicki-Garcia *et al.,* 1989, 1990) between the position of the VSC in the hyphoid model and the position of the Spitzenkörper in real hyphae. This coincidence supports the notion that the organized accumulation of apical vesicles, or Spitzenkörper, so typical of higher fungi (McClure *et al.,* 1968, Girbardt, 1957, 1969; Grove and Bracker, 1970; Howard, 1981; Roberson and Fuller, 1988), is a manifestation of an essential feature of hyphal morphogenesis, namely, the existence of a center for the final distribution of vesicles responsible for tip growth.

E. Morphogenetic Transitions

One of the virtues of the model is that it explains morphogenetic transitions. Basically, a morphogenetic conversion would occur if there were a significant change in the ratio V/N, that is, in the speed of the VSC relative to the number of vesicles produced. An increase in this ratio would produce elongated shapes, a decrease would create spheroidal shapes. The basic model covers a wide range of shapes from a perfect sphere to a highly elongated narrow tube.

VI. MODELING MYCELIAL–YEAST DIMORPHISM

The phenomenon of mycelial–yeast dimorphism fits well within the scope of the proposed model. Accordingly, the dimorphic capacity of a fungus may depend on a single critical factor: the ability of the cell to displace its vesicle-generating or vesicle-releasing apparatus (VSC) during vegetative growth with two substantially different rates relative to the number of wall-making vesicles produced. For instance, the simulation shown in Fig. 2 may be used to illustrate the conversion of a spherical yeast cell into hyphae, as observed in the dimorphism of *Mucor* spp. (Fig. 5)

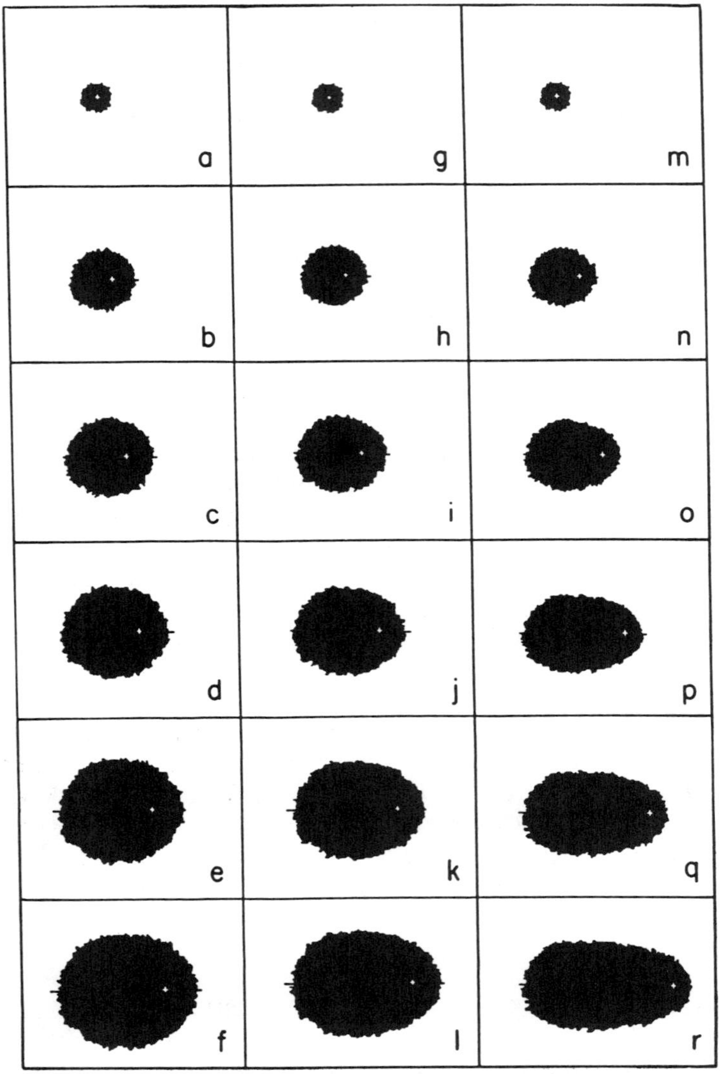

Fig. 4. Computer simulation of yeast-cell morphogenesis. The production of ovoidal/ellipsoidal shapes typical of ordinary yeast cells can be simulated by invoking a limited displacement of the VSC and a high rate of vesicle generation. Each vertical sequence started with a spherical bud generated by 4000 "vesicles" released from the VSC (+). Subsequently, 8000 "vesicles" were released per frame. The three different yeast shapes of increasing cylindricity were generated by varying the rate of advance of the VSC *(V)* in a 1.0 : 1.5 : 2.0 ratio (left, middle, and right, respectively). Reproduced, with permission, from Bartnicki-Garcia *et al.* (1990).

(Bartnicki-Garcia, 1963; Cihlar, 1985) or *Paracoccidioides brasiliensis* (San-Blas and San-Blas, 1984).

A. Yeast Cell Morphology and Budding

If the VSC is held stationary, a spherical cell will be formed (Fig. 2, a–h). If the VSC is displaced very slowly, relative to the rate of vesicle production, an ellipsoidal cell will be produced (Fig. 4). The faster the VSC moves, the more elongated the shape of the resulting cells. In all these instances, yeast-like morphologies are simulated by invoking a high rate of vesicle generation with only a limited displacement of the VSC.

After a yeast cell has grown to "adult" size, the ensuing budding process can be simulated by programming the VSC to advance linearly at high rate for a short time (Fig. 7, frames 4–5; Figs. 8 and 9). This maneuver generates a distinct protuberance (bud) on the mother cell. Shortly after the bud has emerged, V is returned to its original value to produce a spherical or an ellipsoidal daughter cell. An alternative scheme, yielding the same result but adding a biologically significant feature, simulates the budding process by invoking the replication of the VSC when the cell reaches adult size. The new VSC would then start moving at a relatively high speed to generate the bud and then stop or slow down its rate of

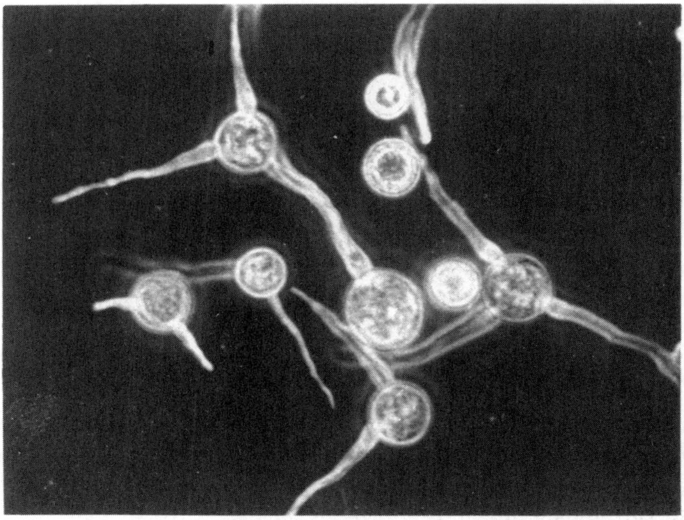

Fig. 5. M–Y dimorphism of *Mucor rouxii.* Hyphae arising from yeast cells exposed to air. Yeast cells were first grown anaerobically under an atmosphere of CO_2.

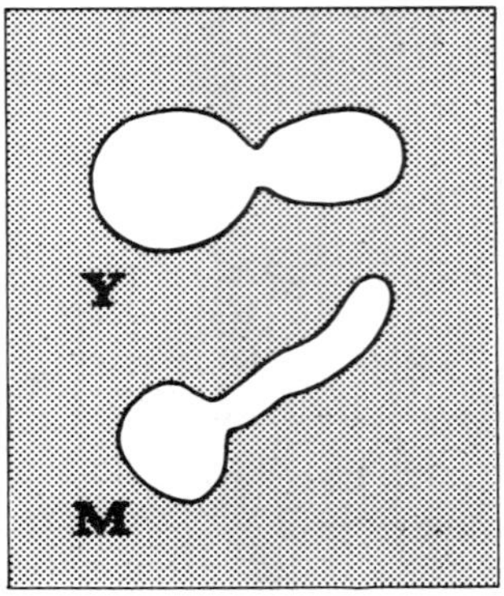

Fig. 6. M–Y dimorphism of *Candida albicans.* Y, Budding yeast cell; M, hyphal tube emerging from yeast cell (drawn from scanning electron micrographs in Soll, 1985).

advance to produce a larger sphere or an ellipsoidal daughter cell, respectively. The old VSC may continue functioning with the same parameters to enlarge the mother cell or may cease to function altogether.

B. Hyphal Morphology and Branching

As discussed earlier, tubular shapes are created when the VSC is displaced continuously along a linear axis at a relatively high speed. Assuming the rate of vesicle production is constant, the faster the VSC moves, the narrower is the hypha that is produced. To simulate the sprouting of a new hypha from a parental cell, the same process invoked for yeast-type budding is initiated. The VSC in the parental cell is programmed to move rapidly to produce a protuberance (Fig. 7, frames 4–5). The key difference with yeast budding is that V is not returned to a slower value but maintained at the same high rate before, during, and after emergence of the tube from the mother cell (Figs. 7 and 9). To simulate the formation of a branch, a new VSC needs to be created at a given distance from the primary tip and programmed to move toward the lateral wall of the hypha.

C. Computer Simulation of *Candida albicans* Dimorphism

The dimorphism exhibited by *Candida albicans* is shown in Fig. 6. Basically, the yeast form consists of ellipsoidal cells that multiply by budding. The hyphal form is commonly obtained by exposing yeast cells to different variables (high temperature, acid pH, and/or certain nutrients, e.g., *N*-acetyl glucosamine) (Shepherd *et al.*, 1985; Soll, 1985). A simulta-

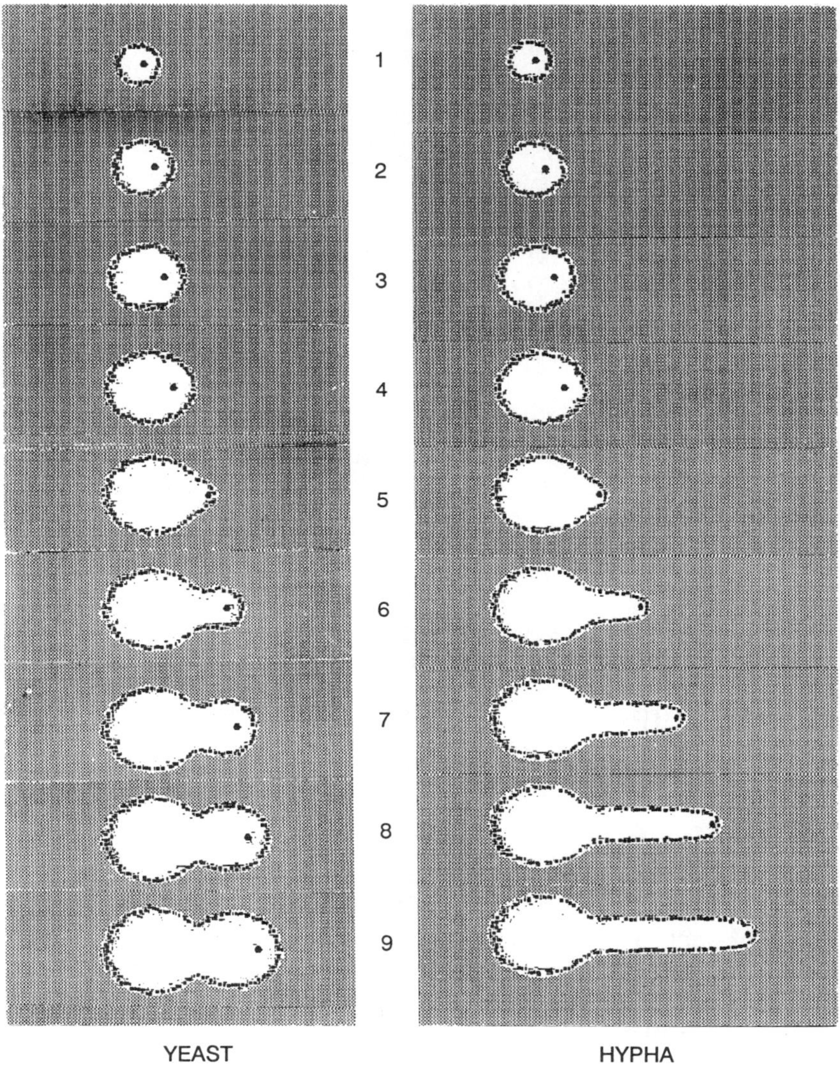

Fig. 7. Computer simulation of dimorphism in *Candida albicans.* Both yeast and hyphal shapes were grown simultaneously at the same rate (10,000 vesicles/frame). Each frame (1–9) indicates a unit of time. The black dot shows the position of the VSC; the different rates of VSC displacement used to generate these shapes are illustrated in Fig. 9. To generate the budding yeast shape, the speed of the VSC was increased fourfold between frames 4 and 5 only. The hyphal shape was generated by increasing the speed of the VSC fourfold after frame 4 and maintaining it thereafter. Details of bud emergence between frames 4 and 5 are shown in Fig. 8. See text for additional explanation.

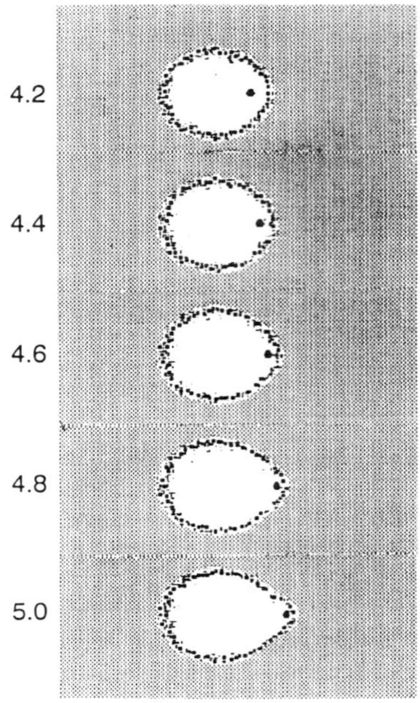

Fig. 8. Computer simulation of bud emergence. Stepwise detail of bud emergence between frames 4 and 5 in the yeast sequence in Figure 7. Each frame represents 0.2 time units (see Fig. 9, a–d). During this sequence the VSC speed was increased fourfold over that in frame 4. Each frame was generated by the release of the 2000 vesicles. After frame 5 the VSC was returned to its original value.

neous computer simulation of the two types of morphological development is shown in Fig. 7. In this dual sequence, both cells were grown at the same rate, that is, the same number of vesicles were released per unit time, N was constant, and shape was regulated by controlling V, the rate of movement of the VSC (Fig. 9). To obtain a yeast morphology, an ellipsoidal mother cell was produced by a VSC that was programmed to advance very slowly; when the cell reached "adult" size, the VSC was accelerated fourfold for a brief period of time. This maneuver caused the appearance of a protuberance or bud on the mother cell; afterward, the rate of displacement of the VSC was returned to its original value to generate the daughter ellipsoidal bud. For the hyphal morphology, a mother yeast cell was first produced as just described. Germ tube emergence was also equally programmed but, after the incipient tube (bud) had emerged, the rate of displacement of the VSC was not lowered to the initial value but main-

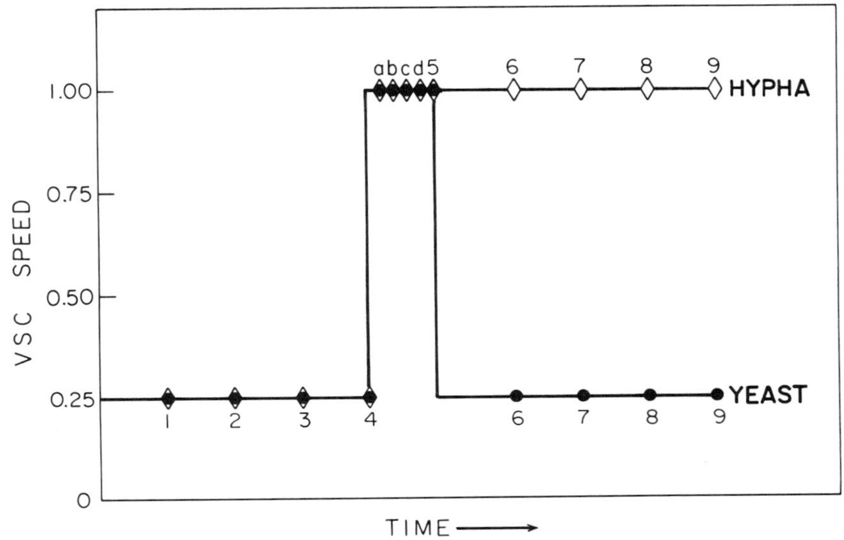

Fig. 9. Changes in the relative rate of displacement of the VSC (V) used to generate the shapes shown in Figs. 7 and 8. Numbers 1–9 correspond to the images shown in Fig. 7. Letters a–d correspond to the images taken at intervals of 0.2 time units shown in Fig. 8.

tained at the same high rate. This caused the protuberance to elongate into a typical hyphal shape.

VII. PREDICTING THE KEY MOLECULAR EVENTS IN DIMORPHISM

The successful simulation of dimorphism described earlier suggests that the same basic morphogenetic principles may operate in real cells, namely, that the dual morphology would result from the intracellular position and movement of the source of wall-building vesicles. If the source is kept stationary, or if it is displaced very slowly, yeast cells will ensue; if the source moves rapidly, hyphal tubes will be generated. Accordingly, we predict that the critical molecular event(s) responsible for dimorphism is that directly involved in the position or movement of the ultimate source of wall-making vesicles. From our current and admittedly rudimentary knowledge of fungal cell biology, we speculate that elements of the cyto-skeleton are the most likely candidates for regulating the displacement of the biological equivalent of the VSC. As previously discussed (Bartnicki-Garcia *et al.*, 1989, 1990), there are two different ways to generate the linear displacement of the VSC: pulling or pushing mechanisms based respectively in microfilaments anchoring VSC to the cell surface (apical pole of the tube) or a scaffolding of microtubules advancing the VSC (pushing mechanism). There is considerable cytological and inhibitor evidence to support the involvement of microfilaments in hyphal developmental tips and other wall-growing regions of fungal cells (Grove and Sweigard, 1980; Adams and Pringle, 1984; Hoch and Staples, 1985; Anderson and Soll, 1986; Runeberg and Raudaskoski, 1986; Tucker *et al.*, 1986; Heath, 1987). Therefore the ultimate molecular event responsible for dimorphism may be a single specific change in a macromolecule that is, or controls, the cytoskeletal element responsible for VSC movement.

Studies on the effects of cytoskeleton inhibitors support its role in fungal dimorphism. Anderson and Soll (1986) and Yokoyama *et al.* (1990) found that actin microfilaments rather than microtubules are essential for hyphal development in *Candida albicans.* However, it would seem premature to discount the role of microtubules since studies on other fungi indicate that microtubules do play a role in hyphal development (Howard and Aist, 1977).

The aforementioned prediction of a specific difference in the cytoskeleton being responsible for the dimorphic behavior of fungi may explain why some previous biochemical studies of dimorphic fungi failed to detect major differences in composition or metabolism between the yeast and hyphal phenotypes. Conversely, the major biochemical differences recorded for other dimorphic fungi are probably circumstantial effects caused by the drastically different environmental conditions employed to elicit dimorphism. Conceivably, the cytoskeletal organization needed to propel the VSC is highly susceptible to internal conditions of the cytoplasm. Accordingly, it should not be surprising that a wide variety of seemingly unrelated external factors may affect dimorphism.

VIII. CONCLUSIONS

1. Cytological and biochemical studies support the concept of vesicles being responsible for delivering to the cell surface all macromolecular ingredients needed to make the cell wall. Autoradiographic studies reinforce the notion that the pattern of vesicle discharge determines the shape of a fungal cell and therefore the key to morphogenesis resides in the mechanisms that orchestrate the movement of wall-destined vesicles.

2. A vesicle-based computer simulation of morphogenesis, and the surprisingly simple mathematical function derived from it, has provided new

insight into the mechanism of fungal morphogenesis (Bartnicki-Garcia *et al.,* 1989, 1990). Foremost was the realization that a simpler solution than was previously anticipated could explain the apical growth of fungal hyphae. By simply advancing its vesicle-generating or vesicle-releasing apparatus (VSC) in a continuous linear fashion, the cell would automatically establish the polarized pattern of surface expansion typical of a hypha.

3. The cellular basis of dimorphism probably revolves around a single factor controlling the displacement of the VSC or vesicle-generating or vesicle-releasing apparatus. Elements of the cytoskeleton are the likely candidates to perform this function. A stationary or slow-moving VSC would give rise to spherical or ellipsoidal yeast cells, respectively, and a fast-moving VSC would produce hyphal tubes.

4. We predict that the molecular basis of dimorphism resides in the macromolecules that comprise or regulate the mechanism for the displacement of the vesicle supply center of the cell.

ACKNOWLEDGMENTS

This study was supported in part by a grant from the NIH (GM-33513).

REFERENCES

Adams, A. E. M., and Pringle, J. R. (1984). Relationship of actin and tubulin distribution to bud growth in wild-type and morphogenetic-mutant *Saccharomyces cerevisiae. J. Cell. Biol.* **98,** 934–945.

Anderson, J. M., and Soll, D. R. (1986) Differences in actin localization during bud and hypha formation in the yeast *Candida albicans. J. Gen. Microbiol.* **132,** 2035–2047.

Au-Young, J., and Robbins, P. W. (1990). Isolation of a chitin synthase gene (ChS1) from *Candida albicans* by expression in *Saccharomyces cerevisiae. Mol. Microbiol.* **4,** 197–207.

Ballou, C. (1976). Structure and biosynthesis of the mannan component of the yeast cell envelope. *Adv. Microbial Physiol.* **14,** 93–157.

Barstow, W. E., and Lovett, J. S. (1974). Apical vesicles and microtubules in rhizoids of *Blastocladiella emersonii:* Effects of actinomycin D and cycloheximide on development during germination. *Protoplasma* **82,** 103–117.

Bartnicki-Garcia, S. (1963). Symposium on the biochemical bases of morphogenesis in fungi. III. Mold–yeast dimorphism of *Mucor. Bacteriol. Rev.* **27,** 293–304.

Bartnicki-Garcia, S. (1970). Cell wall composition and other biochemical markers in fungal phylogeny. *In* "Phytochemical Phylogeny" (J. B. Harborne, ed.), pp. 81–103. Academic Press, London.

Bartnicki-Garcia S. (1973). Fundamental aspects of hyphal morphogenesis. *In* "Microbial Differentiation" (J. M. Ashworth and J. E. Smith, eds.), pp. 245–267. Cambridge University Press, Cambridge.

Bartnicki-Garcia S. (1981). Cell wall construction during spore germination in Phycomycetes. *In* "The Fungal Spore: Morphogenetic Controls" (G. Turian and H. R. Hohl, eds.), pp. 533–556. Academic Press, London.

Bartnicki-Garcia, S. (1984). Kingdoms with walls. *In* "Structure, Function, and Biosynthesis of Plant Cell Walls" (W. M. Dugger and S. Bartnicki-Garcia, eds.), pp. 1–18. American Society of Plant Physiology, Rockville, Maryland.

Bartnicki-Garcia, S. (1990). Role of vesicles in apical growth and a new mathematical model of hyphal morphogenesis. *In* "Tip Growth in Plant and Fungal Cells" (I. B. Heath, ed.), pp. 211–232. Academic Press, San Diego.

Bartnicki-Garcia, S., and Bracker, C. E. (1984). Unique properties of chitosomes. *In* "Microbial Cell Wall Synthesis and Autolysis" (C. Nombela, ed.), pp. 101–112. Elsevier Science Publishers, Amsterdam.

Bartnicki-Garcia, S., and Lippman, E. (1969). Fungal morphogenesis: Cell wall construction in *Mucor rouxii*. *Science* **165,** 302–304.

Bartnicki-Garcia, S., and Lippman, E. (1977). Polarization of cell wall synthesis during spore germination of *Mucor rouxii*. *Exp. Mycol.* **1,** 230–240.

Bartnicki-Garcia, S., Bracker, C. E., Reyes, E., and Ruiz-Herrera, J. (1978). Isolation of chitosomes from taxonomically diverse fungi and synthesis of chitin microfibrils *in vitro*. *Exp. Mycol.* **2,** 173–192.

Bartnicki-Garcia, S., Bracker, C. E., Lippman, E., and Ruiz-Herrera, J. (1984). Chitosomes from the wall-less *slime* mutant of *Neurospora crassa*. *Arch. Microbiol.* **139,** 105–112.

Bartnicki-Garcia, S., Hergert, F., and Gierz, G. (1989). Computer simulation of fungal morphogenesis and the mathematical basis for hyphal (tip) growth. *Protoplasma* **153,** 46–57.

Bartnicki-Garcia, S., Hergert, F., and Gierz, G. (1990). A novel computer model for generating cell shape: Application to fungal morphogenesis. *In* "Biochemistry of Cell Walls and Membranes of Fungi" (P. J. Kuhn, A. P. J. Trinci, M. J. Jung, M. W. Goosey, and L. G. Copping, eds.), pp. 43–60. Springer-Verlag, Berlin.

Biely, P., Kovarik, J., and Bauer, S. (1973). Cell wall formation in yeast. An electron microscopic autoradiographic study. *Arch Mikrobiol.* **94,** 365–371.

Bracker, C. E., Ruiz-Herrera, J., and Bartnicki-Garcia, S. (1976). Structure and transformation of chitin synthetase particles (chitosomes) during microfibril synthesis *in vitro*. *Proc. Nat. Acad. Sci. U.S.A.* **73,** 4570–4574.

Bulawa, C. E., Slater, M., Cabib, E., Au-Young, J., Sburlati, A., Adair W. L., and Robbins P. W. (1986). The *S. cerevisiae* structural gene for chitin synthase is not required for chitin synthesis *in vivo*. *Cell* **46,** 213–225.

Cabib, E. (1987). The synthesis and degradation of chitin. *In* "Advances in Enzymology and Related Areas of Molecular Biology" (A. Meister, ed.), pp. 59–101. John Wiley & Sons, New York.

Cabib, E., and Roberts, R. (1982). Synthesis of the yeast cell wall and its regulation. *Ann. Rev. Biochem.* **51,** 763–793.

Cihlar, R. L. (1985). Morphogenesis and dimorphism of *Mucor*. *In* "Gene Manipulations in Fungi" (J. W. Bennett and L. L. Lasure, eds.), pp. 449–467. Academic Press, Orlando, Florida.

Collinge, A. J., and Trinci, A. P. J. (1974). Hyphal tips of wild type and spreading colonial mutants of *Neurospora crassa*. *Arch. Microbiol* **99,** 353–368.

Flores-Martinez, A., Lopez-Romero, E., Martinez, J. P., Bracker, C. E., Ruiz-Herrera, J. and Bartnicki-Garcia, S. (1990). Protein composition of purified chitosomes from *Mucor rouxii*. *Exp. Mycol.* **14,** 160–168.

Girbardt, M. (1957). Der Spitzenkörper von *Polystictus versicolor*. *Planta* **50,** 47–59.

Girbardt, M. (1969). Die Ultrastruktur der Apikalregion von Pilzhyphen. *Protoplasma* **67,** 413–441.

Gooday, G. W. (1971). An autoradiographic study of hyphal growth of some fungi. *J. Gen. Microbiol.* **67,** 125–133.

Gooday, G. W., and Trinci, A. P. J. (1980). Wall structure and biosynthesis in fungi. *In* "The Eukaryotic Microbial Cell," 30th Symposium of the Society for General Microbiology (G. W. Gooday, D. Lloyd, and A. P. J. Trinci, eds.), pp. 207–251. Cambridge University Press, Cambridge.

Grove, S. N., and Bracker, C. E. (1970). Protoplasmic organization of hyphal tips among fungi: Vesicles and Spitzenkörper. *J. Bacteriol.* **104,** 989–1009.

Grove, S. N., and Sweigard, J. A. (1980). Cytochalasin A inhibits spore germination and hyphal tIp growth in *Gilbertella persicaria. Exp. Mycol.* **4,** 239–250.

Grove, S. N., Bracker, C. E., and Morre, D. J. (1970). An ultrastructural basis for hyphal tip growth in *Pythium ultimum. Amer. J. Bot.* **57,** 245–266.

Hanseler, E., Nyhlen, L. E., and Rast, D. M. (1983). Isolation and properties of chitin synthetase from *Agaricus bisporus* mycelium. *Exp. Mycol.* **7,** 17–30.

Heath, I. B. (1987). Preservation of a labile cortical array of actin filaments in growing hyphal tips of the fungus *Saprolegnia ferax. Eur. J. Cell Biol.* **44,** 10–16.

Heath, I. B., Gay, J. L., and Greenwood, A. D. (1971). Cell wall formation in the saprolegniales: Cytoplasmic vesicles underlying developing walls. *J. Gen. Microbiol.* **65,** 225–232.

Hoch, H. C., and Staples, R. C. (1985). The microtubule cytoskeleton in hyphae of *Uromyces phaseoli* germlings: Its relationship to the region of nucleation and to the F-actin cytoskeleton. *Protoplasma* **124,** 112–122.

Howard, R. J. (1981). Ultrastructural analysis of hyphal tip cell growth in fungi: Spitzenkörper, cytoskeleton and endomembranes after freeze-substitution. *J. Cell Sci.* **48,** 89–103.

Howard, R. J., and Aist, J. R. (1977). Effects of MBC on hyphal tip organization, growth and mitosis of *Fusarium acuminatum,* and their antagonism by D_2O. *Protoplasma* **92,** 195–210.

Howard, R. J., and Aist, J. R. (1979). Hyphal tip cell ultrastructure of the fungus *Fusarium:* Improved preservation by freeze substitution. *J. Ultrastruct. Res.* **66,** 224–234.

Leal-Morales, C. A., Bracker, C. E., and Bartnicki-Garcia, S. (1988). Localization of chitin synthetase in cell-free homogenates of *Saccharomyces cerevisiae:* Chitosomes and plasma membrane. *Proc. Nat. Acad. Sci. U.S.A.* **85,** 8516–8520.

Maresca, B., and Kobayashi, G. S. (1989). Dimorphism in *Histoplasma capsulatum:* A model for the study of cell differentiation in pathogenic fungi. *Microbiol Rev.* **53,** 186–209.

McClure, W. K., Park, D., and Robinson, P. M. (1968). Apical organization in the somatic hyphae of fungi. *J. Gen. Microbiol.* **50,** 177–182.

Orlean, P. (1987). Two chitin synthases in *Saccharomyces cerevisiae. J. Biol. Chem.* **262,** 5732–5739.

Reid, I. D., and Bartnicki-Garcia, S. (1976). Cell-wall composition and structure of yeast cells and conjugation tubes of *Tremella mesenterica. J. Gen. Microbiol.* **96,** 35–50.

Roberson, R. W., and Fuller, M. S. (1988). Ultrastructural aspects of the hyphal tip of *Sclerotium rolfsii* preserved by freeze substitution. *Protoplasma* **146,** 143–149.

Robertson, N. F. (1965). Presidential address: The fungal hypha. *Trans. Brit. Mycol. Soc.* **48,** 1–8.

Ruiz-Herrera, J. (1984). The role of chitosomes in the apical growth of fungi. *In* "Microbial Cell Wall Synthesis and Autolysis" (C. Nombela, ed.), pp. 113–120. Elsevier Science Publishers, Amsterdam.

Runeberg, P., and Raudaskoski, M. (1986). Cytoskeletal elements in the hyphae of the homobasidiomycete *Schizophyllum commune* visualized with indirect immunofluorescence and NBD phallacidin. *Eur. J. Cell Biol.* **41,** 25–32.

San-Blas, G., and San-Blas, F. (1984). Molecular aspects of fungal dimorphism. *CRC Crit. Rev. Microbiol.* **11,** 101–127.

Saunders, P. T., and Trinci, A. P. J. (1979). Determination of tip shape in fungal hyphae. *J. Gen. Microbiol.* **110,** 469–473.

Selitrennikoff, C. P. (1983). Cell wall assembly of *Neurospora crassa:* Lack of evidence for preexisting cell wall acting as primer or template. *Dev. Biol.* **97,** 245–249.

Shepherd, M. G., Poulter, R. T. M., and Sullivan, P. A. (1985). *Candida albicans:* Biology, genetics, and pathogenicity. *Ann. Rev. Microbiol.* **39,** 579–614.

Soll, D. R. (1985). *Candida albicans. In* "Fungal Dimorphism. With Emphasis on Fungi Pathogenic for Humans" (P. J. Szaniszlo, and J. L. Harris, eds.), pp. 167–195. Plenum Press, New York.

Szaniszlo, P. J., and Harris, J. L. (1985). "Fungal Dimorphism. With Emphasis on Fungi Pathogenic for Humans." Plenum Press, New York.

Tucker, B. E., Hoch, H. C., and Staples, R. C. (1986). The involvement of F actin in *Uromyces* cell differentiation. The effects of cytochalasin E and phalloidin. *Protoplasma* **135,** 88–101.

Vermeulen, C. A., and Wessels, J. G. H. (1984). Ultrastructural differences between wall apices of growing and nongrowing hyphae of *Schizophyllum commune. Protoplasma* **120,** 123–131.

Wessels, J. G. H. (1986). Cell wall synthesis in apical hyphal growth. *Int. Rev. Cytology* **104,** 37–79.

Wessels, J. G. H., and Sietsma, J. H. (1981). Fungal cell walls: A survey. *In* "Plant Carbohydrates II. Encyclopedia of Plant Physiology" (W. Tanner and F. A. Loewus, eds.), pp. 352–394. Springer-Verlag, Berlin.

Yokoyama, K., Kaji, H., Nishimura, K., and Miyaji, M. (1990). The role of microfilaments and microtubules in apical growth and dimorphism of *Candida albicans. J. Gen. Microbiol.* **136,** 1067–1075.

II

Tools of the Trade

3

Cloning and Analysis of Fungal Genes

WILLIAM E. TIMBERLAKE
Departments of Genetics and Plant Pathology
University of Georgia
Athens, Georgia

I. INTRODUCTION

Mishra and Tatum (1973) first reported transformation of a filamentous fungus. In these early experiments, the investigators treated an inositol-requiring *Neurospora crassa* strain with DNA from a wild-type strain and obtained inositol-independent colonies. Although it seems likely that these colonies indeed arose by transformation, at the time molecular techniques were lacking to prove that the strains had incorporated exogenously added DNA. Direct demonstration of fungal transformation came a few years later from the experiments of Hinnen *et al.* (1978) and Beggs (1978) with the yeast *Saccharomyces cerevisiae* and soon thereafter from the experiments of Case *et al.* (1980) with *N. crassa.* During the 1980s, transformation systems were developed for many fungal species. Initially, the species studied were the traditional favorites of the fungal geneticist, for example, *N. crassa* and *Aspergillus nidulans* (Ballance *et al.,* 1983; Tilburn *et al.,* 1983; Yelton *et al.,* 1984) and a great deal was learned about the types of transformation vectors and selection procedures that could be reliably employed with these Euascomycetes. Similarly, information was accumulated concerning the nature of transformation events and how newly incorporated DNA molecules behaved during growth and reproduction. With this developing knowledge, more and more fungal species with particularly interesting biological features or of medical, agricultural, or industrial importance were transformed. Thus, representatives from all major fungal classes have been shown to be amenable to transformation. This has justifiably led to excitement about the prospects for solving some long standing and fascinating problems in fungal biology.

The knowledge gained from numerous transformation experiments has led to some generalizations about the fates of transforming DNA. This information can be used to develop strategies for the manipulation of fungal genomes in new and experimentally useful ways. In this chapter I attempt to summarize how transformation can be exploited to help us understand better how the fungal genome is organized and expressed and how genes control the biological activities of this group of organisms. I have not tried to provide a comprehensive review of the now extensive literature on the isolation of genes from fungi or on fungal transformation, but have chosen to highlight methods and approaches that have proved to be of particular experimental utility. The reader may wish to consult the excellent review of fungal transformation systems by Rambosek and Leach (1987) and other chapters in this volume for more complete details about transformation protocols and the use of selective genetic markers.

II. VECTORS FOR FUNGAL TRANSFORMATION

Numerous plasmids for fungal transformation have been constructed and are readily available (Rambosek and Leach, 1987). The basic components of these plasmids are (1) a gene that can be used for selection of fungal transformants and (2) bacterial plasmid sequences that can be used for selection and propagation of the plasmid in *Escherichia coli.* Sometimes sequences for selection of plasmids in yeast or another alternative host are also included. The number of different selective markers available for fungi is now fairly large, and these selective genes often can be used in many different species, thus providing substantial flexibility in designing plasmids for specific purposes.

Selective markers can be divided into at least three functional groups: genes that complement pre-existing mutations and lead to prototrophic growth (auxotrophic markers); genes that provide a new function and lead to drug resistance or growth on a previously nonutilizable nutrient source (drug resistance or added-function markers); and DNA fragments that give rise to selectable mutations when integrated into the genome of the recipient strain at specific locations (mutagenic markers). Each type of marker has specific uses, and in some cases combinations of markers can be used to good advantage.

A. Auxotrophic Markers

Many fungal genes with known functions have been cloned and characterized since the *Neurospora crassa qa-2* gene was first isolated by virtue of its expression in *Escherichia coli* (Vapnek *et al.,* 1977) and used for transformation of a *qa-2*$^{-}$ *Neurospora* strain. Among the most frequently used genes for transformation of auxotrophic strains of filamentous fungi, in addition to the *qa-2* gene, are the *N. crassa* genes *pyr-4,* encoding orotidine-5′-phosphate decarboxylase (Ballance *et al.,* 1983), *trp-1,* encoding a trifunctional enzyme involved in tryptophan biosynthesis from chorismate (Schechtman and Yanofsky, 1983), and *am,* encoding NADP-specific glutamate dehydrogenase (Kinnaird *et al.,* 1982), and the *Aspergillus nidulans* genes *amdS,* encoding acetamidase (Hynes *et al.,* 1983; Tilburn *et al.,* 1983), *trpC,* equivalent to *Neurospora trp-1* (Yelton *et al.,* 1983, 1984; Mullaney *et al.,* 1985), *argB,* encoding ornithine carbamoyl transferase (Berse *et al.,* 1983; Upshall *et al.,* 1986), and *pyrG,* equivalent to *Neurospora pyr-4* (Oakley *et al.,* 1987). The popularity of these genes as selective markers is due to (1) their early cloning in the development of fungal transformation systems, (2) the ready availability of the corresponding mutant strains, and (3) the simplicity and dependability of selection schemes for transformants. In addition, these genes have been sequenced and engineered into a variety of plasmids, facilitating their use. However, numerous other genes have been cloned that have been or could be used for transformation when appropriate mutant recipient strains exist or can be obtained readily.

Selective marker genes from one species can often be used to transform other species, at least within the same subclass. Thus, the first demonstration of transformation in *A. nidulans* (series Plectomycetidae) involved use of the *N. crassa* (series Pyrenomycetidae) *pyr-4* gene (Ballance *et al.,* 1983). Similarly, the *A. nidulans argB* gene has been used to transform the Pyrenomycete *Magnaporthe grisea* (Chumley *et al.,* 1985) and the *A. nidulans niaD* gene, encoding nitrate reductase, has been used to transform at least seven other fungal species (Daboussi *et al.,* 1989; Malardier *et al.,* 1989). Often in cases in which a gene from one species fails to complement a mutation in a different species, it has been possible to clone the corresponding gene from the target species. For example, the basidiomycetes *Schizophyllum commune* and *Coprinus cinereus* have been transformed with genes encoding enzymes of tryptophan biosynthesis that were cloned by their sequence similarity with previously cloned fungal genes (Muñoz-Rivas *et al.,* 1986; Binninger *et al.,* 1987; Skrzynia *et al.,* 1989). The *Aspergillus niger* and *Penicillium chrysogenum trpC* genes have been isolated (Kos *et al.,* 1988; Picknett *et al.,* 1989). Similarly, *Aspergillus oryzae* and *P. chrysogenum* have been transformed with their own *pyrG* genes (Mattern *et al.,* 1987; Cantoral *et al.,* 1988). In most instances, therefore, it should be possible to find suitable genes for complementation of auxotrophic mutations if the appropriate mutant strains are available or can be produced.

The *A. nidulans amdS* gene has been of particular value in devising transformation systems for other fungi. Many other species have limited ability, if any, to utilize acetamide as a nitrogen or carbon source. Thus, *amdS* serves as an added function marker. Expression of the *amdS* gene, encoding acetamidase, permits conversion of acetamide to acetate and ammonia and provides a direct selection for transformants by their ability to grow on acetamide (Hynes and Davis, 1986). Genomic integration of multiple gene copies leads to more vigorous growth of the transformed strain, providing an additional useful selection criterion.

Auxotrophic selective markers have several advantages in transformation experiments. First, they can often be used to direct chromosomal integration of plasmids to homologous sites. Second, selections against auxotrophic strains are frequently very effective, leading to low backgrounds of growth from nontransformed cells. Third, complementation by a single copy of a plasmid is usually complete, so there is little or no selection for transformants containing multiple plasmid copies. The principle disadvantages of auxotrophic markers are that the necessary mutant recipient strains may not be readily available and that these genes may not promote the desired genomic integration events.

B. Drug Resistance Markers

Transformation of many fungal species with plasmids containing metabolic genes may not be feasible, or will at least be quite difficult, because appropriate auxotrophic mutant strains do not already exist. For example, with polyploid or obligately pathogenic species it is unlikely that auxotrophic mutants will ever be obtained. Fortunately, several dominant or semidominant drug resistance genes have found use in fungal transformation and do not require mutant recipient strains. For example, benomyl resistance genes (altered β-tubulin genes) have been isolated from *Neurospora crassa (Bml)* and *Aspergillus nidulans* (*BenA*^{R}) and incorporated into plasmids for fungal transformation (Orbach *et al.,* 1986; Dunne and Oakley, 1988). Antibiotic resistance genes isolated from bacteria have also been engineered for use with filamentous fungi. Most notably, because of the general sensitivity of fungi to the corresponding antibiotics, hygromycin and bleomycin resistance genes have been fused with fungal transcription signals and used for transformation of a number of different species (Yoder *et al.,* 1986; Austin *et al.,* 1990).

Drug resistance markers have the advantage that they can be used to transform most fungal species, even those that have not previously been subjected to laboratory investigation. In addition, they can be used to transform better characterized fungi without carrying out crosses or other manipulations to obtain strains with the correct genotypes. Finally, they can be used to retransform previously transformed strains that have lost their auxotrophic markers in the initial transformation. These markers have the disadvantage that the selections often give high backgrounds and the drugs are fairly expensive. In some instances, a strong selective pressure is seen for strains carrying multiple copies of the resistance marker. Finally, the drug resistance genes cannot easily be used to direct plasmid integration to preselected chromosomal sites.

C. Mutagenic Markers

In yeast and at least some filamentous fungi, plasmids carrying DNA fragments from the recipient species often integrate into the chromosome by homologous recombination (see Section III, C). As illustrated in Fig. 1, homologous integration of a circular plasmid carrying an intragenic DNA fragment leads to a duplication in which two partially deleted copies of the gene are separated by plasmid DNA sequences. Such an event most often leads to a loss of gene function that, in some cases, can be used for selection of the integration event and thus for transformation. For example, integration of an internal fragment of the *Aspergillus nidulans pyrG* gene, as shown in Fig. 1, is expected to lead to loss of orotidine-5′-phosphate decarboxylase, which can be selected for by resistance to 5-fluoroorotic acid (Boeke *et al.,* 1984; Dunne and Oakley, 1988; Grimm *et al.,* 1988). This approach may be of use for direct selection of transformants or for secondary selection of transformants in which a desired integration event has occurred. Thus, for example, disruption of the *A. nidulans yA* or *wA* genes, the products of which are needed for development of conidial pigmentation, has been used to screen for plasmid integration events at these loci following an initial selection for prototrophy due to transformation with a covalently linked or unlinked (i.e., co-transformed) marker (Mayorga and Timberlake, 1990; R. Aramayo, T. Adams, and W. Timberlake, unpublished data).

Mutagenic markers have the advantage that they can be used to transform strains that carry wild-type copies of the selective gene. However, the disruption events they mediate must be frequent in order to be detected without direct selection, and the number of loci that provide a powerful reverse selection procedure such as the one used with orotidine-5′-phosphate decarboxylase inactivation are limited.

III. TRANSFORMATION EVENTS

Transforming DNAs may exist extrachromosomally or be integrated into the genome of the recipient strain at homologous or heterologous sites.

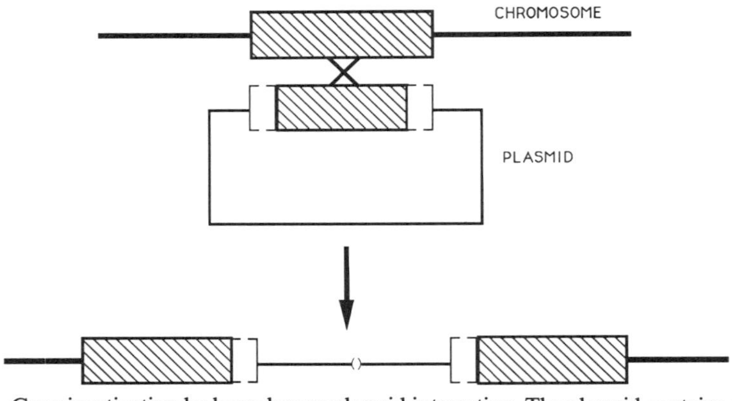

Fig. 1. Gene inactivation by homologous plasmid integration. The plasmid contains a DNA fragment from within a gene, shown as a hatched rectangle. The deleted regions are shown by brackets [ ]. Chromosomal integration by a single homologous recombination event (X), leads to formation of a duplication of two partially deleted copies of the gene separated by plasmid DNA sequences. Most often this leads to gene inactivation.

In addition, one or more copies of added DNAs may be present in transformed strains. The variety of ways in which transforming DNA molecules can reside in cells opens up many avenues for manipulating fungal genomes.

A. Autonomously Replicating Plasmids

A notable deficiency in filamentous fungal transformation systems is the lack of nuclear plasmids that are maintained extrachromosomally. Such plasmids have been used extensively in yeast, where they have made possible many experimental approaches and expedited numerous others. Some yeast extrachromosomal plasmids contain replication and stability functions derived from the endogenous yeast plasmid 2 μ (Beggs, 1978; Hicks *et al.,* 1978; Struhl *et al.,* 1979). Others possess ARS, sequences obtained from the yeast genome that mediate autonomous plasmid replication (Hsiao and Carbon, 1979; Stinchcomb *et al.,* 1979; Struhl *et al.,* 1979). The latter class of plasmids is poorly maintained in cells. Their instability through mitosis results in the accumulation of significant numbers of nontransformed cells in cultures after a few generations. Addition of centromeric DNA (CEN sequences) to ARS plasmids stabilizes them so they are transmitted efficiently through mitosis and meiosis and also reduces their copy number to about one per cell (Clarke and Carbon, 1980; Stinchcomb *et al.,* 1982). Addition of telomeric DNA to CEN plasmids allows maintenance of the plasmids in linear form (Szostak and Blackburn, 1982). This has led to the development of yeast vectors for cloning very large DNA fragments, referred to as yeast artificial chromosomes or YACs (Burke *et al.,* 1987). No analogous technology yet exists for even the best studied filamentous fungi.

It has been argued that the coencytial nature of filamentous fungi makes selection for unstable replicating plasmids difficult (Buxton and Radford, 1984; Timberlake and Marshall, 1989). DNA sequences that stabilize plasmids may exist and would help circumvent this problem (Turner *et al.,* 1985), although nuclear plasmids analogous to yeast 2 μ have not been found in filamentous fungi. Centromeres are an obvious possibility because it is expected that centromeric DNA would, on the one hand, stabilize plasmids by providing an attachment site for mitotic and meiotic spindles and, on the other hand, inhibit chromosomal integration which would result in formation of unstable dicentric chromosomes (Clarke and Carbon, 1980). However, yeast centromeres are inactive in filamentous fungi (Boylan *et al.,* 1986) and centromeres have not yet been isolated and characterized from filamentous species. If fungal centromeres are not too large and complex, it can be anticipated that, once isolated, they will be of immediate use in construction of stably replicating circular and linear plasmids. In the interim, most desired types of molecular genetic studies can be done with filamentous fungi through use of integrative transformation, although the approaches may be more tedious than they would be with reasonably sophisticated replicating plasmid vectors.

B. Integration at Heterologous Sites

Plasmids can integrate into the genome at either homologous or heterologous sites. For the purposes of this discussion, I define homologous sites as those with known identity between plasmid and genomic DNA sequences. The ratio of the two types of integration events varies according to the plasmid used and the species and genetic strain being transformed. Heterologous integration is presumably favored with plasmids containing low amounts of DNA homologous to the target genome. However, even with plasmids having no known homology with the target genome, small stretches of target–plasmid DNA identity could provide nucleation sites to initiate homologous recombination reactions. If these sites are small, integration would be expected to occur at many genomic locations, thus approximating random integration. For example, a given 6-bp sequence is expected to occur approximately once every 4096 bp (4^6) by random chance, assuming an unskewed distribution of base pairs. This assumption is certainly unwarranted given the known species-specific nonrandom compositions of genomes (Arnold *et al.,* 1988), but is adequate to illustrate the point. The probability *(P)* of encountering at least one copy of a particular 6-bp sequence within this interval *(N)* is

$$\begin{aligned} P &= 1 - [1 - (1/N)]^{N-5} \\ &= 1 - [1 - (1/4096)]^{4091} \\ &= 0.63 \end{aligned}$$

Thus, a circular plasmid of 5000 bp, containing 5000 individual 6-bp sequences, would be virtually assured of having at least one perfect 6-bp match with the DNA over the 4096-bp interval. As the size of sites of identity mediating recombination increases, of course, the probability of encountering a site by random chance decreases. The existence of more than about 1000 dispersed integration sites in the genome would experimentally approximate a large number of random sites. Considering an average fungal genome of 2.6×10^7 bp (Timberlake, 1978; Krumlauf and Marzluf, 1979), this equates to approximately one sequence per 26 kb. The probability of *not* encountering a particular 12-bp sequence within a 26-kb region is 0.9984516, but the probability of *not* encountering any of the 12-bp sequences contained in a 5000-bp plasmid is only 4.3×10^{-4}. Thus, homologous recombination between random 12-bp sequences present in a transforming plasmid and the genome could approximate random distribution. However, there is only about a 60% chance that a 5000-bp plasmid will have a 14-bp region of sequence identity over the 26-kb interval, so if regions of identity > 14 are required to mediate recombination, integration clearly becomes nonrandom.

To my knowledge, no systematic studies have been undertaken to determine if integration of plasmids at heterologous sites actually occurs by homologous recombination between short DNA sequences. Simple homologous recombination events involving circular plasmid DNA and the genome will lead to formation of directly repeated sequences (see Fig. 1). Diallinas and Scazzocchio (1989) showed that heterologous integration of a plasmid into the *Aspergillus nidulans* genome was not random but showed site preference. However, they did not determine the sequence of DNA at the junctions between integrated plasmids and the genome. Such sequence determination should clarify the mechanism(s) underlying heterologous plasmid integration. Even without this information, however, heterologous integration may be useful for cloning fungal genes, as discussed in Section IV, B.

C. Integration at Homologous Sites

Circular plasmids can integrate into the genome by single crossover events as shown in Fig. 1 (Hinnen *et al.,* 1978; Tilburn *et al.,* 1983; Yelton *et al.,* 1984). Formation of plasmid co-integrants by homologous recombination preceding chromosomal recombination can result in integration of multiple plasmid copies arranged in head to tail arrays (Fig. 2A). Integration of a single plasmid molecule can be followed by integration of additional plasmids to form tandemly repeated arrays (Fig. 2B). Tandem arrays can be formed at heterologous sites by similar processes. In some instances, transformation of a mutant strain leads to no detectable change in the genome. Such transformants presumably arise from simple gene conversion or double crossover events (Fig. 3A; Hinnen *et al.,* 1978; Yelton, *et al.,* 1984).

Linear molecule can also integrate into the genome at homologous sites (Miller *et al.,* 1985). Often this occurs by a process formally equivalent to a double crossover event as shown in Fig. 3A. However, as shown in Fig. 3B, linear molecules may also first be circularized after entering the cell and subsequently be integrated by a single crossover event, leading to the formation of a precise tandem duplication of the integrated sequence.

Generally, integrated plasmids are stably maintained during vegetative growth and through the asexual reproductive cycle. For example, a tandem duplication of the *Aspergillus nidulans argB* locus was stable for > 50

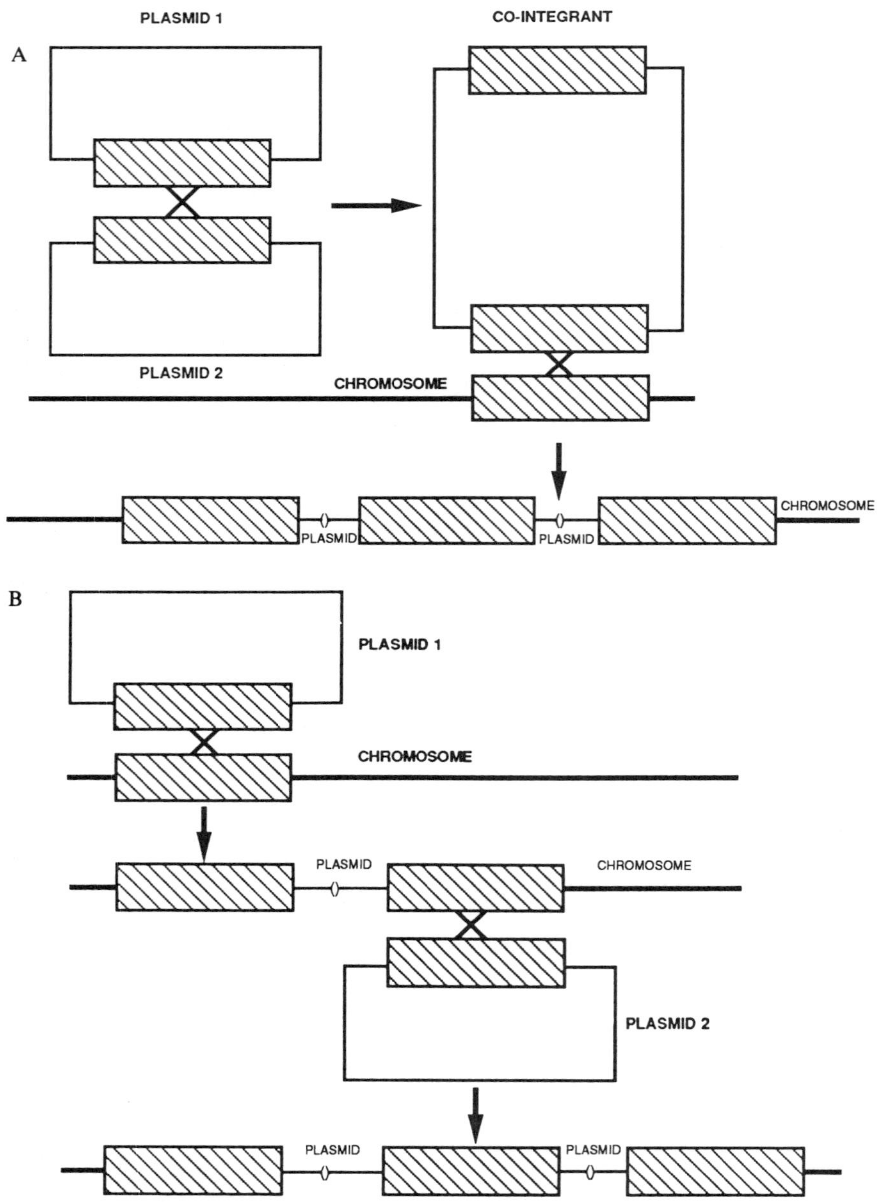

Fig. 2. Integration of multiple plasmid copies. (A) Chromosomal integration of plasmid co-integrants. Recombination between plasmids leads to formation of dimers (shown) and higher order multimers (not shown). Recombination of multimeric plasmids with homologous sequences in the chromosome leads to formation of head-to-tail arrays of integrated plasmid DNA. (B) Repeated integration of plasmids. One molecule of the plasmid integrates into the genome by homologous recombination, leading to the formation of a tandem duplication of chromosomal sequences separated by plasmid DNA. Subsequent integration of additional plasmid molecules leads to formation of head-to-tail arrays of plasmid DNA.

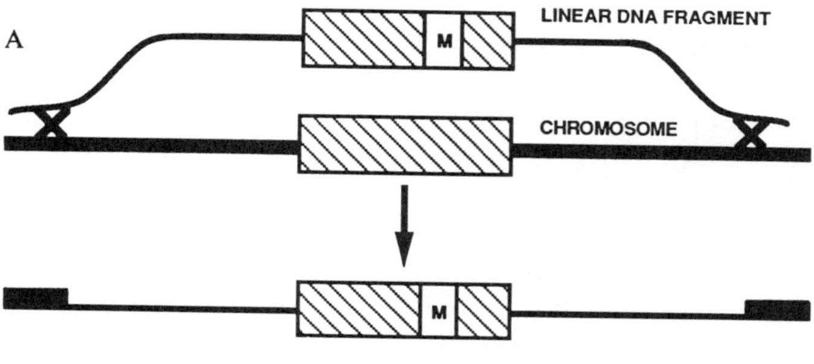

Fig. 3. Homologous integration of linear DNA fragments. (A) Integration without circularization. A linear DNA fragment containing a mutant allele of the target gene (M) recombines with the genome to yield a mutant strain. (B) Integration with circularization. A linear DNA fragment containing a mutant allele of the target gene is first ligated in the cell to form a circle. Homologous integration of the circle results in formation of a precise tandem duplication of the cloned region.

generations of hyphal growth in liquid submerged culture (Boylan *et al.,* 1986). Similarly, many targeted disruptions of the *A. nidulans* conidial pigmentation loci *yA* and *wA,* that lead to the formation of tandem duplications, are stable during extended growth in Petri dishes where assays based on changes in conidial coloration make detection of rare reversion events possible (Mayorga and Timberlake, 1990; R. Aramayo and W. Timberlake, unpublished results). By contrast, integrated DNA sequences are often very unstable through the sexual cycle. Tandemly repeated sequences are frequently lost after selfing or outcrossing in *A. nidulans*

(Tilburn *et al.,* 1983; Yelton *et al.,* 1985; Kelly and Hynes, 1987). During selfing, loss of sequences may be due to unequal crossing over, because progeny are also obtained with increased numbers of the reiterated sequences. Loss of sequences could also be explained by intrachromosomal recombination. In contrast to *A. nidulans* DNA, in which little or no cytosine methylation occurs, introduced DNA sequences in *N. crassa* are frequently hypermethylated. In this species, sequence duplications resulting from transformation are eliminated at high frequency during sexual reproduction by a process referred to as repeat-induced point mutation (RIP) (Selker *et al.,* 1987; Selker and Garrett, 1988; see also Chapter 11). During RIP, both the original copy and the introduced copy of the duplicated sequence are left functionally inactivated (Cambareri *et al.,* 1989). Neither hypermethylation nor RIP appears to occur in the related pyrenomycete, *Sordaria macrospora* (LeChevanton *et al.,* 1989), so it is unclear how widespread this phenomenon is.

Homologous integration events form the basis for many of the experimental approaches summarized in the following sections. In combination with heterologous integration events, they make it possible to adopt numerous strategies for cloning fungal genes and subjecting them to functional analysis.

IV. GENE CLONING TECHNIQUES

Before beginning an analysis of gene structure, function, or regulation, it is of course useful to have a cloned gene in hand. For those fungi with well-developed genetic systems, the experimental strategies made possible by cloned genes provide an extremely valuable adjunct to more traditional genetic approaches. For those fungi with less well-developed genetic systems, or even lacking useful genetic systems, cloned genes may provide the only practical way of approaching a problem at the molecular level. Since the advent of recombinant DNA technology, many clever procedures have been developed to clone desired genes (Sambrook *et al.,* 1989). A few techniques that have proved to be of general use for cloning interesting genes from filamentous fungi are described in the following sections.

A. Cloning Genes by Their Regulatory Properties

One extremely powerful way of cloning specific genes or sets of genes involves "plus–minus" hybridization screening of clone banks, or screening with probes prepared by "subtractive" or "cascade" hybridization. One of the first examples of clone selection by plus–minus screening involved the isolation of galactose–inducible genes from yeast (St. John and Davis, 1979). In these experiments, radioactively labeled probes were prepared from RNA isolated from yeast cells grown in the presence of either glucose or galactose as the sole carbon source. Filter replicas of Petri dishes containing plaques of recombinant lambda clones were hybridized individually with these probes, and clones were selected that gave strong signals with the galactose probe and weak or negligible signals with the glucose probe. The selected clones were subsequently shown to contain genes encoding regulated enzymes of galactose metabolism. These genes have been of great use in studying the molecular control of the yeast galactose regulon (Guarente, 1987).

The approach used to clone galactose-inducible genes can be extended to a variety of physiological or biological situations. The only requirement is that mRNA can be isolated from cells that differ from one another in some way. The success of the procedure, of course, depends on how different the mRNA populations are. Sequences that are absent from cells under one set of conditions and present at high levels under another set of conditions will be easily detected. Sequences that are always present but are moderately induced or repressed will be more difficult to detect reliably. In addition, care must be taken when choosing the type of clone bank to screen. In clones with fairly short genomic DNA inserts, it is unlikely that the sequences will contain multiple genes with different regulatory properties. However, in clones with longer inserts, for example, those present in many lambda banks and all cosmid and YAC clone banks, it is probable that each insert will contain many differently regulated genes. Thus, these clones may yield positive signals with most probes because there is a significant probability that each insert will contain at least one constitutively expressed gene. Sequences present at fairly low abundance (<0.01% of mRNA) usually will be difficult to detect by hybridization with mixed probes.

cDNA libraries are valuable for plus–minus selections. In this case, a cDNA bank is constructed by using mRNA from the cell type that is expected to contain the transcript(s) of interest. Construction of cDNA banks in bacteriophage, plasmid, or hybrid (phagemid) vectors is now very straightforward, and kits for doing so are commercially available. The proportion of clones in these collections that belong to the class of interest will be determined by the prevalence of the mRNA in the population rather than by the representation of the corresponding gene in the genome. The cDNA collection can then be hybridized with radiolabeled probes synthesized from a reference mRNA (e.g., mRNA from uninduced cells) and the mRNA from which the collection was synthesized. Clones showing differential hybridization are then picked the way they are from genomic clone collections.

A more complex approach is based on depleting a nucleic acid population of sequences that are not of interest, thereby enriching the population for sequences that are of interest. For example, Timberlake (1980) used cascade hybridization to enrich for cDNAs corresponding to mRNAs that accumulate preferentially during conidiophore development in *Aspergillus nidulans.* These cDNA probes were subsequently used to isolate lambda clones containing developmentally regulated genes (Zimmerman *et al.,* 1980). In these experiments, a radiolabeled cDNA was synthesized from mRNA isolated from conidiating cultures. The cDNA was then hybridized with a 20-fold excess of mRNA from vegetative cells and the hybrids were removed by hydroxyapatite chromatography. The remaining cDNA was then challenged with a 50-fold and then a 100-fold excess of vegetative mRNA to remove residual vegetative cDNAs. This represents a stringent removal of sequences from the population. The cDNA was finally hybridized with mRNA from either purified conidia or conidiating cultures to obtain stage-specific sequences. These cDNAs were subsequently shown to identify clones containing developmentally regulated genes with a high degree of specificity.

Many modifications of this basic approach are possible, and reagent kits for conducting subtractive hybridization are available commercially. In one modification, subtracted cDNA populations are prepared and then used to make cDNA clones instead of hybridization probes for previously constructed recombinant DNA libraries (Sargent and Dawid, 1983). Other modifications include reducing the number of hybridization steps by using greater mRNA excesses in preliminary hybridizations (Raper and Timberlake, 1985); using the biotin–avidin interaction to remove hybrids from hybridization reactions (Wieland *et al.,* 1989); using cDNA banks constructed in vectors containing bacteriophage promoters to make large amounts of RNA *in vitro;* and exploiting the polymerase chain reaction (PCR) (Saiki *et al.,* 1988) to increase the amount of nucleic acid available (Weiland *et al.,* 1989).

In addition, of course, it is possible to prepare subtracted probes by cross-hybridization of genomic DNAs that differ in one or more sequences. For example, DNA from a strain containing a deletion can be used to remove all of the sequences except those corresponding to the deletion from DNA from an isogenic strain lacking the deletion. Alternatively, mRNA can be isolated from the deletion strain and hybridized to cDNA made from mRNA isolated from the wild-type reference strain. Many other variations of the original depletion hybridization procedure are possible, making this general approach adaptable to many different situations.

Other strategies for cloning genes by their regulatory properties are fairly obvious, but have not yet been extensively explored in fungal systems. One example involves construction of transformation plasmids containing the *Escherichia coli lacZ* or other reporter gene fused to a nonregulated promoter. The host–vector system is chosen so the plasmid integrates at numerous chromosomal sites. Insertion of the plasmid in the vicinity of a regulated enhancer or silencer should lead to up or down regulation of the promoter, which can be detected by staining colonies with a chromogenic β-galactosidase substrate. The plasmid and adjacent sequences can then be cloned into cosmid or bacteriophage vectors by standard procedures. This general approach, modeled after the *lacZ* gene fusion technology developed for bacteria (Casadaban *et al.*, 1980), has been used in higher eukaryotic organisms and is referred to as "enhancer trapping" (O'Kane and Gehring, 1987; Bellen *et al.*, 1989; Bier *et al.*, 1989). In a related approach, genomic DNA fragments can be inserted upstream of a promoterless reporter gene in a suitable transformation plasmid. The resultant collection can then be transformed into a recipient strain and the transformants can be tested for the desired regulatory pattern. Plasmids then can be recovered from candidate clones and retested. The inserted sequences are used as hybridization probes to obtain the corresponding genes from standard genomic libraries. Modifications of this general approach could be of use for isolating sequences that respond to specific *trans*-acting factors, especially if regulated expression can be detected in a heterologous system (e.g., *Neurospora crassa* elements detected by their regulated activity in yeast).

B. Cloning Genes by Function

One of the principal advantages of filamentous fungi as experimental organisms is that genes can often be identified by the interesting properties of strains carrying their mutant alleles. Thus, there is frequently a need to clone a gene that has been detected by mutation. This cloning can be accomplished in many species by genetic complementation of mutations with recombinant DNA clones containing wild-type DNA sequences. At least two types of problems have been encountered in efforts to clone genes by complementation. First, transformation frequencies with fungi have often been low (<100 transformants/μg transforming DNA), making it difficult to screen large numbers of transformants. Second, for some fungi, DNA rearrangements occur either during cloning into some *Escherichia coli* strains or upon reintroduction into the fungus. Both of these problems largely have been overcome by a variety of approaches.

Johnstone *et al.*, (1985) constructed an *Aspergillus nidulans* cloning vector containing the *argB* gene as the selective marker. A bank of plasmids, containing ~10 kb *A. nidulans* genomic DNA inserts, transformed recipient strains at a frequency high enough to allow detection of complementation of a developmentally abnormal mutant strain (a *brlA*$^-$ strain). Plasmid DNA was recovered from the transformants by restriction enzyme digestion, recircularization, and transformation of *E. coli,* and was shown to contain *brlA* gene sequences. Yelton *et al.,* (1985) constructed an *A. nidulans* cosmid cloning vector containing the *trpC* gene as the selective marker. Since the cloning capacity of this cosmid was 40 kb, fewer clones needed to be screened for a high probability of success. A bank of cosmids containing 35–40 kb inserts was used to transform a yellow-spored *(yA2), trpC801* mutant strain. In 1000 *trpC*$^+$ transformants, 3 produced wild-type green spores. Cosmids were recovered from these transformants by treating DNA *in vitro* with a bacteriophage lambda packaging extract and using the resultant defective phage particles to transduce *E. coli* cells to ampicillin resistance. The recovered cosmids contained the *yA* gene and the cloned inserts were unrearranged (Yelton *et al.,* 1985; O'Hara and Timberlake, 1989; Aramayo and Timberlake, 1990).

The mechanisms that lead to successful recovery of cosmid DNA by this method are unknown. Yelton *et al.,* (1985) found that the complementing cosmids had integrated into the *A. nidulans* genome at the *yA* locus, probably by a single homologous recombination event, resulting in the duplication of the DNA insert, with the copies separated by vector sequences. Since the vector contained a single *cos* site, this DNA was not expected to serve as a substrate for lambda packaging. It is possible that multimeric cosmids that would serve as packaging substrates were present in the DNA population but were undetectable by Southern blotting due to their large size, low prevalence, or both. It is also possible that the packaging extracts employed in these studies were recombination proficient and led to formation of multimeric molecules during the packaging reaction itself. Recovery of cosmids by packaging and transduction, although usually quite efficient, has not always been successful, in which case cosmids have been recovered by transformation of *E. coli* with unmodified DNA from the transformant (Andrianopoulos and Hynes, 1988).

Numerous genes have been cloned from *A. nidulans* by complementation of mutations, and methods have been developed for rapidly mapping mutation-complementing DNA fragments (Timberlake *et al.,* 1985). This general approach likely will be useful for cloning genes of known function from related ascomycetes and perhaps from even more distantly related fungi. The very large number of mutant *A. nidulans* strains available makes this species particularly suitable for this purpose. In addition, plasmids containing drug-resistance genes make it unnecessary to cross selective markers into strains prior to transformation. An interesting example of how *A. nidulans* can be used to identify heterologous genes is the cloning of PDA from the plant-pathogenic pyrenomycete *Nectria haematococca*

(Weltring *et al.*, 1988). PDA is a disease-determinant gene in this organism, encoding pisatin demethylating activity. Pisatin, a phytoalexin, is produced by pea plants at sites of infection. PDA inactivates the toxin, thereby permitting fungal growth. *A. nidulans,* which does not produce PDA, was transformed with a cosmid library containing inserts from a PDA-producing *N. haematococca* strain, and colonies were screened for PDA production. One such colony was found, a cosmid was recovered in *E. coli,* and the cosmid subsequently shown to contain the PDA gene. Similarly, Katz and Hynes (1989a,b) introduced the 2-pyrrolidinone utilization pathway from *A. nidulans* into *Aspergillus terreus,* which lacks the pathway. These experiments led to the identification of a new *A. nidulans* gene, *lamB,* required for 2-pyrrolidinone utilization, that was initially identified molecularly by its physical linkage to *lamA.*

Cloning vectors have also been developed for *Neurospora crassa.* Some problems have been encountered with DNA rearrangements in recovered clones, in part because the bacterial recipient cells used restrict methylated DNA. This has not been a problem with *A. nidulans* and some other fungi because these organisms contain little or no methylated DNA. In *Neurospora,* by contrast, DNA sequences, especially sequences introduced by transformation, are frequently methylated (Cambareri *et al.*, 1989). However, *E. coli* strains that are tolerant of methylated DNA are now readily available and should largely circumvent this problem (Orbach *et al.*, 1986, 1988).

An alternative to direct selection and recovery involves use of ordered collections of genomic clones constructed with drug resistance cosmids (Vollmer and Yanofsky, 1986; Mann *et al.*, 1988). DNA from dichotomously divided pools of clones is tested for its ability to complement a mutation until a single complementing clone is identified ("sib selection"). Due to the high transformation frequencies achievable with *N. crassa* drug resistance plasmids, only small amounts of DNA are needed to conduct these screens. Once DNA has been isolated from the many subcollections required for this approach, the method is very rapid and efficient.

For any clone isolated by mutation complementation, it is imperative to show formally that the cloned DNA sequence actually corresponds to the target gene. Suppressors of various sorts may be cloned instead of the gene itself. For example, Mann *et al.* (1988, 1989) inadvertently cloned the *N. crassa pho-4*$^{+}$ gene, encoding a phosphate-repressible phosphate permease, because of the ability of that gene to suppress a mutation in the *nuc-1* regulatory gene. Confirmation of clone identity usually involves demonstration that the cloned DNA fragment is completely linked to the target locus and that mutations induced by the cloned fragment give the expected phenotype and are *trans*-noncomplementing (see, for example, Yelton *et al.,* 1985; Mann *et al.,* 1988; O'Hara and Timberlake, 1989; Mayorga and Timberlake, 1990; Tilburn *et al.,* 1990).

Two potential cloning strategies related to genetic complementation involve loss or gain of genetic function in a wild-type background. The first of these strategies stems from the observation of Hynes *et al.* (1988) that transcriptional regulators can be titrated away by introduction of multiple copies of sequences to which they bind. Titration of transcription factors often leads to a readily identifiable alteration in phenotype. Because vectors are available that permit selection for integration of multiple plasmid copies, for example, plasmids containing the *amdS* or bleomycin-resistance genes, it should be possible to identify specific DNA sequences by their ability to titrate away regulatory factors. Genetic function also can be lost due to insertion of transforming DNA into a locus. Fungal transformation plasmids exist that have no homology with target genomes but still transform at high frequencies. These plasmids integrate into the genome at many positions (see Section III, B). Transformed strains can be examined for desired mutations that may have occurred as a result of plasmid insertion. Plasmids containing genomic sequences adjacent to the point of insertion can be recovered by digesting DNA with a restriction enzyme that cuts once within the plasmid, religating, and transforming *E. coli* cells. Alternatively, genomic libraries can then be made from the mutants and clones containing the plasmid, and adjacent chromosomal sequences can be isolated by colony or plaque filter hybridization. Subsequently, standard reverse genetic procedures can be used to confirm the identity of the cloned genomic region.

The second cloning strategy is based on the finding that either overexpression or misscheduled expression of many regulatory and some nonregulatory genes leads to growth inhibition or other readily detectable changes in phenotype (Adams *et al.,* 1988; Mirabito *et al.,* 1989; Marshall and Timberlake, 1990). Because vectors containing regulatable promoters are available (e.g., the *A. nidulans alcA* promoter; Gwynne *et al.,* 1987), it is possible to make recombinant DNA banks in which genomic or cDNA sequences are fused downstream of the promoter. Transformants can be selected for the primary transformation marker under conditions that inactivate the regulatable promoter, and then screened under conditions that activate the promoter. Another variation of this strategy involves identification of genes by complementation of mutations following induced expression of the cDNA clones. Many other novel approaches to gene cloning are possible with the currently available fungal transformation technology and will undoubtedly contribute to the continued advancement of our understanding of these organisms.

C. Cloning Genes by Chromosomal Position

It is not always practical to screen the large number of clones necessary to find a particular clone. For example, in a search for fungal avirulence genes, the screening procedure might involve testing individual transformants for their ability to infect several nearly isogenic hosts, a very space, labor, and time intensive process. Similarly, the phenotypes of transformants in the desired class might easily be confused with phenotypes of transformants in undesired classes. An increasingly useful way to reduce the number of clones to be screened and to be sure that a targeted chromosomal region is completely evaluated involves chromosome walking.

Several technological advances have made chromosome walking over fairly large distances straightforward, especially in fungi, in which repeated DNA sequences represent a low proportion of the genome and therefore interfere minimally with walking procedures. For some fungi, extensive genetic maps already exist and it is usually easy to map a new gene of interest between two known markers. If the flanking markers have not been cloned, they can be by use of the procedures just outlined. Chromosome walks are then initiated in both directions from the flanking markers until contiguous clones representing the region bounded by the markers are obtained. The number of directions involved in the walk can be reduced if additional tightly linked marker genes for which there are functional assays are present. Efficient methods for rapid chromosome walking are being developed in numerous laboratories as part of the human genome sequencing initiative. These methods, in conjunction with clone banks that have very long inserts in cosmid or YAC vectors, speed up the process of selecting overlapping clones from contiguous chromosomal regions. Given the small size of fungal genomes and the close spacing of fungal genes, cloning a gene by chromosomal position is a viable alternative strategy. For fungi that have less well-developed genetic maps, it may be possible to produce RFLP maps that cover the entire genome, and then to use clones defining flanking RFLPs as starting points for walks.

Ordered chromosomal clone banks are becoming a valuable genetic resource. Ordered *Neurospora crassa* and *Aspergillus nidulans* cosmid banks have been deposited with the Fungal Genetics Stock Center (FGSC) for general distribution. As genes are identified in these banks and that information is incorporated into databases, these ordered collections become increasingly useful. It therefore benefits the entire fungal genetics community if clones are related to these collections and the information is transmitted to the FGSC for incorporation into the databases. Recently it has become possible to separate fungal chromosomes physically by pulsed field electrophoretic techniques (see Chapter 4). Fractionated chromosomes are being used to generate chromosome-specific sublibraries in two ways. First, isolated chromosomes are radiolabeled and used as hybridization probes for existing banks. This approach has been used to assign over 3000 *A nidulans* cosmid clones to specific chromosomes (H. Brody and W. E. Timberlake, unpublished results). In the second, isolated chromosomal DNA is used directly to construct cosmid or YAC libraries. Chromosome specific sublibraries greatly reduce the effort involved in constructing along contiguous sequences of DNA called "contigs." Efforts are underway to establish "contig maps" for the complete genomes of *A. nidulans* and *N. crassa.* These maps would make cloning by chromosomal position almost instantaneous; one would simply pick overlapping clones from microtiter dishes that connect two markers.

D. Cloning Genes by Sequence Similarity

A well-proved procedure for cloning specific genes of interest is by sequence similarity with previously cloned genes of related function from other organisms. This approach (the "clone by phone" approach) simply involves obtaining cloned DNA fragments and using them as hybridization probes, usually at low stringency, for Southern blots or filter replicas of clone banks. As the number of genes cloned increases, the probability that a gene related to a new target gene will already be available becomes higher and higher. Most university and industrial laboratories have ready access to the GenBank and EMBL DNA databases, making the search for homologous genes simple and straightforward. The requirement of many scientific journals that a database access number be included with manuscripts should assure that new sequences are entered in a timely fashion. Numerous fungal genes have been cloned by this approach and, due to its rapidity and simplicity, it is one that should be explored before undertaking more complicated cloning procedures.

V. TARGETED GENOMIC INSERTIONS

The ability to direct integration of transforming DNA to preselected chromosomal sites is of great value for the analysis of gene structure, function, and regulation, and is possible with many fungi. For example, genetic regulatory elements that have been subjected to *in vitro* mutagenesis can be tested for activity *in vivo* by reintroducing the sequences at a single genomic site. The activities of one mutant element can then be compared directly with those of other mutant elements without concern about interference from chromosomal position effects. Similarly, the functions of cloned genes with unknown activities can be examined by site-directed mutagenesis. Once inactivated, these previously unknown genes can be mapped by their tight linkage to the transformation marker. A number of procedures have proved useful for directing integration of transforming DNA molecules to preselected sites. These procedures can be applied both to making targeted mutations, by replacing chromosomal elements with novel constructs, and to repeatedly inserting constructs at a single preselected locus. A few such procedures that have been used extensively are outlined in the following two sections.

A. Targeted Mutations

It is often important to determine the phenotype of a mutant in which a gene under study has lost all function. For example, genes cloned by many of the procedures summarized in the previous sections could have unknown functions that might be inferred from the phenotypes of mutants. Similarly, existing mutations used to select wild-type copies of a gene of interest might not result in complete loss of gene function and null mutations could be highly informative. Directed inactivation of preselected genes can be accomplished by transformation with linear DNA molecules (Rothstein, 1983; Miller *et al.,* 1985) or circular plasmids (Scherer and Davis, 1979; Botstein and Davis, 1982; Miller *et al.,* 1985).

The first strategy is illustrated in Fig. 4. First, a DNA fragment containing the gene to be inactivated is cloned and a restriction map is prepared. A transformation marker is then introduced at a restriction site occurring within the target gene. If a single restriction site is used, the mutation will be a simple insertion. If separated restriction sites are used, the resident fragment will be partially or entirely deleted and replaced with the marker. In either case, the sites are chosen so a linear DNA fragment containing the target gene with inserted transformation marker and as much flanking DNA as possible can be excised from the vector used to make the construct. The use of different restriction sites at each end of the linear fragment may reduce the possibility of recircularization of the DNA molecule during transformation, which can lead to undesired transformation events (Miller *et al.,* 1985). The linear DNA fragment is then used to transform an appropriate recipient strain. Genomic integration of the linear molecule at the homologous site occurs by a process formally equivalent to a double crossover event (Fig. 4), leading to eviction of the resident wild-type gene and insertion of the mutated version and the selective marker.

The second gene inactivation strategy is illustrated in Fig. 1. A DNA fragment contained entirely within the coding region of the target gene is

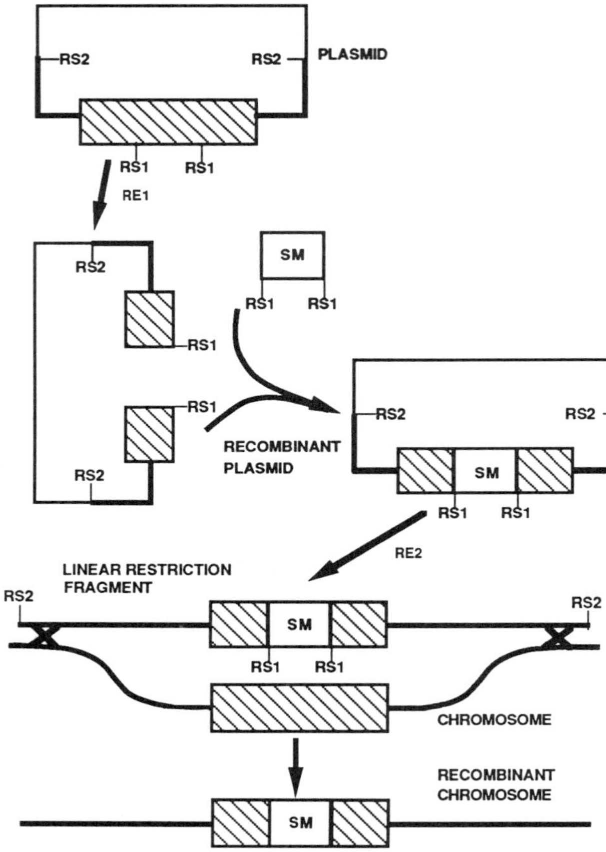

Fig. 4. Directed gene inactivation with linear DNA fragments. A plasmid containing the gene to be inactivated is digested with restriction endonuclease 1 (RE1) to delete internal sequences defined by restriction site 1 (RS1). A selective marker gene (SM) is ligated into the RS1 sites to produce an intermediate plasmid. The plasmid is cut with restriction enzyme 2 (RE2) to produce the linear restriction fragment containing genomic DNA. This fragment is then used to transform cells, and transformants containing the recombinant chromosome, as evidenced by the presence of the marker gene (SM), are selected.

cloned into a circular plasmid containing a transformation marker. Integration of the plasmid by a single homologous recombination event between the target gene and the cloned internal sequence leads to formation of a tandem duplication in which one copy of the gene lacks the 5′ end and the other copy lacks the 3′ end.

Neither of these procedures is completely efficient in filamentous fungi.

By using Southern blot analysis, transformants containing the desired events can be distinguished from those that have undesired events. The frequency of desired events may be increased in the first procedure by including as much flanking DNA as possible, presumably by increasing the probability of initiation of a homologous recombination event. The frequency of desired events may be increased in the second procedure by linearization of circular plasmids by cutting within the target sequence, presumably by providing recombinogenic single stranded DNA ends.

An alternative approach to gene inactivation has been adopted in *Neurospora crassa* because, in this species, the frequency of homologous integration events is fairly low. In this organism, introduced duplicated sequences are eliminated at high frequency during the sexual phase by RIP (see Chapter 11). Thus, genes can be inactivated efficiently simply by introducing an additional copy of the target gene and conducting a cross. It is not clear how widespread the RIP phenomenon is, so its potential for use in other organisms cannot yet be evaluated.

In *Saccharomyces cerevisiae,* it is possible to make a directed mutation in an essential gene in a diploid and then prove that the gene is essential by tetrad analysis (Botstein and Davis, 1982; Struhl, 1983). This approach has limited applicability in filamentous fungi because, in many species in which tetrad analysis is possible (for example, *N. crassa*), stable vegetative diploids cannot be formed. In other species (for example, *Aspergillus nidulans*), in which stable diploids can be formed, tetrad analysis is not practical. Two alternative approaches have been adopted with *A. nidulans.* In the first strategy, one gene copy is disrupted in a vegetative diploid by site-directed mutagenesis. The diploid is then "haploidized" (Clutterbuck, 1974) by treatment with the antitubulin antibiotic benomyl, and haploid segregants are tested for the presence of the mutated gene and the transformation marker. Absence of the mutated and marker genes in all segregants accompanied by normal segregation of other markers is considered *a priori* evidence that the gene is essential. By contrast, recovery of the disrupted allele in a haploid segregant proves that it is not essential. In the second approach, a haploid strain is transformed with a plasmid that disrupts the gene under study (Osmani *et al.,* 1988). Gene disruption can be tolerated in primary transformants because they are most often heterokaryotic, containing both transformed and untransformed nuclei. In many fungi, the asexual spores are uninucleate, or multinucleate but homokaryotic, so masked lethal phenotypes may be exposed by inducing asexual sporulation in the primary transformants, as long as the lethality is not expressed prior to spore formation. If the mutation leads to cell death any time during spore germination, it can be detected by observing abortive growth of germinating spores form the heterokaryon. In *N. crassa,* premeiotic mutation of essential genes by RIP leads to 2:2 segregation of viable versus inviable ascospores in many asci (Selker *et al.,* 1987).

B. Directed Insertions

It is frequently important to examine transformants that are isogenic except for known differences in the introduced DNA sequences. Thus, techniques for reproducibly directing the insertion of plasmids to a single chromosomal site are desirable so strains will not vary according to the chromosomal positions of integrated plasmids. Several such techniques have been developed for filamentous fungi, in which integration of plasmids at heterologous sites is common. In one approach, plasmids are used that contain an internal restriction fragment from a gene whose inactive state lends itself to a convenient screening or selection procedure. Integration of the plasmid by recombination with this homologous site inactivates the gene (Fig. 1), so directed integrants can be readily selected. For example, in *Aspergillus nidulans* two genes, *wA* and *yA,* whose products are required for conidial pigmentation, have been cloned and molecularly mapped (Yelton *et al.,* 1985; O'Hara and Timberlake, 1989; Aramayo and Timberlake, 1990; Mayorga and Timberlake, 1990; Tilburn *et al.,* 1990). Homologous integration of plasmids containing internal fragments from these genes leads to formation of white- or yellow-spored colonies that can be easily selected from a background of wild-type green-spored colonies. Similarly, inclusion of an internal fragment from the *pyrG* gene in a circular transformation plasmid will sometimes lead to recombinational inactivation of this gene, an event that can be selected for by resistance of the *pyrG*$^-$ strains to 5-fluoroorotic acid. This approach can be generalized to many genes in different fungal species.

A converse approach requires the existence of a suitable recipient strain that contains a mutant allele of the gene used for selection. Transformation of this strain with a plasmid containing a different mutant allele of the same gene will lead to formation of a wild-type allele only by homologous recombination, as illustrated in Fig. 5. Ideally the mutations in the selective gene are as widely separated as possible to provide the maximum interval over which recombination can occur. In a related approach, shown in Fig. 6, a linear DNA fragment containing one mutant allele coupled to a sequence to be inserted is used to transform a strain containing a different mutant allele (M. Case, R. Geever, and N. Giles, personal communication). Transformants containing a wild-type allele often contain the linked sequence.

Efficient double selection schemes are also possible. For example, in *A. nidulans,* the *alcA* gene, encoding catabolic alcohol dehydrogenase, can be

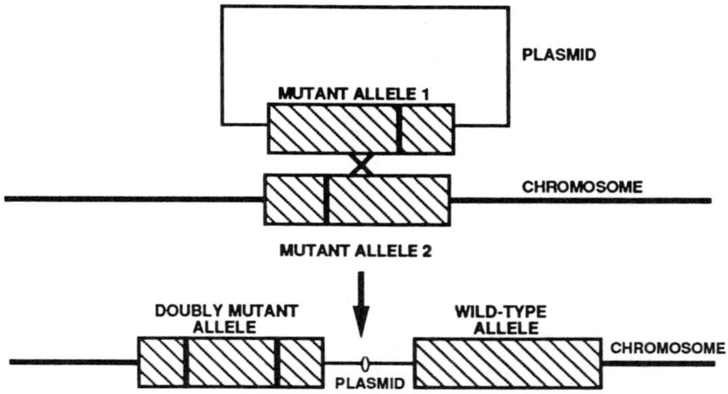

Fig. 5. Targeted plasmid integration. If the recipient strain contains a mutation in the gene to be utilized for selection, then a circular plasmid containing a different mutant allele can integrate and give rise to a wild-type allele only by a homologous recombination event (X) occurring between the positions of the mutations, indicated by the vertical bars. The recombination event shown leads to formation of a doubly mutant allele (two vertical bars) and a wild-type allele that provides a selection for this integration event.

inserted into the *argB* gene, thereby inactivating it, in an *alcA*$^-$ strain. The resultant strain will be *alcA*$^+$ and use ethanol as a sole carbon source; thus, it will be sensitive to allyl alcohol. The strain will also require arginine for growth. Transformation of this strain with a linear DNA fragment containing the *argB* gene and flanking sequences can lead to displacement of the *alcA*$^+$ allele and reconstitution of the *argB*$^+$ allele. Such transformants can be selected directly by arginine prototrophy or allyl alcohol resistance.

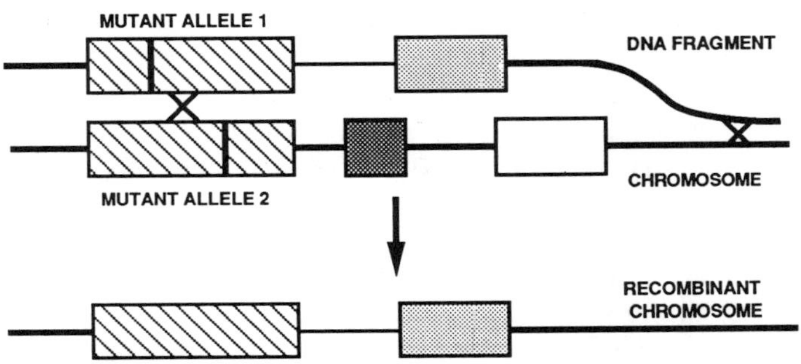

Fig. 6. Extended gene conversion. A linear DNA fragment containing one mutant allele of a selective marker gene and associated sequences is used to transform a recipient strain containing a different mutant allele of the selective marker gene. Recombination events between the mutant sites and at a distal site (X) lead to incorporation of the exogenous DNA and formation of a recombinant chromosome.

This general approach can be modified in numerous ways to adapt it to specific experimental requirements.

VI. ANALYSIS OF GENE FUNCTION AND REGULATION

The transformation technology currently available for many filamentous fungi opens up new ways to study the mechanisms controlling gene expression and the functions and activities of genes. First, promoters and associated regulatory elements can be fused to heterologous "reporter" genes or DNA sequences so their activities can be studied without interference from the endogenous genes. Second, genes can be mutated *in vitro* and the activities of the new alleles can be studied in the presence or absence of the activity of the wild-type allele. Two general techniques for these types of analyses are summarized in the following sections.

A. Promoter Analysis with Reporter Genes

Reporter genes have been used extensively in bacterial and higher eukaryotic molecular genetics. The β-galactosidase *(lacZ)* and β-glucuronidase *(GUS)* genes have been traditional favorites because of the ease and sensitivity with which they can be assayed either *in situ* or *in vitro.* Although both genes are useful in fungi, the *lacZ* gene has been most often employed. van Gorcom *et al.* (1985) showed that fusion of the *Aspergillus nidulans trpC* promoter/regulatory (p/r) sequences to *lacZ* led to production of β-galactosidase in transformed strains. The novel β-galactosidase was not subject to glucose repression, in contrast with the endogenous enzyme. Subsequently, Hamer and Timberlake (1987) used *trpC*(p/r)–*lacZ* fusion genes to carry out a sequential deletion analysis of the *trpC* promoter. In addition, they showed that fusion of a sporulation-specific promoter to *lacZ* led to developmental regulation of β-galactosidase. As shown in Fig. 7, *lacZ* fusion genes may involve joining a gene's promoter, upstream regulatory sequences, and part of the transcribed and translated region, or different genetic elements in different combinations. Thus, for example, the activity of an enhancer element can be assessed by fusion to a heterologous promoter region.

The ability to detect β-galactosidase by cytochemical staining also has been used to establish cell specificity of gene expression (Adams and Timberlake, 1990; Aguirre *et al.,* 1990). Construction of appropriate fusion genes is currently quite simple because of the numerous lacZ constructs available (Casadaban *et al.,* 1983) and the development of efficient procedures for making oligonucleotide-directed mutations in plasmids (Kunkel, 1985).

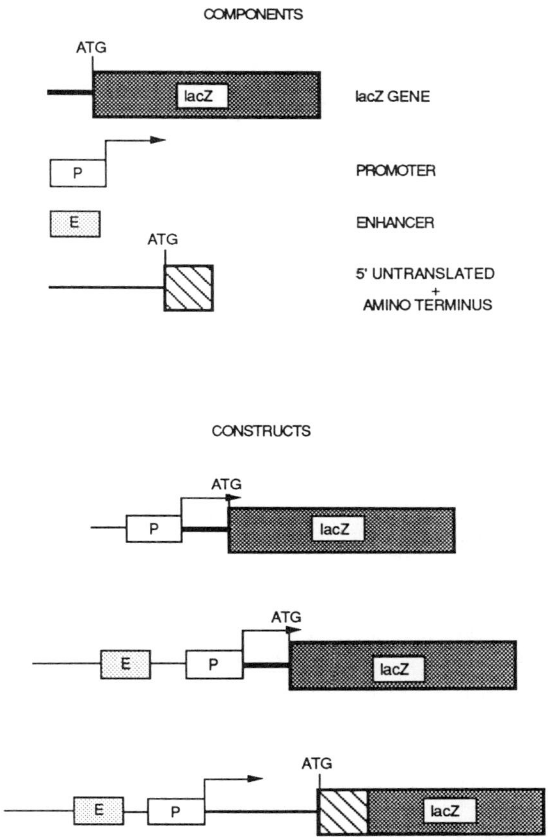

Fig. 7. Reporter gene fusions. The genetic elements *(top)* can be fused in various combinations *(bottom)* to examine different aspects of gene regulation.

Caution must be exercised in interpreting the results of experiments with reporter genes. For example, little is known about the effects of growth and development conditions on the stability of β-galactosidase or its mRNA, so variations in enzyme level under different conditions may reflect parameters other than transcription. Similarly, nothing is known about the effects on growth and development of high-level expression of the reporter genes. Nevertheless, this experimental approach is extremely valuable for making a relatively quick assessment of the ability of selected sequences to control gene activity. In many instances, it is important to follow up initial experiments with more carefully controlled studies of the associated regulatory elements and the actual gene or of the regulatory elements fused to other reporter genes.

Genes can, of course, be tagged in different ways to avoid use of reporter genes. For example, they can be altered by introducing deletions or insertions that make it possible to distinguish experimentally between the resident copy of the gene and an introduced copy by S1 nuclease or primer extension analysis. A problem sometimes encountered is that the modification leads to mRNA destabilization, making it impossible to detect the novel species. In such cases, the endogenous gene copy may be deleted or otherwise modified. As shown by these examples, many strategies for studying genes are possible because of available fungal transformation systems.

B. Functional Analysis of Genes by Controlled Expression

In some instances, misscheduled or ectopic expression of a gene provides as much information about its function as does its inactivation by mutation. The strategy for causing incorrect expression of a regulated gene involves fusion of the gene to a promoter that permits controlled expression, and subsequent introduction of the fusion gene into a recipient strain that contains or lacks the endogenous copy of the gene. The number of conveniently regulatable fungal promoters is currently quite limited, and only the *Aspergillus nidulans alcA* promoter [*alcA*(p)] has been used extensively (Adams *et al.,* 1988; Osmani *et al.,* 1988; Mirabito *et al.,* 1989; Marshall and Timberlake, 1990). This promoter is subject to glucose repression and is strongly induced by a variety of alcohols. The promoter can be readily fused to a gene by standard recombinant DNA techniques; the fusion gene can then be reintroduced into the organism by transformation. After selection of appropriate transformants, which are grown on medium containing glucose as the carbon source to repress the gene, they can be transferred to medium containing an alcohol as the sole carbon source to induce the gene. Alcohol induction occurs rapidly and large numbers of cells can be induced synchronously, thus facilitating the isolation of macromolecules for analysis. It can be expected that additional highly regulated fungal promoters will become available in the future.

VII. RELATIONSHIP OF CLASSICAL GENETICS AND TRANSFORMATION

Although the experimental approaches made possible in fungi by DNA-mediated transformation offer many ways to attack problems, classical genetics continues to play an important role in analysis of biological problems in fungi. For example, genetics is used to prove that a particular gene has been cloned by complementation, rather than a suppressor of the mutation employed in the screen. Similarly, crosses are employed to demonstrate that a mutant phenotype thought to be caused by a tranformation event is actually due to insertion of DNA sequences at a particular locus. New mutations made by reverse genetics need to be checked to see if they are located in previously recognized genes or in new genes. Genetic construction of strains before and after using them for transformation is essential for many experimental strategies. Thus, even though modern transformation technology makes many new systems accessible to experimentation, traditional genetic organisms such as *Saccharomyces cerevisiae, Schizosaccharomyces pombe, Aspergillus nidulans, Neurospora crassa,* and *Schizophyllum commune* can be expected to continue to be most useful in addressing many basic questions in fungal biology. Other species with tractable, although perhaps less well-characterized, genetic systems, such as *Nectria haematococca, Magnaporthe grisea,* and *Ustilago maydis,* will be logical choices for more specialized studies. However, some species with no, or essentially intractable, genetic systems have unique biological features that warrant considerable study, for example, many obligate pathogens and industrial fungi. Development of DNA-mediated transformation systems for this last group of fungi will certainly have a large effect on the ability to advance our knowledge of the activities of filamentous fungi.

ACKNOWLEDGMENTS

I wish to thank Meg Marshall, Alex Andrianopoulos, and Jonathan Arnold for critical reviews of the manuscript, and Kathy Vinson for her invaluable assistance in preparing the manuscript. Our studies with *A. nidulans* have been supported by grants from the NIH, NSF, and USDA.

REFERENCES

Adams, T. H., Boylan, M. T., and Timberlake, W. E. (1988). *brlA* is necessary and sufficient to direct conidiophore development in *Aspergillus nidulans. Cell* **54,** 353–362.

Adams, T. H., and Timberlake, W. E. (1990). Upstream repressors modulate transcription from Inr elements in *Aspergillus* developmental regulatory genes. *Mol. Cell. Biol.* **10,** 4912–4919.

Aguirre, J., Adams, T. H., and Timberlake, W. E. (1990). Spatial control of developmental regulatory genes in *Aspergillus nidulans. Exp. Mycol.* **14,** 290–293.

Andrianopoulos, A., and Hynes, M. J. (1988). Cloning and analysis of the positively acting regulatory gene *amdR* from *Aspergillus nidulans. Mol. Cell. Biol.* **8,** 3532–3541.

Aramayo, R., and Timberlake, W. E. (1990). Sequence and molecular structure of the *Aspergillus nidulans yA* gene. *Nuc. Acids Res.* **18,** 3415.

Arnold, J., Cuticchia, A. J., Newsome, D. A., Jennings, W. W., III, and Ivarie, R. (1988). Mono- through hexanucleotide composition of the sense strand of yeast DNA: A Markov chain analysis. *Nuc. Acids Res.* **16,** 7145–7158.

Austin, B., Hall, R. M., and Tyler, B. M. (1990). Optimized vectors and selection for transformation of *Neurospora crassa* and *Aspergillus nidulans* to bleomycin and phleomycin resistance. *Gene.* **93,** 157–162.

Ballance, D. J., Buxton, F. P., and Turner, G. (1983). Transformation of *Aspergillus nidulans* by the orotidine-5′-phosphate decarboxylase gene of *Neurospora crassa. Biochem. Biophys. Res. Commun.* **112,** 284–289.

Beggs, J. D. (1978). Transformation of yeast by a replicating hybrid plasmid. *Nature (London)* **275,** 104–109.

Bellen, H. J., O'Kane, C. J., Wilson, C., Grossniklaus, U., Pearson, R. K., and Gehring, W. J. (1989). P-element-mediated enhancer detection: A versatile method to study development in *Drosophila. Genes Dev.* **3,** 1288–1300.

Berse, B., Dmochowska, A., Skrzypek, M., Weglenski, P., Bates, M. A., and Weiss, R. L. (1983). Cloning and characterization of the ornithine carbamoyl transferase gene from *Aspergillus nidulans. Gene* **25,** 109–117.

Bier, E., Vaessin, H., Shepherd, S., Lee, K., McCall, K., Barbel, S., Ackerman, L., Carretto, R., Uemura, T., Grell, E., Jan, L. Y., and Jan, Y. N. (1989). Searching for pattern and mutation in the *Drosophila* genome with a P-*lacZ* vector. *Genes Dev.* **3,** 1273–1287.

Binninger, D. M., Skrzynia, C., Pukkila, P. J., and Casselton, L. A. (1987). DNA-mediated transformation of the basidiomycete *Coprinus cinereus. EMBO J.* **64,** 835–840.

Boeke, J. D., LaCroute, F., and Fink, G. R. (1984). A positive selection for mutants lacking orotidine-5′-phosphate decarboxylase activity in yeast: 5-fluoro-orotic acid resistance. *Mol. Gen. Genet.* **197,** 345–346.

Botstein, D., and Davis, R. W. (1982). Principles and practice of recombinant DNA research with yeast. *In* "The Molecular Biology of the Yeast Saccharomyces. II. Metabolism and Gene Expression" (J. N. Strathern, E. W. Jones, and J. R. Broach, eds)., pp. 607–637. Cold Spring Harbor Laboratory Press, Cold Spring Harbor, New York.

Boylan, M. T., Holland, M. J., and Timberlake, W. E. (1986). *Saccharomyces cerevisiae* centromere CEN11 does not induce chromosome instability when integrated into the *Aspergillus nidulans* genome. *Mol. Cell. Biol.* **6,** 3621–3625.

Burke, D. T., Carle, G. F., and Olson, M. V. (1987). Cloning of large segments of exogenous DNA into yeast by means of artificial chromosome vectors. *Science* **236,** 806–812.

Buxton, F. P., and Radford, A. (1984). The transformation of mycelial spheroplasts of *Neurospora crassa* and the attempted isolation of an autonomous replicator. *Mol. Gen. Genet.* **196,** 339–344.

Cambareri, E. B., Jensen, B. C., Schabtach, E., and Selker, E. V. (1989). Repeat-induced G–C to A–T mutations in *Neurospora. Science* **244,** 1571–1575.

Cantoral, J. M., Barredo, J. L., Alvarez, E., Diez, B., and Martin, J. F. (1988). Nucleotide sequence of the *Penicillium chrysogenum pyrG* (orotidine-5′-phosphate decarboxylase) gene. *Nuc. Acids Res.* **16,** 8177.

Casadaban, M. J., Chou, J., and Cohen, S. N. (1980). *In vitro* gene fusions that join an enzymatically active β-galactosidase segment to amino-terminal fragments of exogenous proteins: *Escherichia coli* plasmid vectors for the detection and cloning of translational initiation signals. *J. Bacteriol.* **143,** 971–980.

Casadaban, M. J., Martinez-Arias, A., Shapira, S. K., and Chou, J. (1983). β-Galactosidase gene fusions for analyzing gene expression in *Escherichia coli* and yeast. *In* "Methods in Enzymology" (R. Wu, L. Grossman, and K. Moldave, eds.), Vol. 100, pp. 293–308. Academic Press, San Diego.

Case, M. E., Schweizer, M., Kushner, S. R., and Giles, N. H. (1980). Efficient transformation of *Neurospora crassa* by utilizing hybrid plasma DNA. *Proc. Natl. Acad. Sci. U.S.A.* **76,** 5259–5263.

Chumley, F. G., Parsons, K. A., and Valent B. (1985). Transformation and gene cloning in *Magnaporthe grisea. J. Cell. Biochem. Suppl.* **9C** 197.

Clarke, L., and Carbon, J. (1980). Isolation of a yeast centromere and construction of functional small circular chromosomes. *Nature (London)* **287,** 504–509.

Clutterbuck, A. J. (1974). *Aspergillus nidulans. In* "Handbook of Genetics 1" (R. C. King, ed.), pp. 447–510. Plenum, New York.

Daboussi, M. J., Djeballi, A., Gerlinger, C., Blaiseau, P. L., Bouvier, I., Cassan, M., Lebrun, M. H., Parisot, D., and Brygoo, Y. (1989). Transformation of seven species of filamentous fungi using the nitrate reductase gene of *Aspergillus nidulans. Curr. Genet.* **15,** 453–456.

Diallinas, G., and Scazzocchio, C. (1989). A gene coding for the uric acid–xanthine permease of *Aspergillus nidulans:* Inactivational cloning, characterization, and sequence of a *cis*-acting mutation. *Genetics* **122,** 341–350.

Dunne, P. W., and Oakley, B. R. (1988). Mitotic gene conversion, reciprocal recombination and gene replacement at the *benA,* beta-tubulin, locus of *Aspergillus nidulans. Mol. Gen. Genet.* **213,** 339–345.

Gems, D. H., Johnstone, I. L., and Clutterbuck, A. J. (1991). An autonomously replicating plasmid transforms *Aspergillus nidulans* at high frequency. *Gene* **98,** 61–68.

Grimm, C., Kohli, J., Murray, J., and Maundrell, K. (1988). Genetic engineering of *Schizosaccharomyces pombe:* A system for gene disruption and replacement using the *ura4* gene as a selectable marker. *Mol. Gen. Genet.* **215,** 81–86.

Guarente, L. (1987). Regulatory proteins in yeast. *Ann. Rev. Genet.* **21,** 425–452.

Gwynne, D. I., Buxton, F. P., Sibley, S., Davies, R. W., Lockington, R. A., Scazzocchio, C., and Sealey-Lewis, H. M. (1987). Comparison of *cis*-acting control regions of two coordinately controlled genes involved in ethanol utilization in *Aspergillus nidulans. Gene* **51,** 205–216.

Hamer, J. E., and Timberlake, W. E. (1987). Functional organization of the *Aspergillus nidulans trpC* promoter. *Mol. Cell. Biol.* **7,** 2352–2359.

Hicks, J. B., Hinnen, A., and Fink, G. R. (1978). Properties of yeast transformation. *Cold Spring Harbor Symp. Quant. Biol.* **XLIII,** 1305–1313.

Hinnen, A., Hicks, J. B., and Fink, G. R. (1978). Transformation of yeast. *Proc. Natl. Acad. Sci. U.S.A.* **75,** 1929–1933.

Hsiao, C-H., and Carbon, J. (1979). High-frequency transformation of yeast by plasmids containing the cloned yeast *ARG4* gene. *Proc. Natl. Acad. Sci. U.S.A.* **76,** 3829–3833.

Hynes, M. J., and Davis, M. A. (1986). The *amdS* gene of *Aspergillus nidulans:* Control by multiple regulatory signals. *BioEssays* **5,** 123–128.

Hynes, M. J., Corrick, C. M., and King, J. A. (1983). Isolation of genomic clones containing the *amdS* gene of *Aspergillus nidulans* and their use in the analysis of structural and regulatory mutations. *Mol. Cell. Biol.* **3,** 1430–1439.

Hynes, M. J., Corrick, C. M., Kelly, J. M., and Littlejohn, T. G. (1988). Identification of the sites of action for regulatory genes controlling the *amdS* gene of *Aspergillus nidulans. Mol. Cell. Biol.* **8,** 2589–2596.

Johnstone, I. L., Hughes, S. G., and Clutterbuck, A. J. (1985). Cloning an *Aspergillus nidulans* developmental gene by transformation. *EMBO J.* **4,** 1307–1311.

Katz, M. E., and Hynes, M. J. (1989a). Characterization of the *amdR* controlled *IamA* and *IamB* genes of *Aspergillus nidulans. Genetics* **122,** 331–339.

Katz, M. E., and Hynes, M. J. (1989b). Gene function identified by interspecific transformation. *Gene* **78,** 167–171.

Kelly, J. M., and Hynes, M. J. (1987). Multiple copies of the *amdS* gene of *Aspergillus nidulans* cause titration of *trans*-acting regulatory proteins. *Curr. Genet.* **12,** 21–31.

Kinnaird, J. H., Kinsey, J. A., Eaton, M., and Fincham, J. R. S. (1982). Cloning of the *am* (glutamate dehydrogenase) gene of *Neurospora crassa* through the use of a synthetic DNA probe. *Gene* **20,** 387–396.

Kos, T., Kuijvenhoven, A., Hessing, H. G. M., Pouwels, P. H., and van den Hondel, C. A. M. J. J. (1988). Nucleotide sequence of the *Aspergillus niger trpC* gene: Structural relationship with analogous genes of other organisms. *Curr. Genet.* **13,** 137–144.

Krumlauf, R., and Marzluf, G. A. (1979). Characterization of the sequence complexity and organization of the *Neurospora crassa* genome. *Biochem.* **18,** 3705–3713.

Kunkel, T. A. (1985). Rapid and efficient site-specific mutagenesis without phenotypic selection. *Proc. Natl. Acad. Sci. U.S.A.* **82,** 488–492.

LeChevanton, L., Leblon, G., and Lebilcot, S. (1989). Duplications created by transformation in *Sordaria macrospora* are not inactivated during meiosis. *Mol. Gen. Genet.* **218,** 390–396.

Malardier, L., Daboussi, M. U., Julien, J., Roussel, F., Scazzocchio, C., and Brygoo, Y. (1989). Cloning of the nitrate reductase gene *(niaD)* of *Aspergillus nidulans* and its use for transformation of *Fusarium oxysporum. Gene* **78,** 147–156.

Mann, B. J., Akins, R. A., Lambowitz, A. M., and Metzenberg, R. L. (1988). The structural gene for a phosphorus-repressible phosphate permease in *Neurospora crassa* can complement a mutation in positive regulatory gene *nuc-1. Mol. Cell. Biol.* **8,** 1376–1379.

Mann, B. J., Bowman, B. J., Grotelueschen, J., and Metzenberg, R. L. (1989). Nucleotide sequence of *pho-4*$^+$, encoding a phosphate-repressible phosphate permease of *Neurospora crassa. Gene* **83,** 281–289.

Marshall, M. A., and Timberlake, W. E. (1991). *Aspergillus wetA* regulates spore specific gene expression. *Mol. Cell. Biol.* **11,** 55–62.

Mattern, I. E., Unkles, S., Kinghorn, J. R., Pouwels, P. H., and van den Hondel, C. A. M. J. J. (1987). Transformation of *Aspergillus oryzae* using the *A. niger pyrG* gene. *Mol. Gen. Genet.* **210,** 460–461.

Mayorga, M. E., and Timberlake, W. E. (1990). Isolation and molecular characterization of the *Aspergillus nidulans wA* gene. *Genetics.* in press.

Miller, B. L., Miller, K. Y., and Timberlake, W. E. (1985). Direct and indirect gene replacements in *Aspergillus nidulans. Mol. Cell. Biol.* **5,** 1714–1721.

Mirabito, P. M., Adams, T. H., and Timberlake, W. E. (1989). Interactions of three sequentially expressed genes control temporal and spatial specificity in *Aspergillus* development. *Cell* **57,** 859–868.

Mishra, N. C., and Tatum, E. L. (1973). Non-Mendelian inheritance of DNA-induced inositol independence in *Neurospora. Proc. Natl. Acad. Sci. U.S.A.* **70,** 3875–3879.

Mullaney, E. J., Hamer, J. E., Roberti, K. A., Yelton, M. M., and Timberlake, W. E. (1985). Primary structure of the *trpC* gene from *Aspergillus nidulans. Mol. Gen. Genet.* **199,** 37–45.

Muñoz-Rivas, A. M., Specht, C. A., Ullrich, R. C., and Novotny, C. P. (1986). Isolation of the DNA sequence coding indole-3-glycerol phosphate synthetase and phosphoribosylanthranilate isomerase of *Schizophyllum commune. Curr. Genet.* **10,** 909–913.

Oakley, B. R., Rinehart, J. E., Mitchell, B. L., Oakley, C. E., Carmona, C., Gray, G. L., and May, G. S. (1987). Cloning, mapping and molecular analysis of the *pyrG* orotidine-5′-phosphate decarboxylase gene of *Aspergillus nidulans. Gene* **61,** 385–400.

O'Hara, E. B., and Timberlake, W. E. (1989). Molecular characterization of the *Aspergillus nidulans yA* locus. *Genetics* **129,** 249–254.

O'Kane, C. J., and Gehring, W. J. (1987). Detection *in situ* of genomic regulatory elements in *Drosophila. Proc. Natl. Acad. Sci. U.S.A.* **84,** 9123–9127.

Orbach, M. J., Porro, E. B., and Yanofsky, C. (1986). Cloning and characterization of the gene for β-tubulin from a benomyl-resistant mutant of *Neurospora crassa* and its use as a dominant selection marker. *Mol. Cell. Biol.* **6,** 2452–2461.

Orbach, M. J., Schneider, W. P., and Yanofsky, C. (1988). Cloning of methylated transforming DNA from *Neurospora crassa* in *Escherichia coli. Mol. Cell. Biol.* **8,** 2211–2213.

Osmani, S. A., Pu, R. T., and Morris, N. R. (1988). Mitotic induction and maintenance by overexpression of a G2-specific gene that encodes a potential protein kinase. *Cell* **53,** 237–244.

Picknett, T. M., Saunders, G., and Holt, G. (1989). Expression of the *trpC* gene from *Penicillium chrysogenum* in *Aspergillus nidulans. Biotech. Appl. Biochem.* **11,** 464–470.

Rambosek, J., and Leach, J. (1987). Recombinant DNA in filamentous fungi: Progress and prospects. *CRC Crit. Rev. Biotechnol.* **6,** 357–393.

Raper, C. A., and Timberlake, W. E. (1985). Identification of a polyadenylated transcript correlated with development in the sexual cycle of *Schizophyllum. Exp. Mycol.* **9,** 351–355.

Rothstein, R. J. (1983). One-step gene disruption in yeast. *In* "Methods in Enzymology" (R. Wu, L. Grossman, and K. Moldave, eds.), Vol. 101, pp. 202–211. Academic Press, San Diego.

Saiki, R. K., Gelfand, D. H. Stoffel, S., Scharf, S. J., Higuchi, R., Horn, G. T., Mullis, K. B., and Erlich, H. A. (1988). Primer-directed enzymatic amplification of DNA with a thermostable DNA polymerase. *Science* **239,** 487–491.

St. John, T. P., and Davis, R. W. (1979). Isolation of galactose-inducible sequences from *Saccharomyces cerevisiae* by differential plaque hybridization. *Cell* **16,** 443–452.

Sambrook, J., Fritsch, E. F., and Maniatis, T. (1989). "Molecular Cloning" (C. Nolan, ed.). Cold Spring Harbor Laboratory Press, Cold Spring Harbor, New York.

Sargent, T. D., and Dawid, I. G. (1983). Differential gene expression in the gastrula of *Xenopus laevis. Science* **222,** 135–139.

Schechtman, M. G., and Yanofsky, C. (1983). Structure of the trifunctional *trp-1* gene from *Neurospora crassa* and its aberrant expression in *Escherichia coli. J. Mol. Appl. Genet.* **2,** 83–99.

Scherer, S., and Davis, R. W. (1979). Replacement of chromosome segments with altered DNA sequences constructed *in vitro. Proc. Natl. Acad. Sci. U.S.A.* **76,** 4951–4955.

Selker, E. U., and Garrett, P. W. (1988). DNA sequence duplications trigger gene activation in *Neurspora crassa. Proc. Natl. Acad. Sci. U.S.A.* **85,** 6870–6874.

Selker, E. U., Cambareri, E. B., Jensen, B. C., and Haack, K. R. (1987). Rearrangement of duplicated DNA in specialized cells of *Neurospora. Cell* **51,** 741–752.

Skrzynia, C., Binninger, D. M., Alspaugh, J. A., II, and Pukkila, P. J. (1989). Molecular characterization of *TRP1,* a gene coding for tryptophan synthetase in the basidiomycete *Coprinus cinereus. Gene* **81,** 73–82.

Stinchcomb, D. T., Struhl, K., and Davis, R. W. (1979). Isolation and characterization of a yeast chromosomal replicator. *Nature (London)* **282,** 39–43.

Stinchcomb, D. T., Mann, C., and Davis, R. W. (1982). Centromeric DNA from *Saccharomyces cerevisiae. J. Mol. Biol.* **158,** 157–179.

Struhl, K. (1983). The new yeast genetics. *Nature (London)* **305,** 391–397.

Struhl, K., Stinchcomb, D. T., Scherer, S., and Davis, R. W. (1979). High-frequency transformation of yeast: Autonomous replication of hybrid DNA molecules. *Proc. Natl. Acad. Sci. U.S.A.* **76,** 1035–1039.

Szostak, J. W., and Blackburn, E. H. (1982). Cloning yeast telomeres on linear plasmid vectors. *Cell* **29,** 245–255.

Tilburn, J., Scazzocchio, C., Taylor, G. G., Zabicky-Zissman, J. H., Lockington, R. A., and

Davies, R. W. (1983). Transformation by integration in *Aspergillus nidulans. Gene* **26,** 205–221.

Tilburn, J., Roussel, F., and Scazzocchio, C. (1990). Insertional inactivation and cloning of the *wA* gene of *Aspergillus nidulans. Genetics.* **126,** 81–90.

Timberlake, W. E. (1978). Low repetitive DNA content in *Aspergillus nidulans. Science* **202,** 973–975.

Timberlake, W. E. (1980). Developmental gene regulation in *Aspergillus nidulans. Dev. Biol.* **78,** 497–510.

Timberlake, W. E., and Marshall, M. A. (1989). Genetic engineering in filamentous fungi. *Science* **244,** 1313–1317.

Timberlake, W. E., Boylan, M. T., Cooley, P. M., Mirabito, P. M., O'Hara, E. B., and Willett, C. E. (1985). Rapid identification of mutation-complementing restriction fragments from *Aspergillus nidulans* cosmids. *Exp. Mycol.* **9,** 351–355.

Turner, G., Ballance, D. J., Ward, M., and Beri, R. K. (1985). Development of cloning vectors and a marker for gene replacement techniques in *Aspergillus nidulans. In* "Molecular Genetics of Filamentous Fungi" (W. E. Timberlake, ed.), pp. 15–28. Alan R. Liss, New York.

Upshall, A., Gilbert, T., Saari, G., O'Hara, P. J., Weglenski, P., Berse, B., Miller, K., and Timberlake, W. E. (1986). Molecular analysis of the *argB* gene of *Aspergillus nidulans. Mol. Gen. Genet.* **204,** 349–354.

van Gorcom, R. F. M., Pouwels, P. H., Goosen, T., Visser, J., van den Broek, H. W. J., Hamer, J. E., Timberlake, W. E., and van den Hondel, C. A. M. J. J. (1985). Expression of the *Escherichia coli* β-galactosidase fusion gene in *Aspergillus nidulans. Gene* **40,** 99–106.

Vapnek, D., Hautala, J. A., Jacobson, J. W., Giles, N. H., and Kushner, S. R. (1977). Expression in *Escherichia coli* K-12 of the structural gene for catabolic dehydroquinase of *Neurospora crassa. Proc. Natl. Acad. Sci. U.S.A.* **74,** 3508–3512.

Vollmer, S. J., and Yanofsky, C. (1986). Efficient cloning of genes of *Neurospora crassa. Proc. Natl. Acad. Sci. U.S.A.* **83,** 4869–4873.

Weltring, K-M., Turgeon, B. G., Yoder, O. C., and VanEtten, H. D. (1988). Isolation of a phytoalexin-detoxification gene from the plant pathogenic fungus *Nectria haematococca* by detecting its expression in *Aspergillus nidulans. Gene* **68,** 335–344.

Wieland, I., Bolger, G., Asouline, G., and Wigler, M. (1989). A method for difference cloning: Gene amplification following subtractive hybridization. *Proc. Natl. Acad. Sci. U.S.A.* **87,** 2720–2724.

Yelton, M. M., Hamer, J. E., de Souza, E. R., Mullaney, E. J., and Timberlake, W. E. (1983). Developmental regulation of the *Aspergillus nidulans trpC* gene. *Proc. Natl. Acad. Sci. U.S.A.* **80,** 7576–7580.

Yelton, M. M., Hamer, J. E., and Timberlake, W. E. (1984). Transformation of *Aspergillus nidulans* by using a *trpC* plasmid. *Proc. Natl. Acad. Sci. U.S.A.* **81,** 1470–1474.

Yelton, M. M., Timberlake, W. E., and van den Hondel, C. A. M. J. J. (1985). A cosmid for selecting genes by complementation in *Aspergillus nidulans:* Selection of the developmentally regulated *yA* locus. *Proc. Natl. Acad. Sci. U.S.A.* **82,** 834–838.

Yoder, O. C., Weltring, K., Turgeon, B. G., and Van Etten, H. D. (1986). Technology for molecular cloning of fungal virulence genes. *In* "Biology and Molecular Biology of Plant–Pathogen Interactions" (J. A. Bailey, ed.), pp. 371–384. Springer-Verlag, Berlin.

Zimmerman, C. R., Orr, W. C., Leclerc, R. F., Barnard, E. C., and Timberlake, W. E. (1980). Molecular cloning and selection of genes regulated in *Aspergillus* development. *Cell* **21,** 709–715.

NOTE ADDED IN PROOF

Gems *et al.* (1991) have recently presented convincing evidence for the existence of an autonomously replicating plasmid that transforms *Aspergillus nidulans, A. oryzae,* and *A. niger* at high frequency. The plasmid, called ARp1, is maintained extrachromosomally at about 10 copies per nucleus, is mitotically unstable, and can be rescued directly in *E. coli.* Although ARp1 and its initial derivatives are not ideally suited as cloning vectors, it is likely that derivatives with desirable properties can be made.

4

Molecular Karyotype Analysis of Fungi

DANIEL Z. SKINNER
USDA-ARS and Department of Agronomy
Kansas State University
Manhattan, Kansas

ALLEN D. BUDDE
USDA-ARS Plant Disease Resistance Research Unit
Madison, Wisconsin

SALLY A. LEONG
USDA-ARS Plant Disease Resistance Research Unit and
Department of Plant Pathology
University of Wisconsin
Madison, Wisconsin

MORE GENE MANIPULATIONS
IN FUNGI

I. HISTORICAL OVERVIEW

Traditionally, determining the number of chromosomes present in a fungus has been done by direct microscopic examination or by establishing genetic linkage relationships between known genes. Neither of these procedures is particularly rapid or straightforward and the results obtained from such analyses frequently have yielded different results for many fungi. Much of this variation is due to the difficulty of observing the small condensed chromosomes of fungi or to the small number of genetic markers available for most fungal species.

A new experimental approach to karyotype analysis, called pulsed-field gel electrophoresis (PFGE), involves subjecting intact chromosomes to a pulsed field in an agarose gel matrix. Intact chromosomes migrate through the gel matrix with a velocity dependent on their size and three-dimensional structure. The result is a resolution of the genome into chromosomal bands, that is, an "electrophoretic karyotype." PFGE promises to assist researchers in making great strides in the determination of both chromosome number and size.

Electrophoretic karyotyping was first developed by Schwartz and Cantor (1984) using a PFGE device consisting of two perpendicularly oriented, alternately pulsing electrical fields, at least one of which was inhomogeneous. PFGE allowed separation of large DNA molecules up to 2000 kb (2 megabasepairs, Mgb) in an agarose matrix. In the best case, DNA from lysed *Saccharomyces cerevisiae* cells was separated into 11 distinct zones of resolution. Hybridization of cloned single-copy genes to Southern transfers of the gels showed unique bands of hybridization, indicating that each band was composed largely of a single chromosome. It is now generally accepted that the resolved bands represent intact chromosomes.

PFGE suffers from variable migration of DNA molecules in different lanes that is caused by the variation in electrical field strength across the gel. This makes lane-to-lane comparisons and accurate size determination extremely difficult. This problem was addressed to some extent by Carle and Olson (1984), who developed the orthogonal-field-alternation gel electrophoresis (OFAGE) system. OFAGE applies two alternately pulsing inhomogeneous fields to the DNA, resulting in much more consistent migration of DNA in adjacent lanes than is seen with PFGE. However, the angle between the electrical fields varies from the top to the bottom of the gel, causing DNA molecules to migrate at different rates in different positions in the gel. This results in the characteristically curved migration of DNA in the outer lanes seen in OFAGE gels (Carle and Olson, 1984). Nonetheless, OFAGE electrophoretic karyotypes of *S. cerevisiae* (Carle and Olson, 1984; Carle *et al.*, 1986), *Candida* spp. (Magee and Magee,

1987), and *Ustilago maydis* (Kinscherf and Leong, 1988) have been constructed.

Factors that impact on the resolution of DNA molecules in PFGE and OFAGE have been described in a series of articles (Cantor *et al.,* 1988; Mathew *et al.,* 1988a,b,c). Since these studies were limited to chromosomes with a molecular size of less than 2 Mgb, these data apply directly only to fungi with chromosomes in that size range.

The OFAGE karyotype of *S. cerevisiae* developed by Carle and Olson (1984) was composed of 11 bands and, therefore, did not represent a significant advance over the results of Schwartz and Cantor (1984). To obtain further resolution of chromosomes, the same authors, along with M. Frank, developed another technique: field-inversion gel electrophoresis or FIGE (Carle *et al.,* 1986). FIGE differs from PFGE and OFAGE by the application of only a single electrical field to the system. The field is periodically inverted—the forward portion of the cycle lasting longer than the reverse—yielding a net forward migration. FIGE has a peculiar bimodal distribution of molecular mobilities. In the original example, *S. cerevisiae* chromosomes less than 0.5 Mgb migrated at about the same velocity as the largest 2 chromosomes (~2 Mgb), whereas the chromosomes of intermediate size (0.6–0.7 Mgb) did not migrate at all. This phenomenon was overcome by using a switching interval ramp, that is, the length of the forward and backward pulses was gradually increased over time (Carle *et al.,* 1986). By successively applying OFAGE and FIGE technologies, the author accomplished resolution of the *S. cerevisiae* chromosome complement into 14 distinct bands. Magee *et al.* (1988) used a similar approach to resolve as many as 11 chromosomes of *C. albicans,* an improvement over the 5 zones resolved by FIGE (Lott *et al.,* 1987).

Although these results were impressive and productive, and the methods used represented an improvement over PFGE, the lack of uniformity of electrical fields across lanes still made lane-to-lane comparisons difficult. Two systems have been developed which largely overcome this problem. Contour-clamped homogeneous electric field (CHEF) gel electrophoresis and transverse alternating-field electrophoresis (TAFE) were described virtually simultaneously (Chu *et al.,* 1986; Gardiner *et al.,* 1986).

The CHEF system employs a hexagonal array of 24 electrodes that surround the gel and produce a contour-clamped homogeneous electrical field that alternates between two orientations at an angle of 120° (Chu *et al.,* 1986). The dynamics of DNA migration in the CHEF system are quite complex and have been thoroughly characterized by Vollrath and Davis (1987), who identified four regions of separation in CHEF gels. Region 1 is an area of good resolution containing the smallest DNA molecules and mobility appears to be a linear function of molecular size; Region 2, near the middle of the gel, is characterized by poor resolution; Region 3 is another area of good resolution where mobility and size are again linearly related; and Region 4 is an area of unresolved molecules of the largest size (Vollrath and Davis, 1987). The molecular weight of molecules in region 3 can be adjusted by altering the electrical field strength, the switching interval, and the agarose concentration (Vollrath and Davis, 1987). Using this system, Chu *et al.* (1986) resolved the *S. cerevisiae* karyotype into 15 bands and Vollrath and Davis (1987) resolved the 3 chromosomes of *Schizosaccharomyces pombe.* The CHEF system also has been used to develop electrophoretic karyotypes of *Aspergillus nidulans* (Brody and Carbon, 1989), *S. pombe* and *Histoplasma capsulatum* (Steele *et al.,* 1989), *C. albicans* (Magee *et al.,* 1988), *Neurospora crassa* (Orbach *et al.,* 1988), *Phytophthora megasperma* (Howlett, 1989), *Colletotrichum gloeosporoides* (Masel *et al.,* 1990), *Magnaporthe grisea* (Orbach, 1989; Skinner *et al.,* 1990), *Ustilago hordei* (McCluskey and Mills, 1990), *U. maydis* (Budde and Leong, 1990) and other species from the order Ustilaginales (see Fig. 1). The karyotypes of *M. grisea* and *U. maydis,* developed in our laboratory, will be discussed in detail in the following sections. A report on the *Coprinus* karyotype is included in this volume in Chapter 6.

The TAFE system utilizes an arrangement of electrodes that orients the electrical field transversely to a vertical gel. The DNA molecules actually traverse a three-dimensional path through the gel (Gardiner *et al.,* 1986; Gardiner and Patterson, 1988; Stewart *et al.,* 1988). This system produces straight lanes and has yielded karyotypes of *S. cerevisiae* and *S. pombe* (Gardiner and Patterson, 1988; Stewart *et al.,* 1988) and *Cephalosporium acremonium* (Skatrud and Queener, 1989).

Various other systems have been developed and have been reviewed elsewhere (Anand, 1986; Lai *et al.,* 1989). To our knowledge, none of these have been used to resolve karyotypes of fungi other than yeast. The distinguishing features of these other systems are the manner in which alternating electrical fields are produced and the path the DNA is made to traverse. We note here the programmable autonomously controlled electrode (PACE) gel electrophoresis system (Clark *et al.,* 1988). The PACE system uses a hexagonal array of 24 electrodes, as does the CHEF system. However, each electrode is independently controlled, allowing the investigator to generate virtually an unlimited number of electrophoresis conditions including field strength, reorientation angles, and pulse times. The PACE system represents the state of the art, and will undoubtedly lead to improvements in electrophoretic karyotypes. It is however, expensive to construct, requiring two power supplies, a personal computer, and a 24-point electrode system.

An inexpensive alternative to the PACE system has been described by

Schwartz *et al.* (1989). Simply named ED ("electrophoresis device"), this system is of simple design, yet allows flexibility to alter field shape and reorientation angle. ED has yielded good karyotypes of *S. cerevisiae* and *S. pombe* (Schwartz *et al.,* 1989), and probably will prove quite useful to investigators with limited funds.

Examples of published fungal electrophoretic karyotypes are given in Table 1. In addition to the four *Candida* species listed, Magee and Magee (1987) examined *C. tropicalis* and *C. parapsilosis,* but were able to resolve

TABLE I

Fungal Electrophoretic Karyotypes Recently Determined by PFGE

Genus and species	Haploid number of chromosomes[a]	Size range (Mgb)	Reference
Aspergillus nidulans	8	2.9–5.0	Brody and Carbon, 1989
Candida albicans	7	1.1–>2.2	Magee and Magee, 1987
C. stellatoidea	9	—	Magee and Magee, 1987
C. glabrata	10	—	Magee and Magee, 1987
C. guilliermondii	6	—	Magee and Magee, 1987
Cephalosporium acremonium	8	2.9–5.0	Skatrud and Queener, 1989
Colletotrichum gloeosporoides type A	5	2.0–>6	Masel *et al.,* 1990
type B	3	4.7–>6	Masel *et al.,* 1990
Fusarium oxysporum			
f. sp. conglutinans	8	2.2–6.2	Kistler and Momol, 1990
f. sp. raphani	11	0.6–6.4	Kistler and Momol, 1990
Histoplasma capsulatum	7	2.0–>5.7	Steele *et al.,* 1989
Magnaporthe grisea	6	3.4–9.5	Skinner *et al.,* 1990
Neurospora crassa	7	4.0–12.6	Orbach *et al.,* 1988
Phenaerchateae chrysosporium	7	2.0–4.8	Cullen, 1990
Phytophthora megasperma	13–14	1.4–?	Howlett, 1989
Saccharomyces cerevisiae	15	0.24–2.2	Chu *et al.,* 1986[b]
S. uvarum	14–15	0.2–2.0	Jager and Philippsen, 1989
Schizosaccharomyces pombe	3	3.0–9.0 (3.5–5.7)[c]	Smith *et al.,* 1987
Tilletia caries	12–14	0.88–4.0	Patil *et al.,* 1990
T. controversa	12–16	0.88–3.9	Patil *et al.,* 1990
Ustilago hordei	15–19	0.17–3.15	McCluskey and Mills, 1990
U. maydis	20	0.3–>2.0	Kinscherf and Leong, 1988

[a] Strains with differing chromosome numbers can be found in most species of fungi

[b] A number of *Saccharomyces cerevisiae* karyotypes have been published (see text). The reference indicated was the first to demonstrate resolution of 15 chromosomes.

[c] Physical mapping supports the sizes in parentheses (Fan *et al.,* 1989); the larger size estimates predate the physical mapping data.

only one or two chromosomes from the mass of unresolved DNA, respectively. Therefore these two species are not included on the list.

In addition to *Saccharomyces uvarum* (Table 1), Jager and Philippsen (1989) presented OFAGE-derived karyotypes of five other *Saccharomyces* species. Since these investigators reported the number of bands resolved under only one set of electrophoresis conditions and did not claim to have resolved all the chromosomes, we have not included those results in Table 1. However the chromosome sizes of all six *Saccharomyces* species studied (*S. uvarum, S. exiguus, S. unisporus, S. dairensis, S. kluyveri,* and *S. servazzii*) were similar, and were in the size range of *S. cerevisiae* chromosomes. Jager and Philippsen also reported 20 chromosomes for one *S. cerevisiae* strain.

In general, fungal chromosome sizes may be categorized as *S. cerevisiae*-like (<1–2 Mgb) or *S. pombe*-like (>2 Mgb). In our laboratory, we have worked extensively with both classes. All of the *Ustilago* species we have examined have chromosomes in the size range of *S. cerevisiae*. *Magnaporthe grisea* has chromosomes greater than 2 Mgb in length. Our findings for each of these systems are discussed subsequently.

II. CHROMOSOME LENGTH POLYMORPHISMS

Chromosome length polymorphisms are common in fungi, for example, Magee and Magee (1987) studied five isolates of *Candida albicans* and found that none of them had the same karyotype. In a subsequent study, 11 different chromosomal bands were resolved from various *C. albicans* strains. Since *C. albicans* is thought to be diploid (2N = 14), the resolution of 11 chromosomes indicated length polymorphisms among four sets of homologues large enough to allow separation into the component chromosomes (Magee *et al.,* 1988).

Chromosome length polymorphisms in *Saccharomyces cerevisiae* are inherited in a 1 : 1 ratio, indicating stable transmission of the polymorphism (Ono and Ishino-Arao, 1988). Other polymorphisms do not segregate 1 : 1, but result in nonparental chromosome lengths in progeny. These polymorphisms are thought to represent chromosomes with one, or more, structural alterations, and concomitant recombination between homologous chromosomes having regions of inhomogeneity. These results demonstrate that chromosome plasticity is common in fungi, and that new gene combinations apparently are created by physical rearrangement without detriment to the fungal isolate.

McCluskey and Mills (1990) have reported a unique karyotype for each of the 14 races of *Ustilago hordei*. Progeny analyses of intraracial crosses for four of the races indicated that the karyotype was typically conserved for each race, although the progeny of one tetrad showed a 2:2 ratio for one chromosome-sized band.

Translocation-derived chromosome length polymorphisms have been described in *S. cerevisiae* (Schwartz and Cantor, 1984), *Neurospora crassa* (Orbach *et al.,* 1988), *Ustilago maydis* (Kinscherf and Leong, 1988), and *Aspergillus nidulans* (Brody and Carbon, 1989). Chromosome length polymorphisms found in *U. maydis* (Kinscherf and Leong, 1988) and *Magnaporthe grisea* (D. Z. Skinner, unpublished results) are discussed in the following sections.

III. SUPERNUMERARY CHROMOSOMES

Supernumerary, or "B," chromosomes are small chromosomes that are not necessary for the survival of the organism. Supernumerary chromosomes are extremely common in both plants and animals; when the subject was reviewed in 1982, over 1000 species of plants and "more than" 260 species of animals had been reported as having at least some individuals with supernumerary chromosomes (Jones and Rees, 1982). Perhaps because of the generally small size of fungal chromosomes, supernumerary chromosomes in fungi were not discovered until the development of electrophoretic karyotyping techniques. Supernumerary chromosomes have now been reported in *Nectria haematococca* (Miao, 1990), *Magnaporthe grisea* (Orbach, 1989), *Schizosaccharomyces pombe* (Niwa *et al.,* 1986), and *Colletotrichum gloeosporoides* (Masel *et al.,* 1990).

Two pathotypes of *C. gloeosporoides* attack *Stylosanthes* (a forage legume). Karyotypes of isolates from the two pathotypes are substantially different (Table 1), but can be broadly classified as having a constant number of large (>2 Mgb) chromosomes within each type, and a variable number of "mini-chromosomes," less than 0.6 Mgb in length (Masel *et al.,* 1990). Because the mini-chromosomes apparently are dispensable, they probably represent supernumerary chromosomes.

In *M. grisea,* field isolates have from one to four small chromosomes ranging in size from 0.9 Mgb to 2 Mgb (Orbach, 1989). These minichromosomes are absent from laboratory strains, probably because they are not transmitted to progeny in meiosis (B. Valent, personal communication), and are clearly supernumerary.

To date, no functional genes have been detected on any of the supernumerary chromosomes of *M. grisea* or *C. gloeosporoides.* However, functional genes do occur on supernumerary chromosomes of *S. pombe* (Niwa *et al.,* 1986) and on those of the plant pathogen *N. haematococca* (Miao, 1990).

The stability of supernumerary chromosomes has been investigated in a *S. pombe* strain carrying a 0.5 Mgb mini-chromosome that was isolated by γ irradiation of an unstable aneuploid disomic for chromosome III (Niwa *et al.,* 1986). Genetic analysis and probing studies of Southern transfers of PFGE gels demonstrated that the mini-chromosome was derived from the pericentric region of chromosome III, which contains the gene *Ade6*. Interestingly, this mini-chromosome, at one copy per genome, was highly stable and behaved like a natural chromosome both in mitosis and meiosis. In contrast, multiple copies of the same chromosome were mitotically unstable and frequently failed to segregate in meiosis.

The pisatin demethylase *(Pda)* gene of *N. haematococca,* which is involved in pathogenicity, is located on a 1.6 Mgb chromosome. In some crosses between two *Pda*$^+$ isolates, unexpected *Pda*$^-$ progeny could be recovered. Southern analysis revealed that the *Pda*$^-$ progeny no longer carried DNA sequences homologous to the *Pda* gene. Karyotype analysis using CHEF gels revealed that they no longer carried the intact 1.6 Mgb chromosome. Southern analysis of *Pda*$^+$ karyotypes confirmed that the *Pda* gene is located on the 1.6 Mgb chromosome. Fungal isolates lacking the 1.6 Mgb chromosome were not pathogenic but were fully viable; hence, the 1.6 Mgb chromosome is dispensable, and is a supernumerary chromosome. A second gene involved in pathogenic ability, *Mak*1, is also located on the 1.6 Mgb supernumerary chromosome (Miao, 1990). The evolutionary significance of two pathogenicity-related genes occurring on the same supernumerary chromosome is not known. However, this finding reveals a heretofore unknown source of variation used by some fungi, and indicates that the chromosomes of fungi are dynamic molecules subject to variations caused by mechanisms that are not fully understood. An analogous situation that is much better characterized, is the existence of plasmids in bacteria. Many pathogenicity genes are found on plasmids; perhaps the best understood is the Ti plasmid of *Agrobacterium tumefaciens.* (See Coplin, 1989, for a review).

IV. METHODS OF CHROMOSOME PREPARATION

Different methods are used to prepare yeast and filamentous fungal chromosomes for PFGE. The element common to all procedures is the need to handle the very large and very fragile DNA molecules without breaking them.

Initial research efforts addressed this problem in yeasts by embedding intact cells in agarose, enzymatically removing the cell walls, and removing the plasma membranes and other cellular components with a mixture of detergents and proteolytic enzymes at elevated temperatures (Schwartz and

Cantor, 1984). Filamentous fungi traditionally have been handled by first enzymatically generating protoplasts, embedding protoplasts in agarose, and treating the agarose-embedded protoplasts with detergents and proteolytic enzymes at elevated temperatures. Several studies of yeasts and yeast-like fungi also employed an initial protoplasting step prior to embedding in agarose.

Recent reports indicate that protoplasting is not necessary (McCluskey and Mills, 1990), and that proteolytic enzymes and elevated temperatures also are not necessary (Miao and Van Etten, 1990). Thus, techniques are being simplified empirically so electrophoretic karyotypes may be quickly and routinely produced. These developments will facilitate karyotype studies of large numbers of isolates, as is necessary in population biology investigations.

V. ELECTROPHORESIS CONDITIONS

The electrophoretic parameters employed may be broadly classified for chromosomes that are "yeast-like" or "non-yeast-like." Yeast-like chromosomes ($\leq$2 Mgb) may be resolved in approximately 24 hr with strong electrical current flow, high agarose concentrations, and relatively short pulse switching intervals. For example, CHEF electrophoresis of species from the Ustilaginales (Fig. 1) is carried out under conditions of 1.0% Seakem LE agarose in 0.5× TBE, 140 V, a switching interval of 60 sec for 17 hr followed by 140 sec for 8 hr at 14°C.

Non-yeast-like chromosomes ($>$2 Mgb) require extended periods of electrophoresis with low current flow, reduced agarose concentrations, and long pulse switching intervals for resolution. For example, we carry out CHEF electrophoresis of *Magnaporthe grisea* (Fig. 2) under conditions of 0.65% Seakem LE agarose in 0.5× TBE, 35 V, a switching interval of 90 min for 5 days, followed by 60 min for 2 days at 14°C.

VI. ASSIGNMENT OF DNA MARKERS TO CHROMOSOMES

The preparation of molecular karyotypes by PFGE allows direct gene mapping. Any cloned DNA segment may be assigned to a chromosome by Southern hybridization studies. We employed this technique to assign linkage groups of restriction fragment length polymorphism (RFLP) markers in *Magnaporthe grisea* to their chromosomes of origin (Skinner *et al.*, 1990). A similar technique may be used to assign cloned genes to chromosomes and thereby discover linkage relationships. This mapping

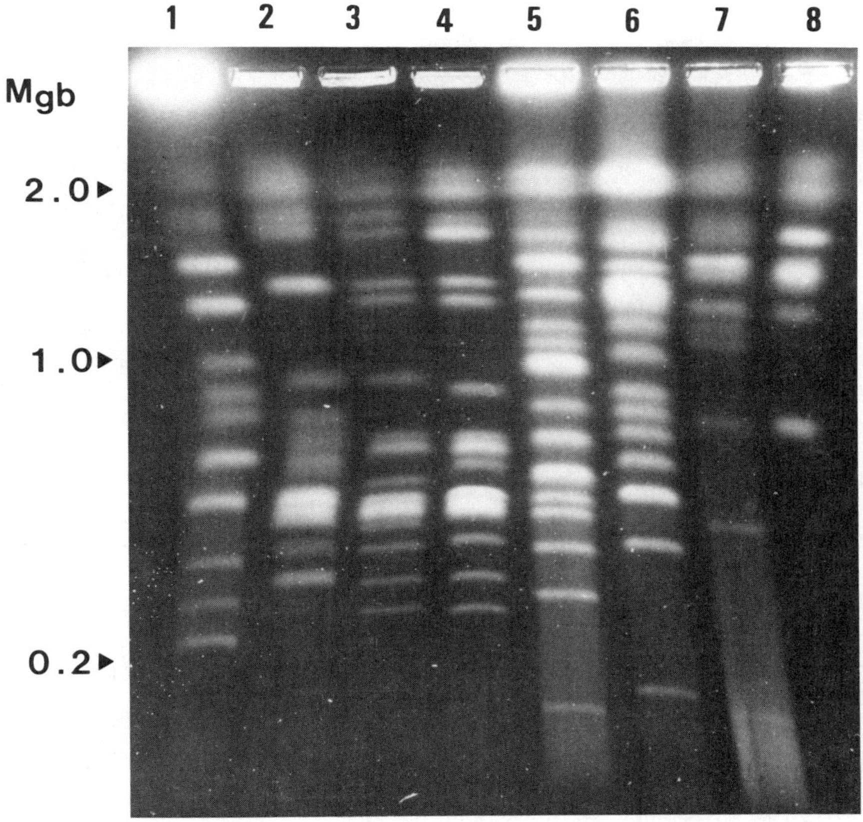

Fig. 1. CHEF electrophoretic karyotypes of species from the order Ustilaginales. Lanes: 1, yeast standard (G2); 2, *Sphacelotheca reiliana;* 3, *Ustilago maydis;* 4, *Ustilago longissima;* 5, *Ustilago hordei;* 6, *Ustilago nigra;* 7, *Ustilago koleri;* 8, *Cryptococcus neoformans.* Conditions were 1.0% Seakem LE agarose in 0.5× TBE, 14°C, a switching interval of 60 sec for 17 hr followed by 140 sec for 8 hr at 140 V. (Unit fabricated following schematics supplied by R. Davis.)

technique is particularly useful when studying fungi lacking a sexual phase. For example, Magee *et al.* (1988) determined the chromosomal origin of 14 genes in *Candida albicans,* and Skatrud and Queener (1989) established that 2 key genes in the cephalosporin C pathway are on different chromosomes in *Cephalosporium acremonium.* Hence, electrophoretic karyotyping is proving valuable in extending the knowledge of linkage relationships in genetically intractable fungi.

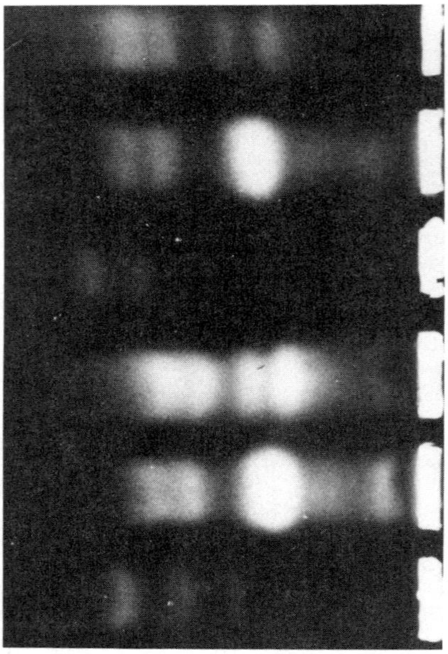

Fig. 2. CHEF electrophoresis of *Magnaporthe grisea*. Lanes: 1 and 4, Guy11; 2 and 5, 2539; 3 and 6, *Schizosaccharomyces pombe* standard. Conditions were 0.65% Seakem LE agarose in 0.5× TBE, 14°C, a switching interval of 90 min for 5 days followed by 60 min for 2 days at 35 V. (Unit fabricated following schematics supplied by R. Davis.)

VII. PREPARATION OF CHROMOSOME-SPECIFIC PROBES

The resolution of chromosomes in a gel creates the possibility of constructing chromosome-specific libraries. The search for genes or RFLP markers on a particular chromosome will be streamlined if the analysis can be limited to a library of a single chromosome. We have attempted to construct chromosome-specific plasmid libraries of *Magnaporthe grisea* DNA using the following protocol. Chromosomes resolved on CHEF gels are excised from the gel, the DNA is digested with *Hin*dIII while in the gel, and the fragments are recovered with "Gene-Clean" (Bio101, La Jolla, California). The DNA is size-fractionated in 0.7% low-melting-temperature agarose, and fragments greater than 2 kb are again recovered with "Gene-Clean." These fragments are ligated into pUC18 and transformed into *Escherichia coli* strain DH5α. A similar approach should be possible for constructing cosmid libraries, except that "Gene-Clean" is not well suited for the recovery of large fragments. Other procedures, such as electroelution or agarase treatment of isolated chromosome bands, could be used for that purpose.

Plasmid libraries obtained by the procedure just described were enriched for fragments from the targeted chromosomes, but were contaminated with fragments from other chromosomes. The DNA from each resolved chromosomal band on pulsed-field gels was not pure, indicating that the resolution of large chromosomes by the methods we employed was not complete. As materials and techniques are refined, resolution should improve and allow for construction of truly chromosome-specific libraries. Small chromosomes have yielded DNA of sufficient purity to be used as chromosome-specific probes (Niwa *et al.,* 1986; Kinscherf and Leong, 1988). In our laboratory, a gel-purified chromosome was used as a probe to document a translocation event in *Ustilago maydis.*

VIII. PHYSICAL MAPPING

The resolution of very large DNA fragments leads to the possibility of determining the physical distance between two genetically linked markers, or determining the physical size of tandemly repeated genes. The principle of the former technique is to digest genomic DNA with an infrequently cutting enzyme, resolve the fragments on a pulsed-field gel, transfer the DNA to a solid support, and probe with two markers known to be genetically linked. If they hybridize to the same fragment of known size, an immediate indication of the maximum distance separating them is obtained. Other enzymes and partial digests may then be used to further refine and confirm the estimate. This procedure has been used with *Schizosaccharomyces pombe* to generate a detailed physical map of the genome of the fungus (Fan *et al.,* 1989). This method also has been employed to define the size of a known coding region, the human HLA gene complex, for example, (Lawrance *et al.,* 1987). Likewise, CHEF gels have been used to analyze the sizes of ribosomal DNA regions of various isolates of *Neurospora crassa.* Frequent changes in the size of the nucleolus organizer region are seen in this species (Butler and Metzenberg, 1989).

Recently a technique referred to as "Achilles heel cleavage" (Koob and Szybalski, 1990) was developed that permits cleavage of complex genomes at a single site (Fig. 3). A symmetric *lacO* sequence was introduced into one of the yeast chromosomes. Agarose beads containing yeast chromosomes were treated with *Hha*I methylase in the presence of *lac* repressor. The *Hha*I site (5′GCGC3′) in the *lacO* sequence was protected from methylation by the presence of the *lac* repressor. Following denaturation of the *lac* repressor, the DNA was cleaved specifically at the *lacO* sequence

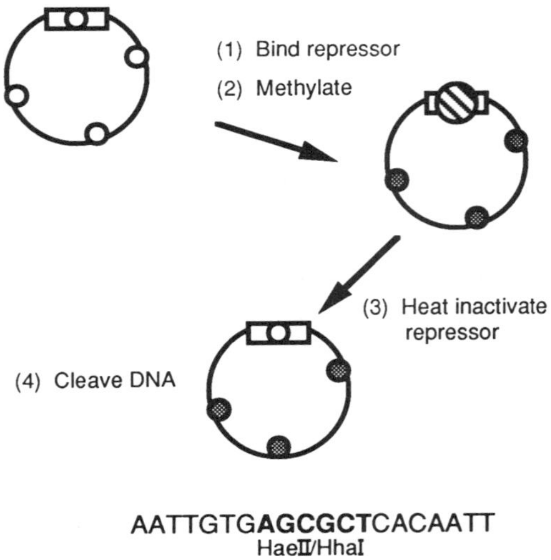

Fig. 3. "Achilles heel cleavage" of plasmid DNA (Koob and Szybalski, 1990). A plasmid carrying a symmetric *lacO* sequence (boxed region of plasmid depicts 20 bp sequence indicated at the bottom of the figure) and four sites for *Hha*I(5′GCGC3′)/*Hae*II($5'{}^{\mathrm{A}}_{\mathrm{G}}\mathrm{GCGC}^{\mathrm{T}}_{\mathrm{C}}3'$) (open circles) is treated with *Hha*I methyltransferase in the presence of *lac* repressor (hatched circle). *Hha*I sites that are not protected by repressor are methylated (shaded circles). *Lac* repressor is heat inactivated and the single unmethylated *Hae*II site of the *lacO* sequence is cleaved with *Hae*II.

with *Hae*II ($5'{}^{\mathrm{A}}_{\mathrm{G}}\mathrm{GCGC}^{\mathrm{T}}_{\mathrm{C}}3'$). By targeting the *lacO* sequence to RFLP loci throughout the genome of *Magnaporthe grisea,* we hope to assess the relationship of physical to genetic distance in various sectors of the fungal genome. Other applications include facilitation of long-range restriction mapping analysis and the isolation of discrete regions of the fungal genome.

IX. USTILAGINALES

The chromosomes of *Ustilago maydis* have proved to be quite amenable to PFGE analysis. Protocols are available for protoplasting sporidia. The chromosomes are similar in size to yeast chromosomes, which allows for application of electrophoretic conditions similar to those used to separate yeast chromosomes. Kinscherf and Leong (1988) exploited these traits to create an electrophoretic karyotype for *U. maydis.* The laboratory strain #518 obtained from Dr. Robin Holiday (CSIRO, North Ryde, Sydney, Australia) was used as a reference standard. This strain has been in culture for over 20 years and is currently being used by many of the laboratories working with this fungus. By employing a number of pulse times and different electrical currents, the minimum number of chromosomes present in #518 was estimated to be 20, with evidence of chromosome size polymorphisms among strains gathered from different areas of the world. One strain, F200, had an apparent chromosomal translocation that could be documented by Southern hybridization analysis of DNA transfers of the OFAGE gels. Southern hybridization analysis was also used to assign 7 genes to chromosomes of the OFAGE electrophoretic karyotype of the reference strain (Kinscherf and Leong, 1988). This work was further expanded to 13 genes by Budde and Leong (1990). As more genes become available, the number of chromosomal assignments will increase, and these assignments will aid in establishing linkage relationships.

Other species in the order Ustilaginales, including *U. longissima, U. hordei, U. nigra, U. koleri, Sphacelotheca reiliana,* and *Cryptococcus neoformans,* have been karyotyped by CHEF electrophoresis (Fig. 1). All of these species have chromosomes of yeast-like size; however, the number of chromosomes varies considerably from the 20 in *U. maydis* to a minimum number of 7 chromosomes in *C. neoformans. U. maydis, U. longissima,* and *S. reiliana* have very similar karyotypes; in fact, these karyotypes do not vary more than those of the different strains of *U. maydis* described by Kinscherf and Leong (1988). These species were analyzed because of the ease with which they are protoplasted. Sample preparation via protoplasting has been one of the biggest obstacles to overcome for those species with yeast-sized chromosomes, since the electrophoretic equipment currently allows routine karyotyping of chromosomes in this size range. The discovery that chromosomes of some fungal cells can be analyzed after embedding spores directly in agarose, thus avoiding protoplasting altogether (McCluskey *et al.,* 1990), should extend PFGE analysis to fungi that cannot be readily protoplasted.

X. *MAGNAPORTHE GRISEA*

In our attempts to separate the chromosomes of *Magnaporthe grisea,* we routinely resolve a karyotype of four distinct bands; the largest band appears to be a multiplet (Fig. 2). Hybridization of DNA probes to Southern transfers of CHEF gels has shown that the band of highest molecular weight is composed of at least two chromosomes. In one instance, a malfunction of our CHEF equipment caused the DNA to migrate at an angle, and the largest band was resolved into two chromosomes (Skinner *et al.,* unpublished). This result supports the theory that inhomogeneous electrical fields are important for good resolution of large DNA molecules

(Schwartz and Cantor, 1984). Others claim that a reorientation angle of greater than 90° is an important consideration. The equipment Schwartz and Cantor used to produce the inhomogeneous field also introduced an obtuse angle, and in a recent report Schwartz *et al.* (1989) state only that "field gradients can sharpen bands a little." During our analysis, resolution was achieved when the chromosomal DNA was subjected to at least one inhomogeneous electrical field that resolved the two largest chromosomes. Normally functioning equipment (homogeneous fields) under the same conditions did not yield the same resolution (Fig. 2). Therefore, our results suggest that inhomogeneity of fields may be important for resolution of large molecules. More research with the PACE system should advance the study of large chromosomes (>5 Mgb).

The chromosomes of the three *M. grisea* isolates we studied varied in size; one of the isolates had an obvious chromosome length polymorphism caused by a translocation. RFLP probes demonstrated that approximately 2 Mgb of chromosome 3 (as defined by a reference isolate) had translocated to chromosome 4 in another isolate.

We also identified a small interstitial translocation in *M. grisea* that was too small to cause a detectable chromosome polymorphism. A RFLP probe hybridized to single polymorphic *Hin*dIII fragments in parental isolates. However, of two complete asci examined, one was parental ditype and the other was nonparental ditype, in which four progeny had both fragments and four had neither fragment, indicating independent assortment. Hybridization of the probe to a Southern transfer of a CHEF gel of the parents and one of the progeny revealed that the probe bound to a different chromosome in each parent and bound to both chromosomes in one of the descendents. This provided conclusive evidence that a translocation was involved. How many other translocations of this kind occur in *M. grisea* is unknown, but the finding for this RFLP probe demonstrates that the level of genetic complexity in *M. grisea* is high, and that movement about the genome of discrete regions of DNA does occur. Perkins (1985) stated that "insertional translocations are common" in *Neurospora* species. Evidence is accumulating that the chromosomes of fungi, and perhaps other organisms, are not invariable passive carriers of genetic information, but are dynamic molecules subject to alterations by mechanisms currently not understood.

XI. SUMMARY

The development of pulsed-field and molecular karyotyping technology has led to rapid advances in the understanding of genetic diversity at the chromosomal level in fungi. The detection of translocations and variations in chromosome number, previously an arduous task, is nearly a routine process with this technique. The assignment of cloned genes to chromosomal locations is accomplished without the need for crosses and progeny tests; hence, the development of genetic maps in previously intractable fungi is now possible. Currently, the primary limitation of pulsed-field technology is that only relatively small chromosomes can be resolved. Using the current support matrices (various formulations of agarose), the upper size limit of resolvable DNA molecules is approximately 10 Mgb. New formulations of agarose may increase this limit to some extent, but extensive increases in the size of resolvable DNA molecules probably will not occur until a new support matrix is developed. Resolution of mammalian or plant chromosomes, for example, is not possible with the current technology. The smallest human chromosome is estimated to be approximately 30 Mgb (Anand, 1986), about three times larger than the resolution limit. Advances in this field will also result in more precise separation of fungal chromosomes, enabling the investigator to obtain pure preparations of individual chromosomes, leading to rapid advances in studies targeted to defined regions of the genome.

ACKNOWLEDGMENTS

The authors acknowledge The Rockefeller Foundation, the National Institutes of Health, and the USDA for support of the projects from which much of our experience with PFGE has been derived.

REFERENCES

Anand, R. (1986). Pulsed field gel electrophoresis: A technique for fractionating large DNA molecules. *Trends Genet.* **2,** 278–283.

Brody, H., and Carbon, J. (1989). Electrophoretic karyotype of *Aspergillus nidulans. Proc. Natl. Acad. Sci. U.S.A.* **86,** 6260–6263.

Budde, A. D., and Leong, S. A. (1990). *Ustilago maydis,* causal agent of corn smut (n = 20). *In* "Genetic Maps," (S. J. O'Brien, ed.) 5th Ed., pp. 3.92–3.93. Cold Spring Harbor Laboratory Press, Cold Spring Harbor, New York.

Butler, D. K., and Metzenberg, R. L. (1989). Premeiotic change of nucleolus organizer size in *Neurospora. Genetics* **122,** 783–791.

Cantor, C. R., Gaal, A., and Smith, C. L. (1988). High-resolution separation and accurate size determination in pulsed-field gel electrophoresis of DNA. 3. Effect of electrical field shape. *Biochemistry* **27,** 9216–9221.

Carle, G. F., and Olson, M. V. (1984). Separation of chromosomal DNA molecules from yeast by orthogonal-field-alternation gel electrophoresis. *Nuc. Acids Res.* **12,** 5647–5664.

Carle, G. F., Frank, M., and Olson, M. V. (1986). Electrophoretic separations of large DNA molecules by periodic inversion of the electric field. *Science* **232,** 65–68.

Chu, G., Vollrath, D., and Davis, R. W. (1986). Separation of large DNA molecules by contour-clamped homogeneous electric fields. *Science* **234,** 1582–1585.

Clark, S. M., Lai, E., Birren, B. W., and Hood, L. (1988). A novel instrument for separating large DNA molecules with pulsed homogeneous electric fields. *Science* **241,** 1203–1205.

Coplin, D. L. (1989). Plasmids and their role in the evolution of plant pathogenic bacteria. *Ann. Rev. Phytopathol.* **27,** 187–212.

Cullen, D. C. (1990). Genomic organization of lignin peroxidase genes for *Phenaerchateae chrysosporium.* Submitted.

Fan, J.-B., Chikashige, Y., Smith, C. L., Niwa, O., Yanagida, M., and Cantor, C. R. (1989). Construction of a *Not*1 restriction map of the fission yeast *Schizosaccharomyces pombe* genome. *Nuc. Acids Res.* **17,** 2801–2818.

Gardiner, K., and Patterson, D. (1988). Transverse alternating electrophoresis. *Nature (London)* **331,** 371–372.

Gardiner, K., Lass, W., and Patterson, D. (1986). Fractionation of large mammalian DNA restriction fragments using vertical pulsed-field gradient gel electrophoresis. *Somatic Cell Molec. Genet.* **12,** 185–195.

Howlett, B. (1989). An electrophoretic karyotype for *Phytophthora megasperma. Exp. Mycol.* **13,** 199–202.

Jager, D., and Philippsen, P. (1989). Many yeast chromosomes lack the telomere-specific Y′ sequence. *Mol. Cell. Biol.* **9,** 5754–5757.

Jones, R. N., and Rees, H. (1982). "B Chromosomes." Academic Press, London.

Kinscherf, T. G., and Leong, S. A. (1988). Molecular analysis of the karyotype of *Ustilago maydis. Chromosoma* **96,** 427–433.

Kistler, H. C., and Momol, (1990). Cell fusion and cytoplasmic reassortment in *Fusarium oxysporum.* Submitted.

Koob, M., and Szybalski, W. (1990). Cleaving yeast and *Escherichia coli* genomes at a single site. *Science* **250,** 271–273.

Lai, E., Birren, B. W., Clark, S. M., Simon, M. I., and Hood, L. (1989). Pulsed field gel electrophoresis. *Biotechniques* **7,** 34–42.

Lawrance, S. K., Smith, C. L., Srivastava, R., Cantor, C. R., and Weissman, S. M. (1987). Megabase-scale mapping of the HLA gene complex by pulsed-field gel electrophoresis. *Science* **235,** 1387–1390.

Lott, T. J., Boiron, P., and Reiss, E. (1987). An electrophoretic karyotype for *Candida albicans* reveals large chromosomes in multiples. *Mol. Gen. Genet.* **209,** 170–174.

McCluskey, K., and Mills, D. (1990). Identification and characterization of chromosome-length polymorphisms among strains representing fourteen races of *Ustilago hordei. MP-MI* **3,** 366–373.

McCluskey, K., Russell, B., and Mills, D. (1990). Electrophoretic karyotyping without the need for generating protoplasts. *Curr. Genet.* In press.

Magee, B. B., and Magee, P. T. (1987). Electrophoretic karyotypes and chromosome numbers in *Candida* species. *J. Gen. Microbiol.* **133,** 425–430.

Magee, B. B., Koltin, Y., Gorman, J. A., and Magee, P. T. (1988). Assignment of cloned genes to the seven electrophoretically separated *Candida albicans* chromosomes. *Mol. Cell. Biol.* **8,** 4721–4726.

Masel, A., Braithwaite, K., Irwin, J., and Manners, J. (1990). Highly variable molecular karyotypes in *Colletotrichum gloeosporoides* pathogenic on *Stylosanthes.* sp. *Curr. Genet.* **18,** 81–86.

Mathew, M. K., Smith, C. L., and Cantor, C. R. (1988a). High-resolution separation and accurate size determination in pulsed-field gel electrophoresis of DNA. 1. DNA size standards and the effect of agarose and temperature. *Biochemistry* **27,** 9204–9210.

Mathew, M. K., Smith, C. L., and Cantor, C. R. (1988b). High-resolution separation and accurate size determination in pulsed-field gel electrophoresis of DNA. 2. Effect of pulse time and electric field strength and implications for models of the separation process. *Biochemistry* **27,** 9210–9216.

Mathew, M. K., Hui, C-F., Smith, C. L., and Cantor, C. R. (1988c). High-resolution separation and accurate size determination in pulsed-field gel electrophoresis of DNA. 4. Influence of DNA topology. *Biochemistry* **27,** 9222–9226.

Miao, V. P. W. (1990). Genes for phytoalexin detoxification in *Nectria haematococca:* Role in virulence on chickpea, and location on *B* chromosomes. Ph. D. Dissertation. Department of Plant Pathology, Cornell University, Ithaca, New York.

Miao, V. P. W., and Van Etten, H. D. (1990). Preparing chromosomes from protoplasts of a filamentous fungus for pulsed-field gel electrophoresis. *Biotechniques* **8,** 393.

Niwa, O., Matsumoto, T., and Yanagida, M. (1986). Construction of a mini-chromosome by deletion and its mitotic and meiotic behaviour in fission yeast. *Mol. Genet.* **203,** 397–405.

Ono, I., and Ishino-Arao, Y. (1988). Inheritance of chromosome length polymorphisms in *Saccharomyces cerevisiae. Curr. Genet.* **14,** 413–418.

Orbach, M. J. (1989). Electrophoretic characterization of the *Magnaporthe grisea* genome. *Fung. Genet. Newslett.* **36,** 14.

Orbach, M. J., Vollrath, D., Davis, R. W., and Yanofsky, C. (1988). An electrophoretic karyotype of *Neurospora crassa. Mol. Cell. Biol.* **8,** 1469–1473.

Patil, S. S., Mills, D., and Vance, C. 1990.

Perkins, D. D. (1985). Aspects of the *Neurospora* genome. In "Molecular Genetics of Filamentous Fungi," W. E. Timberlake, ed. pp. 277–294. Alan R. Liss, New York.

Schwartz, D., and Cantor, C. R. (1984). Separation of yeast chromosome-sized DNAs by pulsed field gradient gel electrophoresis. *Cell* **37,** 67–75.

Schwartz, D. C., Smith, L. C., Baker, M., and Hsu, M. (1989). ED: Pulsed electrophoresis instrument. *Nature (London)* **342,** 575–576.

Skatrud, P. L., and Queener, S. W. (1989). An electrophoretic molecular karyotype for an industrial strain of *Cephalosporium acremonium. Gene* **78,** 331–338.

Skinner, D. Z., Leung, H. L., and Leong, S. A. (1990). Genetic map of the blast fungus *Magnaporthe grisea* (n = 6). *In* "Genetic Maps," (S. J. O'Brien, ed.) 5th Ed., pp. 3.82–3.83. Cold Spring Harbor Laboratory Press, Cold Spring Harbor, New York.

Smith, C. L., Matsumoto, T., Niwa, O., Klco, S., Fan, J-B., Yanagida, M., and Cantor, C. R. (1987). An electrophoretic karyotype for *Schizosaccharomyces pombe* by pulsed field gel electrophoresis. *Nuc. Acids Res.* **15,** 4481–4489.

Steele, P. E., Carle, G. F., Kobayashi, S., and Medoff, G. (1989). Electrophoretic analysis of *Histoplasma capsulatum* chromosomal DNA. *Mol. Cell. Biol.* **9,** 983–987.

Stewart, G., Furst, A., and Avdalovic, N. (1988). Transverse alternating field electrophoresis (TAFE). *Biotechniques* **6,** 68–73.

Vollrath, D., and Davis, R. W. (1987). Resolution of DNA molecules greater than 5 megabases by contour-clamped homogeneous electric fields. *Nuc. Acids Res.* **15,** 7865–7876.

III

Molecular Morphogenesis

5

Cell Cycle and Tubulin Mutations in Filamentous Fungi

STEPHEN A. OSMANI
Department of Cell Biology
Baylor College of Medicine
Houston, Texas

BERL R. OAKLEY
Department of Molecular Genetics
Ohio State University
Columbus, Ohio

I. INTRODUCTION

In this chapter we will discuss the contributions that molecular and genetic analyses of the filamentous fungi have made to our understanding of how cells regulate their cell cycle and undergo mitosis. We will focus on *Aspergillus nidulans,* the organism in which the great majority of these studies have been done. The first part of the chapter will review the mutations that block the cell cycle (the *nim* and *bim* genes). The second part will discuss tubulin mutations.

II. CELL CYCLE MUTATIONS AND CELL CYCLE REGULATION

The isolation of mutants defective in cell cycle progression is a powerful tool for the study of cell division. Using this approach, preconceived theories about which genes are important for cell cycle progression are not required. The genetic system itself identifies the genes. Mutants defective in cell cycle progression have been obtained in a variety of organisms. However, by far the largest number of these mutants has been isolated from three different *ascomycete* fungi: the yeasts *Saccharomyces cerevisiae* (Hartwell *et al.,* 1970, 1974) and *Schizosaccharomyces pombe* (Nurse *et al.,* 1976) and the filamentous fungus *Aspergillus nidulans* (Morris, 1976). Well-advanced classical and molecular genetic methodologies have been developed for each of these fungi, so analysis of mutants and subsequent molecular cloning and manipulation of the genes involved is straightforward. The major contributions to cell cycle research made by studies of these fungal systems are a testament to the power of this approach. Several excellent reviews have been written about the two yeast systems (Hartwell and Weinert, 1989; Murray and Kirschner, 1989; Nurse, 1990).

A. Isolation of the *nim* and *bim* Genes of *Aspergillus nidulans*

Genes that are required for cell cycle progression are expected to be essential for life. It is, therefore, necessary to identify such genes by the isolation of conditional mutations. Traditionally, cell cycle mutations in fungi have been conditional at low temperature (cold sensitive, cs^-) or high temperature (designated either temperature sensitive, ts^-, or heat sensitive, hs^-, by different workers). At permissive temperatures, these mutants are able to grow; at restrictive temperatures they cannot complete the nuclear division cycle, so they cannot form colonies.

Conditional mutants fall into two categories; those that cause a generalized inability to traverse the cell cycle (as would result from mutation of an essential intermediary metabolism enzyme) and those that specifically affect the cell cycle. In the 1970s, Morris isolated numerous temperature-sensitive mutants of *Aspergillus* and, by a "brute-force" screen, determined which mutants had a phenotype indicative of a cell-cycle-specific arrest

(Morris, 1976). To identify mutations in cell-cycle-specific genes, the percentage increase or decrease in both the spindle and chromosome mitotic indices of these strains was determined after they were shifted to the restrictive temperature. Mutations that caused a decrease in mitotic index at the restrictive temperature were given the gene symbol *nim,* standing for *n*ever *i*n *m*itosis. Some of the *nim* mutations subsequently were shown to block the cell cycle specifically in either S phase (*nim*C, *nim*G, *nim*K, and *nim*Q) or G_2 phase (*nim*A, *nim*B, *nim*E, *nim*T, and *nim*U) (Bergen *et al.,* 1984). Although *Aspergillus* does have a discrete G_1 phase in its cycle (Bergen and Morris, 1983), no G_1-specific mutants were isolated. Mutations that caused an increase in mitotic index at the restrictive temperature were designated *bim,* standing for *b*locked *in* *m*itosis (*bim*A, *bim*B, *bim*C, *bim*D, *bim*E, *bim*F, and *bim*G). The *bim* mutants block various stages of mitosis.

With the advent of a transformation system for *Aspergillus nidulans* (Ballance *et al.,* 1983; Tilburn *et al.,* 1983), the possibility of cloning the wild-type copies of cell-cycle-specific genes by complementation of the mutant phenotypes (Nasmyth and Reed, 1980) became a reality. Several of the *bim* mutants and one of the *nim* mutants have now been cloned.

B. The *nim*A Gene

The first cell-cycle-specific gene cloned from *Aspergillus nidulans* by the complementation strategy was *nim*A (Osmani *et al.,* 1987). The temperature-sensitive mutation *nim*A5 causes a specific block at the restrictive temperature in the G_2 phase of the cell cycle. This G_2 block is readily reversible, that is mitosis is resumed simply by returning cells to the permissive temperature (Oakley and Morris, 1983). This observation indicates that *nim*A function is required, possibly in a regulatory capacity, specifically in late G_2 at a time when cells are poised to enter mitosis. Thus, the *nim*A gene is of particular interest when studying mitotic regulation.

The deduced amino acid sequence of *nim*A cDNA shows that this gene encodes a protein with a theoretical molecular mass of 79kDa (designated NIMA) that contains the motifs typical of the catalytic site of protein kinases (Osmani *et al.,* 1988b). The catalytic capacity of NIMA has been established biochemically. NIMA-specific immunoprecipitates have protein kinase activity that will phosphorylate the artificial substrate β-casein (Osmani *et al.,* 1991).

The levels of *nim*A mRNA (Osmani *et al.,* 1987) and NIMA protein kinase activity (Osmani *et al.,* 1991) are under cell cycle regulation; there is a positive correlation between mitosis and high levels of *nim*A. By introducing extra copies of the *nim*A gene under the control of an inducible promoter into the genome of *Aspergillus,* it is possible to greatly increase the level of *nim*A mRNA and NIMA protein kinase activity. This has the effect of promoting mitotic events and causing an arrest of nuclei in a mitotic state, even if cells are blocked in S phase before inducing the NIMA protein kinase. Thus, overexpression of NIMA profoundly affects the cell cycle by promoting premature mitotic events (Osmani *et al.,* 1988b).

Because *nim*A is required for progression into mitosis and encodes a cell-cycle-specific protein kinase that, when overexpressed, is able to influence mitotic events, we propose that *nim*A plays an essential regulatory role during the G_2–M–G_1 transitions of the cell cycle. According to our model, NIMA catalyses cell-cycle-specific phosphorylation of target proteins during G_2, changing the biochemical properties of the target proteins and helping to manifest spindle formation and chromosome condensation. The target proteins are then dephosphorylated to allow completion of mitosis and entry into G_1 (Osmani *et al.,* 1988b).

It is well documented that numerous proteins are reversibly phosphorylated during mitosis and it is widely believed that a conserved phosphorylation cascade promotes the initiation of mitosis in all eukaryotes (Murray and Kirschner, 1989; Nurse, 1990). The NIMA protein kinase is probably part of this phosphorylation cascade. Data obtained with the monoclonal antibody MPM2 suggest that, in fact, the NIMA protein kinase is required when the mitotic phosphorylation cascade has already been activated during G_2. MPM2, raised against mitotic human cells, detects mitosis-specific phosphoproteins (Davis *et al.,* 1983). The epitope detected by this antibody, by either Western blotting or immunofluorescence, is removed by treatment with phosphatase (Vandre *et al.,* 1984). In all species tested, this antibody reacts in a cell-cycle-dependent fashion with the microtubule-organizing center responsible for establishing the mitotic spindle (Vandre *et al.,* 1984). The spindle-pole body (SPB) of *Aspergillus* (the mammalian equivalent is the centrosome) is reactive to MPM2 only during G_2 and mitosis (Engle *et al.,* 1988). There is no reactivity during G_1 or S phase. This pattern of staining is thought to reflect the phosphorylation cascade occurring during G_2. SPBs in *nim*A mutants that are arrested in G_2 are fully reactive to the MPM2 antibody, showing that *nim*A function is not required for the phosphorylations detected by the MPM2 antibody during G_2 (Engle *et al.,* 1988). This observation also suggests that *nim*A functions at a time during mitotic initiation after the mitotic phosphorylation cascade has begun.

More work on the NIMA protein kinase is required to ascertain how it relates to the other protein kinases involved in mitotic regulation. The most widely studied cell-cycle-specific protein kinase is the *cdc2* gene of fission yeast and its homologues in other systems. The *cdc2* gene encodes the catalytic subunit of a protein complex that phosphorylates histone H1 in a cell-cycle-specific fashion. The *cdc2* kinase is also present in maturation promotion factor (MPF), a universally active entity thought to regulate meiosis and mitosis in all eukaryotes. In addition to *cdc2,* MPF contains a cell-cycle-regulated protein called cyclin. The activity of the *cdc2* protein kinase is regulated through the cell cycle by its association with cyclin and by its state of phosphorylation. During interphase, *cdc2* protein is phosphorylated at Tyr 15. At the transition into mitosis, Tyr 15 is dephosphorylated, allowing activation of the kinase. (See Murray and Kirschner, 1989, and Nurse, 1990, for reviews and references.) Preliminary data indicate that *nim*A is required for the initiation of mitosis after tyrosine dephosphorylation of *cdc2* (A. H. Osmani, S. L. M. McGuire, and S. A. Osmani, unpublished observations).

C. The *bim*G Gene

The importance of reversible phosphorylation as a regulatory mechanism during mitosis is further highlighted by the discovery that one of the *Aspergillus bim* genes (*bim*G) encodes a protein homologous to a mammalian phosphoprotein phosphatase (Doonan and Morris, 1989). The phosphorylation cascade that occurs during G_2 is thought to promote the initiation of mitosis. Much of the phosphorylation that occurs during G_2 and mitosis is subsequently reversed when cells exit mitosis and enter a new cycle of division. The shift from a phosphorylated to a dephosphorylated state involves phosphoprotein phosphatase enzymes. Therefore, reduction in protein phosphatase activity may prevent proper progress out of mitosis. Studies on *bim*G11 indicate that this is indeed the case.

At the restrictive temperature, *bim*G11 causes an increase in mitotic index and a very marked modification of reactivity toward the MPM2 antibody. As noted earlier, the MPM2 monoclonal antibody reacts with mitosis-specific phosphoproteins in a cell-cycle-specific manner in higher and lower eukaryotes. In *bim*G11 strains, both the degree and distribution of MPM2 immunofluorescent staining of nuclei is dramatically affected. The MPM2 antibody is normally only reactive with cells that are in G_2 or mitosis. Approximately 20% of cells stain in asynchronous cultures of wild-type strains at all temperatures. At the permissive temperature for *bim*G11, the MPM2 staining index is greater than 40%, increasing to greater than 80% at the restrictive temperature. In addition to reacting more frequently with nuclei, MPM2 stains nucleolar material in *bim*G11 strains at the restrictive temperature, something never seen in wild-type cells at any temperature. These observations strongly suggest that *bim* G11 causes a major increase in the level of some of the phosphoproteins involved in mitosis (Doonan and Morris, 1989).

The *bim*G gene has been cloned by complementation. The gene encodes a 37kDa protein that has 86% identity with the catalytic subunit of mammalian phosphoprotein phosphatase type 1 (PP1) (Doonan and Morris, 1989). This homology is greater than that found between the β-tubulin (May *et al.,* 1987) and histone H2A genes (May and Morris, 1987) of *Aspergillus* and their respective highly conserved counterpart genes in mammalian cells. The level of PP1 was measured in extracts derived from *bim*G11 strains grown at the restrictive temperature. The results indicate that a temperature-sensitive PP1 is produced in this strain. In addition, a cDNA encoding the mammalian PP1 fully complements *bim*G11 in *Aspergillus* (J. H. Doonan *et al.,* unpublished observations). Thus, the defects caused by the *bim*G11 mutation are due to a decrease in protein phosphatase activity, which consequently increases the net phosphorylation state of proteins in the cell.

Eukaryotic protein phosphatases were once thought to be restricted to regulating major metabolic pathways (Cohen, 1989). However, they are rapidly being established as key components of cell cycle regulation (Cyert and Thorner, 1989; Gould and Nurse, 1989).

D. The *bim*E Gene

The *bim*E gene is defined by a temperature-sensitive mutation (*bim*E7) that blocks mitosis at the restrictive temperature (Morris, 1976). *bim*E7 causes the most rapid, extensive, and prolonged mitotic arrest of all the *bim* mutations presently available. When viewed with the electron microscope, *bim*E7 cells form a mitotic spindle that appears normal, and chromosome condensation occurs as in wild-type mitosis (B. R. Oakley and N. R. Morris, unpublished observations). An informative feature of the phenotype of *bim*E7 is that this mutation can induce mitotic events in cells that have been arrested in interphase, either in S phase or G_2 (Osmani *et al.,* 1988a). These effects are due to the loss of *bim*E function.

The *bim*E gene has been cloned by complementation and encodes a large 240kDa protein that contains three potential membrane-spanning domains (Engle *et al.,* 1990). Molecular disruption of *bim*E yields the same phenotype as growth of *bim*E7 at the restrictive temperature (Osmani *et al.,* 1988a). Normally, no mitotic events take place if the cell cycle is arrested in interphase. Loss of *bim*E function, however, makes cells undergo mitotic events even when they are blocked in interphase. Its normal function must be, therefore, to play some role in preventing mitosis from occurring during interphase.

The eukaryotic cell cycle is an ordered sequence of events. In normal cells, completion of one event precedes subsequent events. The common analogy is to a biochemical pathway, in which the product of one catalytic reaction forms the substrate for the next step in the pathway. If the substrate is not formed in one of the steps, the pathway is blocked at this point (Hartwell and Weinert, 1989). The properties of the somatic cell cycle are analogous. For instance, if DNA replication is halted, the cell cycle arrests in S phase and does not progress into G_2 or mitosis. One could, therefore, argue that the product of DNA synthesis (replicated DNA) is an essential "substrate" for mitosis. Data obtained utilizing the *bim*E7 mutation of *Aspergillus* has, however, challenged this model for the cell cycle (Osmani *et al.,* 1988a; Hartwell and Weinert, 1989).

If the cell cycle runs merely as an ordered series of events, as does a biochemical pathway, cells blocked in interphase should not be able to initiate downstream events. However, loss of *bim*E function does allow cells to bypass interphase blocks and enter mitosis. Similarly, overexpression of the NIMA protein kinase initiates mitotic events when cells are in interphase (Osmani *et al.,* 1988b). These observations, coupled with similar observations in several other systems, strongly argue that the "biochemical pathway" hypothesis is not an adequate explanation for cell cycle control (Osmani *et al.,* 1988a; Hartwell and Weinert, 1989).

How, then, is temporal control of the cell cycle achieved? In our opinion, temporal controls revolve around signals generated by certain key elements of cell-cycle progression. Hartwell and Weinert (1989) called these "checkpoints." A cellular sensing mechanism detects these signals and relays this information to the next step in the cycle. If, for any reason, one key element (e.g., DNA replication) is not completed, a signal is not generated, and downstream events are halted. Each step in the cell cycle, therefore, does not necessarily produce a substrate for the next step but generates a signal to prompt the next step. By interfering with the signaling or response systems that control the cell cycle (e.g., by inactivating *bim*E or increasing NIMA activity), events of the cell cycle can be made to happen out of order.

E. The *bim*C Gene

The *bim*C3 mutation of *Aspergillus* causes a distinctive defect in mitotic progression. At the restrictive temperature, this mutant exhibits nuclei with condensed chromatin and a very small mitotic spindle (Morris, 1976). Immunofluorescence studies indicate that this small spindle is, in fact, a monopolar spindle structure (Enos and Morris, 1990) and electron microscopy shows that, in blocked nuclei, the SPB has replicated but the daughter SPBs have not separated (B. R. Oakley and N. R. Morris, unpublished observations). Thus, the mitotic defect caused by *bim*C3 is characterized by an inability to separate SPBs at the very beginning of mitosis. This is the earliest mitotic block known in *Aspergillus*.

The *bim*C gene has been cloned by complementation. Analysis of the *bim*C DNA sequence reveals that its gene product is related to the mechanochemical enzyme kinesin (Enos and Morris, 1990). Kinesin was first identified in higher eukaryotes as a protein with ATPase activity that generates forces when interacting with microtubules. Functionally, the kinesin molecule has two domains. The amino-terminal region binds to microtubules and hydrolyzes ATP. This region is thought to be the actual "motor" part of the molecule that can bind, and then move along, microtubules. This motor domain is attached to a tail that has the potential to bind to another molecule, the "cargo." The kinesin "motor" transports the "cargo" along the microtubule, fueled by ATP hydrolysis. Indeed, kinesin was first isolated from neuronal tissue by its ability to move beads along microtubules. (See Vale and Goldstein, 1990, for review and references.)

Of the amino acid residues found in the motor portion of *Drosophila* kinesin, 40% are also found in the *bim*C gene product. The rest of the primary amino acid sequence shows no similarity between the two proteins. Therefore, *bim*C has similarity to the microtubule-interactive ATP-hydrolyzing motor portion of kinesin (Enos and Morris, 1990), suggesting the attractive possibility that *bim*C plays a role in mitosis as a mechanochemical enzyme involved in generating the force required during mitosis. In short, *bim*C could be a mitotic "motor." The phenotype of *bim*C3 is consistent with this possibility. If mitosis were initiated under conditions of a nonfunctioning mitotic motor, one might expect to see chromosome condensation and activation of the mitotic microtubule-organizing centers, coupled with a halt in mitotic progression. The force required to move the spindle poles apart and to organize and move the mitotic spindle would be absent. This is precisely the defect caused by *bim*C3 (Enos and Morris, 1990).

If a *bim*C DNA probe is used to probe *Aspergillus* genomic DNA at low stringency, more than one band of hybridization is observed, suggesting the presence of several kinesin-like molecules in addition to *bim*C in *Aspergillus*. *bim*C3 specifically affects mitosis and has no apparent effect on other cytoplasmic motility systems. It is, therefore, a distinct possibility that a family of kinesin molecules exists that contains a common motor portion and a variable tail region that specifies the cargo to be moved (Enos and Morris, 1990). A protein related to kinesin has been isolated from yeast (Maluh and Rose, 1990) and *Drosophila* (Endow *et al.*, 1990). These proteins appear to differ from the *bim*C gene product. Recent studies have localized cytoplasmic dynein (another mechanochemical enzyme) to the kinetochore of higher eukaryotic cells (Pfarr *et al.,* 1990; Steuer *et al.,* 1990). The isolation of these other motor-like proteins further supports the notion that more than one motor molecule could be involved in driving various aspects of the mitotic process.

III. TUBULIN MUTATIONS

Another attractive genetic approach aimed at understanding the molecular basis of cell division is the isolation and study of specific genes whose products are known to be involved in the cell cycle. This approach has been most successfully employed in *Aspergillus* with α and β tubulin. α and β tubulin form a heterodimer called, simply, tubulin that assembles to form microtubules. Microtubules constitute the major component of the mitotic apparatus and are essential to mitosis and many other forms of intracellular motility. Tubulin mutations have been used to study microtubule-mediated processes and to isolate other genes whose products interact with α and β tubulin. Recently one of these tubulin interacting genes was found to encode a third member of the tubulin superfamily of proteins, γ tubulin (Oakley and Oakley, 1989).

Tubulins are both regulated by, and part of, the cell cycle regulatory apparatus. The assembly of mitotic and cytoplasmic microtubules is regulated in a cell-cycle-specific fashion. Thus, the machinery that regulates the cell cycle must, directly or indirectly, act on tubulin. Tubulin mutations, or antimicrotubule agents that prevent the formation of microtubules, block the cell cycle at mitosis. This implies that microtubules are somehow involved in the signaling process that stipulates that the transition from mitosis to G_1 should take place.

A. β-Tubulin Mutations

The identification of β-tubulin mutations grew out of attempts to understand the mechanism of resistance to, and mechanism of action of, the commercially important fungicide, benomyl. The mechanism of action of benomyl and related compounds has been the subject of an extensive and excellent review (Davidse, 1986). Early clues about the mode of action of benomyl came from the work of several groups who showed that the chemical inhibited nuclear division (Clemons and Sisler, 1971; Davidse, 1973; Hammerschlag and Sisler, 1973). The arrest of nuclear division was similar to that seen in other organisms treated with the antimicrotubule agent colchicine, suggesting that benomyl might bind to tubulin and inhibit the assembly or function of the microtubules of the mitotic apparatus.

Van Tuyl (1977) investigated the genetics of benomyl resistance by isolating and characterizing mutations, in *Aspergillus nidulans* and other fungi, that conferred resistance to benomyl and related compounds. The *A. nidulans* mutations were mapped to three loci: *ben*A, *ben*B, and *ben*C. Of 28 mutations characterized, 26, including all the mutations that conferred resistance to high concentrations of benomyl, mapped to *ben*A. Thus *ben*A was identified as the major locus involved in benomyl resistance in *A. nidulans.*

Davidse and associates hypothesized that benomyl resistance might be due to the mutation of an α- or β-tubulin gene, so tubulin no longer bound benomyl. Davidse and Flach (1977, 1978) tested this hypothesis by carrying out binding studies with methyl benzimidazole-2-yl carbamate (MBC; sometimes called carbendazim), the active breakdown product of benomyl in aqueous solution. MBC bound to an *A. nidulans* protein with characteristics typical of tubulin. Additionally, the MBC binding affinity of the tubulin-like protein correlated with sensitivity. Binding was greatest in a strain that carried a *ben*A mutation that conferred MBC supersensitivity, lower in a wild-type strain, and lowest in a strain carrying a *ben*A mutation that conferred resistance. These data suggested strongly that MBC bound to the tubulin dimer and that the MBC-resistance mutations altered α or β tubulin to reduce MBC binding.

Sheir-Neiss and colleagues (1978) used a combination of genetic experiments and two-dimensional gel electrophoresis to demonstrate unequivocally that *ben*A was the structural gene for β tubulin. This was the first identification, in any organism, of a gene that encoded a microtubule protein and demonstrated that β-tubulin mutations could be isolated as benomyl-resistance mutations.

These seminal studies left little doubt that the fungicidal activity of benomyl on *A. nidulans* is due to its binding to β tubulin and inhibiting the function of microtubules. Subsequently, the benomyl-resistance genes identified in several other filamentous fungi and yeasts turned out to be β tubulin genes as well (e.g., Neff *et al.,* 1983; Hiraoka *et al.,* 1984; Orbach *et al.,* 1986; Panaccione *et al.,* 1988).

Both α and β tubulin are highly conserved proteins. Thus, low stringency hybridization with cloned tubulin genes can be used to identify other tubulin genes within and among species. The two β-tubulin genes of *A. nidulans, ben*A and *tub*C, were cloned by low stringency hybridization using a chicken β-tubulin cDNA probe (May *et al.,* 1985, 1987). The *ben*A gene was subsequently used to clone the β-tubulin gene of *Neurospora crassa* (Orbach *et al.,* 1986), designated *tub-2. tub-2,* in turn, was used to clone two β-tubulin genes from *Colletotrichum graminicola* (Panaccione *et al.,* 1988; Panaccione and Hanau, 1990).

In principle, understanding the β-tubulin mutations that confer resistance to benomyl should reveal a great deal about the interactions of benomyl and other benzimidazole antimicrotubule agents with β tubulin. Orbach and co-workers (1986) isolated benomyl-resistant mutants of *N. crassa* and cloned the *tub-2* allele from a benomyl-resistant strain. This allele carried a single mutation, a phenylalanine-to-tyrosine substitution at amino acid (aa) 167, and it was used to transform a non-benomyl-resistant strain to benomyl resistance. This result demonstrated that in *N. crassa,* as in *A. nidulans,* benomyl resistance is conferred by a mutation in a β-tubulin gene. It also implicated a specific region of the protein in benomyl binding.

β-tubulin mutations that alter resistance to benomyl have been extensively analyzed in *A. nidulans* (Jung and Oakley, 1990; M. K. Jung, I. B. Wilder, and B.R. Oakley, unpublished observations). Twenty *ben*A alleles that alter benomyl resistance have been cloned and sequenced. Eighteen of the alleles confer resistance to benomyl, MBC, thiabendazole (TBZ), and nocodazole (another benzimidazole antimicrotubule agent), and two alleles confer resistance to TBZ and supersensitivity to the other compounds. These mutations are all single amino acid substitutions. Twelve mutations are substitutions at aa 6, two are substitutions at aa 165, four are substitutions at aa 198, and two are substitutions at aa 200. The single benomyl resistance mutation identified from *N. crassa* is a substitution at aa 167 of the β tubulin encoded by the *tub-2* gene (Orbach *et al.,* 1986). (The amino acid sequences encoded by *ben*A and the *N. crassa tub-2* genes are sufficiently similar that the aa numbers correspond perfectly in this region. Thus aa 167 of *ben*A corresponds to aa 167 of *tub-2.*)

Data from binding studies (reviewed by Davidse, 1986) indicate only one binding site for benomyl and related compounds on β tubulin. Sequences of mutant alleles indicate that the areas around amino acids 6, 165, and 198 of β tubulin are all important to benomyl binding. Recent models for the structure of β tubulin (Mandelkow *et al.,* 1985a,b; Sternlicht *et al.,* 1987) suggest that the β-tubulin molecule is composed of several large loops, each of which contains alpha helical or beta pleated-sheet regions. At least three of these loops are thought to interact to form the GTP-binding site. It is, likewise, possible that the regions around amino acids 6, 165, and 198 are all part of a single binding site created by the folding together of three regions of the protein.

The two mutations at aa 165, *ben*A16 and 19, are particularly interesting because they simultaneously confer resistance to TBZ and supersensitivity to benomyl and MBC (Van Tuyl, 1977), and nocodazole (Jung and Oakley, 1990). These are closely related compounds, but benomyl, MBC, and nocodazole have a methylcarbamate at position 2 whereas TBZ has a 4-thiazole group (Fig. 1). MBC and TBZ are, moreover, identical except

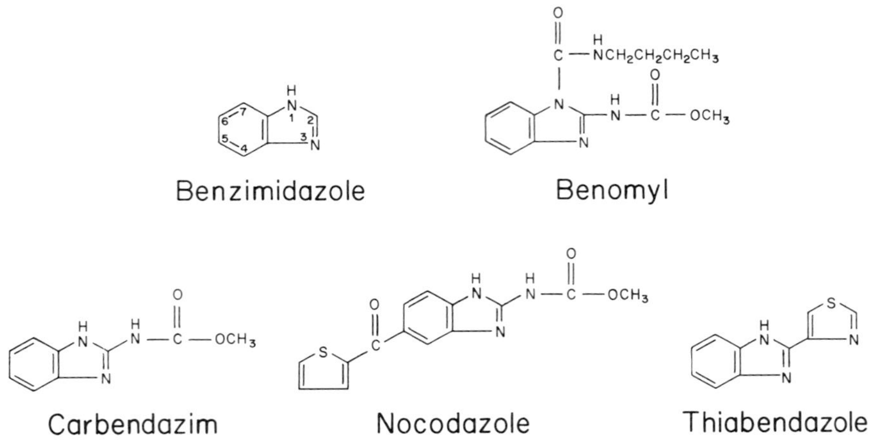

Fig. 1. Structures of benzimidazole antimicrotubule agents.

for the R2 group. These and additional data have led Davidse and Flach (1978) to propose that *ben*A16 and 19 alter a region of β tubulin that is involved in the binding of the R2 group. The fact that both *ben*A16 and 19 are alanine to valine substitutions at aa 165 (Jung and Oakley, 1990) suggests that the region around aa 165 is involved in the binding of the R2 group.

Amino acid 165 of β tubulin shows phylum-specific variation, that is, the wild-type fungal β-tubulin genes sequenced to date have alanine at aa 165, metazoan β tubulins have asparagine at aa 165, and plants and algae have leucine or methionine at aa 165 (references in Jung and Oakley, 1990). These phylum-specific variations at aa 165 may explain, at least partially, the different sensitivities of organisms to the benzimidazole antimicrotubule agents. The resistance of plants to benomyl, TBZ, and nocodazole, for example, might be caused by a failure of β tubulins with leucine or methionine at aa 165 to bind to these compounds. On a practical level one might be able to alter the R2 group of benzimidazoles to devise new compounds that specifically inhibit the functioning of microtubules of specific organisms. For example, it might be possible to design compounds that inhibit microtubule function in parasitic protozoans, but not in the host organism (June and Oakley, 1990).

B. Uses of Tubulin Mutations

Tubulin mutations have had wide application as dominant selectable markers in developing transformation systems in a variety of fungal species and, hence, have been an important tool in the application of modern genetic technology to filamentous fungi. Nutritional markers give strong selections for transformation; however, they are not available in most species and, where available, have to be crossed into each strain prior to transformation. Dominant selectable markers transform virtually any strain. Orbach and colleagues (1986) first showed that a benomyl-resistant β-tubulin allele can be used as a dominant selectable marker in *Neurospora crassa,* establishing a paradigm that has been followed in many other fungi. In some, cases benomyl-resistant β-tubulin alleles from *Aspergillus nidulans* or *N. crassa* have been used to transform other fungi to benomyl resistance (e.g., Dickman, 1988; Henson *et al.,* 1988; Panaccione *et al.,* 1988; Dickman and Partridge, 1989; Goettel *et al.,* 1990). In other cases, benomyl-resistant mutants of the fungus of interest have been isolated and used to transform non-benomyl-resistant strains (e.g., Panaccione *et al.,* 1988).

Tubulin mutations have also been important tools in basic research on microtubule-mediated functions of the cell. Since microtubules are required for mitosis and other essential cellular functions, mutational analysis has relied on the use of conditional lethal mutations. The first conditional tubulin mutations were identified by Morris and co-workers (1979), who found that three of the *ben*A alleles isolated by Van Tuyl (1977) were moderately restricted for growth at high temperatures. The first tightly heat-sensitive (hs^-) tubulin mutations were isolated by Oakley and Morris (1981). Benomyl-resistant mutants were selected and then screened for blocked growth at high and low temperatures. No tight cold-sensitive (cs^-) mutants were isolated; however, several moderately hs^- mutants, and one mutant that caused a virtually complete block of growth at high temperatures, were isolated. The tight hs^- mutation was mapped to the *ben*A locus and designated *ben*A33 (Oakley and Morris, 1981).

At restrictive temperatures, *ben*A33 inhibited nuclear division and nuclear migration along the hyphae and caused a substantial increase in the mitotic index. These results were not surprising because earlier studies had demonstrated that the β tubulin encoded by the *ben*A gene was required for nuclear movement (Oakley and Morris, 1980). What was surprising about *ben*A33, however, was that it did not prevent microtubule assembly at restrictive temperatures. Rather, indirect evidence suggested that *ben*A33 inhibited microtubule disassembly at restricted temperatures (Oakley and Morris, 1981), implying that microtubule disassembly is required for the successful completion of mitosis.

The simultaneous inhibition of mitosis and nuclear migration by *ben*A33 implies, moreover, that both the mitotic apparatus and the cytoplasmic microtubules that are required for nuclear migration contain the β

tubulin encoded by *ben*A. Many morphological studies in fungi and other organisms have demonstrated independent temporal and spatial regulation of different classes of microtubules. The microtubules of the mitotic apparatus, for example, assemble at the time that cytoplasmic microtubules disassemble. At the end of mitosis, the mitotic microtubules disassemble while the cytoplasmic microtubules reassemble. If the different classes of microtubules had different isotypes of α and/or β tubulin with slightly different characteristics, these differences might provide the specificity necessary for differential regulation of assembly and disassembly (Fulton and Simpson, 1976). The results obtained with *ben*A33 and similar results obtained with other organisms (reviewed by Raff, 1984; Cleveland and Sullivan, 1985) demonstrate that this is not generally the case.

C. α- and γ-Tubulin Mutations

The first α-tubulin mutation in filamentous fungi and, indeed, in any organism was isolated by Morris and colleagues (1979) in *Aspergillus nidulans.* As described previously, three of the *ben*A alleles isolated by Van Tuyl (1977) caused a partial inhibition of growth at high temperatures. The inhibition was sufficient to allow isolation of revertants of these alleles as fast-growing sectors of colonies incubated at a high temperature. In 14 of 18 revertants characterized, suppression of the heat sensitivity conferred by the *ben*A alleles was due to extragenic suppressor mutations (Morris *et al.,* 1979). Using a combination of genetics and two-dimensional gel electrophoresis, Morris and co-workers (1979) were able to show that one of these mutations was in an α-tubulin gene that they designated *tub*A. The single *tub*A allele, *tub*A1, conferred supersensitivity to a variety of antimicrotubule agents, suggesting that it destabilized microtubules (Morris *et al.,* 1979). This work not only identified the first α-tubulin gene, but also showed that isolation and analysis of revertants of β-tubulin mutations could be used to identify other genes whose products interact with β tubulin.

Oakley and colleagues (1985, 1987) and Weil and co-workers (1986) carried out an extensive isolation and analysis of revertants of *ben*A33. Seventeen cs$^-$ *tub*A alleles were identified. All but one of these alleles conferred lethality when crossed into a wild-type *ben*A background. One allele, *tub*A4, however, conferred weak cold sensitivity and extreme sensitivity to a variety of antimicrotubule agents in a wild-type *ben*A background.

When cs$^-$ revertants of *ben*A33 were analyzed, most of the mutations that suppressed the *ben*A33 phenotype mapped to the *ben*A and *tub*A loci (Oakley *et al.,* 1985, 1987). Four of the mutations, however, mapped to two previously undiscovered loci that were designated *mip*A and *mip*B (*mip,* for *m*icrotubule-*i*nteracting *p*rotein) (Weil *et al.,* 1986). Oakley and

Oakley (1989) cloned and sequenced the *mip*A gene and found that it encoded a new member of the tubulin superfamily of proteins, a protein that was not α or β tubulin but was closely related to both. Oakley and Oakley proposed the name γ tubulin for this protein. The discovery of γ tubulin was quite unexpected, since two decades of biochemical studies on microtubules had failed to reveal its existence.

Disruption of the *mip*A gene blocks nuclear division, and causes a virtually complete loss of mitotic microtubules and a substantial loss of cytoplasmic microtubules (Oakley *et al.,* 1990). γ tubulin, thus, is essential for microtubule assembly and function. Immunofluorescence microscopy with a γ-tubulin-specific antibody showed that γ tubulin is a component of the SPB. Based on these and additional data, Oakley and co-workers (1990) postulated that γ tubulin functions *in vivo* to nucleate the assembly of microtubules from the SPB.

IV. CONCLUSION AND PROSPECTS

When the first cell cycle mutations of *Aspergillus nidulans* were isolated, few foresaw the value of fungi for studying mitosis, microtubules, and the regulation of the cell cycle. Since the isolation of these mutants, the development of molecular genetic methods in general, and cloning by complementation in particular, has made it possible to clone virtually any gene for which one has a selectable marker or an appropriate probe. The cloning of cell cycle genes is permitting the construction of a catalog of the genes and proteins required for the cell cycle. Although such a catalog is necessary for understanding the cell cycle, it is not sufficient. One must be able to understand the functions of the products of these genes. As we have discussed, in a remarkable number of cases the predicted amino acid sequences of the proteins encoded by the cloned genes have given important clues about the functions of the proteins. The tools are now in hand, moreover, to determine the functions of the proteins *in vivo.* Cloned genes can be used to disrupt the chromosomal copy, can be overexpressed *in vivo,* or can be mutated and then expressed *in vivo.* In addition, the cloned gene can be used to generate specific antibodies to the encoded protein, which can then be used to isolate the gene product.

It is also clear from many studies in many organisms that remarkably similar proteins are involved in the regulation of the cell cycle in filamentous fungi, yeasts, and vertebrates. By studying the cell cycle in fungi, we have learned about cell cycle regulation in all eukaryotic cells.

Much remains to be discovered. We have, at best, only a rudimentary understanding of how microtubule assembly is regulated, how chromosomes move, and how the cell cycle is regulated. Most of the known loci that are required for progression through the cell cycle have not yet been cloned and many such loci have yet to be identified. Many specific questions remain to be answered about proteins that have already been identified. For example, how does the *nim*A protein kinase interact with other protein kinases involved in the regulation of the G_2 to M transition? Is γ tubulin a substrate for the *cdc2*- or *nim*A-encoded protein kinase?

In spite of the remaining questions, progress has been gratifyingly rapid in recent years. The promise that genetic approaches could help us understand mitosis and the cell cycle has been fully realized. Moreover, we now have the techniques in hand to bring about even greater progress in understanding mitosis and the cell cycle in years to come.

ACKNOWLEDGMENTS

We wish to thank Dr. John H. Doonan for sharing unpublished data. This work was supported by National Institute of Health grants GM42564 to S.A.O. and GM31837 to B.R.O..

REFERENCES

Ballance, D. J., Buxton, F. P., and Turner, G. (1983). Transformation of *Aspergillus nidulans* by the orotidine-5′-phosphate decarboxylase gene of *Neurospora crassa*. *Biochem. Biophys. Res. Comm.* **112,** 284–289.

Bergen, L. G., and Morris, N. R. (1983). Kinetics of the nuclear division cycle of *Aspergillus nidulans*. *J. Bacteriol.* **156,** 155–160.

Bergen, L. G., Upshall, A., and Morris, N. R. (1984). S-phase, G_2, and nuclear division mutants of *Aspergillus nidulans*. *J. Bacteriol.* **159,** 114–119.

Clemons, G. P., and Sisler, H. D. (1971). Localization of the site of action of a fungitoxic benomyl derivative. *Pest. Biochem. Physiol.* **1,** 32–43.

Cleveland, D. W., and Sullivan, K. F. (1985). Molecular biology and genetics of tubulin. *Ann. Rev. Biochem.* **54,** 331–365.

Cohen, P. (1989). The structure and regulation of protein phosphatase. *Ann. Rev. Biochem.* **58,** 453–508.

Cyert, M. S., and Thorner, J. (1989). Putting it on and taking it off: Phosphoprotein phosphatase involvement in cell cycle regulation. *Cell* **57,** 891–893.

Davidse, L. C. (1973). Antimitotic action of methyl-benzimidazole-2-yl carbamate (MBC) in *Aspergillus nidulans*. *Pest. Biochem. Physiol.* **3,** 317–325.

Davidse, L. C. (1986). Benzimidazole fungicides: Mechanism of action and biological impact. *Ann. Rev. Phytopathol.* **24,** 43–65.

Davidse, L. C., and Flach, W. (1977). Differential binding of methyl benzimidazole-2-yl carbamate to fungal tubulin as a mechanism of resistance to this antimitotic agent in mutant strains of *Aspergillus nidulans*. *J. Cell Biol.* **72,** 174–193.

Davidse, L. C., and Flach, W. (1978). Interaction of thiabendazole with fungal tubulin. *Biochim. Biophys. Acta* **543,** 82–90.

Davis, F. M., Tsao, T. Y., Fowler, S. K., and Rao, P. N. (1983). Monoclonal antibodies to mitotic cells. *Proc. Natl. Acad. Sci. U.S.A.* **80,** 2962–2930.

Dickman, M. B. (1988). Whole cell transformation of the alfalfa fungal pathogen *Colletotrichum trifolii. Curr. Genet.* **14,** 241–246.

Dickman, M. B., and Partridge, J. E. (1989). Use of molecular markers for monitoring fungi involved in stalk rot of corn. *Theor. Appl. Genet.* **77,** 535–539.

Doonan, J. H., and Morris, N. R. (1989). The *bim*G gene of *Aspergillus nidulans,* required for completion of anaphase, encodes a homolog of mammalian phosphatase 1. *Cell* **57,** 987–996.

Endow, S. A., Henikoff, S., and Soler-Niedziela, L. (1990). Mediation of meiotic and early mitotic chromosome segregation in *Drosophila* by a protein related to kinesin. *Nature (London)* **343,** 81–83.

Engle, D. B., Doonan, J. H., and Morris, N. R. (1988). Cell cycle modulation of MPM-2-specific spindle-pole body phosphorylation in *Aspergillus nidulans. Cell Motil. Cytoskeleton* **10,** 432–437.

Engle, D. B., Osmani, S. A., Osmani, A. H., Rosborough, S., Xiang, X., and Morris, N. R. (1990). A negative regulator of mitosis in *Aspergillus* is a putative membrane spanning protein. *J. Biol. Chem.* **256,** 16132–16137.

Enos, A. P., and Morris, N. R. (1990). Mutation of a gene that encodes a kinesin-like protein blocks nuclear division in *A. nidulans. Cell* **60,** 1019–1027.

Fulton, C., and Simpson, P. A. (1976). Selective synthesis and utilization of flagellar tubulin. The multi-tubulin hypothesis. *In* "Cell Motility" (R. Goldman, T. Pollard, and J. Rosenbaum, eds.), pp. 987–1005. Cold Spring Harbor Laboratory Press, Cold Spring Harbor, New York.

Goettel, M. S., Leger, R. J. S., Bhairi, S., Jung, M. K., Oakley, B. R., Roberts, D. W., and Staples, R. C. (1990). Pathogenicity and growth of *Metarhizium anisopliae* stably transformed to benomyl resistance. *Curr. Genet.* **17,** 129–132.

Gould, K. L., and Nurse, P. (1989). Tyrosine phosphorylation of the fission yeast cdc2$^+$ protein kinase regulates entry into mitosis. *Nature (London)* **342,** 39–45.

Hammerschlag, R. S., and Sisler, H. D. (1973). Benomyl and methyl-2-benzimidazole carbamate (MBC): Biochemical, cytological, and chemical aspects of toxicity to *Ustilago maydis* and *Saccharomyces cerevisiae. Pest. Biochem. Physiol.* **3,** 42–54.

Hartwell, L. H., and Weinert, T. A. (1989). Checkpoints: Controls that ensure the order of cell cycle events. *Science* **246,** 629–634.

Hartwell, L. H., Culotti, J., and Reid, B. (1970). Genetic control of the cell division cycle in yeast. I. Detection of mutants. *Proc. Natl. Acad. Sci. U.S.A.* **66,** 352–359.

Hartwell, L. H., Culotti, J., Pringle, J. R., and Reid, B. J. (1974). Genetic control of the division cycle in yeast. A model to account for the order of cell cycle events is deduced from the phenotypes of yeast mutants. *Science* **183,** 46–51.

Henson, J. M., Blake, N. K., and Pilgeram, A. L. (1988). Transformation of *Gaeumannomyces graminis* to benomyl resistance. *Curr. Genet.* **14,** 113–117.

Hiraoka, Y., Toda, T., and Yanagida, M. (1984). The *nda3* gene of fission yeast encodes beta tubulin: A cold-sensitive *nda3* mutation reversibly blocks spindle formation and chromosome movement in mitosis. *Cell* **39,** 349–358.

Jung, M. K., and Oakley, B. R. (1990). Identification of an amino acid substitution in the *ben*A β-tubulin gene of *Aspergillus nidulans* that confers thiabendazole resistance and benomyl super-sensitivity. *Cell Motil. Cytoskeleton* **17,** 87–94.

Maluh, P. B., and Rose, M. D. (1990). KAR3, a kinesin-related gene required for yeast nuclear fusion. *Cell* **60,** 1029–1041.

Mandelkow, E.-M., Herrmann, M., and Rühl, U. (1985a). Tubulin domains probed by limited proteolysis and subunit-specific antibodies. *J. Mol. Biol.* **185,** 311–327.

Mandelkow, E.-M., Kirchner, K., and Mandelkow, E. (1985b). Domains, antigenicity, and substructure of tubulin. *In* "Microtubules and Microtubule Inhibitors" (M. De Brabander and J. De May, eds.), pp. 31–47. Elsevier Science, Amsterdam.

May, G. S., and Morris, N. R. (1987). The unique histone H2A gene of *Aspergillus nidulans* contains three introns. *Gene* **58,** 59–66.

May, G. S., Gambino, J., Weatherbee, J. A., and Morris, N. R. (1985). Identification and functional analysis of beta-tubulin genes by site specific integrative transformation in *Aspergillus nidulans. J. Cell Biol.* **101,** 712–719.

May, G. S., Tsang, M. L.-S., Smith, H., Fidel, S., and Morris, N. R. (1987). *Aspergillus nidulans* β-tubulin genes are unusually divergent. *Gene* **55,** 231–243.

Morris, N. R. (1976). Mitotic mutants of *Aspergillus nidulans. Genet. Res. Camb.* **26,** 237–254.

Morris, N. R., Lai, M. H., and Oakley, C. E. (1979). Identification of a gene for α-tubulin in *Aspergillus nidulans. Cell* **16,** 437–442.

Murray, A. W., and Kirschner, M. W. (1989). Dominoes and clocks: The union of two views of the cell cycle. *Science* **246,** 614–621.

Nasmyth, K. A., and Reed, S. I. (1980). Isolation of genes by complementation in yeast: Molecular cloning of a cell-cycle gene. *Proc. Natl. Acad. Sci. U.S.A.* **77,** 2119–2123.

Neff, N. F., Thomas, J. H., Grisafi, P., and Botstein, D. (1983). Isolation of the β-tubulin gene from yeast and demonstration of its essential function *in vivo. Cell* **33,** 211–219.

Nurse, P. (1990). Universal control mechanism regulating onset of M-phase. *Nature (London)* **344,** 503–508.

Nurse, P., Thuriaux, P., and Nasmyth, K. (1976). Genetic control of the cell division cycle in the fission yeast *Schizosaccharomyces pombe. Mol. Gen. Genet.* **146,** 167–178.

Oakley B. R., and Morris, N. R. (1980). Nuclear movement is β-tubulin-dependent in *Aspergillus nidulans. Cell* **19,** 255–262.

Oakley, B. R., and Morris, N. R. (1981). A β-tubulin mutation in *Aspergillus nidulans* that blocks microtubule function without blocking assembly. *Cell* **24,** 837–845.

Oakley, B.R., and Morris, N. R. (1983). A mutation in *Aspergillus nidulans* that blocks the transition from interphase to prophase. *J. Cell Biol.* **96,** 1155–1158.

Oakley, B. R., Oakley, C. E., Kniepkamp, K. S., and Rinehart, J. E. (1985). Isolation and characterization of cold-sensitive mutations at the *ben*A, β-tubulin, locus of *Aspergillus nidulans. Mol. Gen. Genet.* **201,** 56–64.

Oakley, B. R., Oakley, C. E., and Rinehart, J. E. (1987). Conditionally lethal *tub*A α-tubulin mutations in *Aspergillus nidulans. Mol. Gen. Genet.* **208,** 135–144.

Oakley, B. R., Oakley, C. E., Yoon, Y., and Jung, M. K. (1990). γ-tubulin is a component of the spindle-pole-body that is essential for microtubule function in *Aspergillus nidulans. Cell* **61,** 1289–1301.

Oakley, C. E., and Oakley, B. R. (1989). Identification of γ-tubulin, a new member of the tubulin superfamily encoded by *mip*A gene *Aspergillus nidulans. Nature (London)* **338,** 662–664.

Orbach, M. J., Porro, E. B., and Yanofsky, C. (1986). Cloning and characterization of the gene for β-tubulin from a benomyl-resistant mutant of *Neurospora crassa* and its use as a dominant selectable marker. *Mol. Cell. Biol.* **6,** 2452–2461.

Osmani, A. H., O'Donnell, K., Pu, R. T., and Osmani, S. A. (1991). Activation of the *nim*A protein kinase plays a unique role during mitosis that cannot be bypassed by absence of the *bim*E checkpoint. *EMBO J.* In Press.

Osmani, S. A., May, G. S., and Morris, N. R. (1987). Regulation of the mRNA levels of *nim*A, a gene required for the G_2–M transition in *Aspergillus nidulans. J. Cell Biol.* **104,** 1495–504.

Osmani, S. A., Engle, D. B., Doonan, J. H., and Morris, N. R. (1988a). Spindle formation and chromatin condensation in cells blocked at interphase by mutation of a negative cell cycle control gene. *Cell* **52,** 241-51.

Osmani, S. A., Pu, R. T., and Morris, N. R. (1988b). Mitotic induction and maintenance by overexpression of a G_2-specific gene that encodes a potential protein kinase. *Cell* **53,** 237-244.

Panaccione, D. G., and Hanau, R. M. (1990). Characterization of two divergent β-tubulin genes from *Colletotrichum graminicola. Gene* **86,** 163-170.

Panaccione, D. G., McKiernan, M., and Hanau, R. M. (1988). *Colletotrichum graminicola* transformed with homologous and heterologous benomyl-resistant genes retains expected pathogenicity to corn. *Molecular Plant-Microbe Interactions* **1,** 113-120.

Pfarr, C. M., Coue, M., Grissom, P. M., Hays, T. S., Porter, M. E., and McIntosh, J. R. (1990). Cytoplasmic dynein is localized to kinetochores during mitosis. *Nature (London)* **345,** 263-265.

Raff, E. C. (1984). Genetics of microtubule systems. *J. Cell Biol.* **99,** 1-10.

Sheir-Neiss, G., Lai, M. H., and Morris, N. R. (1978). Identification of a gene for β-tubulin in *Aspergillus nidulans. Cell* **15,** 639-649.

Sternlicht, H., Yaffe, M. B., and Farr, G. W. (1987). A model of the nucleotide-binding site in tubulin. *FEBS Lett.* **214,** 226-235.

Steuer, E. R., Wordeman, L., Schroer, T. A., and Sheetz, M. P. (1990). Localization of cytoplasmic dynein to mitotic spindles and kinetochores. *Nature (London)* **345,** 266-268.

Tilburn, J., Scazzocchio, C., Taylor, G. G., Zabicky-Zissman, J. H., Lockington, R. A., and Davies, W. R. (1983). Transformation by integration in *Aspergillus nidulans. Gene* **26,** 205-221.

Vale, R. D., and Goldstein, S. B. (1990). One motor, many tails: An expanding repertoire of force-generating enzymes. *Cell* **60,** 883-885.

Vandre, D. D., Davis, F. M., Rao, P. N., and Borisy, G. G. (1984). Phosphoproteins are components of mitotic microtubule organizing centers. *Proc. Natl. Acad. Sci. U.S.A.* **81,** 4439-4443.

Van Tuyl, J. M. (1977). Genetics of fungal resistance to systemic fungicides. *Meded. Landbouwhogesch. Wageningen* **2,** 1-137.

Weil, C. F., Oakley, C. E., and Oakley, B. R. (1986). Isolation of *mip* (microtubule interacting protein) mutations of *Aspergillus nidulans. Mol. Cell. Biol.* **6,** 2963-2968.

6

Molecular Genetics of the Agaric *Coprinus cinereus*

PATRICIA J. PUKKILA
Department of Biology and Curriculum in Genetics
University of North Carolina
Chapel Hill, North Carolina

LORNA A. CASSELTON
School of Biological Sciences
Queen Mary and Westfield College
University of London
London, England

MORE GENE MANIPULATIONS
IN FUNGI

I. INTRODUCTION

The basidiomycete fungus *Coprinus cinereus* has a typical hymenomycete life cycle with distinct monokaryotic and dikaryotic phases, culminating in the development of the highly differentiated mushroom fruit body that provides a platform from which to disperse the sexual spores. This life cycle is easily manipulated in the laboratory and, as a result, it is possible to select for mutations in genes affecting a wide spectrum of metabolic and developmental functions. The hymenomycetes include a number of valuable edible species as well as others of potential industrial and agricultural importance that are far less tractable to genetic analysis. *C. cinereus* provides an important model system in which to look at the structure and regulation of basidiomycete genes. With the advent of a simple and efficient transformation system, it is also a useful host for testing the expression of genes from other species.

In this chapter, we describe the traditional genetic and cytological techniques as well as more recently developed molecular techniques that make *C. cinereus* such as attractive organism to study. Apart from a general interest in the organization of the basidiomycete genome, there are three fundamental problems which are uniquely interesting to study in *C. cinereus*. The first is the complex and unusual mating type system first described by Bensaude in 1918 (see Casselton, 1978). This system is referred to only briefly in this chapter. The genes that constitute the *A* mating type factor have been cloned (Casselton *et al.*, 1989; May *et al.*, 1991). They are considered in detail elsewhere in this volume (Novotny *et al.*, Chapter 10). The second is mitochondrial transmission and the opportunity for mitochondrial gene recombination in an organism in which no specialized sex cells are involved in mating. The third is chromosome structure and function, with particular emphasis on meiotic recombination and chromosome pairing.

A note on nomenclature: *Coprinus cinereus* (Shaeff, ex Fr.) S. F. Gray has been identified variously as *C. lagopus, C. fimentarius,* and *C. macrorhizus f. microsporus* Hongo, but it is now generally accepted that the reports in the literature do refer to the same species (North, 1987).

II. GENETIC MANIPULATION OF THE LIFE CYCLE

A. Monokaryons and Dikaryons

The monokaryon (or homokaryon) is the primary mycelium that develops from a single sexual spore. It has uninucleate cells, is sexually sterile, but produces abundant uninucleate asexual spores, the oidia. Mutant isolation techniques developed for other fungi have been equally successful in *Coprinus cinereus.* Filtration enrichment following mutagen treatment of oidial suspensions has provided a wide variety of auxotrophic mutants. Selection for resistance to metabolic inhibitors such as fluoroacetate (Casselton and Casselton, 1974), 5-fluoroindole (Veal and Casselton, 1985), and cycloheximide (North, 1982) has isolated mutants altered in metabolic regulation and ribosome function. Some 100 different gene mutations have been described, and recombination analysis has identified 10 linkage groups (North, 1987). Oidia are also a source of highly competent protoplasts for DNA-mediated transformation.

For sexual reproduction, compatible monokaryons must fuse to form a dikaryon, which has two genetically different haploid nuclei in each cell, maintained by a complex mode of cell division that generates the characteristic clamp connection at all cross walls. The dikaryotic cell is effectively diploid and, for genetic analysis, offers an unambiguous test of dominance, recessiveness, and complementation. On occasion it is important to resolve the dikaryon into its two nuclear components. This can be done by germinating the thick-walled chlamydospores that form on the underside of the mycelium and often produce monokaryotic hyphae (Lewis, 1961).

Given the right temperature and light stimuli, the fruit bodies develop on the dikaryon. The dikaryotic nuclei eventually fuse in the specialized basidia that develop on the gill surfaces and the meiotically produced nuclei enter the four basidiospores. When it is important to analyze a single meiotic event, the tetrad of spores can be removed by micromanipulation.

B. Heterokaryons and Diploids

Compatibility between monokaryons is determined by the mating type genes. *Coprinus cinereus* is a tetrapolar species with two unlinked mating type factors, *A* and *B,* that each derive many specificities from a combination of two closely linked multiallelic genes termed α and β (see Novotny *et al.,* Chapter 10). Monokaryons are compatible when *A* and *B* are different. Hyphal fusion triggers a series of events, regulated in part by *A* and in part by *B,* that converts the monokaryon into the dikaryon. Matings between monokaryons with one common factor are incompatible but can generate sterile heterokaryons if suitable forcing markers are present. Such heterokaryons have been useful for two purposes: isolation of mating type gene mutations (Day, 1963; Haylock *et al.,* 1980; Swamy *et al.,* 1984) and somatic diploids (Casselton, 1965).

Conversion of a sterile heterokaryon into a fertile dikaryon is the result of mutation in one of the genes of the common factor. Such mutations always give universal compatibility and constitutive expression of that part of the dikaryon morphogenetic sequence regulated by the mutated factor. A combination of mutations in both factors gives a mycelium that resembles a true dikaryon, with binucleate cells and clamp connections, and is self-fertile. Unlike a true dikaryon, it also produces uninucleate oidia. Thus, it is possible to switch on the entire life cycle from a single haploid asexual spore. This mutant mycelium opens another dimension in mutant selection. It is possible to use conventional mutagen treatment of oidia to induce mutations in developmentally regulated genes that, whether recessive or dominant, will be expressed in this homothallic background (Kanda and Ishikawa, 1986).

Somatic diploidy is not a normal stage in the life cycle of *C. cinereus* but, like many other fungi, there is a low level of somatic recombination. Buller demonstrated that recombinant nuclei could arise in the dikaryon, an event correctly interpreted by Quintanilha (1937) and called the "Buller phenomenon." Stable diploid monokaryons can be selected from heterokaryons with common *A* or common *A* and *B* factors using a technique based on the one originally described for *Aspergillus nidulans* (Roper, 1952). Diploids were first used to interpret dosage effects of recessive translational suppressor genes (Todd and Casselton, 1973) and compatibility control by the *B* factor (Casselton and Lewis, 1966), and can reveal unusual nuclear gene interactions (North, 1982).

III. DNA-MEDIATED TRANSFORMATION

A. Transformation Systems

DNA-mediated transformation of protoplasts of *Coprinus cinereus* was first reported by Binninger and colleagues (1987). A cloned tryptophan synthetase gene was used to complement a nonrevertible lesion in the resident gene of the host strain. The original technique was based on that used for *Aspergillus nidulans* (Ballance *et al.*, 1983). The protoplast procedure of Kiguchi and Yanagi (1985) required little modification (Casselton and de la Fuente Herce, 1989). It is now possible to recover more than 1000 transformants per μg cosmid or plasmid DNA.

Several cloned genes have been isolated and are listed in Table I. Not all genes transform as efficiently as the *trp-1* gene. Work with the *trp-2* gene suggests that important regulatory sequences may be lost during cloning. Then expression becomes dependent on a chance integration close to other regulatory sequences. The *trp-2* gene was originally isolated on a 3.5 kb *Eco*RI fragment that gave only 50–100 transformants per μg DNA. Addition of 1.0 kb of flanking sequence increased this frequency threefold (A. Kispert and L. A. Casselton, unpublished observations). Even so, transfor-

TABLE I.

Cloned Genes of *Coprinus cinereus*

Gene	Function	Reference
trp-1	Tryptophan synthetase	Skrzynia *et al.* (1989)
trp-2	PRA isomerase, indole glycerol phosphate synthase, anthranilate synthetase	Kispert and Casselton (unpublished observations)
trp-3	Anthranilate synthetase	Burrows, Stratmann, Elliott, and Casselton (unpublished observations)
trp-3^{iar}	Feedback resistant *trp-3*	Burrows, Stratmann, Elliott, and Casselton (unpublished observations)
trp-4	Phosphoribosyl transferase	Burrows and Casselton (unpublished observations)
ade-8	Adenine biosynthesis	Casselton *et al.* (1989)
pab-1	*p*-Aminobenzoic acid biosynthesis	Casselton *et al.* (1989)
acu-7	Isocitrate lyase	Mellon *et al.* (1987)
acs-1	Acetyl-CoA synthetase	Maconochie and Casselton (unpublished observations)
gdh-1	Glutamate dehydrogenase	Marmeisse, Casselton, and Debaud (unpublished observations)
rad-9	Repair deficient	M. Zolan (personal communication)
A42α3β3	*A* Mating type factor	Casselton *et al.* (1989)
A6mut	Self-compatible mutant *A6*	Casselton *et al.* (1989)
A43	*A* Mating type factor	May *et al.* (1991)
rDNA	Ribosomal RNA repeat	Wu *et al.* (1983)

mation frequency with this gene continues to be much lower than with the *trp-1* gene.

Stable transformants arise by integration of transforming DNA. Linear DNA fragments transform efficiently; identification of gene sequences within longer genomic sequences such as cosmid inserts can be made by simply digesting the DNA with different restriction enzymes and using the entire digest or gel-purified fragments as the transforming DNA.

B. Fate of Transforming DNA

Integration of plasmid transforming DNA occurs predominantly at nonhomologous sites within the genome. An analysis of 93 Trp$^+$ transformants transformed with the *trp-1* gene showed that only 4 resulted from homologous integration and 1 from a gene replacement. Multiple integrations were common and often represented tandemly duplicated copies of the whole plasmid (Binninger *et al.*, 1987). Transformed copies of both the *trp-1* and *acu-7* genes were stable through mitosis and meiosis (Binninger *et al.*, 1987; Mellon *et al.*, 1987).

Nonhomologous integration results in major position effects. Co-transformation of a *trp-1 acu-7* double mutant with separate plasmids, each containing one of the functional genes, showed a variable expression of the unselected *acu-7* gene (Mellon and Casselton, 1988). No correlation could be found between isocitrate lyase activity and gene copy number; as many as eight copies of the gene gave only 25% of the enzyme activity found in an untransformed wild-type strain. The expression of a transformed gene is determined by the site of integration. In regions that are developmentally regulated, the gene will be expressed poorly or not at all in the monokaryotic transformant. Several Acu$^-$ transformants were, in fact, shown to contain undisrupted but unexpressed copies of the *acu-7* gene. This evidence highlights one of the problems faced in testing the expression of manipulated genes and the need to be able to target transforming DNA.

This problem has been addressed (Binninger *et al.*, 1991). The frequency of targeted transformation (5%) could not be enhanced by the presence of single or double strand breaks, single-stranded DNA, or additional flanking homology. However, the observed frequency of homologous replacements and insertions has allowed targeted modifications of the *Coprinus cinereus* genome to be sought and recovered (T. Freedman, C. Skrzynia, and P. J. Pukkila, unpublished observations).

C. Gene Cloning by Transformation

Transformation followed by sib-selection is now an effective way of isolating cloned genes from genomic libraries. Three of the metabolic genes listed in Table I *(trp-3, ade-8,* and *pab-1*) were isolated from cosmid libraries by selecting directly for expression in the appropriate auxotrophic mutant. A cosmid vector containing the *trp-1* gene as a selectable marker has been constructed by Le Chevanton (Fig. 1). Ordered libraries of individual chromosomes are now available in this vector, making it possible to screen for genes that cannot be selected directly, for example, the mating type gene (May *et al.*, 1991) and *rad* genes (M. Zolan, personal communication).

Because transforming DNA integrates nonhomologously, it can be a source of gene mutation and offers a potentially valuable technique for isolating developmentally regulated genes that would be difficult to screen for in the conventional way (see Section VI). Once integrated into a gene, the exogenous DNA not only induces a mutant phenotype but offers a

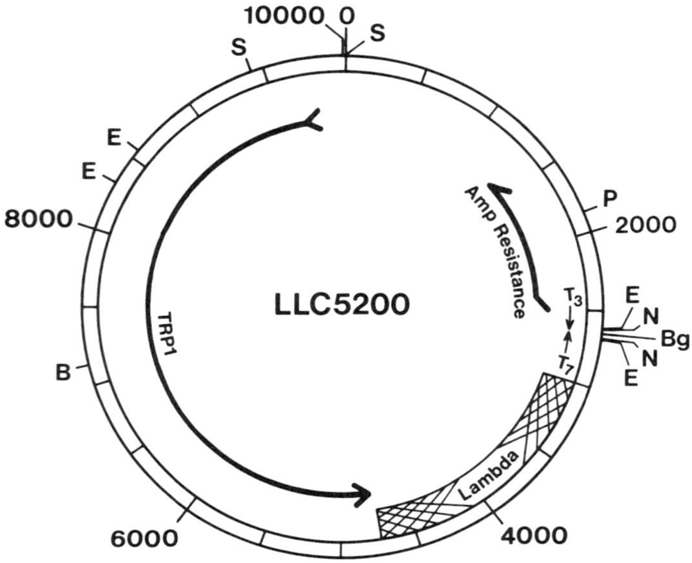

Fig. 1. Schematic diagram illustrating major features of the cosmid cloning vector LLC5200. Abbreviations are: T3, transcriptional promoter from bacteriophage T3; T7, transcriptional promoter from bacteriophage T7; B, *Bam*HI; Bg, *Bgl*II; E, *Eco*RI; N, *Not*I, P; *Pst*I; S, *Sal*I. The vector contains the 1.9-kb *Sal*I–*Pst*I fragment of pBR329 (Covarrubias and Bolivar, 1982) ligated to the 3.1-kb *Sal*I–*Pst*I fragment of pWE15 (Wahl *et al.*, 1987). The *Bgl*II cloning site was added using the appropriate linkers. The *trp-1* gene was introduced by blunt end ligation of the 5-kb *Bgl*II fragment of pCc1001 (Skrzynia *et al.*, 1989) into an *Sph*I site.

molecular marker for recovering the gene. Mutations in the *A* and *B* incompatibility factors can be used to switch on the entire developmental sequence, from haploid oidium to meiotically produced sexual spore. Transformation of a triple mutant, *Amut Bmut pab-1,* with DNA from a cosmid clone containing the *pab-1* gene induced a variety of mutants that produce abnormal fruit bodies (L. A. Casselton and S. Taylor, unpublished observations). Southern blot analysis confirmed that the integration event generated novel-sized restriction fragments that contain sequences from the disrupted gene fused to the transforming DNA. Plasmid integration also causes developmental gene mutations and is potentially more useful, since plasmid rescue techniques (Ballance and Turner, 1986) should recover the plasmid and the flanking sequences from the disrupted gene in a single step.

IV. THE MITOCHONDRIAL GENOME

A. Genome Variation

Coprinus cinereus has a comparatively small mitochondrial genome. The standard genome contains approximately 42 kb (Casselton and Economou, 1985; Weber *et al.,* 1986) and is termed H because it was first isolated from the wild-type strain H9. Screening of 10 separately isolated strains from such widely dispersed locations as the United Kingdom, the United States, Czechoslovakia, Poland, Japan, Singapore, and Java revealed that 7 had the standard H genome, whereas the other 3 had different variants. These are designated J, P, and S (Economou *et al.,* 1987; Economou and Casselton, 1989).

Physical maps of the genomes are shown in Fig. 2. All the variation is found in the region that encodes the cytochrome oxidase subunit 1, *co-1,* and the large ribosomal RNA subunit gene, L-RNA (or *rnl*). J and H DNAs are identical in size and differ only in alternative 1.2 kb insertions. The J insertion is an optional intron in the *co-1* gene, whereas the H insertion appears to be in a nonessential region of the 3′ flanking region of the gene because it is deleted in the P and S genomes. The P and S genomes

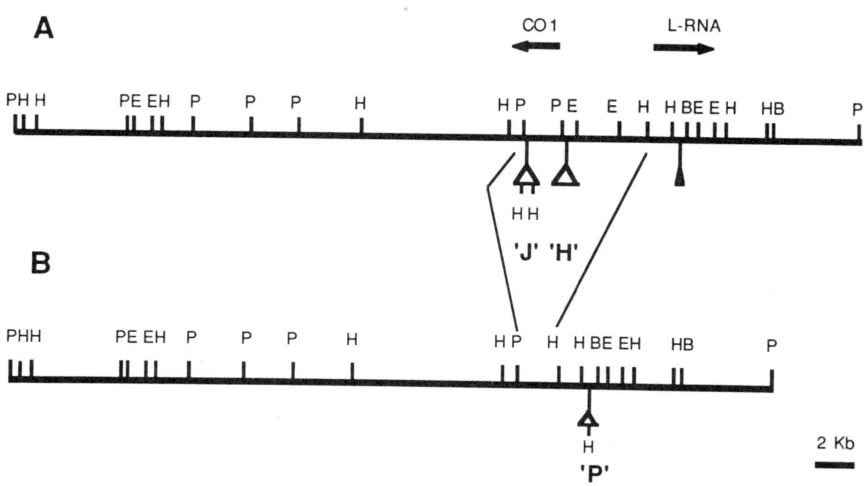

Fig. 2. Linearized physical maps of the *C. cinereus* mitochondrial DNA. (A) H and J genomes; (B) P and S genomes. The positions of the H, J, and P insertions and the *co-1* and L-RNA genes are indicated. Arrows indicate the 5′ ends of the genes. Maps are based on Weber *et al.* (1986) and Economou and Casselton (1989). E, P, H, and B represent *Eco*RI, *Pst*I, *Hin*dIII, and *Bam*HI sites, respectively.

are 37 and 36 kb respectively, lack the J insertion, and have a 4.4 kb deletion in the region between the *co-1* and L-RNA genes. The difference between P and S genomes is a polymorphism within the 3′ end of the L-RNA gene that is large enough to represent an optional intron and is in a similar position to the intron found in the corresponding genes of *Saccharomyces cerevisiae, Neurospora crassa,* and *Aspergillus nidulans* (Hahn *et al.,* 1979; Dujon, 1980; Lazarus *et al.,* 1980). It is particularly interesting that the L-RNA gene in the H and J genomes represents a third form of this gene, with only a 300 bp insertion at the same site. This, perhaps, is the result of an incomplete excision of the larger sequences present in the P genome.

A feature of the genome that is unusual is the opposing orientation of the *co-1* and L-RNA genes, which must therefore be transcribed from different strands of the DNA.

B. Mitochondrial Inheritance

The mitochondria of *Coprinus cinereus* show typical maternal inheritance, despite the fact that there is no sexual differentiation in the hymenomycetes and a compatible interaction occurs between purely vegetative monokaryotic hyphal cells. Even though there is widespread breakdown of septa to allow migration of nuclei during dikaryon formation, there is no corresponding exchange of mitochondria. After nuclear exchange, at which time mated monokaryons have recognizably different mitochondria, it can be seen that two distinct dikaryons are formed, each of which has the mitochondrial type of the recipient monokaryon (Casselton and Condit, 1972). In a similar way, mitochondrial mosaics were seen to arise in monokaryon × dikaryon matings (May and Taylor, 1988). By resolving reciprocally formed dikaryons, it is possible to recombine nuclear and cytoplasmic genomes, a technique used to show that the small P and S genomes do not cause any respiratory deficiency in different nuclear backgrounds.

Recombination between mitochondrial genomes does occur but is restricted to the zone where monokaryotic cells fuse. Recombinant genomes rapidly segregate, so cells never maintain mixed populations (Baptista-Ferreira *et al.,* 1983). Fruit bodies that arise on fusion zones have either of the parental mitochondrial DNAs, but not both (May and Taylor, 1988). Why this rapid segregation occurs and how one mitochondrial type is selected is not known.

Mutations in two genes have been used to monitor mitochondrial recombination: the *acu-10* mutation, which gives a defective cytochrome oxidase phenotype, and *cap-1,* which confers resistance to chloramphenicol and is assumed to map in the L-RNA gene. Spontaneous *cap-1* mutations occur readily in the H, J, and S genomes, and a variety of resistant phenotypes can be distinguished. No *cap-1* mutations have been detected in the P genome. This has been attributed to the putative intron in the P L-RNA gene, which could impose sequence conservation in the region where such mutations are expected to occur (Dujon, 1980). Recombination studies using both mitochondrial gene mutations as markers showed that the H and J insertions exhibit a property characteristic of many group I introns, which Dujon (1989) has called mobile; they show undirectional conversion, so recombinants always inherit both insertions (Economou *et al.,* 1987).

V. CHROMOSOME STRUCTURE

A variety of cytological and molecular methods have elucidated relationships between *Coprinus cinereus* chromosome structure and function. Cytological studies have been confined largely to meiotic chromosomes. These chromosomes can be obtained easily due to a lengthy and synchronous meiotic process. In many respects this fungus exhibits a typical meiosis: the condensed chromatin becomes associated with lateral elements of the synaptonemal complex and, following alignment of these elements, a central component forms between them. Telomeres remain anchored in the nuclear envelope throughout meiotic prophase. The metaphase divisions occur without nuclear envelope dissolution (typical of nuclear divisions in fungi).

Spindle-pole body (SPB) behavior has also been examined. The elongate shape of this organelle at zygotene (Lu, 1978; Holm *et al.,* 1981) is thought to result from fusion of two SPBs during karyogamy (which occurs immediately prior to leptotene/zygotene in this fungus). Both light and electron microscopy have been used to monitor SPB duplication during diplotene and subsequent separation prior to metaphase I (Raju and Lu, 1973b). Using electron microscopy, centromere regions can be detected at pachytene and become increasingly prominent as prophase progresses (Holm *et al.,* 1981). In particularly favorable preparations, a primary constriction (the centromere) is apparent using light microscopy as well (Lu and Raju, 1970). Numerous microtubules radiate from each SPB. Some of these appear to span the distance between the duplicated SPBs (particularly in the central zone of the spindle, which is largely free of chromatin), while others appear to be attached both to a SPB and to a chromosome. Insertion of more than one microtubule into a single centromere has been described (Lu, 1967). Genetic methods also have been used to analyze chromosome segregation in *C. cinereus.* Several complementation groups have been defined that exhibit failure to complete metaphase I (Tani *et al.,* 1977; Kanda *et al.,* 1989a). Surprisingly, in six mutants examined, premeiotic DNA replication failed to occur, although chromosome synapsis and SPB duplication seemed normal (Kanda *et al.,* 1990). In addition, four complementation groups have been defined recently that are resistant to benomyl (a fungicide that acts by binding to and inhibiting the function of microtubules), and defects in nuclear migration and spore formation in strains homozygous for some alleles were noted (Kamada *et al.,* 1989). The synchrony of the meiotic divisions (Fig. 3) may facilitate analyses of the assembly and regulation of the spindle apparatus.

A. Comparison of Molecular and Cytological Karyotypes

Due to the genome size (37,500 kb; Dutta, 1974), chromosome number (13), and degree of meiotic chromosome condensation, three techniques that provide complementary information can be applied to analysis of the *Coprinus cinereus* karyotype: light microscopic analysis of squashes or spreads, electron microscopic analysis of spreads or serial sections, and gel electrophoretic analysis of intact chromosomal DNAs.

Lu and Raju (1970) used light microscopy and iron–hematoxylin staining to describe 12 bivalents in *C. cinereus.* Subsequently, Pukkila and Lu

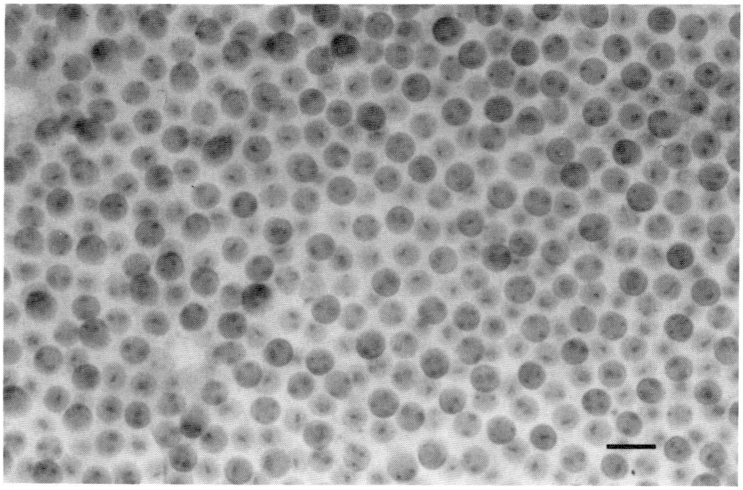

Fig. 3. Light micrograph illustrating surface views of *C. cinereus* basidia fixed and stained with iron–haematoxylin. Of the 250 nuclei that can be scored, over 90% are in metaphase or telophase of the first meiotic division. Bar indicates 10 μm.

(1985) adapted techniques developed by Moses (1977; Dresser and Moses, 1979) to disperse *C. cinereus* nuclear contents and stain the lateral elements of the synaptonemal complexes with silver. The relative lengths of 13 bivalents were reported, and the length of the largest chromosome was approximately five times the length of the smallest, as illustrated in Fig. 4. The data were derived from a dikaryon collected in the wild, resolved into its two component monokaryons, and recrossed (as in Fig. 4), and from two additional monokaryotic strains of different geographic origin. Some differences were noted between this karyotype and that reported from serial sectioning and three-dimensional reconstruction (Holm *et al.,* 1981; Rasmussen *et al.,* 1981). Both light and electron microscopy proved sensitive to gross chromosomal rearrangements (translocations). The ease with which these rearrangements were revealed in standard lab strains suggests their prevalence in *C. cinereus* populations.

More recently, electrophoretic methods have been applied to examining the *C. cinereus* karyotype. Both OFAGE (orthogonal field alternation gel electrophoresis) and CHEF (contour-clamped homogeneous electric field) methods, in which the DNA molecules must alter their orientation during electrophoresis in response to changes in orientation of the electric fields (Carle and Olson, 1984; Chu *et al.,* 1986), have been used with success. Under typical conditions (6 day run, 20 min pulse time, 6 V/cm), 10 bands are completely resolved, and intensity differences suggest that some of these bands contain more than one chromosome. The patterns are highly reproducible from gel to gel, but strikingly divergent from strain to strain

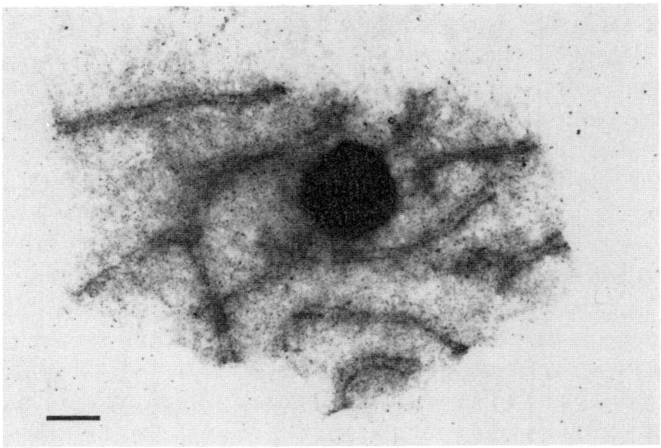

Fig. 4. Electron micrograph of silver-stained chromosomes of *C. cinereus.* A strain 144X × 144Y fruiting body was used. Bar indicates 1 μm.

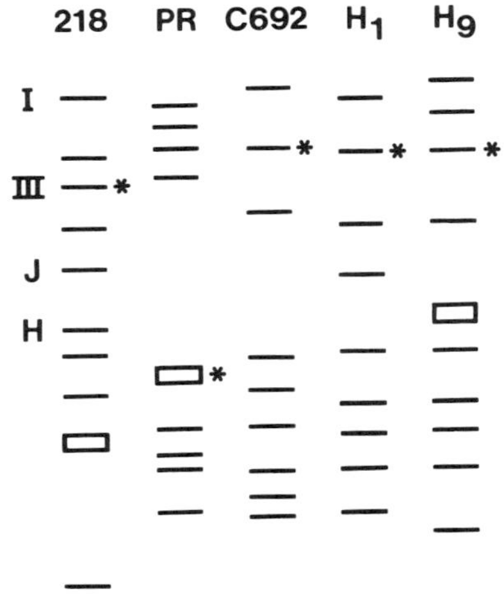

Fig. 5. Tracing of chromosomal-sized DNAs from *C. cinereus* strains following separation by the CHEF method. The DNAs range in size from (approximately) 1 to 5 megabases. Strains are designated at the top (PR is strain PR2301). Linkage groups in 218 are designated along the left side according to North (1987) and were identified using the following cloned genes (see Table I): I, *ade-8, pab-1,* and *A* mating type factor; III, *trp-1;* J, rDNA; H, *acu-7.* Asterisks indicate the chromosome that contains the *trp-1* gene in each strain. Open boxes indicate bands that are not resolved.

(Fig. 5). In fact, with the exception of strains that have been extensively and recently backcrossed, each strain examined has revealed a unique karyotype. The extensive chromosome length polymorphisms (CLPs) make it difficult to identify a presumed linkage group by size alone. Several cloned probes have been used to facilitate the analysis, as illustrated in Fig. 5, lane 218 (a Trp^- strain used extensively in transformation studies). Linkage group III likely is involved in the translocation between chromosomes 3 and 5 that was mapped by light and electron microscopy (Holm *et al.,* 1981; Pukkila and Lu, 1985), since the cloned *trp-1* probe hybridizes to bands of very different sizes in these two strains (Fig. 5, lanes PR and C692), but additional CLPs are also obvious. Single spore isolates from the original Helm collection (Fig. 5, lanes H1 and H9) also exhibit several CLPs. It appears that the extensive natural polymorphism coupled with the requirement for outbreeding makes it likely that natural matings occur between strains harboring many chromosome aberrations. Heterozygosities for structural rearrangements lead to a marked loss in fertility in many eukaryotes (Perkins and Barry, 1977; Gillies, 1989). It will be important to ascertain if essential or dispensable regions of the genome are involved in the observed *C. cinereus* CLPs, and to monitor spore viability in crosses with known heterozygosities.

B. Gene Organization

Although detailed study of gene organization in basidiomycetes is just beginning, three features of studies with *Coprinus cinereus* deserve mention.

First, both classical genetic analysis and less extensive molecular studies indicate that genes coordinately expressed (Yashar and Pukkila, 1985) or of related function (Tilby, 1976) are not tightly linked in the genome, suggesting that *cis*-acting regulatory elements must be present in each sequence. It was somewhat surprising to find that a sequence of 321 nucleotides (nt) upstream from the translation initiation site was sufficient for expression of the tryptophan synthetase gene at a variety of chromosome locations (Skrzynia *et al.,* 1989), although regulation may have been impaired in the transformants. The upstream region lacked CAAT and TATAAA elements; demarcation of the precise regulatory signals must await more extensive deletion analysis. Some regulatory signals may be unique to basidiomycetes, since some genes from filamentous ascomycetes are expressed poorly or not at all in *C. cinereus* (Casselton and de la Fuente Herce, 1989).

Second, sequence analysis has revealed the presence of numerous short introns within basidiomycete genes. It has been proposed that introns facilitated the rearrangement of protein structural domains during the course of evolution (Gilbert, 1978). Sequence analysis of the multifunctional protein tryptophan synthetase (TSase) of *C. cinereus* is of unusual interest (Skrzynia *et al.,* 1989) because several structural and functional domains have been defined and the three-dimensional structure of this enzyme complex from *Salmonella typhimurium* has been described (Hyde *et al.,* 1988). The *C. cinereus* coding region contains nine small interruptions (44–65 nt); sequences at their 5′ and 3′ boundaries match the consensus sequence 5′-GTRNGT . . . YAG-3′ that is typical of splicing signals in filamentous fungi (Gurr *et al.,* 1987). The TSase protein consists of two domains, α and β, separated by a "connector" of variable length and sequence. The α domain consists of an eightfold repeat of β strand alternating with α helix. Although there are four introns within this domain, the exons so defined do not correspond to multiples of the α/β motif. There are no introns in the connector region between the two protein domains, and the introns within the β domain also fail to correlate with boundaries of known structural or functional motifs.

The introns display a striking uniformity in size (44–65 nt), although sequence relatedness could not be detected. It will be of interest to determine if splicing efficiency is influenced by intron size and if differential splicing is used as a regulatory mechanism in this fungus.

Finally, the cluster of repeated genes that encode rRNAs has some unusual features. Unlike previously studied eukaryotic genes, the genes encoding the 5 S rRNA are found within the repeating unit that contains the other three rRNA genes (5.8 S, 17–18 S, and 25–28 S); all four genes are transcribed in the same direction (Cassidy *et al.,* 1984). This novel gene organization is found in other members of the Agaricales. However, in one member of the genus *Coprinus,* the 5 S gene is present in an inverted orientation in all members of the tandem array (Cassidy and Pukkila, 1987). Each 5 S gene lies within "spacer" DNA sequences located between a 26 S and an 18 S gene. These spacer DNA copies are usually homogeneous within a species, but highly diverged among species. This study was the first to suggest that inversions can contribute to divergence in spacer regions.

C. DNA Methylation

Postreplicative modification of DNA occurs in fungi; the addition of a methyl group to certain cytosine residues has received most intensive study. Such cytosine methylation has been associated with instabilities in some filamentous fungi. In *Ascobolus immersus,* transient loss of gene expression has been observed in DNA duplications that are substrates for DNA methylation (Goyon and Faugeron, 1989; Faugeron *et al.,* 1990), while in *Neurospora crassa* both DNA methylation and extensive point mutation in duplicated sequences have been analyzed (Selker *et al.,* 1987; Cambareri *et al.,* 1989).

The nuclear genome of *Coprinus cinereus* is extensively methylated (Zolan and Pukkila, 1985), although no phenotypic consequence of this modification has been observed. Faithful transmission of methylated tracts through meiosis, without influence from the homologous chromosome, was seen in several crosses (Zolan and Pukkila, 1985, 1986). However, the length of the methylated tracts was not always constant; it varied most often by simultaneous *de novo* methylation of several adjacent sites. This *de novo* methylation occurred throughout the life cycle of the fungus, and the methylated regions were transmitted without modification through meiosis.

De novo methylation of transforming DNAs has also been seen (T. Freedman and P. J. Pukkila, unpublished observations), although no loss of gene expression was observed. Unlike in *A. immersus* and *N. crassa,* duplications in *C. cinereus* (either linked or unlinked) are stable through vegetative growth, asexual spore formation, and meiosis (Binninger *et al.,* 1987).

VI. FRUIT BODY FORMATION AND THE MEIOTIC PROCESS

The natural synchrony of basidial development in the genus *Coprinus* was first emphasized by Buller (1909, 1924). He pointed out that deliquescence was necessary to facilitate spore discharge from the vertical nonrigid gills of *Coprinus* species. The process of autodigestion also placed limits on the number of separate generations of basidia that could develop and bear spores in such fruit bodies. Many morphogenetic changes occur in close coordination in developing fruit bodies. Three major processes appear to be under separate, although coordinated, controls: meiosis and spore formation, cap expansion and autolysis, and stipe expansion. Four major questions have been the focus of considerable experimentation.

1. What structural changes accompany differentiation of the tissues of the fruit body?
2. What metabolic changes are associated with fruit body differentiation?
3. What signals in the environment regulate fruit body formation?
4. What are the genetic controls of *C. cinereus* development?

A. Developmental Studies

Matthews and Niederpruem (1972) reviewed the earlier work of Brefeld and Levine and developed methods to observe hyphal anastomoses that preceded fruit body formation. This striking change in hyphal growth pattern from an invasive to a cooperative mode underlies both this initial stage as well as subsequent stages in primordium differentiation. Individual hyphal cells exhibit a limited range of behaviors (reviewed in Reijnders and Moore, 1985): they can become partitioned only by transverse cross-walls, they can enlarge in length and width (which can produce a variety of shape changes), and they can coordinate their behavior with adjacent cells to form well-demarcated tissues (e.g., the central stalk, the sheets of basidia organized into gills, and the outer covering of veil cells). At this time, little is known about how the three-dimensional fruit body emerges using only these few processes.

Metabolic changes are obvious during the fruiting process. Madelin (1956a) demonstrated that developing primordia draw upon metabolites in the mycelium in an uneven way, so scattered clumps of primordia develop using resources provided by the entire mycelial colony. The key metabolic features appear to be the provision of energy sources to allow DNA, RNA, protein, and carbohydrate accumulation at defined times and locations (Matthews and Niederpruem, 1973; McLaughlin, 1974; Kamada *et al.*, 1976); modifications of cell-wall components to permit localized wall expansion (Kamada and Takemaru, 1977a,b, 1983; McLaughlin, 1977;

Kamada *et al.*, 1980; Yoon and McLaughlin, 1980; reviewed in Gooday, 1985); regulation of osmotic balance during cell expansion (Stewart and Moore, 1974; Ewaze *et al.*, 1978; reviewed in Moore *et al.*, 1979; Moore, 1984); and production of lytic enzymes for wall modifications and autolysis (Iten and Matile, 1970; Miyake *et al.*, 1980; Kamada *et al.*, 1985). Many of these processes can occur in isolated tissues of the fruit body (Gooday, 1974; Kamada *et al.*, 1982; Chiu and Moore, 1988).

Extensive studies have been done on the environmental controls of fruiting. *Coprinus cinereus* requires a brief light exposure in order to fruit; this stimulus cannot be transferred to portions of the colony that do not receive light exposure (Madelin, 1956b). There is also a stage at which light appears to be inhibitory, particularly above 28°C (Tsusue, 1969; Lu, 1972, 1974a,b; Kamada *et al.*, 1978), and the required dark-induced factor appears to be diffusible (Kamada and Tsuji, 1979). Since development is tightly coordinated with environmental changes, the availability of any desired stage can be controlled by culturing *C. cinereus* under a defined light–dark regime and simply varying the time of day at which the light cycle begins (Morimoto and Oda, 1973). Detailed cytological studies have indicated that meiocyte development occurs in a regular and predictable manner under these conditions (Raju and Lu, 1970; Morimoto and Oda, 1974). The signal transduction pathway between the results of light stimulation and sexual morphogenesis may involve cAMP (Uno *et al.*, 1974; reviewed in Uno and Ishikawa, 1982). More recently Swamy and coworkers (1985) reported that high levels of cAMP (as well as high levels of cAMP-binding proteins and cAMP-dependent protein kinases) were found in normal dikaryons as well as in strains with mutational bypasses of the normal control of sexual morphogenesis (e.g., strains that fruit as monokaryons or in self-fertile *Amut Bmut* strains), but not in semi-compatible heterokaryons.

A variety of mutagens and selective schemes have been used to acquire collections of dominant (Takemaru and Kamada, 1972; Gibbins and Lu, 1982), recessive (Zolan *et al.*, 1988), and conditional (Kamada *et al.*, 1984) developmental mutants. Spontaneous variants (Day, 1959; Uno and Ishikawa, 1971) have also proved useful. The earliest mutants revealed the existence of several developmental pathways that could operate independently of each other, including meiosis and basidiospore formation, cap expansion and autolysis, and stipe elongation. Mutants blocked at several steps in these pathways were obtained. Progress in this area was greatly facilitated by the isolation and use of *Amut Bmut* strains. In such strains, haploid cells can be mutagenized and screened for developmental defects after the formation of homokaryotic fruit bodies (Swamy *et al.*, 1984; Kanda and Ishikawa, 1986). Even tight recessive mutations that prevent the formation of primordia (Kanda *et al.*, 1989b) or viable basidiospores

(Kanda *et al.,* 1989a; A. B. Maynard and P. J. Pukkila, unpublished observations) can be obtained and analyzed, because *Amut Bmut* strains will donate nuclei to monokaryotic strains, enabling recovery of the new mutations in strains that have conventional mating types. Now DNA mediated transformation can be used to isolate cloned sequences that can complement the most interesting mutants.

B. Meiosis and Genetic Recombination

Lu (1967) first recognized that synchrony of basidial development also extended to synchrony of the meiotic process. *Coprinus cinereus* is particularly well suited to comparative studies using a variety of cytological techniques, because small gill samples can be removed at different times from the same fruit body without any apparent effects on development. A variety of simple staining methods have been used, with success, for light microscopic monitoring of chromosome behavior during the meiotic process, for example, hematoxylin (Lowry, 1963; Lu and Raju, 1970; Moore and Pukkila, 1985), Giemsa (Aist, 1969; Tani *et al.,* 1977), silver (Pukkila and Lu, 1985), and 4,6-diamidino-2-phenyl indole (DAPI) (Kanda *et al.,* 1989a). Under standardized conditions, karyogamy and meiotic prophase take 12 hr, the meiotic divisions take 4 hr, and sterigma formation, spore maturation, and release take 16 hr (Raju and Lu, 1970; reviewed in Pukkila *et al.,* 1984). Since there are approximately 10 million synchronously developing basidia in each fruit body, abundant material for cytological, biochemical, and biophysical studies of the meiotic process can be obtained.

Lu and co-workers have used environmental perturbations such as temperature (Lu, 1969), ionizing radiation (Raju and Lu, 1973a), hydroxyurea (Raudaskoski and Lu, 1980), cycloheximide (Lu, 1984), and divalent cations (Lu and Xiu-Yu, 1988) to alter meiotic recombination frequencies at particular stages. Meiotic synchrony has also been exploited to demonstrate that meiotic cells contain a distinct form of DNA polymerase (polymerase b) (Sakaguchi and Lu, 1982). Of particular interest was the report of an endonuclease, specific to meiotic cells, that introduces single-strand nicks into supercoiled DNA (Lu *et al.,* 1988). If the gene encoding this enzyme can be isolated, sequenced, and disrupted, the role of this protein in genetic recombination can be evaluated directly.

A complementary approach to analyzing mechanisms underlying chromosome pairing and genetic recombination was taken by Zolan and colleagues (1988). They identified mutant strains that were unusually sensitive to ionizing radiation, and showed that two of the genes thus defined were required for completion of meiosis. One gene *(rad-3)* was of particular interest, since chromosome condensation and pairing in *rad-3;1* homozygotes was aberrant, although surviving spores exhibited normal levels of recombination. It is important to develop direct assays for genetic exchange in this system, to enable independent monitoring of crossing over and chromosome pairing, even in cells that have not completed these processes. The use of the transformation system to clone and characterize genes critical for the meiotic process, in conjunction with the genetic and biochemical approaches discussed in the preceding sections, should provide considerable advances in our understanding of meiotic mechanisms.

ACKNOWLEDGMENTS

We thank our colleagues for providing preprints of their work, and S. Whitfield for preparing the figures. The work from our laboratories described in this chapter was supported by the ACS, NIH, and NSF (P.J.P.), as well as the SERC and the Gatsby Charitable Foundation (L.A.C.).

REFERENCES

Aist, J. R. (1969). The mitotic apparatus in fungi, *Ceratocystis fagacearum* and *Fusarium oxysporum*. *J. Cell Biol.* **40,** 120–135.

Ballance, D. J., and Turner, G. (1986). Gene cloning in *Aspergillus nidulans*. Isolation of the isocitrate lyase gene *(acuD)*. *Mol. Gen. Genet.* **202,** 271–275.

Ballance, D. J., Buxton, F. P., and Turner, G. (1983). Transformation of *Aspergillus nidulans* by the orotidine-5′-phosphate decarboxylase gene of *Neurospora crassa*. *Biochem. Biophys. Res. Commun.* **112,** 284–289.

Baptista-Ferreira, J. L. C., Economou, A., and Casselton, L. A. (1983). Mitochondrial genetics of *Coprinus*. Recombination of mitochondrial genomes. *Curr. Genet.* **7,** 405–407.

Binninger, D. M., Skrzynia, C., Pukkila, P. J., and Casselton, L. A. (1987). DNA-mediated transformation of the basidiomycete *Coprinus cinereus*. *EMBO J.* **6,** 835–840.

Binninger, D. M., LeChevanton, L., Skrzynia, S., Shubkin, C. D. and Pukkila, P. J. (1991). Targeted transformation in *Coprinus cinereus*. *Mol. Gen. Genet.* (in press).

Buller, A. H. R. (1909). "Researches on Fungi," Vol. I. Longmans, Green, London.

Buller, A. H. R. (1924). "Researches on Fungi," Vol. III. Longmans, Green, London.

Cambareri, E. B., Jensen, B. C., Schabtach, E., and Selker, E. U. (1989). Repeat-induced G–C to A–T mutations in *Neurospora*. *Science (Washington, D.C.)* **244,** 1571–1575.

Carle, G. F., and Olson, M. V. (1984). Separation of chromosomal DNA molecules from yeast by orthogonal-field-alternation gel electrophoresis. *Nucl. Acids Res.* **12,** 5647–5664.

Casselton, L. A. (1965). The production and behavior of diploids of *Coprinus lagopus*. *Genet. Res.* **6,** 190–208.

Casselton, L. A. (1978). Dikaryon formation in higher basidiomycetes. *In* "The Filamentous Fungi. Developmental Mycology" (J. D. Smith and D. R. Berry, eds), Vol. III, pp. 275–297. Edward Arnold, London.

Casselton, L. A., and Casselton, P. J. (1974). Functional aspects of fluoroacetate resistance in *Coprinus* with special reference to acetyl-CoA synthetase activity. *Mol. Gen. Genet.* **132,** 255–264.

Casselton, L. A., and Condit, A. (1972). A mitochondrial mutant of *Coprinus lagopus*. *J. Gen. Microbiol.* **72,** 521–527.

Casselton, L. A., and de la Fuente Herce, A. (1989). Heterologous gene expression in the basidiomycete fungus *Coprinus cinereus*. *Curr. Genet.* **16,** 35–40.

Casselton, L. A., and Economou, A. (1985). Dikaryon formation. *In* "Developmental Biology of Higher Fungi" (D. Moore, L. A. Casselton, D. A. Wood, and J. C. Frankland, eds.), British Mycological Society Symposium, Vol. 10, pp. 213–229. Cambridge University Press, Cambridge.

Casselton, L. A., and Lewis, D. (1966). Compatibility and stability of diploids in *Coprinus lagopus*. *Genet. Res.* **8,** 61–72.

Casselton, L. A., Mutasa, E. S., Tymon, A., Mellon, F. M., Little, P. F. R., Taylor, S., Bernhagen, J., and Stratmann, R. (1989). The molecular analysis of basidiomycete mating type genes. *In* "Proceedings of the EMBO–Alko Workshop on Molecular Biology of Filamentous Fungi" (H. Nevalainen and M. Penttila, eds.), Vol. 6, pp. 139–148. Foundation for Biotechnical and Industrial Fermentation Research, Helsinki.

Cassidy, J. R., Moore, D., Lu, B.C., and Pukkila, P. J. (1984). Unusual organization and lack of recombination in the ribosomal RNA genes of *Coprinus cinereus*. *Curr. Genet.* **8,** 607–613.

Cassidy, J. R., and Pukkila, P. J. (1987). Inversion of 5 S ribosomal RNA genes within the genus *Coprinus*. *Curr. Genet.* **12,** 33–36.

Chiu, S. W., and Moore, D. (1988). Evidence for developmental commitment in the differentiating fruit body of *Coprinus cinereus*. *Trans. Br. Mycol. Soc.* **90,** 247–253.

Covarrubias, L., and Bolivar, F. (1982). Construction and characterization of new cloning vehicles. VI. Plasmid pBR329, a new derivative of pBR328 lacking the 482-basepair inverted duplication. *Gene* **17,** 79–89.

Chu, G., Vollrath, D., and Davis, R. W. (1986). Separation of large DNA molecules by contour-clamped homogeneous electric fields. *Science (Washington, D.C.)* **234,** 1582–1585.

Day, P.R. (1959). A cytoplasmically controlled abnormality of the tetrads of *Coprinus lagopus*. *Heredity* **13,** 81–87.

Day, P. R. (1963). Mutations affecting the *A* mating-type factor in *Coprinus lagopus*. *Genet. Res.* **4,** 323–325.

Dresser, M. E., and Moses, M. J. (1979). Silver staining of synaptonemal complexes in surface spreads for light and electron microscopy. *Exp. Cell Res.* **121,** 416–419.

Dujon, B. (1980). Sequence of the intron and flanking exons of the mitochondrial 21 S rRNA gene of yeast strains having different alleles at the ω and *rib-1* loci. *Cell* **20,** 185–197.

Dujon, B. (1989). Group I introns as mobile genetic elements: Fact and mechanistic speculations—A review. *Gene* **82,** 91–113.

Dutta, S. K. (1974). Repeated DNA sequences in fungi. *Nucl. Acids Res.* **11,** 1411–1419.

Economou, A., and Casselton, L. A. (1989). Polymorphism of the mitochondrial L-RNA gene of the basidiomycete *Coprinus cinereus*. *Curr. Genet.* **16,** 41–46.

Economou, A., Lees, V., Pukkila, P. J., Zolan, M. E., and Casselton, L. A. (1987). Biased inheritance of optional insertions following mitochondrial genome recombination in the basidiomycete fungus *Coprinus cinereus*. *Curr. Genet.* **11,** 513–519.

Ewaze, J. O., Moore, D., and Stewart, G. R. (1978). Co-ordinate regulation of enzymes involved in ornithine metabolism and its relation to sporophore morphogenesis in *Coprinus cinereus*. *J. Gen. Microbiol.* **107,** 343–357.

Faugeron, G., Rhounim, L., and Rossignol, J-L. (1990). How does the cell count the number of ectopic copies of a gene in the premeiotic inactivation process acting in *Ascobolus immersus*? *Genetics* **124,** 585–591.

Gibbins, A. M. V., and Lu, B. C. (1982). An ameiotic mutant of *Coprinus cinereus* halted prior to pre-meiotic S-phase. *Curr. Genet.* **5,** 119–126.

Gilbert, W. (1978). Why genes in pieces? *Nature (London)* **271,** 501.

Gillies, C. B. (1989). "Fertility and Chromosome Pairing." CRC Press, Boca Raton, Florida.

Gooday, G. W. (1974). Control of development of excised fruit bodies and stipes of *Coprinus cinereus. Trans. Br. Mycol. Soc.* **62,** 391–399.

Gooday, G. W. (1985). Elongation of the stipe of *Coprinus cinereus. In* "Developmental Biology of Higher Fungi" (D. Moore, L.A. Casselton, D. A. Wood, and J. C. Frankland, eds.), British Mycological Society Symposium Vol. 10, pp. 311–331. Cambridge University Press, Cambridge.

Goyon, C., and Faugeron, G. (1989). Targeted transformation of *Ascobolus immersus* and *de novo* methylation of the resulting duplicated DNA sequences. *Mol. Cell. Biol.* **9,** 2818–2827.

Gurr, S. J., Unkles, S. E., and Kinghorn, J. R. (1987). The structure and organization of nuclear genes of filamentous fungi. *In* "Gene Structure in Eukaryotic Microbes" (J. R. Kinghorn, ed.), pp. 93–139. IRL Press, Oxford.

Hahn, U., Lazarus, C. M., Lunsdorf, H., and Kuntzel, H. (1979). Split gene for mitochondrial 24 S ribosomal RNA of *Neurospora crassa. Cell* **17,** 191–200.

Haylock, R. W., Economou, A., and Casselton, L. A. (1980). Dikaryon formation in *Coprinus cinereus:* Selection and identification of *B* factor mutants. *J. Gen. Microbiol.* **121,** 17–26.

Holm, P. B., Rasmussen, S. W., Zickler, D., Lu, B. C., and Sage, J. (1981). Chromosome pairing, recombination nodules and chiasma formation in the basidiomycete *Coprinus cinereus. Carlesberg Res. Commun.* **46,** 305–346.

Hyde, C. C., Ahmed, S. A., Padlan, E. A., Miles, E. W., and Davies, D. R. (1988). Three-dimensional structure of the tryptophan synthase $\alpha_2\beta_2$ multienzyme complex from *Salmonella typhimurium. J. Biol. Chem.* **263,** 17857–17871.

Iten, W., and Matile, P. (1970). Role of chitinase and other lysosomal enzymes of *Coprinus lagopus* in the autolysis of fruiting bodies. *J. Gen. Microbiol.* **61,** 301–309.

Kamada, T., and Takemaru, T. (1977a). Stipe elongation during basidiocarp maturation in *Coprinus macrorhizus:* Mechanical properties of stipe cell wall. *Plant Cell Physiol.* **18,** 831–840.

Kamada, T., and Takemaru, T. (1977b). Stipe elongation during basidiocarp maturation in *Coprinus macrorhizus:* Changes in polysaccharide composition of stipe cell wall during elongation. *Plant Cell Physiol.* **18,** 1291–1300.

Kamada, T., and Takemaru, T. (1983). Modifications of cell-wall polysaccharides during stipe elongation in the basidiomycete *Coprinus cinereus. J. Gen. Microbiol.* **129,** 703–709.

Kamada, T., and Tsuji, M. (1979). Darkness-induced factor affecting basidiocarp maturation in *Coprinus macrorhizus. Plant Cell Physiol.* **20,** 1445–1448.

Kamada, T., Miyazaki, S., and Takemaru, T. (1976). Quantitative changes of DNA, RNA, and protein during basidiocarp maturation in *Coprinus macrorhizus. Trans. Mycol. Soc. Japan* **17,** 451–460.

Kamada, T., Kurita, R., and Takemaru, T. (1978). Effects of light basidiocarp maturation in *Coprinus macrorhizus. Plant Cell Physiol.* **19,** 263–275.

Kamada, T., Fujii, T., and Takemaru, T. (1980). Stipe elongation during basidiocarp maturation in *Coprinus macrorhizus:* Changes in activity of cell wall lytic enzymes. *Trans. Mycol. Soc. Japan* **21,** 359–367.

Kamada, T., Hamada, Y., and Takemaru, T. (1982). Autolysis *in vitro* of the stipe cell wall in *Coprinus macrorhizus. J. Gen Microbiol.* **128,** 1041–1046.

Kamada, T., Katsuda, H., and Takemaru, T. (1984). Temperature-sensitive mutants of

Coprinus cinereus defective in hyphal growth and stipe elongation. *Curr. Microbiol.* **11,** 309–312.

Kamada, T., Fujii, T., Nakagawa, T., and Takemaru, T. (1985). Changes in (1 → 3)-beta-glucanase activities during stipe elongation in *Coprinus cinereus. Curr. Microbiol.* **12,** 257–260.

Kamada, T., Sumiyoshi, T., Shindo, Y., and Takemaru, T. (1989). Isolation and genetic analysis of resistant mutants to the benzimidazole fungicide benomyl in *Coprinus cinereus. Curr. Microbiol.* **18,** 215–218.

Kanda, T., and Ishikawa, T. (1986). Isolation of recessive developmental mutants in *Coprinus cinereus. J. Gen. Appl. Microbiol.* **32,** 541–543.

Kanda, T., Goto, A., Sawa, K., Arakawa, H., Yasuda, Y., and Takemaru, T. (1989a). Isolation and characterization of recessive sporeless mutants in the basidiomycete *Coprinus cinereus. Mol. Gen. Genet.* **216,** 526–529.

Kanda, T., Ishihara, H., and Takemaru, T. (1989b). Genetic analysis of recessive primordiumless mutants in the basidiomycete *Coprinus cinereus. Bot. Mag. Tokyo* **102,** 561–564.

Kanda, T., Arakawa, H., Yasuda, Y., and Takemaru, T. (1990). Basidiospore formation in a mutant of the incompatibility factors and mutants that arrest at meta-anaphase I in *Coprinus cinereus. Expt. Mycol.* **14,** 218–226.

Kiguchi, T., and Yanagi, S. O. (1985). Intraspecific heterokaryon and fruit body formation in *Coprinus macrorhizus* by protoplast fusion of auxotrophic mutants. *Appl. Microbiol. Biotechnol.* **22,** 121–127.

Lazarus, C. M., Lunsdorf, H., Hahn, P., Stepien, P. P. and Kuntzel, H. (1980). Physical map of *Aspergillus nidulans* mitochondrial genes coding for ribosomal RNA and intervening sequence in the large rRNA cistron. *Mol. Gen. Genet.* **177,** 389–397.

Lewis, D. (1961). Genetical analysis of methionine suppressors in *Coprinus. Genet. Res.* **2,** 141–155.

Lowry, R. J. (1963). Aceto-iron-hematoxylin for mushroom chromosomes. *Stain Technology* **38,** 199–200.

Lu, B. C. (1967). Meiosis in *Coprinus lagopus:* A comparative study with light and electron microscopy. *J. Cell Sci.* **2,** 529–536.

Lu, B. C. (1969). Genetic recombination in *Coprinus.* I. Its precise timing as revealed by temperature treatment experiments. *Can. J. Genet. Cytol.* **11,** 834–847.

Lu, B. C. (1972). Dark dependence of meiosis at elevated temperatures in the basidiomycete *Coprinus lagopus. J. Bacteriol.* **111,** 833–834.

Lu, B. C. (1974a). Meiosis in *Coprinus.* V. The role of light on basidiocarp initiation, mitosis, and hymenium differentiation in *Coprinus lagopus. Can. J. Bot.* **52,** 299–305.

Lu, B. C. (1974b). Meiosis in *Coprinus.* VI. The control of the initiation of meiosis. *Can. J. Genet. Cytol.* **16,** 155–164.

Lu, B. C. (1978). Meiosis in *Coprinus.* VIII. A time-course study of the fusion and division of the spindle pole body during meiosis. *J. Cell Biol.* **76,** 761–766.

Lu, B. C. (1984). The cellular program for the formation and dissolution of the synaptonemal complex in *Coprinus. J. Cell Sci.* **67,** 25–43.

Lu, B. C., and Raju, N. B. (1970). Meiosis in *Coprinus.* II. Chromosome pairing and the lampbrush diplotene stage of meiotic prophase. *Chromosoma (Berlin)* **29,** 305–316.

Lu, B. C., and Xiu-Yu, L. (1988). Correlation of endonuclease activity at meiotic prophase and cofactor-induced genetic recombination in the basidiomycete *Coprinus cinereus. Genome* **30,** 380–386.

Lu, B. C., Wong, W., Fanning, L., and Sakaguchi, K. (1988). Purification and characterization of an endonuclease from fruiting caps of basidiomycete *Coprinus cinereus. Eur. J. Biochem.* **174,** 725–732.

McLaughlin, D. J. (1974). Ultrastructural localization of carbohydrate in the hymenium and subhymenium of *Coprinus*. *Protoplasma* **82,** 341–364.

McLaughlin, D. J. (1977). Basidiospore initiation and early development in *Coprinus cinereus*. *Am. J. Bot.* **64,** 1–16.

Madelin, M. F. (1956a). Studies on the nutrition of *Coprinus lagopus* Fr., especially as affecting fruiting. *Ann. Botany N.S. (London)* **20,** 306–330.

Madelin, M. F. (1956b). The influence of light and temperature on fruiting of *Coprinus lagopus* Fr. in pure culture. *Ann. Botany N.S. (London)* **20,** 467–481.

Matthews, T. R., and Niederpruem, D. J. (1972). Differentiation in *Coprinus lagopus*. I. Control of fruiting and cytology of initial events. *Arch. Microbiol.* **87,** 257–268.

Matthews, T. R., and Niederpruem, D. J. (1973). Differentiation in *Coprinus lagopus*. II. Histology and ultrastructural aspects of developing primordia. *Arch. Microbiol.* **88,** 169–180.

May, G., LeChevanton, L., and Pukkila, P. J. (1991). Molecular analysis of the *Coprinus cinereus* mating type *A* factor demonstrates an unexpectedly complex structure. *Genetics* (In press).

May, G., and Taylor, J. H. (1988). Patterns of mating and mitochondrial DNA inheritance in the Agaric basidiomycete *Coprinus cinereus*. *Genetics* **118,** 213–220.

Mellon, F. M., and Casselton, L. A. (1988). Transformation as a method of increasing gene copy number and gene expression in the basidiomycete fungus *Coprinus cinereus*. *Curr. Genet.* **14,** 451–456.

Mellon, F. M., Little, P. F. R., and Casselton, L. A. (1987). Gene cloning and transformation in the basidiomycete fungus *Coprinus cinereus:* Isolation and expression of the isocitrate lyase gene *(acu-7)*. *Mol. Gen. Genet.* **210,** 352–357.

Miyake, H., Takemaru, T., and Ishikawa, T. (1980). Sequential production of enzymes and basidiospore formation in fruiting bodies of *Coprinus macrorhizus*. *Arch. Microbiol.* **126,** 201–205.

Moore, D. (1984). Developmental biology of the *Coprinus cinereus* carpophore: Metabolic regulation in relation to cap morphogenesis. *Exp. Mycol.* **8,** 283–297.

Moore, D., and Pukkila, P. J. (1985). *Coprinus cinereus:* An ideal organism for studies of genetic and developmental biology. *J. Biol. Educ.* **19,** 31–40.

Moore, D., Elhiti, M. M. Y., and Butler, R. D. (1979). Morphogenesis of the carpophore of *Coprinus cinereus*. *New Phytol.* **83,** 695–722.

Morimoto, N., and Oda, Y. (1973). Effects of light on fruit-body formation in a basidiomycete, *Coprinus macrorhizus*. *Plant Cell Physiol.* **14,** 217–225.

Morimoto, N., and Oda, Y. (1974). Photo-induced karyogamy in a basidiomycete, *Coprinus macrorhizus*. *Plant Cell Physiol.* **15,** 183–186.

Moses, M. J. (1977). Synaptonemal complex karyotyping in spermatocytes of Chinese hamster *(Cricetulus griseus)*. I. Morphology of the autosomal complement in spread preparations. *Chromosoma (Berlin)* **60,** 99–125.

North, J. (1982). A dominance modifier for cycloheximide resistance in *Coprinus cinereus*. *J. Gen. Microbiol.* **128,** 2747–2753.

North, J. (1987). Linkage map of *Coprinus cinereus* (Schaeff. ex Fr.) S. F. Gray, Ink cap. *In* "Genetic Maps 1987" (S. J. O'Brian, ed.), Vol. 4, pp. 340–345. Cold Spring Harbor Laboratory Press, Cold Spring Harbor, New York.

Perkins, D. D., and Barry, E. G. (1977). The cytogenetics of *Neurospora*. *Adv. Genetics* **19,** 133–285.

Pukkila, P. J., and Lu, B. C. (1985). Silver staining of meiotic chromosomes in the fungus, *Coprinus cinereus*. *Chromosoma (Berlin)* **91,** 108–112.

Pukkila, P. J., Yashar, B. M., and Binninger, D. M. (1984). Analysis of meiotic development in *Coprinus cinereus. In* "Controlling Events in Meiosis" (C. W. Evans and H. G. Dickinson, eds.) Symposia of the Society for Experimental Biology, Vol. 38, pp. 177–194. Company of Biologists Limited, Cambridge.

Quintanilha, A. (1937). Contributions a l'etude genetique du phenomene de Buller. *Compt. Rend. Acad. Sci., Paris* **205,** 745–747.

Raju, N. B., and Lu, B. C. (1970). Meiosis in *Coprinus.* III. Timing of meiotic events in *C. lagopus* (sensu Buller). *Can. J. Bot.* **48,** 2183–2186.

Raju, N. B., and Lu, B. C. (1973a). Genetic recombination in *Coprinus.* III. Influence of gamma-irradiation and temperature treatments on meiotic recombination. *Mutat. Res.* **17,** 37–48.

Raju, N. B., and Lu, B. C. (1973b). Meiosis in *Coprinus.* IV. Morphology and behavior of spindle pole bodies. *J. Cell Sci.* **12,** 131–141.

Rasmussen, S. W., Holm, P. B., Lu, B. C., Zickler, D., and Sage, J. (1981). Synaptonemal complex formation and distribution of recombination nodules in pachytene trivalents of triploid *Coprinus cinereus. Carlsberg Res. Commun.* **46,** 347–360.

Raudaskoski, M., and Lu, B. C. (1980). The effect of hydroxyurea on meiosis and genetic recombination in the fungus *Coprinus lagopus. Can. J. Genet. Cytol.* **22,** 41–50.

Reijnders, A. F. M., and Moore, D. (1985). Developmental biology of agarics—An overview. *In* "Developmental Biology of Higher Fungi" (D. Moore, L. A. Casselton, D. A. Wood, and J. C. Frankland, eds.) British Mycological Society Symposia, Vol. 10, pp. 581–595. Cambridge University Press, Cambridge.

Roper, J. A. (1952). Production of heterozygous diploids in filamentous fungi. *Experientia* **8,** 14–15.

Sakaguchi, K., and Lu, B. C. (1982). Meiosis in *Coprinus:* Characterization and activities of two forms of DNA polymerase during meiotic stages. *Mol. Cell. Biol.* **2,** 752–757.

Selker, E. U., Cambareri, E. B., Jensen, B. C., and Haack, K. R. (1987). Rearrangement of duplicated DNA in specialized cells of *Neurospora. Cell* **51,** 741–752.

Skrzynia, C., Binninger, D. M., Alspaugh, J. A., II, and Pukkila, P. J. (1989). Molecular characterization of *TRP1,* a gene coding for tryptophan synthetase in the basidiomycete *Coprinus cinereus. Gene* **81,** 73–82.

Stewart, G. R., and Moore, D. (1974). The activities of glutamate dehydrogenases during mycelial growth and sporophore development in *Coprinus lagopus (sensu* Lewis). *J. Gen Microbiol.* **83,** 73–81.

Swamy, S., Uno, I., and Ishikawa, T. (1984). Morphogenetic effects of mutations at the *A* and *B* incompatibility factors in *Coprinus cinereus. J. Gen. Microbiol.* **130,** 3219–3224.

Swamy, S., Uno, I., and Ishikawa, T. (1985). Regulation of cyclic AMP metabolism by the incompatibility factors in *Coprinus cinereus. J. Gen. Microbiol.* **131,** 3211–3217.

Takemaru, T., and Kamada, T. (1972). Basidiocarp development in *Coprinus macrorhizus.* I. Induction of developmental variations. *Bot. Mag. Tokyo* **85,** 51–57.

Tani, K., Kuroiwa, T., and Takemaru, T. (1977). Cytological studies on sporeless mutant in the basidiomycete *Coprinus macrorhizus. Bot. Mag. Tokyo* **90,** 235–245.

Tilby, M. J. (1976). Tryptophan biosynthesis in *Coprinus lagopus:* A genetic analysis of mutants. *J. Gen. Microbiol.* **93,** 126–132.

Todd, N. K., and Casselton, L. A. (1973). Non-complementation and recessiveness as properties of missense suppressor genes in the fungus *Coprinus. J. Gen. Microbiol.* **77,** 197–207.

Tsusue, Y. M. (1969). Experimental control of fruit-body formation in *Coprinus macrorhizus. Dev. Growth Differentiation* **11,** 164–178.

Uno, I., and Ishikawa, T. (1971). Chemical and genetical control of induction of monokaryotic fruiting bodies in *Coprinus macrorhizus. Mol. Gen. Genet.* **113,** 228–239.

Uno, I., and Ishikawa, T. (1982). Biochemical and genetic studies on the initial events of fruitbody formation. *In* "Basidium and Basidiocarp" (K. Wells and E. K. Wells, eds.), pp. 113–123. Springer-Verlag, New York.

Uno, I., Yamaguchi, M., and Ishikawa, T. (1974). The effect of light on fruiting body formation and adenosine 3′,5′-cyclic monophosphate metabolism in *Coprinus macrorhizus. Proc. Natl. Acad. Sci. U.S.A.* **71,** 479–483.

Veal, D., and Casselton, L. A. (1985). Regulation of tryptophan metabolism in *Coprinus cinereus:* Isolation and characterisation of mutants resistant to 5-fluoroindole. *Arch. Microbiol.* **142,** 157–163.

Wahl, G. M., Lewis, K. A., Ruiz, J. C., Rothenberg, B., Zhao, J., and Evans, G. A. (1987). Cosmid vectors for rapid genomic walking, restriction mapping, and gene transfer. *Proc. Natl. Acad. Sci. U. S. A.* **84,** 2160–2164.

Weber, C. A., Hudspeth, E. S., Moore, G. P., and Grossman, L. I. (1986). Analysis of the mitochondrial and nuclear genomes of two basidiomycetes, *Coprinus cinereus* and *Coprinus sterorarius. Curr. Genet.* **10,** 515–525.

Wu, M. M. J., Cassidy, J. R., and Pukkila, P. J. (1983). Polymorphisms in DNA of *Coprinus cinereus. Curr. Genet.* **7,** 385–392.

Yashar, B. M., and Pukkila, P. J. (1985). Changes in polyadenylated RNA sequences associated with fruiting body morphogenesis in *Coprinus cinereus. Trans. Br. Mycol. Soc.* **84,** 215–226.

Yoon, K. S., and McLaughlin, D. J. (1980). Fluorescence microscopic examination of basidiosporogenesis in *Boletus rubinellus* and *Coprinus cinereus. Mycologia* **72,** 1150–1158.

Zolan, M. E., and Pukkila, P. J. (1985). DNA methylation in *Coprinus cinereus. In* "Molecular Genetics of Filamentous Fungi" (W. Timberlake, ed.), UCLA Symposia on Molecular and Cellular Biology, pp. 333–344. Liss, New York.

Zolan, M. E., and Pukkila, P. J. (1986). Inheritance of DNA methylation in *Coprinus cinereus. Mol. Cell. Biol.* **6,** 195–200.

Zolan, M. E., Tremel, C. J., and Pukkila, P. J. (1988). Production and characterization of radiation-sensitive meiotic mutants of *Coprinus cinereus. Genetics* **120,** 379–387.

7

Regulatory Circuits in *Aspergillus nidulans*

MERYL A. DAVIS
MICHAEL J. HYNES
Department of Genetics
The University of Melbourne
Parkville, Victoria, Australia

I. GENETIC ANALYSIS OF REGULATORY CIRCUITS

Modern molecular genetic techniques have been very successful tools for analyzing mechanisms of gene regulation in eukaryotes. Research on a wide range of systems has provided strong evidence for common mechanisms of transcriptional control in eukaryotes, ranging from *Saccharo-*

MORE GENE MANIPULATIONS
IN FUNGI

myces cerevisiae to the higher vertebrates, and includes studies of genes regulated by environmental, metabolic, and developmental signals. Although there has been much work of significance using systems in which no detailed genetics work is possible, comprehensive approaches to understanding all the regulatory components involved have been available only when classical genetic methodology can be applied. This limitation has led to a research emphasis on organisms such as *S. cerevisiae, Drosophila melanogaster, Caenorhabditis elegans,* and *Arabidopsis thaliana.*

Among the filamentous fungi, genetic analysis has long been applied to the ascomycetes *Neurospora crassa* and *Aspergillus nidulans.* A variety of systems for the study of gene regulation in these fungi has been developed using the genetic approach, facilitating the application of molecular techniques to these systems in recent years.

The review by Arst and Scazzocchio (1985) in *Gene Manipulations in Fungi* was written at a time when molecular data was just becoming available (Green and Scazzocchio, 1985). That review should be consulted for a comprehensive outline of the background to many of the genetic systems to be discussed in this chapter. We will review the progress that has been made in the molecular genetic analysis of some fungal regulatory circuits. Emphasis will be on work with *A. nidulans,* but examples drawn from *N. crassa* will also be considered when appropriate. We will restrict ourselves largely to considering systems of catabolic control, but it should be noted that developmental control circuits are beginning to be understood in *A. nidulans* (e.g., Mirabito *et al.*, 1989); these findings may have implications for more general aspects of the control of primary and secondary metabolism (Adams and Timberlake, 1990).

Most progress has been made in the investigation of the regulation of catabolic pathways in *A. nidulans.* These advances have resulted from the great catabolic diversity of this organism and the ease with which mutant strains can be isolated. *A. nidulans* forms compact colonies on solid media, and mutants increasing or decreasing the utilization of particular substrates are readily recognized. The morphology of the organism allows subtle distinctions to be made between different strains on different media. This property of *A. nidulans* also has been of great utility in the study of the effects of introducing regulatory DNA by transformation (Kelly and Hynes, 1987; Andrianopoulos and Hynes, 1988).

II. PATTERNS OF CATABOLIC GENE REGULATION

Many genes encoding catabolic activities are expressed at elevated levels only under conditions of nutritional stress. A good example is an alkaline extracellular protease that is produced when *A. nidulans* is starved for carbon, nitrogen, sulfur, or phosphorus (Cohen, 1973). Other activities respond to pH changes or changes in oxygen status (Caddick *et al.*, 1986b; Shaffer *et al.*, 1988). These general metabolic controls are mediated by gene products encoded by the so-called wide-domain regulatory genes. Many activities are also specifically inducible by the substrates or early products of the catabolic pathway. The regulatory genes involved have been termed pathway specific (Arst and Scazzocchio, 1985). These patterns of regulation are consistent with an organism capable of exploiting a wide range of substrates for growth; the appropriate activities are synthesized only if the inducer is present and if the products of the pathway (e.g., carbon or nitrogen metabolites) are in short supply. Activities responding solely to the wide-domain controls may represent a final desperate effort to scavenge nutrients from any available substrate.

Other control mechanisms have yet to be subjected to detailed analysis. Under conditions of carbon starvation, some activities specific to nitrogen metabolism are not produced, for example, nitrate reductase, NADP-linked glutamate dehydrogenase, and histidase (Hynes, 1973, 1974; Polkinghorne and Hynes, 1982)—a phenomenon which may be termed reverse carbon catabolite repression. Carbon limitation acts on NADP–glutamate dehydrogenase at the level of transcription (Hawkins *et al.*, 1989). There is evidence for nitrate repression of some nitrogen catabolic activities by a mechanism apparently independent of nitrogen metabolite repression (Cove, 1979; Polkinghorne and Hynes, 1982). Conversely, some enzymes involved in acetate utilization are not induced to high levels under nitrogen limiting conditions (Kelly, 1980).

The fact that many genes involved in catabolic pathways are subject to more than one regulatory mechanism has allowed the isolation of mutations affecting particular control circuits. The general strategy was originally described by Arst and Cove (1973) and has been of fundamental importance in the analysis of the best studied systems; mutations in both *trans*- and *cis*-acting regulatory elements have been isolated (see Arst and Scazzocchio, 1985; Hynes and Davis, 1986; Davis and Hynes, 1989). Progress in the molecular characterization of some of these systems will be described in the rest of this chapter.

III. GENE STRUCTURE AND EXPRESSION IN FILAMENTOUS FUNGI

Sequence analysis of fungal genes has shown them to be organized similarly to genes of other eukaryotes. (See Gurr *et al.*, 1988, for review.) Unlike *Saccharomyces cerevisiae* genes, most fungal genes contain small introns (usually between 50 and 100 bp). Since many genes have been isolated directly as genomic clones, the determination of coding sequences has involved detailed transcription mapping, cDNA sequencing, or both. Intron boundaries nearly always follow the GT–AG rules of Breathnach and Chambon (1981), with a preference for the organization 5′GT*RN*GT . . . *Y*AG3′. There is no strong preference for particular sequences preceding either the GT or YAG splice site. An internal consensus sequence for lariat formation in splicing is evident in all introns and has the sequence GCTAACN. Many genes contain sequences related to the consensus AATAAA in the 3′ noncoding region; this may be involved in polyadenylation of the 3′ terminus. Translation usually begins at the first ATG. A Kozak consensus sequence surrounding this codon is commonly found.

At this time, no clear-cut examples of posttranscriptional control, such as differential splicing or regulation of messenger stability, have been described in *Aspergillus nidulans.* However, Adams and Timberlake (1990) recently have provided evidence for posttranscriptional controls accompanying a major switch involved in conidial development, and Kelly and colleagues (1990) have shown that posttranscriptional control of *alcC* expression in response to anaerobic growth conditions is likely. Nevertheless, most studies of regulation of fungal gene activity have shown, via Northern blot analysis, that control is at the messenger RNA level, and, as for other organisms, it is likely that the major effects of regulatory genes are the modulation of transcription initiation at the promoter.

The analysis of promoter structure and function in eukaryotes is currently an area of intense research (see, for example, Lewin, 1990; Ptashne and Gann, 1990). Although promoter structures in *S. cerevisiae* are somewhat different than those in higher eukaryotes, many of the fundamental mechanisms involved in transcription initiation and its control are similar in all eukaryotes (for reviews, see Ptashne, 1988; Struhl, 1989). The promoter sequences of many filamentous fungal genes are now known. Commonly, the region immediately upstream of transcription start points is pyrimidine rich in the coding strand. These CT blocks may be extensive in some cases, particularly in highly expressed genes lacking apparent TATA boxes (Gurr *et al.*, 1988; Punt *et al.*, 1988; de Graaff and Visser, 1988; Turner *et al.*, 1989). Polypyrimidine sequences also have been found in higher eukaryotic promoters (Hariharan and Perry, 1990).

The majority of promoters in higher eukaryotes contain a TATA box between 25 and 30 bp upstream from the initiation site. This sequence is generally AT rich, containing the fairly strong consensus TATAAA (Bucher and Trifonov, 1986). In contrast, most *S. cerevisiae* promoters contain one or more TATA elements, located anywhere between 40 and 120 bp from the initiation site (Struhl, 1989). This sequence is recognized by the transcription factor TFIID and plays a role in determining levels of expression in response to other regulatory elements. Recent work suggests that there is considerable complexity in these interactions; evidence indicates more than one class of TATA element in *S. cerevisiae* (Harbury and Struhl, 1989) and similar DNA-binding specificities for the TFIID proteins of yeast and other species (Wobbe and Struhl, 1990). Also, there is the possibility that species-specific coactivator molecules interact with TFIID (see Lewin, 1990; Ptashne and Gann, 1990).

AT-rich sequences are found approximately 30–50 bp upstream of the initiation site in some fungal promoters; some of these are consistent with the TATAAA consensus. (See Gurr *et al.*, 1988, for some examples.) Many fungal genes, however, do not have any sequences resembling a TATA element.

It has been suggested that a common CAAT consensus exists approximately 70–90 bp upstream of the initiation site in eukaryotic genes and that this sequence is analogous to the −35 sequence of prokaryotic promoters. This sequence has been found in some fungal promoter sequences (Gurr *et al.*, 1988). However, a survey of 168 higher eukaryotic promoters showed clearly that the occurrence of CAAT sequences in this region is not statistically significant and that the sequence is unlikely to be a core promoter element (Bucher and Trifonov, 1986). The fungal sequence data would appear to support this conclusion.

Sequences near the start site of yeast genes can be important in determining accurate transcription initiation (Chen and Struhl, 1985). In higher eukaryotes, the start site is often located about 30 bp downstream from a TATA element. In at least some higher eukaryotic genes, however, the start site is determined by an initiator (Inr) element approximately 8 bp in length; initiation occurs at an A surrounded by pyrimidine residues (Smale and Baltimore, 1989; Smale *et al.*, 1990). Significantly, *in vitro* studies demonstrate that an Inr sequence can work in conjunction with a TATA element; the start point is determined by the Inr. Furthermore, the experiments suggest that transcription determined by either TATA or Inr elements is responsive to an Sp1-dependent upstream activating sequence; in both cases this is TFIID dependent (Smale *et al.*, 1990). This evidence clearly indicates the possibility that fungal genes may have Inr elements with or without TATA elements and that TFIID may be involved in the expression and regulation of most genes.

Functional analysis of fungal promoters is possible using DNA-mediated transformation following *in vitro* mutagenesis of the 5′ region of cloned structural genes. Commonly, *lacZ* is used as a reporter gene and, where possible, transformants with a single copy of the construct integrated at a defined locus are analyzed (Hamer and Timberlake, 1987; van Gorcom *et al.*, 1986). In the earliest of these studies with the *trpC* gene of *A. nidulans* it was shown that a deletion from −67 to −11 led to heterogeneous initiation sites for transcription without affecting the final level of gene expression (Hamer and Timberlake, 1987). Included in this deletion is an AT-rich element and a CT-rich region 13 bp in length. Similar results for the *oliC* gene of *A. nidulans* have been found: removal of the very large CT blocks affects the start sites but not the level of expression (Turner *et al.*, 1989). A number of fungal genes are being analyzed in this way in various laboratories. It is to be hoped that a clearer picture of fungal promoters will emerge. Unfortunately, this kind of analysis has yet to be reported for a gene whose regulation is reasonably well understood.

Current thinking about mechanisms of transcriptional control is dominated by the concept that regulatory proteins bind to specific DNA sites and interact with the transcriptional apparatus at the promoter to influence transcription initiation (Ptashne, 1988). It is, therefore, important to understand the nature of the fundamental components of promoters. One strategy to achieve this understanding is to see if genes that are co-regulated by a common regulatory gene share promoter elements such as TATA boxes, CT blocks, or possible Inr sequences. Inspection of genes co-regulated by the *A. nidulans amdR* regulatory gene (see Section V,A,5) indicates that these genes all lack long CT blocks but have potential TATA elements (Corrick *et al.*, 1987; Richardson *et al.*, 1989; I. B. Richardson, M. Katz, and M. J. Hynes, unpublished data). Similarly, genes of the *alcR* regulon have potential TATA elements and do not have CT blocks (Gwynne *et al.*, 1987; Pickett *et al.*, 1987; Felenbok *et al.*, 1988). Apart from a tendency for initiation to occur at, or close to, purines within a generally pyrimidine-rich region, there are no obvious highly conserved Inr sequences. Detailed functional analysis of selected regulated promoter regions is required.

Clustering of genes of related function is not uncommon in filamentous fungi (Arst and Scazzocchio, 1985; Giles *et al.*, 1985). Clusters may include all the known genes for a catabolic pathway including those involved in specific regulation, for example, the *prn* cluster of *A. nidulans* (Hull *et al.*, 1989) and the *qa* and *qut* clusters of *Neurospora crassa* and *A. nidulans* (Geever *et al.*, 1989; Lamb *et al.*, 1990). Alternatively, only some of the genes of a pathway may be clustered and the specific regulatory gene may not be closely linked to any of the structural genes (e.g., nitrate utilization in *A. nidulans*; Arst and Scazzocchio, 1985) or to only one of the structural genes (e.g., the *alcR* regulatory gene adjacent to the *alcA* gene but unlinked to the *aldA* gene; Pateman *et al.*, 1983; Lockington *et al.*, 1985). In all cases examined, clustered genes do not encode polycistronic messages.

Gene clusters raise the question of whether there are chromosomal domains that affect the expression of genes within that domain. Such domains would lead to position effects on gene activity. In *A. nidulans,* a very large cluster of genes (the *spoC1* cluster) is expressed during conidiation. Transformation experiments involving insertion of a gene from that cluster elsewhere in the genome show altered regulation and expression of that gene. Integration of the *argB* gene into the cluster leads to elevated expression of *argB* (Miller *et al.*, 1987). This is the clearest case of position effects described in fungi. Recently, elements involved in position effects in higher eukaryotes have been discovered. These elements have been called dominant control regions or locus activating regions. They have been particularly well studied in the human β-globin gene cluster, where they are characterized by tissue-specific DNaseI hypersensitive sites located up to 20 kb from the globin genes (see Talbot *et al.*, 1989; Novak *et al.*, 1990; Townes and Behringer, 1990). These sites confer position-independent expression on *cis*-linked genes in transfected cells and may be involved in mediating attachment of the chromatin to the nuclear scaffold (see Stief *et al.*, 1989). Whether similar elements determine chromosomal domains of gene activity in fungi must still be determined.

There is no obvious pattern in the clustering of co-regulated genes in fungi, and the functional and evolutionary significance of clustering remains obscure. Three genes for penicillin biosynthesis are clustered in *A. nidulans* and *Penicillium chrysogenum* and, notably, also in the prokaryote *Flavobacterium* (MacCabe *et al.*, 1990; D. J. Smith *et al.*, 1990). All the genes for quinate utilization, including the regulatory genes, are clustered in *A. nidulans* and in *N. crassa,* but the gene order is not the same in the two organisms (Geever *et al.*, 1989; Lamb *et al.*, 1990). Structural genes for nitrate reductase and nitrite reductase are adjacent and divergently transcribed in *A. nidulans* but not in *N. crassa* (Johnstone *et al.*, 1990).

Transformation can be used to investigate whether clustering is absolutely required for regulation and expression of the genes. For example, the *lamA* and *lamB* genes of *A. nidulans* are adjacent and divergently transcribed, but are functional when inserted at separate locations in the genome of either *A. nidulans* or *A. terreus* (Katz and Hynes, 1989a,b). The *prn* cluster—encoding genes for L-proline catabolism, including a *trans*-acting regulatory gene, *prnA*—represents an interesting situation in which a genetically defined *cis*-acting region (see Section V,A,5) affects the expression of a gene (*prnC*) separated from it by another gene (*prnB*) (Hull *et al.*, 1989). Since only monocistronic messages have been detected, this effect indicates that the *cis*-acting sequence has enhancer-like properties. Moreover, DNA fragments carrying only *prnC* (and therefore separated from the enhancer region) would not be expressed normally when present in ectopic locations arising either by transformation experiments or by chromosomal rearrangements. This model predicts that breaking up and dispersal of the gene cluster in the genome would be constrained during evolution.

In some cases, clustering may be necessary to optimize the efficiency of regulation and expression of clustered genes. Geever *et al.* (1989) have discussed two possible models for how clustering might affect regulation. Detailed transformation studies using precisely engineered gene rearrangements followed by accurate measurements of gene regulation are required to substantiate either model.

IV. CLONING THE COMPONENTS OF REGULATORY CIRCUITS

Molecular methods, characterized mutants, and a transformation system have provided powerful tools for gene cloning in *Aspergillus nidulans.* A variety of different approaches has been used to clone both regulated structural genes and the regulatory genes themselves (see Table I). The isolation of the *areA* gene (Caddick *et al.*, 1986a) and the *prn* genes (Hull *et al.*, 1989) involved the use of mutants that had chromosome rearrangements within the desired genes. The *areA* gene was cloned by screening a lambda library for sequences present in wild-type DNA but absent from duplication-deficiency progeny arising from crosses using an *areA* inversion mutation. The genes of the *prn* cluster were cloned by screening Southern blots of *prn* mutants with fragments from a wild-type genomic library. A number of *A. nidulans* genes were cloned by differential screening. This method provides an effective way to clone those genes that are highly expressed under some conditions but poorly expressed under others or for which appropriate mutants are available (e.g., *amdS,* Hynes *et al.*, 1983; *alcA* and *aldA,* Lockington *et al.*, 1985). However, except under exceptional circumstances, regulatory genes are not well suited to this approach.

When a transformation system is available, it is possible to clone any gene with mutant alleles that confer a detectable phenotype. Complementation of recessive loss-of-function mutations has provided a means of cloning a number of different regulatory genes (e.g., *amdR,* Andrianopoulos and Hynes, 1988; *creA,* Dowzer and Kelly, 1989) as well as structural genes (e.g., *gatA,* Richardson *et al.*, 1989). It is also possible to clone the mutant, rather than the wild-type, version of the gene, provided the cloned sequence confers a phenotype that is dominant or semidominant over that of the recipient. Since transformation in *A. nidulans* is integrative, relevant sequences are rescued by transformation of genomic DNA into *Escherichia coli.* A different approach employing transformation was used in the cloning of *uapA:* the resident gene was inactivated by integration of a plasmid carrying *amdS* as a selective marker (Diallinas and Scazzocchio,

TABLE I

Examples of Cloning of the Components of Regulatory Circuits from *Aspergillus nidulans*

Gene	Function	Method	Reference
aciA	Unknown	Molecular[a]	Atkinson *et al.*, 1985
acuD	Isocitrate lyase	Complementation of *acuD*$^-$	Ballance and Turner, 1986
acuE	Malate synthase	Molecular	Sandeman and Hynes, 1989
alcA; alcR gene cluster	Alcohol dehydrogenase I and regulatory gene	Molecular	Lockington, *et al.*, 1985; Doy *et al.*, 1985
aldA	Aldehyde dehydrogenase	Molecular	Lockington *et al.*, 1985; Pickett *et al.*, 1987
alcC	Alcohol dehydrogenase III	Complementation of *Saccharomyces cerevisiae adh*$^-$	McKnight *et al.*, 1985
amdR	ω-Amino acid induction	Complementation of *amdR*$^-$	Andrianopoulos and Hynes, 1988
amdS	Acetamidase	Molecular	Hynes *et al.*, 1983
areA	Nitrogen metabolite repression	Molecular	Caddick *et al.*, 1986a
creA	Carbon catabolite repression	Complementation of *creA*$^-$	Dowzer and Kelly, 1989
cmA; niiA; niaD gene cluster	Nitrate catabolism	Complementation of *niiA*$^-$	Johnstone *et al.*, 1990
facA	Acetyl CoA synthetase	Molecular	Sandeman and Hynes, 1989
facB	Acetate induction	Complementation of *facB*$^-$	Katz and Hynes, 1989c
gatA	GABA transaminase	Complementation of *gatA*$^-$	Richardson *et al.*, 1989
gdhA	NADP–glutamate dehydrogenase	NADP–gdh *(Neurospora crassa)*	Gurr *et al.*, 1986
lamA, B gene cluster	2-Pyrrolidinone catabolism	Complementation of *lamA*$^-$	Katz and Hynes, 1989b
prnA, D, B, C gene cluster	Proline catabolism	Molecular	Hull *et al.*, 1989
qutA, B, C, D, E, G, R gene cluster	Quinate catabolism	*qa* genes *(N. crassa)*	Hawkins *et al.*, 1985
uapA	Uric acid permease	Inactivation	Diallinas and Scazzocchio, 1989

[a] Molecular is used to describe those situations in which hybridization using DNA or RNA probes has been employed in the initial isolation of the gene.

1989). McKnight and colleagues (1985) isolated the *alcC* gene of *A. nidulans* by complementing *adh*⁻ mutants of *Saccharomyces cerevisiae* with *A. nidulans* cDNA clones.

When genes are clustered, sometimes several genes, including regulatory genes, can be cloned simultaneously. Examples are *alcA* and *R* (Lockington *et al.*, 1985); *prnA, B, C,* and *D* (Hull *et al.*, 1989); *niiA, crnA,* and *niaD* (Johnstone *et al.*, 1990); the *qut* gene cluster (Hawkins *et al.*, 1985; Beri *et al.*, 1987; Whittington *et al.*, 1987); and *lamA* and *B* (Katz and Hynes, 1989b). As increasing numbers of cloned sequences from other organisms become available, cloning genes from *A. nidulans* by cross-hybridization becomes possible. The *A. nidulans gdhA* gene and the genes of the *qut* cluster were isolated using the relevant genes from *Neurospora crassa* (Hawkins *et al.*, 1985; Gurr *et al.*, 1986).

V. PATHWAY-SPECIFIC REGULATORY CIRCUITS

A. Characterization of Regulatory Genes

Pathway-specific regulatory genes usually have been defined by frequent recessive mutations that lead to loss of inducibility of all the activities of a particular pathway. Less commonly, allelic mutations that result in elevated constitutive expression of the structural genes are also found (Arst and Scazzocchio, 1985). The *amdA* gene was defined initially only by the latter kind of mutation (Arst and Cove, 1973; Hynes, 1978). The strong implication is that these genes produce activator proteins that turn on transcription in the presence of inducer. Nevertheless it is possible that, in some cases, proposed activator genes code for structural proteins (e.g., components of permeases) rather than specific DNA-binding proteins. A number of proposed filamentous fungal activator genes have now been isolated and sequenced (see previous sections). It is encouraging to find that these genes specify proteins with characteristic domains predicted for DNA-binding regulators of gene expression (Fig. 1).

1. DNA Binding Domains

Genes involved in induction of catabolic activities specify proteins with the six-cysteine (C6) motif characteristic of *GAL4* and many other yeast regulatory proteins. The C6 structure may coordinate two zinc atoms rather than a single atom as previously proposed (Pan and Coleman, 1990). In *GAL4* and other yeast C6 proteins, the zinc-binding motif is part of the DNA-binding domain. *In vitro* mutagenesis of the *GAL4* C6 region generates a zinc-dependent phenotype (Johnston, 1987a). Similar experi-

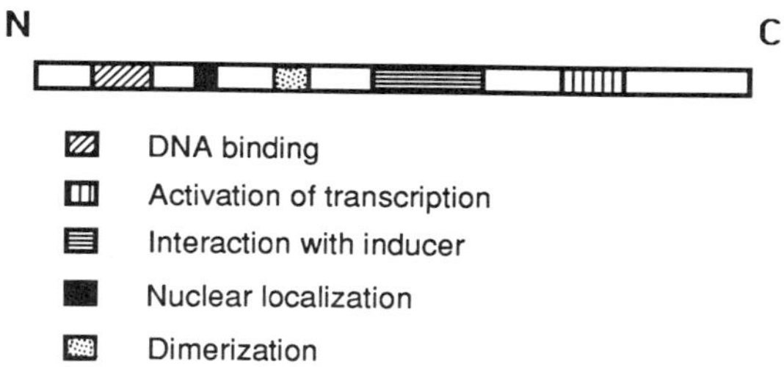

Fig. 1. Hypothetical structure of an activator protein involved in induction of catabolic genes showing the functional domains required for activity. The DNA-binding domain is required for specific binding to *cis*-acting DNA sequences in the 5′ region of regulated genes. Inducer molecules interact either directly or indirectly (e.g., via another protein) with the protein to allow the activation region to contact the transcription apparatus and stimulate transcription initiation.

ments with HAP1 produce variants showing zinc-dependent *in vitro* DNA binding (Kim *et al.*, 1990).

Three filamentous fungal sequences are compared with the *GAL4* C6 sequence in Fig. 2. All of these sequences occur in the N terminus of the proteins. This region is required for DNA binding by protein products of *qa-1F* and *amdR* (see Geever *et al.*, 1989; R. van Heejswyck and M. J. Hynes, unpublished data). A temperature-sensitive *qa-1F* mutation caused by an amino acid substitution in the C6 region results in a protein that exhibits temperature-sensitive DNA binding *in vitro* (Geever *et al.*, 1989).

```
GAL4   CDICRLKKLKCSKEKPKCAKCLKNNWEC
qa-1F  CDQCRAARGKCDGIQPACFPCVSQGRSC
qutA   CDSCRSKKDKCDGAEPICSTCASLSRPC
amdR   CVHCHRRKVRCDARLVGLPCSNCRSAGKTDC
```

Fig. 2. Comparison of some C6 DNA-binding regions of fungal regulatory genes. The *GAL4* gene of *S. cerevisiae* is required for galactose induction of various genes involved in galactose utilization (Johnston, 1987b). The *qa-1F* and *qutA* genes are required for induction of the genes of the quinate utilization cluster in *N. crassa* and *A. nidulans*, respectively (Giles *et al.*, 1985; Geever *et al.*, 1989; Beri *et al.*, 1990). The *amdR* gene regulates the expression of the *amdS, gabA, gatA,* and *lamA* and *B* genes of *A. nidulans* (Andrianopoulos and Hynes, 1990). Other genes from *A. nidulans* known to have similar motifs include the *prnA* gene involved in proline induction of proline catabolic genes, the *alcR* gene required for induction of the *alcA* and *aldA* genes (see Felenbok *et al.*, 1988,1989), and the *facB* gene which regulates the *amdS, facA, acuD,* and *acuE* genes (Katz and Hynes, 1989c; M. E. Katz, S. I. Sapats, and M. J. Hynes, unpublished data).

Sequences flanking the C6 region are likely to be involved in binding to the specific DNA sites recognized by the activator. The C6 region may also be involved in a more general way in DNA binding (Corton and Johnston, 1989). The *GAL4* and *LAC9* proteins of *Saccharomyces cerevisiae* and *Kluveromyces lactis* recognize the same sites; homology extends beyond the zinc coordination region (Wray *et al.*, 1987). The *qa-1F* and *qutA* proteins, which also recognize similar DNA sequences, show extended similarity outside the C6 region (Geever *et al.*, 1989).

It should be noted in passing that there is some evidence in yeast systems for additional functions for the C6 DNA-binding domain. The *HAP1-18* mutation increases activation of the *CYC7* gene without altering the DNA-binding affinity of the activator for this gene (Kim and Guarente, 1989). The *LAC9* protein of *K. lactis* plays a role in glucose repression; amino acid substitutions in the DNA-binding domain of the protein can alter sensitivity to glucose (see Kuger *et al.*, 1990). These observations indicate that the protein domains are not necessarily completely autonomous in function.

Different DNA-binding motifs have been found in some other fungal regulatory proteins not involved in pathway-specific induction. The *brlA* gene encodes a key regulator of conidiation in *Aspergillus nidulans* that has two C2H2 zinc fingers of the TFIIIA type. Both of these fingers are necessary for *brlA* activity (Adams *et al.*, 1988; Adams *et al.*, 1990). The *cpc-1*-encoded protein of *Neurospora crassa,* involved in cross-pathway control of amino acid biosynthesis, has a leucine-zipper motif similar to the one found in yeast *GCN-4* and *jun* oncogenes (Paluh *et al.*, 1988). The deduced protein of the *abaA* gene, involved in the regulation of conidiation, contains a possible leucine-zipper motif (Mirabito *et al.*, 1989). Probable DNA-binding motifs for the products of the wide-domain regulatory genes *areA, creA,* and *cys-3* are described in Section VI.

2. *Activation Domains*

Acidic amphipathic α-helical domains may play a role in the activation of transcription (see Ptashne, 1988). It is thought that these acidic regions contact components of the transcription apparatus (e.g., RNA polymerase or TATA binding proteins) and stimulate initiation of transcription. Alternatively, chromatin structure at the promoter may be modified by contact with histones (see Struhl, 1989).

The *amdR* activator of *Aspergillus nidulans* has four acidic regions in the C-terminal half of the polypeptide. Deletion analysis of the cloned gene followed by *in vivo* complementation studies shows that at least two of these regions play a role in *amdR* function, presumably activation, and that both of these sequences are required for full activity (Andrianopoulos and Hynes, 1990). Comparison of the *qa-1F* and *qutA* activators of *Neurospora crassa* and *A. nidulans* reveals a conserved acidic region at the extreme C terminus of the polypeptide. It is highly likely that this region is analogous to the activator domains of other regulatory proteins (Geever *et al.*, 1989).

3. *Negatively Acting Genes*

Regulatory genes with a negative (repressor) function are less commonly found in fungi. Loss-of-function mutations in such genes would give rise to recessive constitutive phenotypes, whereas rare dominant or semidominant alleles would generate noninducible phenotypes. Such mutations are found in the *qa-1S* and *qutR* genes, affecting quinate utilization in *Neurospora crassa* and *Aspergillus nidulans,* respectively (Giles *et al.*, 1985; Grant *et al.*, 1988; Lamb *et al.*, 1990). It is likely, however, that the products of these genes do not bind directly to DNA but interact with the corresponding activator genes (*qa-1F* and *qutA;* see previous text) to prevent activation of transcription. Inducers presumably inhibit this interaction. This model is consistent with that derived for the *GAL4*–*GAL80* interaction in *Saccharomyces cerevisiae* (Johnston, 1987b). Molecular characterization of *qa-1S* mutations indicates that mutants with a constitutive phenotype can result from frameshift mutations, whereas dominant noninducible mutants contain missense mutations, presumably defective in interaction with the inducer or activator (Geever *et al.*, 1989). Genes of this sort therefore respond to an inducer rather than play a primary role in modulating transcription. There may well be other systems in which this interaction between negative and positive regulatory gene products occurs (see Section VI). Repressor genes may be less readily detected because usually selection is possible only for the rare noninducible mutant phenotype rather than for the constitutive phenotype.

4. *Other Domains*

Sequences involved in nuclear localization have been identified for some DNA binding proteins and are likely to be found in fungal regulatory proteins. No studies directed at identifying such sequences have been reported. Regulatory proteins commonly occur as multimers, particularly dimers. This structure allows recognition of DNA-binding sites with dyad symmetry and provides a rationale for bidirectional gene activation (Struhl, 1989). In addition, the formation of heterodimers between the products of different activator genes potentially increases the diversity of available regulatory activities. Leucine-zipper motifs are thought to be involved in dimer formation. Other kinds of dimerization domains are likely to be found in fungal activator proteins but have yet to be identified.

5. *The Role of Inducer*

In inducible systems, low-molecular-weight inducers result in gene activation. The inducer could stimulate the action of the activator protein in a number of ways. The inducer could be required for specific DNA binding, either by causing an essential conformational change or by promoting dimerization, which may be necessary for DNA binding. Inducer binding could turn an inactive activator into an active form by altering the conformation of the protein, perhaps by unmasking the transcription-activation domains. If a separate repressor protein constituted part of the system, the inducer could prevent the repressor from repressing activator synthesis or from inhibiting activator function. Finally, the inducer could modulate the activity of enzymes that modify the structure of the activator to affect its DNA-binding affinity or transcription-activation capacity.

As noted earlier, induction of the quinate utilization genes is probably mediated by an interaction between an inducer and the product of the repressor gene (*qa-1S* = *qutR*). This interaction is highlighted by the observation that the *qa-1S* repressor protein is homologous to the three C-terminal domains of the *arom* pentafunctional enzyme of *A. nidulans* and *S. cerevisiae* (Anton *et al.*, 1987). The repressor acts to inhibit activation by the activator gene (*qa-1F* = *qutA*) in the absence of inducer (Giles *et al.*, 1985; Grant *et al.*, 1988). This inhibition may result directly from the interaction of repressor and activator. An alternative possibility has been raised in which the gene product acts to phosphorylate the activator protein thereby inhibiting its activity, while a second gene (*qutG* = *qa-x*) dephosphorylates the activator (Anton *et al.*, 1987; Lamb *et al.*, 1990). In this model, the inducer would somehow modulate the kinase–phosphatase activities. Comparison of the *qa-1F* and *qutA* proteins reveals a long conserved central segment which may be involved in interaction with the repressor protein and/or in dimerization (Geever *et al.*, 1989).

The *amdR* activator protein activates the *amdS, gabA, gatA,* and *lamA* and *B* genes in response to ω-amino acid inducers such as β-alanine and GABA. Semidominant *amdR*^{c} mutations can result in high level constitutive expression of the structural genes. The sequence alteration in the *amdR6*^{c} mutant allele is a proline–serine substitution in the middle of the protein, which is likely to result in a marked conformational change in the protein (Andrianopoulos and Hynes, 1990). Addition of inducer is not required for the *amdR* protein to bind to DNA *in vitro* (R. van Heejswyck and M. J. Hynes, unpublished data). It is likely that the role of the inducer in this system is to interact directly with the *amdR* protein to increase its ability to activate transcription. *amdR* mutations exist that differentially alter activation of some of the structural genes, and one allele apparently affects inducer specificity (Arst, 1976). Sequence analysis of these mutations is in progress.

Nitrate induction of nitrate reductase and nitrite reductase involves the activator genes *nirA* and *nit-4* in *Aspergillus nidulans* and *Neurospora crassa,* respectively. In these systems, the nitrate reductase protein itself may be involved in induction (Cove, 1979; Fu and Marzluf, 1988). Whether the inducer interacts directly with the activator protein or whether inducer is recognized by the nitrate reductase protein remains to be determined.

6. Regulation of Regulatory Genes

Northern blot analysis of genes involved in activation of catabolic activities indicates that activator genes can regulate their own expression in response to an inducer (autogenous regulation). This is illustrated in Fig. 3 for the *alcR* gene, which is subject to the same regulation as the structural genes except for a higher basal level of expression (Lockington *et al.*, 1988). The regulation of the *facB* and *prnA* genes is similar (Hull *et al.*, 1989; Katz and Hynes, 1989c). Both the activator (*qa-1F* and *qutA*) and repressor (*qa-1S* and *qutR*) genes of the quinate utilization pathway are also regulated autogenously (Giles *et al.*, 1985; Lamb *et al.*, 1990).

The *alcR, facB,* and *prnA* genes are also subject to carbon catabolite repression (see Section VI,C), raising the possibility that carbon control is mediated by expression of the specific activator gene rather than directly by the structural genes. This possibility was eliminated for the *alcR* regulon by showing that, in transformants in which the *alcR* gene is expressed from a constitutive promoter, the *alcA* and *aldA* genes are still subject to carbon catabolite repression (Felenbok *et al.*, 1989).

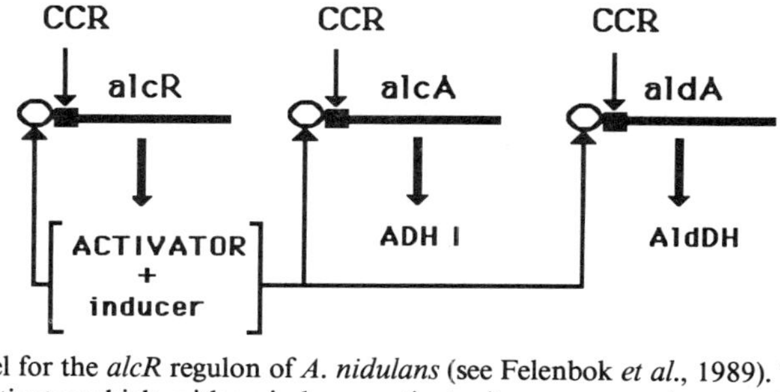

Fig. 3. Model for the *alcR* regulon of *A. nidulans* (see Felenbok *et al.*, 1989). The *alcR* gene encodes an activator which, with an inducer, activates its own expression as well as that of the structural genes encoding alcohol dehydrogenase (ADH1) and aldehyde dehydrogenase (AldDH). Carbon catabolite repression (CCR) also affects expression of all the genes of the regulon.

The expression of some activator genes is apparently not regulated. The *nit-4* gene of *Neurospora crassa* is not induced by nitrate or subject to nitrogen metabolite repression (mediated by the *nit-2* gene; see Section VI,A) but is expressed at a low constitutive level (Fu *et al.*, 1989a). Similarly, the *amdR* gene is constitutively expressed at a low level (Andrianopoulos and Hynes, 1988). Clearly autogenous regulation of activator synthesis by inducer is not always required for pathway-specific induction.

There is evidence that the *facB* gene plays a structural as well as a regulatory role in acetate utilization (J. M. Kelly and M. J. Hynes, unpublished data). Autogenous control could therefore be more important for induction of this structural role than for gene regulation. Conceivably, other activator genes such as *prnA* and *alcR* have dual functions. The possible involvement of the nitrate reductase protein in gene regulation (see Section V,A,5) is perhaps of interest in this regard.

7. Dosage Effects of Regulatory Genes

The ability to generate transformants containing multiple copies of cloned DNA sequences facilitates experiments that assess the effects on gene expression of changing the relative numbers of regulatory elements. Increasing the number of DNA target sites for regulatory proteins provides an opportunity to investigate titration of regulatory proteins (see Section V,B,3). Increasing the dosage of a regulatory gene relative to the controlled structural genes enables studies of the limiting factors for induction.

Transformants with multiple copies of the *amdR* gene show elevated but not high-level constitutive *amdS* expression under inducer-limiting conditions. When the inducer is present, however, expression of the *amdR*-controlled genes is elevated in multicopy *amdR* transformants (Andrianopoulos and Hynes, 1988). These results support a model in which inducer is required for maximal activity of the activator, but the upper level of induction of gene expression is set by the amount of *amdR* protein present, implying that the *amdR*-binding site is not saturated. This model is consistent with the titration and antititration experiments described in Section V,B,3. Similar experiments indicate that the *facB* protein is limiting and requires acetate for efficient activation of *amdS* expression (Katz and Hynes, 1989c). Studies with a multicopy *alcR* transformant have shown that, although *alcR* messenger levels are higher under noninduced conditions than the levels in wild-type strains under induced conditions, *alcA* and *aldA* expression is not constitutive, indicating a requirement for inducer for *alcR* function (Lockington *et al.*, 1988).

Transformants with multiple copies of the activator gene *qutA*, but single copies of the other *qut* genes, do not show abnormal regulation or overexpression of the regulated genes (Beri *et al.*, 1990). However, when equal multiple copies of the *qutA* and *qutE* genes are present, the *qutE* gene is selectively overexpressed, but only upon induction, while regulation and expression of the other genes is normal. In contrast to the systems discussed earlier, *qut* cluster regulation also involves a repressor gene (*qutR*) that is proposed to interact with the *qutA* activator. The results imply either that the *qutR* repressor is present in excess of the *qutA* activator or that the *qutR* product acts catalytically on the activator (Lamb *et al.*, 1990). In addition, it would appear that the *qutA* activator is not limiting for expression under induced conditions, and perhaps all binding sites are occupied by activator (Beri *et al.*, 1990).

B. Identification of *cis*-Acting Sites

A number of different approaches have been used to delineate the *cis*-acting sequences recognized by specific regulatory proteins. In combination, these approaches provide powerful tools for dissecting the specific DNA–protein interactions that regulate gene expression.

1. cis-*Acting Mutations*

Mutations that specifically alter the interaction between regulatory proteins and DNA serve as a focus for identifying sequences likely to be involved in protein recognition. Many of the *cis*-acting mutations isolated have been obtained as suppressor mutations able to circumvent loss-of-function mutations in regulatory genes. The *gabI, nis-5, prn*^{d}, and *uap-100* mutations, for example, were isolated as revertants of *areA* mutants (Arst and Scazzocchio, 1985). In each case, mutations closely linked to structural genes result in a bypass of the need for *areA*-mediated activation. The *uap-100* mutation is a duplication of a 139-bp sequence in the upstream region of the *uapA* gene (Diallinas and Scazzocchio, 1989). This mutation has an up-promoter effect on *uapA* expression, but expression remains dependent on functional *uaY* and *areA* products. The *uap-100* locus may represent a duplication including the binding site for either *uaY, areA,* or both.

The *prn* genes of *Aspergillus nidulans* form a cluster containing both structural genes (*prnB, C, D*) and a regulatory gene (*prnA*). Mutations, designated *prn*^{d}, lie in a putative *cis*-regulatory region between the *prnA* and *D* genes and the *prnB* and *C* genes (Arst and Scazzocchio, 1985). The *prn*^{d} mutations are all single basepair substitutions and are thought to prevent the imposition of carbon catabolite repression, or possibly nitrogen metabolite repression, on expression of the *prnB, C*, and *D* genes (Felenbok *et al.*, 1989).

More detailed studies have been undertaken on the variety of different *cis*-acting mutations, designated *amdI,* that affect expression of the *amdS* gene (reviewed by Hynes and Davis, 1986). The *amdI18* mutation, a single basepair change at −118, leads to an up-promoter effect without altering regulation (Corrick *et al.*, 1987). The *amdI9* mutation, a basepair change at −210, results in increased *facB*-dependent acetate induction, and the *amdI93* mutation abolishes *amdR* induction of *amdS* and has been shown to be due to a 30-bp deletion. Finally, a duplication of sequences −107 to —90 (*amdI66*) leads to increased activation by mutant forms of the *amdA* regulatory product (Hynes *et al.*, 1988). More recently, a mutant carrying three copies of this sequence was isolated and characterized as having a stronger effect on *amdA*-mediated gene expression (Katz *et al.*, 1990). The isolation of mutants with specific alterations in sequences 5′ to the regulated gene can be of great value because the associated change in gene regulation provides a direct indication of the functional importance of particular sequences. However, the rarity of these mutations is a very limiting factor in the analysis of most systems.

2. *Comparison of Coregulated Genes*

A second approach, designed to identify key recognition sequences, is based on the premise that related sequences will be present in the regulatory regions of genes coordinately regulated by a common regulator. This approach has been used in the study of the *alcA, aldA,* and *alcR* genes, all of which are regulated by *alcR* (Gwynne *et al.*, 1987). Although small regions of homology were found between *alcA* and *alcR* or *aldA* and *alcR,* only a hexanucleotide sequence is common to the 5′ regions of all three genes. Since this latter sequence may have a more general promoter function, the comparisons indicate that the *alcR* binding sequences may be degenerate or that *alcR,* like HAP1 of yeast (Pfeifer *et al.*, 1987), may recognize dissimilar sequences. The *qut* gene cluster of *Aspergillus nidulans* similarly illustrates the difficulties in assigning function to sequence homologies detected by computer analysis. Hawkins and co-workers (1988) described three classes of conserved sequence homology in the 5′ regions of two pairs of divergently transcribed *qut* genes. Between each pair of genes, a common pattern of 22-nt–9-nt–16-nt sequence classes was identified. The 16-nt sequence shared 43% homology with the sequence shown to bind the *qa-1F* activator in the analogous *qa* system in *Neurospora crassa* (Baum *et al.*, 1987). However, other sequences with 66% homology to the *qa-1F* site were also detected at various locations relative to the 22-nt sequence. The functional significance of each of these classes of sequences must still be determined.

In the *amdS* system, it has been possible to couple the information derived from *in vivo* mutants (see Section V,B,1) with sequences found in co-regulated genes. Sequence comparisons have indicated that CT-rich sequences around the site of the *amdI9* mutation of the *amdS* gene are shared with the 5′ regions of the *acu* and *facA* genes (*facB* control) (Sandeman, 1988). Sequences homologous to those deleted in part in the *amdI93* mutation are found 5′ to the *gatA* gene and in the intergenic region of the divergently transcribed *lamA* and *B* genes (*amdR* control) (Richardson *et al.*, 1980; I. B. Richardson, M. Katz, and M. J. Hynes, unpublished data). Upstream sequences of the *aciA* gene contain GA-rich sequences similar to those duplicated and triplicated in the *amdI66* and *amdI666* mutations (*amdA* control) (Atkinson *et al.*, 1985; Katz *et al.*, 1990; J. A. Saleeba, unpublished data). Although these putative recognition sequences are reasonably degenerate, sequences from *gatA* will bring about *amdR* regulation of *amdS* (Richardson *et al.*, 1989). However, these differences may result in differing affinities for the regulatory protein as judged by titration studies (see Section V,B,3).

3. Binding of Regulatory Proteins in Vivo *and* in Vitro

Titration analysis provides an *in vivo* assay for the capacity of a DNA sequence to bind a regulatory protein. When regulatory proteins are present in limiting quantity within the cell, multiple copies of a binding sequence may reduce the amount of protein available for control of co-regulated structural genes. Titration analysis has been used successfully to localize regions of the *amdS* gene involved in the binding of the *facB* protein, the *amdR* protein, and the *amdA* protein (Kelly and Hynes, 1987; Hynes *et al.*, 1988). As predicted, multiple copies of the regulatory genes *amdR* and *facB* reverse titration of their gene products (antititration; Andrianopoulos and Hynes, 1988; Katz and Hynes, 1989c).

Specific oligonucleotides have been used to define those sequences necessary and sufficient to titrate the *amdR* protein away from its co-regulated genes. These oligonucleotide sequences are also capable of conferring β-alanine inducibility on an upstream-deleted *amdS–lacZ* fusion gene (T. G. Littlejohn and M. J. Hynes, unpublished data). In the case of *amdR* regulation, the same sequences that bind the protein in *in vivo* titration studies also bind *amdR* protein *in vitro,* as seen by mobility shift experiments (R. van Heejswyck and M. J. Hynes, unpublished data). Miller and colleagues (1987) similarly found evidence suggestive of titration of a negative control factor leading to overexpression of *spoC1* transcripts in multicopy *spoC1C* transformants.

The power of the titration assay is diminished when the regulatory product is abundant. This would appear to be the case for the *areA* protein, a regulatory product involved in the control of over 100 different structural genes (see Section VI,A). Attempts to demonstrate titration of the *areA* product have been successful only in transformants containing in excess of

100 additional copies of the *amdS* gene (Hynes and Andrianopoulos, 1989). The rarity of transformants containing this many gene copies precludes titration analysis in this instance. The success of this approach with *amdR,* in particular, is consistent with the fact that the *amdR* protein is extremely limiting (see Section V,A,7).

To date, there are few published reports of *in vitro* DNA–protein binding studies in *Aspergillus nidulans.* R. van Heeswyck (unpublished data) has demonstrated the presence of protein(s) in nuclear extracts capable of retarding the mobility of CCAAT-containing oligonucleotides. These CCAAT sequences are found in the 5′ region of *amdS,* in the region bound by the *amdR* protein. The two proteins bind independently. The CCAAT motif is thought to be involved in setting the basal levels of *amdS* transcription and may have a regulatory function. As increasing numbers of cloned regulatory gene sequences become available, further information on DNA–protein interactions no doubt will be produced. The overexpression of regulatory genes by the transformation of multiple copies of cloned sequences into *A. nidulans* provides a means of potentially increasing the concentration of regulatory proteins. In addition, expression in *Escherichia coli* may be possible for purification of the protein. However, evidence from studies with the *amdR* protein suggests that the quantity of regulatory protein need not necessarily be a limiting factor.

There are several DNA-binding studies in *Neurospora crassa* that are of considerable interest. Baum and colleagues (1987) have demonstrated *in vitro* binding of the *qa-1F* protein to 16-bp consensus sequences in front of each of the structural genes of the *qa-1F* cluster. Interestingly, transcript inducibility correlates with the proximity of the sequences to the gene rather than with the relative affinity of the *qa-1F* protein for different target sites (Geever *et al.*, 1989). Since related sequences have been found in the 5′ regions of the analogous *qut* genes of *A. nidulans* (Hawkins *et al.*, 1988), comparative binding studies using the *qutA* protein will be significant. The *qa-1F* and *qutA* proteins share three regions of homology, one of which includes the DNA-binding domain (Geever *et al.*, 1989) (see Section V,A,1). The *nit-2* gene of *N. crassa* is analogous in function to the *areA* gene of *A. nidulans* (see Section VI,A). Fu and Marzluf (1990a) have expressed the DNA-binding domain of the *nit-2* protein in *E. coli* for DNA-footprinting studies. The target sequences in their analysis were 5′ sequences of the nitrate reductase (*nit-3*) gene of *N. crassa* and the promoter region between the divergently transcribed nitrate reductase (*niaD*) and nitrite reductase (*niiA*) genes of *A. nidulans.* Similar motifs were protected by the *nit-2* protein in both instances. This finding correlates well with the knowledge that the *nit-2* gene can complement *areA*$^-$ mutants of *A. nidulans* (Davis and Hynes, 1987).

VI. WIDE-DOMAIN CONTROL MECHANISMS

Aspergillus nidulans, like most free-living fungi, is essentially a scavenger, a recycler of available resources. The organism displays incredible metabolic versatility, since it is able to use a wide range of compounds as its nutrient source and has the capacity to tolerate a broad pH range. Global control systems for regulating gene expression are essential to this lifestyle. Limitation of any of the essential nutrients, such as carbon, nitrogen, sulfur, or phosphate, can trigger changes in the levels of transcription of genes involved in procuring these nutrients.

A. Nitrogen Metabolite Repression

The synthesis of many catabolic enzymes is regulated by nitrogen metabolite repression. When readily assimilated ammonium ion or glutamine is unavailable, nitrogen limitation causes cells to express or potentiate the expression of a large number of enzymes involved in catabolizing alternative nitrogen sources.

The key intracellular signal is thought to be the intracellular level of glutamine. Whereas wild-type cells grown in the presence of ammonium ion or glutamine synthesize only basal levels of these catabolic enzymes, mutants lacking a functional NADP–GDH enzyme (*gdhA*) or glutamine synthetase (*glnA*) are insensitive to repression by ammonium ion (see Arst and Scazzocchio, 1985). The NADP–GDH enzyme is the major ammonium ion assimilatory enzyme in *Aspergillus nidulans,* converting ammonium to glutamate and hence to glutamine by the action of glutamine synthetase. However, evidence from studies of nitrogen control in bacteria (see Magasanik, 1988) would caution that a more complex interplay of intracellular molecules may be involved.

Genetic studies clearly establish that the *areA* gene encodes the primary regulatory protein for nitrogen catabolism control. The most common class of *areA* mutants includes those which show a reduced ability to use a wide range of nitrogen sources other than ammonium ion. These mutations are recessive to wild-type, indicative of a positive mode of action for the *areA* gene product. Rarer *areA* mutations appear to subtly alter the conformation of the *areA* protein, differentially affecting its ability to bind to or activate transcription from different promoters. These latter mutations indicate a degree of heterogeneity in the recognition sequences of the *areA* protein for different genes. Arst and Scazzocchio (1985) have described the isolation, phenotypes, and genetics of these mutations in some detail. One very significant *areA* mutation is the *xprD1* mutation, which results in derepressed synthesis of nitrogen catabolic enzymes, and arose from an inversion with a breakpoint within the *areA* gene. This mutation has been exploited to clone the *areA* gene. Crosses between the *xprD1* strains and wild-type strains result in a class of duplication-deficiency progeny that lack part of the *areA* gene and regions centromere-distal to the inversion breakpoint (Arst, 1982). An *areA*-containing clone was identified by hydridizing more weakly to genomic DNA from the duplication-deficiency strain than to wild-type DNA (Caddick *et al.*, 1986a).

The predicted *areA* protein contains a large number of S(T)PXX residues (Kudla *et al.*, 1990), potential phosphorylation sites found predominantly in DNA-binding proteins (Suzuki, 1989). Moreno and Nurse (1990) suggested that these motifs may have a role in the transitory displacement of proteins from the DNA during the cell cycle. Two *areA* mutations, *xprD1* and *areA18,* have been used to define regions of functional importance in the *areA* protein. The inversion event in the *xprD1* mutation deletes the 3′ end of the *areA* gene. Since the *are*A protein in *xpr*D1 strains has full activator function, the C-terminal 124 amino acids are apparently dispensable. The phenotype of this mutant indicates that the sequences 3′ to the breakpoint are, however, crucial for regulation. The *areA18* mutation is a loss-of-function mutation due to a translocation breakpoint within the gene (Rand and Arst, 1977). Rare intracistronic revertants of *areA18* contain further chromosomal rearrangements involving the 3′ portion of this fractured gene. These new arrangements are thought to fuse the 3′ portion to new transcription and translation signals for expression (Arst *et al.*, 1989b). Coding sequences 5′ to the breakpoint are, therefore, also thought to be nonessential for function. The functional limits, as defined by these mutants, indicate that the essential region of the *areA* protein is confined to an internal sequence of 253 residues (Kudla *et al.*, 1990; see Fig. 4).

This essential region of the protein contains sequences capable of forming a zinc finger. This DNA binding motif is flanked on the N-terminal side by an acidic region potentially capable of forming an amphiphatic α helix, a structural arrangement of amino acids implicated in activation of transcription (see Section V,A,2). The zinc finger and adjacent C-terminal sequences are highly basic, and computer analysis suggests that residues outside the zinc finger could form a helix–turn–helix structure (Kudla *et al.*, 1990).

The diversity of mutant alleles available in the *areA* gene is proving invaluable in analyzing the key structural features of the protein. Sequence analysis has shown that several mutants contain sequence changes within the zinc finger region (see Fig. 4). For example, the *areA102* allele differentially affects expression from different promoters, allowing increased growth on some nitrogen sources and reduced or unaltered growth on

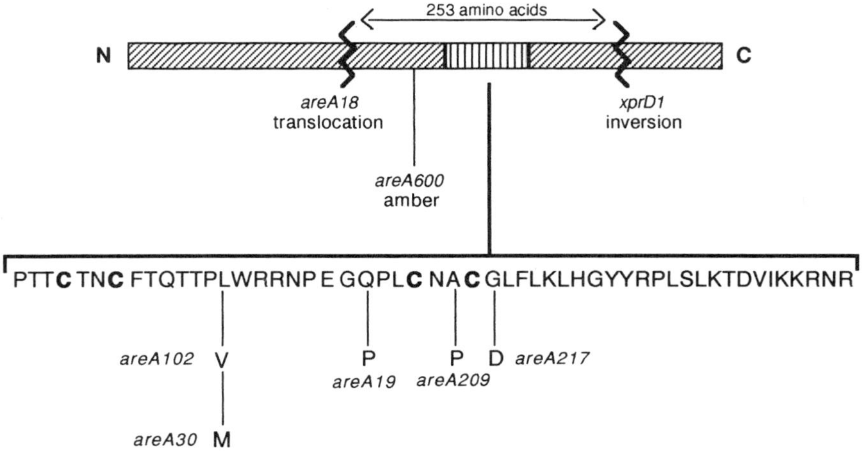

Fig. 4. Sequence changes in a number of *areA* mutants and their effects on function. The *areA18* and *xprD1* mutations define an essential 253-amino-acid region. The *areA18* mutation involves a III–IV reciprocal translocation and leads to a loss of *areA* function. Rare revertants involving further chromosomal rearrangements of the 3′ portion of the gene suggest that sequences 5′ to the *areA18* breakpoint are not essential to function (see text). The *xprD1* mutation is an inversion with one breakpoint in the 3′ end of the *areA* gene. The *areA* product of *xprD1* strains has apparently normal activator function but is insensitive to nitrogen control. Therefore, sequences essential to activator function appear to lie within the region defined by the *areA18* and *xprD1* breakpoints. A nonsense mutation within this region (*areA600*) has a loss-of-function phenotype. Sequences required for response to nitrogen metabolite repression apparently lie 3′ to the *xprD1* breakpoint. A number of *areA* mutations lie within the C2/C2 zinc finger region of the protein. The amino-acid sequence of this region is shown; cysteines are in bold. The *areA19, areA209,* and *areA217* mutations are loss-of-function mutations due to amino-acid substitutions within or adjacent to the zinc finger region. The *areA102* mutation results in an apparent change in specificity of the *areA* protein. The *areA30* mutation is a revertant of *areA102* in which a new substitution at the same site results in altered specificity of the *areA* protein. These mutants and others are described in detail by Kudla and colleagues (1990).

others. This mutation is a conservative leucine–valine amino-acid substitution within the zinc finger. An *areA102* revertant (*areA30*), that produces 'mirror-image' change in phenotype to that of *areA102* strains, is caused by a substitution of a methionine residue for the valine of *areA102* (Kudla *et al.*, 1990). These relatively conservative amino-acid changes at a single position within the finger region suggest that this residue is crucial to the protein's recognition of, and affinity for, promoter sequences. These mutants also indicate the potential heterogeneity of *areA* recognition sequences. However, recent studies in yeast provide evidence that alterations within DNA-binding domains may alter activator function without affecting affinity of the protein for its recognition sequence (Kim *et al.*, 1990). The phenotype of strains carrying the *areA217* mutation indicates that this mutation abolishes *areA* function. The amino-acid substitution in this instance is of the residue immediately adjacent to the finger, indicating that sequences outside the finger itself are also critical for *areA* function.

The zinc finger motif of *areA* is of an unusual type, resembling those found in a tissue-specific transcriptional activator GATA-1 (previously termed GF1, EryF1 or NF-E1) active in vertebrate erythroid cells (see Pevny *et al.*, 1991). The vertebrate proteins contain two related zinc finger structures of a Cys-x-x-Cys-x_{17}-Cys-x-x-Cys form. The *areA* protein contains an identical distribution of cysteine residues and the amino acids internal to, and flanking these cysteines, bear a close resemblance to those of the GATA-1-type protein (Arst *et al.*, 1989a). It is of interest that GATA-1 has been postulated to serve as a general switch protein that activates all erythroid-specific genes (Evans and Felsenfeld, 1989). A targeted gene disruption of the GATA-1 gene has been shown to block erythroid differentiation in chimeric mice (Pevny *et al.*, 1990).

The *nit-2* gene of *Neurospora crassa* and the *areA* gene of *A. nidulans* are genetically analogous. In both instances, loss-of-function mutants are unable to use nitrogen sources other than ammonium (Perrine and Marzluf, 1986). The wild-type *nit-2* gene complements *areA*$^-$ mutants of *A. nidulans,* underlining the conservation of function between these genes in the different filamentous fungi (Davis and Hynes, 1987). Comparison of the deduced amino-acid sequences of the two proteins reveals a considerable conservation of residues throughout much of the protein (Fig. 5). Although *A. nidulans* and *N. crassa* are evolutionarily diverged, the *nit-2* and *areA* proteins show an extremely high conservation of amino-acid sequence at and around the zinc finger region. *In vitro* mutagenesis of conserved residues within and around the zinc finger region of the *nit*-2 gene has shown that these amino acids are essential for DNA binding *in vitro* and activator function *in vivo* (Fu and Marzluf, 1990a). Fu and Marzluf (1990b) have expressed partial *nit-2* proteins in *Escherichia coli* and identified protected sequences 5′ to the nitrate utilization genes of both *N. crassa* and *A. nidulans* (see Section V,B,3). In all cases, the protected sites contain versions of a sequence with the consensus TATCTA. This sequence is very similar to the sequence TATCT, defined as the core recognition sequence of the GATA-1 protein (Evans and Felensfeld, 1989). It is interesting to note that the GLN3 gene of *Saccharomyces cerevisiae,* which has a role in the regulation of nitrogen metabolism, also appears to recognize a similar sequence (Courchesne and Magasanik, 1988; Cooper *et al.*, 1990).

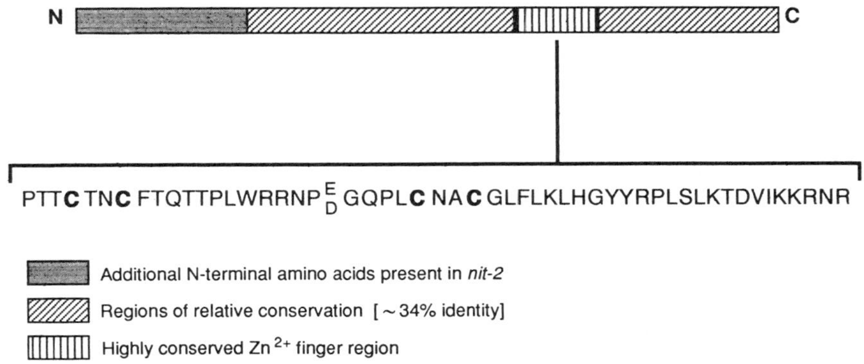

Fig. 5. Comparison of the predicted *areA* protein (Kudla *et al.*, 1990) and *nit-2* protein (Fu and Marzluf, 1990a,b). The *areA* and *nit-2* proteins are the major positively acting regulators of nitrogen metabolite repression in *A. nidulans* and *N. crassa,* respectively (see text). The *nit-2* gene can complement *areA* mutants of *A. nidulans* (Davis and Hynes, 1987). The two proteins share approximately 30% identity over most of their sequence. However, two regions are exceptional. The *nit-2* protein contains additional N-terminal amino acids not present in the *areA* protein. This additional region has an unusual composition and includes three variants of an amino-acid sequence S-S-S-A-N-N-N (Fu and Marzluf, 1990a,b). The other region of interest spans the putative zinc finger region of both proteins. The zinc finger is a C2/C2 type; cysteines are shown in bold. In the 53-amino-acid sequence shown, there is only one difference between *areA* and *nit-2*:*areA* has glutamate at residue 20 of this sequence whereas *nit-2* has aspartate.

The predicted *nit-2* protein of 1036 amino acids is significantly larger than the predicted *areA* protein, primarily due to additional residues at the N terminus (Fig. 5). The function, if any, of these residues is not clear. In the case of *areA,* much of the N-terminal sequence is dispensable to the activator function of the protein (see previous text). It could therefore be argued that these additional N-terminal sequences simply reflect an insertion (in *nit-2*) or a deletion (in *areA*) and are of no functional significance. However, cross-species transformation of *nit-2* into *areA*$^-$ mutants of *A. nidulans* has shown that the transformants, although regaining the ability to use nitrogen sources, are partially derepressed for nitrogen control (Davis and Hynes, 1987). Can differences in the structure/expression of the *areA* and *nit-2* proteins explain this altered regulatory response? Fu and Marzluf (1990c) have commented that repeats of an unusual amino-acid sequence in the N-terminal portion of *nit-2* may reflect earlier duplication events. It is possible that chromosomal rearrangements involving this region of the gene have resulted in different promoters driving *areA* and *nit-2*. The *nit-2* gene is only weakly regulated by the nitrogen status of cells and is not autoregulated (Fu and Marzluf, 1987), whereas *areA* expression may be at least partially autoregulated, requiring sequences at both 3′ and 5′ ends of the gene (Caddick *et al.*, 1989).

The C-terminal 124 amino acids of the *areA* protein (removed by the *xprD1*-associated inversion mutation) are not essential for activator function, but appear to be required for the transition between active and inactive forms. C-terminal deletions of the *nit-2* gene also retain activator function, but their effect on nitrogen repressibility has not been reported (Fu and Marzluf, 1990c). Instead, it has been suggested that nitrogen regulation of gene expression in *N. crassa* involves the product of a second gene, the *nmr* gene. Mutations in the *nmr* gene cause nitrogen derepression, leading to the hypothesis that the *nmr* product has a negative control function (Dunn-Coleman *et al.*, 1981; DeBusk and Ogilvie, 1984). Fu and Marzluf (1990c) have raised the possibility that the *areA* polypeptide is equivalent to a complex of the *nit-2* and *nmr-1* proteins. No recognizable DNA-binding motifs have been found in the sequence of the *nmr* gene (J. L. Young, G. Jarai, Y.-H. Fu, and G. A. Marzluf, unpublished data). Its function is unclear since *nmr* mutants reduce the levels of expression of the *nit-2* gene, a result not immediately reconcilable with their phenotype (Fu and Marzluf, 1987). No mutations equivalent to the *nmr* mutations have been found in *A. nidulans.* The GLN3 control system of yeast similarly involves a second gene, URE2 (Courchesne and Magasanik, 1988). The involvement of hitherto uncharacterized genes in nitrogen control in *A. nidulans* should not be discounted.

Tollervey and Arst (1982) have identified a rare class of *areA*$^-$ suppressor mutations, all of which involve chromosomal rearrangements. This unusual class of mutations is thought to identify an *areA* "pseudogene" (*areB*) that is reactivated by rearrangements that introduce new promoter elements. Isolation of the *areB* gene may provide valuable information about the *areA* gene itself, and about the evolution of sequences within a genome.

B. Carbon Catabolite Repression

Although carbon catabolite repression is well defined in *Escherichia coli* (see Morita *et al.*, 1988, and references therein), little is understood about the process in eukaryotes. Studies in yeast have revealed a large number of genes involved directly, or indirectly, in carbon control (for examples, see Carlson, 1987.) In *Aspergillus nidulans,* mutations in the *creA, creB,* and *creC* genes lead to altered responses to carbon control (Bailey and Arst, 1975; Hynes and Kelly, 1977). Early studies indicated that cyclic AMP was unlikely to be the effector molecule that triggered changes in gene expression and indicated that *creA* specified a negatively acting regulatory protein (Arst and Scazzocchio, 1985). Mutations in the *creA* gene are associated with an abnormal morphology and lead to derepression of many catabolite-repressible functions. These mutations are recessive to the wild-type and have been assumed to represent loss-of-function mutations (see subsequent text).

Cloning of the *creA* gene by complementation of a *creA* mutant has provided new information (Dowzer and Kelly, 1989). The gene produces a transcript of approximately 1.8 kb, the levels of which vary up to twofold depending on the carbon source. Autogenous control may not be involved since the levels and regulation of the transcript are not affected by *creA* mutations. Consistent with negative control, transformants containing multiple copies of the *creA* gene show evidence of tighter carbon catabolite repression than the wild-type (Dowzer and Kelly, 1989). Sequence analysis has indicated a possible zinc finger motif of the C2H2 type in the *creA* predicted protein, suggesting that the *creA* protein may bind DNA directly (C. E. A. Dowzer and J. M. Kelly, unpublished data).

To confirm that the derepressed phenotype of *creA* mutations is characteristic of loss-of-function, Arst and co-workers (1990) used the cloned sequences to demonstrate that the translocation mutation *creA30* has a breakpoint internal to the coding sequences of the gene. However, since messages are detected from both portions of the gene, by analogy with *areA* mutations, the conclusion remains equivocal. Gene disruption experiments currently underway (C. E. A. Dowzer and J. M. Kelly, unpublished data) suggest that a complete loss of *creA* function may be lethal or prevent conidiation. An essential role for *creA* is not inconsistent with the central role of carbon metabolism in development, especially given the abnormal morphology that characterizes *creA* mutant strains. The RGR1 gene recently identified in yeast provides an interesting analogy for *creA*. Mutations in RGR1 prevent catabolite repression and result in an abnormal cell morphology (Sakai *et al.*, 1990). Intriguingly, an RGR1 gene disruption is lethal and C-terminal deletions give phenotypes similar to, but more extreme than, the *in vivo* isolated mutants.

An area of considerable interest that has not yet been addressed is the interrelationship between the carbon and nitrogen control circuits. This, in part, is illustrated by the selection strategy used to isolate *creA* mutants. Many of these mutants were isolated as revertants of *areA*$^-$ mutants able to use acetamide or proline as the sole nitrogen source in the presence of glucose (see Arst and Scazzocchio, 1985). The *areA*$^-$ mutants are unable to use these compounds as nitrogen sources because of the loss of the positively acting *areA* product. Presumably by relieving carbon catabolite repression of the *amdS* and *prn* genes, *creA* mutants are able to catabolize these compounds. If, as the genetic studies suggest, the *creA* mutation abolishes a negative control element, what is replacing the *areA* function? The implication is that a positively acting component activates gene expression when carbon sources are limiting. No mutants of this type have been isolated and it is possible that such mutations are lethal. Some preliminary studies on the *amdS* system suggest that a CCAAT-containing sequence may play a role in activating expression when catabolite repression is released (Hynes *et al.*, 1989). Another topic that needs to be addressed is how the carbon status of the cells influences the ability of the *areA* protein to activate gene expression. In certain systems not regulated by carbon catabolite repression, *areA* is apparently unable to activate expression under carbon-limiting conditions, even if nitrogen is also limiting (reverse catabolite repression; see Section II). Why is nitrogen limitation a trigger for *areA* action only under conditions of carbon catabolite repression? Clearly much remains to be established about the interplay of these global controls and the intracellular conditions necessary for their operation.

C. Other Wide-Domain Controls

The synthesis of catabolic enzymes, particularly those whose activity could contribute to the alleviation of starvation conditions, is regulated by general controls in addition to carbon and nitrogen repression. Limitation for phosphate or sulfur can also lead to enzyme synthesis. Little work has been done on sulfur repression in *Aspergillus nidulans.* A putative loss-of-function mutation in the *suAmeth* gene has been isolated that leads to derepression, suggesting negative control (see Arst and Scazzocchio, 1985). Possibly, *suAmeth* is analogous to *scon* (*scon-1*) of *Neurospora crassa,* mutations in which also lead to a sulfur-derepressed phenotype (Burton and Metzenberg, 1971). The *scon-1* gene regulates the levels of transcript of the *cys-3* gene, the major sulfur regulatory gene of *N. crassa* (Marzluf and Metzenberg, 1968; Fu *et al.*, 1989b). The *cys-3* gene product activates the transcription of a variety of enzymes of sulfur uptake and metabolism, and the predicted *cys-3* protein product contains a leucine zipper with homology to v-*fos,* v-*jun* and GCN4 (Fu *et al.*, 1989b). The levels of *cys-3* transcript are regulated in response to the sulfur status of the cells and may be autoregulated (Paietta *et al.*, 1987). Paietta (1990) has described a new locus, *scon-2,* which may have a role in the sulfur regulatory system of *N. crassa.*

From genetic studies, it appears that *A. nidulans* also has control systems involved in phosphate control (*palcA;* Caddick *et al.*, 1986c), as does *N. crassa* (see Littlewood *et al.*, 1974). There are also mechanisms that sense and regulate enzyme synthesis in response to changes in external and internal pH levels (*pal* and *pac* genes; Caddick *et al.*, 1986b). At this time, no molecular studies have been reported on these systems.

VII. INTERSPECIFIC STUDIES OF GENE EXPRESSION AND REGULATION

It is now possible to transfer genes between different fungal species and study their expression and regulation. This allows direct studies of the evolution of control mechanisms and also provides opportunities to obtain a broader perspective on the logic of regulatory circuits. Filamentous fungi represent a wide range of evolutionary divergent organisms and present an excellent resource for such studies. The well-characterized systems of *Aspergillus nidulans* and *Neurospora crassa* can be used to investigate aspects of gene regulation in species in which genetic analysis is not available, particularly in industrial strains and plant pathogens.

There are now many instances of heterologous expression of particular ascomycete structural genes in other ascomycete species (see van den Hondel, Chapter 18). For example, the *amdS* and *niaD* genes of *A. nidulans* have been used widely as heterologous selection markers for transformation, as has the *N. crassa pyr-4* gene (Fincham, 1989). Furthermore, the complex mechanisms affecting regulation of the *A. nidulans amdS* gene appear to be conserved in *A. niger,* despite the apparent absence of an equivalent gene in this species (Kelly and Hynes, 1985; Davis *et al.*, 1988). Regulated expression of the *A. niger* and *A. nidulans niaD* genes in *Penicillium chrysogenum* has been observed (Whitehead *et al.*, 1989). The *aldA* gene of *A. niger,* when introduced into *A. nidulans,* is regulated by the *alcR* gene (O'Connell and Kelly, 1988). Exchange of the structural genes for acetyl CoA-synthetase between *N. crassa* and *A. nidulans* shows conservation of acetate induction control (Connerton *et al.*, 1990).

A few heterologous expression studies in basidiomycetes have been performed. Corresponding *trp* genes from two basidiomycetes (*Schizophyllum commune* and *Phanerochaete chrysosporium*), but not the ascomycete *A. nidulans,* were expressed in *Coprinus cinereus.* In the same study, a promoter from the basidiomycete *Ustilago maydis* was nonfunctional (Casselton and de la Fuerte Herce, 1989). The glucoamylase promoter from *A. niger* functions in *U. maydis* (Smith et al., 1990). The *acu-7* gene, encoding isocitrate lyase, from *C. cinereus* is capable of complementing the equivalent *acuD* gene of *A. nidulans* but is not regulated by acetate induction (Hynes, 1989). The *A. nidulans* gene, however, is not expressed in *C. cinereus* (Casselton and de la Fuerte Herce, 1989). Two different *S. com-*

mune ade genes complement *P. chrysosporium* adenine auxotrophs and one of these complements a *N. crassa ade-2* mutant (Alic *et al.*, 1990).

Lack of expression or regulation of genes in heterologous species can be due to significant differences in promoter sequences or to specificity differences in the transcription apparatus and associated transcription factors. Recent studies on the transcription factor TFIID from a number of organisms indicate considerable conservation of structure, but also reveal important differences in functional specificity, apparently due to previously undetected transcription factors (for review see Lewin, 1990; Ptashne and Gann, 1990). Lack of regulation could be due to evolutionary divergence of regulatory genes or to a complete absence of regulatory mechanisms. By isolating the components of particular regulatory circuits from a range of fungal species and performing cross-species expression experiments, including switching *cis*-acting sequences between genes of different species, it should be possible to identify the important parameters involved.

Sequence comparison of regulatory genes from different species as a means of determining conserved regions of functional significance in DNA binding proteins has been discussed in Sections V and VI. In particular, the comparison of the *nit-2* and *areA* genes, coupled with cross-species *in vivo* functional studies, illustrates the potential of this approach. Switching putative functional domains between regulatory proteins from different species is also likely to be fruitful.

Similarly, comparison of the 5′ sequences of homologous genes from different fungal species is an approach used to determine likely control elements. Islands of conserved sequence in a background of divergent sequence indicate potential binding sites for regulatory proteins. This method is particularly valuable when there is little information about regulatory mechanisms affecting the gene. *Drosophila* studies have shown the utility of this approach. (For examples, see Kassis *et al.*, 1989; Moses *et al.*, 1990.) Comparisons of the 5′ sequences of the *oliC* genes of *A. nidulans* and *A. niger* reveal homologies that have been used to guide deletion analysis of the promoters (Turner *et al.*, 1989). Sequence comparisons of the promoter regions of the *A. nidulans* and *A. niger gpdA* genes encoding glyceraldehyde-3-phosphate dehydrogenase have identified a conserved sequence (the *gpd* box). Subsequent deletion analysis has shown that the *gpd* box is involved in determining the level of gene expression (Punt *et al.*, 1988; Punt *et al.*, 1990). In contrast, no obvious homologies occur between the *trpC* promoters of *A. nidulans* and *A. niger* (Kos *et al.*, 1988).

In general terms, sequence comparisons coupled with analysis of the expression of *in vitro*-generated promoter mutations in different species will be an excellent method for identifying functional sequences across all fungal taxa.

VIII. PERSPECTIVE

The application of modern molecular techniques has produced a wealth of information on how genes are regulated in eukaryotes, revealing the universality of key elements in the eukaryotic transcription apparatus and highlighting the power of genetics in dissecting components of these systems. In this chapter, we outlined the progress made in analyzing gene regulation in *A. nidulans* using combined genetic and molecular approaches. These fungal systems provide excellent avenues for understanding fundamental mechanisms of eukaryotic gene control as well as the evolution of these mechanisms. The genetically well-defined catabolic regulatory circuits described here provide insights into the strategies fungi employ to coordinate metabolism, and into the multiple controls that regulate gene expression.

Links between developmental switches and metabolic controls have been identified. The control of secondary metabolism, asexual and sexual reproduction, and plant pathogenesis by nutritional signals is now accessible to molecular analysis. It is clear that much progress has been made in understanding the mechanisms of gene regulation in particular systems and in developing molecular technology for fungal research since the previous edition of this book was published. These advances have clearly been rewarding, and have started to yield significant results. Increasingly, however, we will return also to classical genetic methods as new data suggest possibilities for isolating new classes of mutants. We can expect a truly integrated molecular genetic approach to contribute much to our understanding of the control of gene expression.

ACKNOWLEDGMENTS

We thank colleagues who sent us information prior to publication and Cheryl Grant for her superb typing. Research in our laboratory is partially supported by the Australian Research Council.

REFERENCES

Adams, T. H., and Timberlake, W. E. (1990). Developmental repression of growth and gene expression in *Aspergillus*. *Proc. Natl. Acad. Sci. U.S.A.* **87,** 5405–5409.

Adams, T. H., Boylan, M. T., and Timberlake, W. E. (1988). *brlA* is necessary and sufficient to direct conidiophore development in *Aspergillus nidulans*. *Cell* **54,** 353–362.

Adams, T. H., Deising, H., and Timberlake, W. E. (1990). *brlA* requires both zinc fingers to induce development. *Mol. Cell. Biol.* **10,** 1815–1817.

Alic, M., Clark, E. K., Kornejay, J. R., and Gold, M. H. (1990). Transformation of *Phanero-*

chaete chrysosporium and *Neurospora crassa* with adenine biosynthetic genes from *Schizophyllum commune*. *Curr. Genet.* **17,** 305–311.

Andrianopoulos, A., and Hynes, M. J. (1988). Cloning and analysis of the positively acting regulatory gene, *amdR,* from *Aspergillus nidulans*. *Mol. Cell. Biol.* **8,** 3532–3541.

Andrianopoulos, A., and Hynes, M. J. (1990). Sequence and functional analysis of the positively acting regulatory gene *amdR* from *Aspergillus nidulans*. *Mol. Cell. Biol.* **10,** 3194–3203.

Anton, I. A., Duncan K., and Coggins, J. R. (1987). A eukaryotic repressor protein, the *qa-1S* gene product of *Neurospora crassa,* is homologous to part of the *arom* multifunctional enzyme. *J. Mol. Biol.* **197,** 367–371.

Arst, H. N. (1976). Integrator gene in *Aspergillus nidulans*. *Nature (London)* **262,** 231–234.

Arst, H. N. (1982). A near terminal pericentric inversion leads to nitrogen metabolite derepression in *Aspergillus nidulans*. *Mol. Gen. Genet.* **188,** 490–493.

Arst, H. N., and Cove, D. J. (1973). Nitrogen metabolite repression in *Aspergillus nidulans*. *Mol. Gen. Genet.* **216,** 111–141.

Arst, H. N., and Scazzocchio, C. (1985). Formal genetics and molecular biology of the control of gene expression in *Aspergillus nidulans*. *In* "Gene Manipulations in Fungi" (J. W. Bennett and L. L. Lasure, eds.), pp. 309–343. Academic Press, New York.

Arst, H. N., Kudla, B., Martinez-Rossi, N., Caddick, M. X., Sibley, S., and Davies, R. W. (1989a). *Aspergillus* and mouse share a new class of 'zinc finger' protein. *Trends Genet.* **5,** 291.

Arst, H. N., Tollervey, D., and Caddick, M. X. (1989b). A translocation-associated loss-of-function mutation in the nitrogen metabolite repression regulatory gene of *Aspergillus nidulans* can revert intracistronically. *Mol. Gen. Genet.* **215,** 364–367.

Arst, H. N., Tollervey, D., Dowzer, C. E. A., and Kelly, J. M. (1990). An inversion truncating the *creA* gene of *Aspergillus nidulans* results in carbon catabolite derepression. *Mol. Microbiol.* **4,** 851–854.

Atkinson, P. W., King, J. A., and Hynes, M. J. (1985). Identification of the *aciA* gene controlled by the *amdA* regulatory gene in *Aspergillus nidulans*. *Curr. Genet.* **10,** 133–138.

Bailey, C., and Arst, H. N. (1975). Carbon catabolite repression in *Aspergillus nidulans*. *Eur. J. Biochem.* **51,** 573–577.

Ballance, D. J., and Turner, G. (1986). Gene cloning in *Aspergillus nidulans:* Isolation of the isocitrate lyase gene (*acuD*). *Mol. Gen. Genet.* **202,** 271–275.

Baum, J. A., Geever, R., and Giles, N. H. (1987). Expression of *qa-1F* activator protein: Identification of upstream binding sites in the *qa* gene cluster and localization of the DNA-binding domain. *Mol. Cell. Biol.* **7,** 1256–1266.

Beri, R., Whittington, H., Roberts, C. F., and Hawkins, A. R. (1987). Isolation and characterization of the positively acting regulatory gene QUTA from *Aspergillus nidulans*. *Nucl. Acids Res.* **19,** 7991–8001.

Beri, R. K., Grant, S., Roberts, C. F., Smith, M., and Hawkins, A. R. (1990). Selective overexpression of the QUTE gene encoding catabolic 3-dehydroquinase in multicopy transformants of *Aspergillus nidulans*. *Biochem. J.* **265,** 337–342.

Breathnach, R., and Chambon, P. (1981). Organisation and expression of eukaryotic split genes coding for proteins. *Ann. Rev. Biochem.* **50,** 349–383.

Bucher, P., and Trifonov, E. N. (1986). Compilation and analysis of eukaryotic POL II promoter sequences. *Nucl. Acids Res.* **14,** 10009–10026.

Burton, E. G., and Metzenberg, R. L. (1971). Novel mutation causing derepression of several enzymes of sulfur metabolism in *Neurospora crassa*. *J. Bacteriol.* **109,** 140–151.

Caddick, M. X., Arst, H. N., Taylor, L. H., Johnson, R. I., and Brownlee, A. G. (1986a).

Cloning of the regulatory gene *areA* mediating nitrogen metabolite repression in *Aspergillus nidulans. EMBO J.* **5,** 1087–1090.

Caddick, M. X., Brownlee, A. G., and Arst, H. N. (1986b). Regulation of gene expression by pH of the growth medium in *Aspergillus nidulans. Mol. Gen. Genet.* **203,** 346–353.

Caddick, M. X., Brownlee, A. G., and Arst, H. N. (1986c). Phosphatase regulation in *Aspergillus nidulans:* Responses to nutritional starvation. *Genet. Res. Camb.* **47,** 93–102.

Caddick, M. X., Hooley, P., and Arst, H. N. (1989). Transcript analysis of the *Aspergillus nidulans* regulatory gene *areA. J. Cell. Biol.* **13E,** 36.

Carlson, M. (1987). Regulation of sugar utilization in *Saccharomyces* species. *J. Bacteriol.* **109,** 4873–4877.

Casselton, L. A., and de la Fuerte Herce, A. (1989). Heterologous gene expression in the basidiomycete fungus *Coprinus cinereus. Curr. Genet.* **16,** 35–40.

Chen, W., and Struhl, K. (1985). Yeast mRNA initiation sites are determined primarily by specific sequences, not by the distance from the TATA element. *EMBO J.* **4,** 3273–3280.

Cohen, B. L. (1973). Regulation of intracellular and extracellular neutral and alkaline proteases in *Aspergillus nidulans. J. Gen. Microbiol.* **79,** 311–320.

Connerton, I. F., Fincham, J. R. S., Sandeman, R. A., and Hynes, M. J. (1990). Sequence comparison and cross-species expression of the acetyl-CoA synthetase genes of the ascomycete fungi, *Aspergillus nidulans* and *Neurospora crassa. Mol. Microbiol.* **4,** 451–460.

Cooper, T. G., Ferguson, D., Rai, R., and Bysani, N. (1990). The *GLN3* gene product is required for transcriptional activation of allantoin system gene expression in *Saccharomyces cerevisiae. J. Bacteriol.* **172,** 1014–1018.

Corrick, C. M., Twomey, A. P., and Hynes, M. J. (1987). The nucleotide sequence of the *amdS* gene of *Aspergillus nidulans* and the molecular characterization of 5′ mutations. *Gene* **53,** 63–71.

Corton, J. C., and Johnston, S. A. (1989). Altering DNA-binding specificity of GAL4 requires sequences adjacent to the zinc finger. *Nature (London)* **340,** 724–727.

Courchesne, W., and Magasanik, B. (1988). Regulation of nitrogen assimilation in *Saccharomyces cerevisiae. J. Bacteriol.* **170,** 708–713.

Cove, D. J. (1979). Genetic studies of nitrate assimilation in *Aspergillus nidulans. Biol. Rev.* **54,** 291–327.

Davis, M. A., and Hynes, M. J. (1987). Complementation of *areA*$^-$ regulatory gene mutations of *Aspergillus nidulans* by the heterologous regulatory gene *nit-2* gene of *Neurospora crassa. Proc. Natl. Acad. Sci. U.S.A.* **84,** 3753–3757.

Davis, M. A., and Hynes, M. J. (1989). Regulatory genes in *Aspergillus nidulans. Trends Genet.* **5,** 14–19.

Davis, M. A., Cobbett, C. S., and Hynes, M. J. (1988). An *amdS–lacZ* fusion for studying gene regulation in *Aspergillus. Gene* **63,** 199–212.

DeBusk, R. M., and Ogilvie, S. (1984). Regulation of amino acid utilization in *Neurospora crassa:* Effect of *nmr-1* and *ms-5* mutations. *J. Bacteriol.* **160,** 656–661.

de Graaff, L., and Visser, J. (1988). Structure of the *Aspergillus nidulans* pyruvate kinase gene. *Curr. Genet.* **14,** 553–560.

Diallinas, G., and Scazzocchio, C. (1989). A gene coding for the uric acid-xanthine permease of *Aspergillus nidulans:* Inactivational cloning, characterization, and sequence of a *cis*-acting mutation. *Genetics* **122,** 341–350.

Dowzer, C. E. A., and Kelly, J. M. (1989). Cloning the *creA* gene from *Aspergillus nidulans:* A gene involved in carbon catabolite repression. *Curr. Genet.* **15,** 457–459.

Doy, C. H., Pateman, J. A., Olsen, J. E., Kane, H. J., and Creaser, E. H. (1985). Genomic clones of *Aspergillus nidulans* containing *alcA,* the structural gene for alcohol dehydrogenase and *alcR,* a regulatory gene for ethanol metabolism. *DNA* **4,** 105–114.

Dunn-Coleman, N. S., Tomsett, A. B., and Garrett, R. H. (1981). The regulation of nitrate assimilation in *Neurospora crassa:* Biochemical analysis of *nmr-1* mutants. *Mol. Gen. Genet.* **182,** 234–239.

Evans, T., and Felsenfeld, G. (1989). The erythroid-specific transcription factor Eryf1: A new finger protein. *Cell* **58,** 877–885.

Felenbok, B., Sequeval, D., Mathieu, M., Sibley, S., Gwynne, D. I., and Davies, R. W. (1988). The ethanol regulon in *Aspergillus nidulans:* Characterization and sequence of the positive regulatory gene *alcR. Gene* **73,** 385–396.

Felenbok, B., Sophianopolou, V., Mathieu, M., Sequeval, D., Kulmburg, P., Diallinois, G., and Scazzocchio, C. (1989). Regulation of genes involved in the utilization of carbon sources in *Aspergillus nidulans. In* "Proceedings of EMBO–Alko Workshop on Molecular Biology of Filamentous Fungi" (H. Nevalainen and M. Penttila, eds.), Vol. 6, pp. 73–83. Foundation for Biotechnical and Industrial Fermentation Research, Helsinki, Finland.

Fincham, J. R. S. (1989). Transformation in fungi. *Microbiol. Rev.* **53,** 148–170.

Fu, Y.-H., and Marzluf, G. A. (1987). Characterization of *nit-2,* the major nitrogen regulatory gene of *Neurospora crassa. Mol. Cell. Biol.* **7,** 1691–1696.

Fu, Y.-H., and Marzluf, G. A. (1988). Metabolic control and autogenous regulation of *nit-3,* the nitrate reductase structural gene of *Neurospora crassa. J. Bacteriol.* **170,** 657–661.

Fu, Y.-H., and Marzluf, G. A. (1990a). Site-directed mutagenesis of the 'zinc finger' DNA-binding domain of the nitrogen-regulatory protein NIT2 of *Neurospora. Molec Microbiol.* **4,** 1847–1852.

Fu, Y.-H., and Marzluf, G. A. (1990b). *nit-2,* the major positive-acting nitrogen regulatory gene of *Neurospora crassa,* encodes a sequence-specific DNA-binding protein. *Proc. Natl. Acad. Sci. U.S.A.* **87,** 5331–5335.

Fu, Y.-H., and Marzluf, G. A. (1990c). *nit-2,* the major nitrogen regulatory gene of *Neurospora crassa* encodes a protein with a putative zinc finger DNA-binding domain. *Mol. Cell. Biol.* **10,** 1056–1065.

Fu, Y.-H., Kneesi, J. Y., and Marzluf, G. A. (1989a). Isolation of *nit-4,* the minor nitrogen regulatory gene which mediates nitrate induction in *Neurospora crassa. J. Bacteriol.* **171,** 4067–4070.

Fu, Y.-H., Paietta, J. V., Mannix, D. G., and Marzluf, G. A. (1989b). *cys-3,* the positive-acting sulphur regulatory gene of *Neurospora crassa,* encodes a protein with a putative leucine zipper DNA-binding element. *Mol. Cell. Biol.* **9,** 1120–1127.

Geever, R. F., Huiet, L., Baum, J. A., Tyler, B. M., Patel, V. B., Rutledge, B. J., Case, M. E., and Giles, N. H. (1989). DNA sequence, organization and regulation of the *qa* gene cluster of *Neurospora crassa. J. Mol. Biol.* **207,** 15–34.

Giles, N. H., Case, M. E., Baum, J., Geever, R., Huiet, L., Patel, V., and Tyler, B. (1985). Gene organization and regulation in the *qa* (quinic acid) gene cluster of *Neurospora crassa. Microbiol. Rev.* **49,** 338–358.

Grant, S., Roberts, C. F., Lamb, H., Stout, M., and Hawkins, A. R. (1988). Genetic regulation of the quinic acid utilization (*QUT*) gene cluster in *Aspergillus nidulans. J. Gen. Microbiol.* **134,** 347–358.

Green, P. M., and Scazzocchio, C. (1985). A cloning strategy in filamentous fungi. *In* "Gene Manipulations in Fungi" (J. W. Bennett and L. L. Lasure, eds.), pp. 345–353. Academic Press, New York.

Gurr, S. J., Hawkins, A. R., Drainas, C., and Kinghorn, J. R. (1986). Isolation and identification of the *Aspergillus nidulans gdhA* gene encoding NADP-linked glutamate dehydrogenase. *Curr. Genet.* **10,** 761–766.

Gurr, S. J., Unkles, S. E., and Kinghorn, J. R. (1988). The structure and organisation of nuclear genes of filamentous fungi. *In* "Gene Structure in Eukaryotic Microbes" (J. R. Kinghorn, ed.), pp. 93–139. IRL Press, Oxford.

Gwynne, D. I., Buxton, F. P., Sibley, S., Davies, R. W., Lockington, R. A., Scazzocchio, C., and Sealy-Lewis, H. M. (1987). Comparison of the *cis*-acting control regions of two coordinately controlled genes involved in ethanol utilization in *Aspergillus nidulans. Gene* **51,** 205–216.

Hamer, J. E., and Timberlake, W. E. (1987). Functional organization of the *Aspergillus nidulans trpC* promoter. *Mol. Cell. Biol.* **7,** 2352–2359.

Harbury, P., and Struhl, K. (1989). Functional distinctions between yeast TATA elements. *Mol. Cell. Biol.* **9,** 5298–5304.

Hariharan, N., and Perry, R. P. (1990). Functional dissection of a mouse ribosomal protein promoter: Significance of the polypyrimidine initiator and an element in the TATA-box region. *Proc. Natl. Acad. Sci. U.S.A.* **87,** 1526–1530.

Hawkins, A. R., Da Silva, A. J. S., and Roberts, C. F. (1985). Cloning and characterization of three enzymes structural genes *QUTB, QUTC* and *QUTE* from the quinic acid utilization gene cluster in *Aspergillus nidulans. Curr. Genet.* **9,** 305–311.

Hawkins, A. R., Lamb, H. K., Smith, M., Keyte, J. W., and Roberts, C. F. (1988). Molecular organisation of the quinic acid utilization (QUT) gene cluster in *Aspergillus nidulans. Mol. Gen. Genet.* **214,** 224–231.

Hawkins, A. R., Gurr, S. J., Montague, P., and Kinghorn, J. R. (1989). Nucleotide sequence and regulation of expression of the *Aspergillus nidulans gdhA* gene encoding NADP dependent glutamate dehydrogenase. *Mol. Gen. Genet.* **218,** 105–111.

Hull, E. P., Green, P. M., Arst, H. N., and Scazzocchio, C. (1989). Cloning and physical characterization of the L-proline catabolism gene cluster of *Aspergillus nidulans. Mol. Microbiol.* **3,** 553–559.

Hynes, M. J. (1973). The effect of lack of a carbon source on nitrate-reductase activity in *Aspergillus nidulans. J. Gen. Microbiol.* **79,** 155–157.

Hynes, M. J. (1974). The effects of a carbon source on glutamate dehydrogenase activities in *Aspergillus nidulans. J. Gen. Microbiol.* **81,** 165–179.

Hynes, M. J. (1978). Multiple independent control mechanisms affecting the acetamidase of *Aspergillus nidulans. Mol. Gen. Genet.* **161,** 59–65.

Hynes, M. J. (1989). Complementation of an *Aspergillus nidulans* mutation by a gene from the Basidiomycete, *Coprinus cinereus. Exp. Mycology* **13,** 196–198.

Hynes, M. J., and Andrianopoulos, A. (1989). Transformation studies of gene regulation in filamentous fungi. *In* "Genetics and Molecular Biology of Industrial Microorganisms" (C. L. Hershberger, S. W. Queener, and G. Hegeman, eds.), pp. 304–309. American Society for Microbiology, Washington, D.C.

Hynes, M. J., and Davis, M. A. (1986). The *amdS* gene of *Aspergillus nidulans:* Control by multiple regulatory signals. *BioEssays* **5,** 123–128.

Hynes, M. J., and Kelly, J. M. (1977). Pleiotropic mutants of *Aspergillus nidulans* altered in carbon metabolism. *Mol. Gen. Genet.* **150,** 193–204.

Hynes, M. J., Corrick, C. M., and King, J. A. (1983). Isolation of genomic clones containing the *amdS* gene of *Aspergillus nidulans* and their use in the analysis of structural and regulatory mutations. *Mol. Cell. Biol.* **3,** 1430–1439.

Hynes, M. J., Corrick, C. M., Kelly, J. M., and Littlejohn, T. G. (1988). Identification of the sites of action for regulatory genes controlling the *amdS* gene of *Aspergillus nidulans. Mol. Cell. Biol.* **8,** 2589–2596.

Hynes, M. J., Andrianopoulos, A., Davis, M. A., van Heejswick, R., Katz, M. E., Littlejohn, T. G., Richardson, I. B., and Saleeba, J. A. (1989). Multiple circuits regulating the *amdS* gene of *Aspergillus nidulans. In* "Molecular Genetics of Filamentous Fungi" (H. Nevalainen and M. Pentilla, eds.), pp. 53–62. Foundation for Biotechnical and Industrial Fermentation Research, Helsinki, Finland.

Johnston, M. (1987a). Genetic evidence that zinc is an essential co-factor in the DNA binding domain of GAL4 protein. *Nature (London)* **328,** 352–355.

Johnston, M. (1987b). A model fungal gene regulatory mechanism: The *GAL* genes of *Saccharomyces cerevisiae. Microbiol. Rev.* **51,** 458–476.

Johnstone, I. L., McCabe, P. C., Greaves, P., Gurr, S. J., Cole, G. E., Brow, M. A. D., Unkles, S. E., Clutterbuck, A. J., Kinghorn, J. R., and Innis, M. A. (1990). Isolation and characterization of the *crnA–niiA–niaD* gene cluster for nitrate assimilation in *Aspergillus nidulans. Gene* **90,** 181–192.

Kassis, J. A., Desplan, C., Wright, D. K., and O'Farrell, P. H. (1989). Evolutionary conservation of homeodomain-binding sites and other sequences upstream and within the major transcription unit of the *Drosophila* segmentation gene engrailed. *Mol. Cell. Biol.* **9,** 4304–4311.

Katz, M. E., and Hynes, M. J. (1989a). Gene function identified by interspecific transformation. *Gene* **78,** 167–171.

Katz, M. E., and Hynes, M. J. (1989b). Molecular cloning and genetic characterization of the *amdR* controlled *lamA* and *lamB* genes of *Aspergillus nidulans. Genetics* **122,** 331–339.

Katz, M. E., and Hynes, M. J. (1989c). Isolation and analysis of the acetate regulatory gene, *facB,* from *Aspergillus nidulans. Mol. Cell. Biol.* **9,** 5696–5701.

Katz, M. E., Saleeba, J. A., Sapats, S. I., and Hynes, M. J. (1990). A mutation affecting *amdS* expression in *Aspergillus nidulans* contains a triplication of a *cis*-acting regulatory sequence. *Mol. Gen. Genet.* **220,** 373–376.

Kelly, J. M. (1980). Pleiotropic mutants of *Aspergillus nidulans* affected in carbon metabolism. Ph.D. Thesis. La Trobe University, Australia.

Kelly, J. M., and Hynes, M. J. (1985). Transformation of *Aspergillus niger* by the *amdS* gene of *Aspergillus nidulans. EMBO J.* **4,** 475–479.

Kelly, J. M., and Hynes, M. J. (1987). Multiple copies of the *amdS* gene of *Aspergillus nidulans* cause titration of *trans*-acting regulatory proteins. *Curr. Genet.* **11,** 21–31.

Kelly, J. M., Drysdale, M. R., Sealy-Lewis, H. M., Gwyn Jones, I., and Lockington, R. A. (1990). Alcohol dehydrogenase III in *Aspergillus nidulans* is anaerobically induced and posttranscriptionally regulated. *Mol. Gen. Genet.* **222,** 323–328.

Kim, K. S., and Guarente, L. (1989). Mutations that alter transcriptional activation but not DNA binding in the zinc finger of yeast activator HAP1. *Nature (London)* **342,** 200–203.

Kim, K. S., Pfeifer, K., Powell, L., and Guarente, L. (1990). Internal deletions in the yeast transcriptional activator HAP1 have opposite effects at two sequence elements. *Proc. Natl. Acad. Sci. U.S.A.* **87,** 4524–4528.

Kos, T., Kuijvenhoven, A., Hessing, H. G. M., Pouwels, P. H., and van den Hondel, C. A. M. J. J. (1988). Nucleotide sequence of the *Aspergillus niger trpC* gene: Structural relationship with analogous genes of other organisms. *Curr. Genet.* **13,** 137–144.

Kudla, B., Caddick, M. X., Langdon, T., Martini-Rossi, N. M., Bennett, C. F., Sibley, S., Davies, R. W., and Arst, H. N. (1990). The regulatory gene *areA* mediating nitrogen metabolite repression in *Aspergillus nidulans.* Mutations affecting specificity of gene activation alter a loop residue of a putative zinc finger. *EMBO J.* **9,** 1355–1364.

Kuger, P., Godecke, A., and Breunig, K. D. (1990). A mutation in the Zn-finger of the GAL4 homolog LAC9 results in glucose repression of its target genes. *Nucl. Acids Res.* **18,** 745–751.

Lamb, H. K., Hawkins, A. R., Smith, M., Harvey, I. J., Brown, J., Turner, G., and Roberts, C. F. (1990). Spatial and biological characterisation of the complete quinic acid utilization gene cluster in *Aspergillus nidulans. Mol. Gen. Genet.* **223,** 17–23.

Lewin, B. (1990). Commitment and activation at pol II promoters: A tail of protein–protein interactions. *Cell* **61,** 1161–1164.

Littlewood, B. S., Chia, W., and Metzenberg, R. L. (1974). Genetic control of phosphate-metabolizing enzymes in *Neurospora crassa:* Relationships among regulatory mutations. *Genetics* **79,** 419–434.

Lockington, R. A., Sealy-Lewis, H. M., Scazzocchio, C., and Davies, R. W. (1985). Cloning and characterization of the ethanol utilizing regulon in *Aspergillus nidulans. Gene* **33,** 137–149.

Lockington, R., Felenbok, B., Sequeval, D., Mathieu, M., and Scazzocchio, C. (1988). Regulation of *alcR,* the positive regulatory gene of the ethanol utilization regulon of *Aspergillus nidulans. Mol. Microbiol.* **1,** 275–281.

MacCabe, A. P., Riach, M. B. R., Unkles, S. E., and Kinghorn, J. R. (1990). The *Aspergillus nidulans npeA* locus consists of three contiguous genes required for penicillin biosynthesis. *EMBO J.* **9,** 279–287.

McKnight, G. L., Kato, H., Upshall, A., Parker, M. D., Saari, G., and O'Hara, P. J. (1985). Identification and molecular analysis of a third *Aspergillus nidulans* alcohol dehydrogenase gene. *EMBO J.* **4,** 2093–2099.

Magasanik, B. (1988). Reversible phosphorylation of an enhancer binding protein regulates the transcription of bacterial nitrogen utilization genes. *TIBS* **13,** 475–479.

Marzluf, G. A., and Metzenberg, R. L. (1968). Positive control by the *cys-3* locus in regulation of sulfur metabolism in *Neurospora crassa. J. Mol. Biol.* **33,** 423–437.

Miller, B. L., Miller, K. Y., Roberti, K. A., and Timberlake, W. E. (1987). Position-dependent and -independent mechanisms regulate cell-specific expression of the SpoC1 gene cluster of *Aspergillus nidulans. Mol. Cell. Biol.* **7,** 427–434.

Mirabito, P. M., Adams, T. H., and Timberlake, W. E. (1989). Interactions of three sequentially expressed genes control temporal and spatial specificity in *Aspergillus* development. *Cell* **57,** 859–868.

Moreno, S., and Nurse, P. (1990). Substrates for p34^{cdc2}: *In vivo veritas. Cell* **61,** 549–551.

Morita, T., Shigesada, K., Kimizuka, F., and Aiba, H. (1988). Regulatory effect of a synthetic CRP recognition sequence placed downstream of a promoter. *Nucl. Acids Res.* **16,** 7315–7322.

Moses, K., Heberlein, U., and Ashburner, M. (1990). The Adh promoters of *Drosophila melanogaster* and *Drosophila orena* are functionally conserved and share features of sequence structure and nuclease-protected sites. *Mol. Cell. Biol.* **10,** 539–548.

Novak, U., Harris, E. A. S., Forrester, W., Groudine, M., and Gelinas, R. (1990). High level *β*-globin expression after retroviral transfer of locus-activation-region-containing human *β*-globin gene derivatives into murine erythroleukemia cells. *Proc. Natl. Acad. Sci. U.S.A.* **87,** 3386–3390.

O'Connell, M. J., and Kelly, J. M. (1988). Differences in the regulation of aldehyde dehydrogenase genes in *Aspergillus niger* and *Aspergillus nidulans. Curr. Genet.* **14,** 95–103.

Paietta, J. V. (1990). Molecular cloning and analysis of the *scon-2* negative regulatory gene of *Neurospora crassa. Mol. Cell. Biol.* **10,** 5207–5214.

Paietta, J. V., Atkins, R. A., Lambowitz, A. M., and Marzluf, G. A. (1987). Molecular cloning and characterization of the *cys-3* regulatory gene of *Neurospora crassa. Mol. Cell. Biol.* **7,** 2506–2511.

Paluh, J. L., Orbach, M. J., Legerton, T. L., and Yanofsky, C. (1988). The cross-pathway control gene of *Neurospora crassa, cpc-1,* encodes a protein similar to GCN4 of yeast and the DNA-binding domain of the oncogene v-*jun*-encoded protein. *Proc. Natl. Acad. Sci. U.S.A.* **85,** 3728–3732.

Pan, T., and Coleman, J. E. (1990). GAL4 transcription factor is not a "zinc finger" but forms a Zn $(II)_2$ Cys 6 binuclear cluster. *Proc. Natl. Acad. Sci. U.S.A.* **87,** 2077–2081.

Pateman, J. A., Doy, C. H., Olsen, J. E., Norris, U., Creaser, E. H., and Hynes, M. (1983). Regulation of alcohol dehydrogenase (ADH) and aldehyde dehydrogenase (AldDH) in *Aspergillus nidulans. Proc. R. Soc. Lond., Series B* **217,** 243–264.

Perrine, K. G., and Marzluf, G. A. (1986). Amber nonsense mutations in regulatory and structural genes of the nitrogen control circuit of *Neurospora crassa. Curr. Genet.* **10,** 677–684.

Pevny, L., Simon, M. C., Robertson, E., Klein, W. H., Tsai, S-F., D'Agati, V., Orkin, S. H., and Costantini, F. (1991). Erythroid differentiation in chimaeric mice blocked by a targeted mutation in the gene for transcription factor GATA-1. *Nature* **439,** 257–260.

Pfeifer, K., Prezant, T., and Guarente, L. (1987). Yeast HAP1 activator binds to two upstream activation sites of different sequences. *Cell* **49,** 19–27.

Pickett, M., Gwynne, D. I., Buxton, F. P., Elliott, R., Davies, R. W., Lockington, R. A., Scazzocchio, C., and Sealy-Lewis, H. M. (1987). Cloning and characterization of the *aldA* gene of *Aspergillus nidulans. Gene* **51,** 217–226.

Polkinghorne, M. A., and Hynes, M. J. (1982). L-Histidine utilization in *Aspergillus nidulans. J. Bacteriol.* **149,** 931–940.

Ptashne, M. (1988). How eukaryotic transcriptional activators work. *Nature (London)* **335,** 683–689.

Ptashne, M., and Gann, A. A. F. (1990). Activators and targets. *Nature (London)* **346,** 329–334.

Punt, P. J., Dingemanse, M. A., Jacobs-Meijsing, B. J. M., Pouwels, P. H., and van den Hondel, C. A. M. J. J. (1988). Isolation and characterization of the glyceraldehyde-3-phosphate dehydrogenase gene of *Aspergillus nidulans. Gene* **69,** 49–57.

Punt, P. J., Dingemanse, M. A., Kuyvenhoven, A., Soede, R. D. M., Pouwels, P. H., and van den Hondel, C. A. M. J. J. (1990). Functional elements in the promoter region of the *Aspergillus nidulans gpdA* gene, encoding glyceraldehyde-3 phosphate dehydrogenase. *Gene.* **93,** 101–109.

Rand, K. N., and Arst, H. N. (1977). A mutation in *Aspergillus nidulans* which affects the regulation of nitrate reductase and is tightly linked to its structural gene. *Mol. Gen. Genet.* **155,** 67–75.

Richardson, I. B., Hurley, S. K., and Hynes, M. J. (1989). Cloning and molecular characterisation of the *amdR* controlled *gatA* gene of *Aspergillus nidulans. Mol. Gen. Genet.* **217,** 118–125.

Sakai, A., Shimizu, Y., Kondou, S., Chibazakura, T., and Hishinuma, F. (1990). Structure and molecular analysis of RGR1, a gene required for glucose repression of *Saccharomyces cerevisiae. Mol. Cell. Biol.* **10,** 4130–4138.

Sandeman, R. A. (1988). The molecular characterization of genes involved in acetate metabolism in *Aspergillus nidulans.* Ph.D. Thesis. University of Melbourne, Australia.

Sandeman, R. A., and Hynes, M. J. (1989). Isolation of the *facA* (acetyl-CoenzymeA synthetase) and *acuE* (malate synthase) genes of *Aspergillus nidulans. Mol. Gen. Genet.* **218,** 87–92.

Shaffer, P. M., Arst, H. N., Estberg, L., Fernando, L., Tran, L. and Sitter, M. (1988). An asparaginase of *Aspergillus nidulans* is subject to oxygen repression in addition to nitrogen metabolite repression. *Mol. Gen. Genet.* **212,** 337–341.

Smale, S. T., and Baltimore, D. (1989). The initiator as a transcription control element. *Cell* **57,** 103–113.

Smale, S. T., Schmidt, M. C., Berk, A. J., and Baltimore, D. (1990). Transcriptional activation by Sp1 as directed through TATA or initiator: Specific requirement for mammalian transcription factor IID. *Proc. Natl. Acad. Sci. U.S.A.* **87,** 4509–4513.

Smith, D. J., Burnham, M. K. R., Bull, J. H., Hodgson, J. E., Ward, J. M., Browne, P., Brown, J., Barton, B., Earl, A. J., and Turner, G. (1990). β-lactam antibiotic biosynthetic genes have been conserved in clusters in prokaryotes and eukaryotes. *EMBO J.* **9,** 741–747.

Smith, T. L., Gaskell, J., Berka, R. M., Yang, M., Henner, D. J., and Cullen, D. (1990). The promoter of the glucoamylase-encoding gene of *Aspergillus niger* functions in *Ustilago maydis. Gene* **88,** 259–262.

Stief, A., Winter, D. M., Stratling, W. H., and Sippel, A. E. (1989). A nuclear DNA attachment element mediates elevated and position-independent gene activity. *Nature (London)* **341,** 343–345.

Struhl, K. (1989). Molecular mechanisms of transcriptional regulation in yeast. *Ann. Rev. Biochem.* **58,** 1051–1077.

Suzuki, M. (1989). SPXX, a frequent sequence motif in gene regulatory proteins. *J. Mol. Biol.* **207,** 61–84.

Talbot, D., Collis, P., Antoniou, M., Vidal, M., Grosveld, F., and Greaves, D. R. (1989). A dominant control region from the human β-globin locus conferring integration site-independent gene expression. *Nature (London)* **338,** 352–355.

Tollervey, D. W., and Arst, H. N. (1982). Domain-wide, locus-specific suppression of nitrogen metabolite repressed mutations in *Aspergillus nidulans. Curr. Genet.* **6,** 79–85.

Townes, T. M., and Behringer, R. R. (1990). Human globin locus activation region (LAR): Role in temporal control. *Trends Genet.* **6,** 219–223.

Trainor, C. D., Evans, T., Felsenfeld, G., and Boguski, M. S. (1990). Structure and evolution of a human erythroid transcription factor. *Nature (London)* **343,** 92–96.

Turner, G., Brown, J., Kerry-Williams, S., Bailey, A. M., Ward, M., Punt, P. J., and van den Hondel, C. A. M. J. J. (1989). Analysis of the *oliC* promoter of *Aspergillus nidulans. In* "Proceedings of the EMBO-Alko Workshop on Molecular Biology of Filamentous Fungi" (H. Nevalainen and M. Penttila, eds.), Vol. 6, pp. 101–109. Foundation for Biotechnical and Industrial Fermentation Research, Helsinki, Finland.

van Gorcom, R. F. M., Punt, P. J., Pouwels, P. H., and van den Hondel, C. A. M. J. J. (1986). A system for the analysis of expression signals in *Aspergillus. Gene* **48,** 211–217.

Whitehead, M. P., Unkles, S. E., Ramsden, M., Campbell, E. I., Gurr, S. J., Spence, D., van den Hondel, C., Contreras, R., and Kinghorn, J. R. (1989). Transformation of a nitrate reductase deficient mutant of *Penicillium chrysogenum* with the corresponding *Aspergillus niger* and *A. nidulans niaD* genes. *Mol. Gen. Genet.* **216,** 408–411.

Whittington, H., Grant, S., Roberts, C. F., Lamb, H., and Hawkins, A. R. (1987). Identification and isolation of a putative permease gene in the quinic acid utilization (QUT) gene cluster of *Aspergillus nidulans. Curr. Genet.* **12,** 135–19.

Wobbe, C. R., and Struhl, K. (1990). Yeast and human TATA-binding proteins have nearly identical DNA sequence requirements for transcription *in vitro. Mol. Cell. Biol.* **10,** 3859–3867.

Wray, L. V., Witte, M. M., Dickson, R. C., and Riley, M. I. (1987). Characterization of a positive regulatory gene, *LAC9,* that controls induction of the lactose-galactose regulon of *Kluyveromyces lactis:* Structural and functional relationships to *GAL4* of *Saccharomyces cerevisiae. Mol. Cell. Biol.* **7,** 1111–1121.

IV

Meiosis, Mating Types, and Other Mysteries of Fungal Sex

8

Ascomycete Mating Types

N. LOUISE GLASS
Department of Botany and
Biotechnology Laboratory
University of British Columbia
Vancouver, British Columbia
Canada

IAN A. J. LORIMER
Department of Biochemistry and
Biotechnology Laboratory
University of British Columbia
Vancouver, British Columbia
Canada

I. INTRODUCTION

This chapter compares the mating systems of the unicellular yeasts *Saccharomyces cerevisiae* and *Schizosaccharomyces pombe* and the multicellular filamentous fungus *Neurospora crassa*. These fungi are members of the Ascomycotina, but nevertheless are believed to be highly divergent evolutionarily. For example, it has been suggested that the extent of the similarity between *S. pombe* and *S. cerevisiae* is the same as that between each yeast and humans (Russell and Nurse, 1986). Each of these organisms has two haploid mating types that recognize each other and fuse to form a new cell type capable of undergoing karyogamy, meiosis, and sporulation. In each of the three fungi, mating type is under the control of a master mating-type locus (Table I). The focus of review is a comparison of the structure of the mating-type locus in each organism and the manner in which the products of each locus control cell type. A role for chromatin features associated with repression of the silent loci in *S. cerevisiae* and *S. pombe* and with expression of the mating-type locus in *N. crassa* will be presented; however, mating-type switching will not be discussed in detail.

Despite differences in their growth habits, both the coenocytic *N. crassa* and the unicellular yeasts must cope with similar problems during their life cycles. First, each individual must be able to recognize and respond to an individual of the opposite mating type. Presumably, this requires the coordinated regulation of a set of mating-type-specific genes. Second, after the two compatible individuals have recognized each other and fused, they must be able to form a new cell type that is no longer able to fuse with additional partners, but is able to undergo meiosis and sporulation. This requires that the mating-type-specific genes no longer be expressed and that a new regulatory function be present that can trigger sporulation under appropriate conditions. The following sections summarizes the current knowledge of how each organism uses the information present at its mating-type locus to solve these problems.

TABLE I

Functions of the Mating Type Loci in Yeasts and Filamentous Fungi

	S. cerevisiae	*S. pombe*	*N. crassa*
Mating type switching	+	+	−
Mating type control	+	+	+
Sporulation	+	+	+
Meiosis	+	+	+
Heterokaryon incompatibility	−	−	+

II. *SACCHAROMYCES CEREVISIAE*

A. Life Cycle of *Saccharomyces cerevisiae*

Saccharomyces cerevisiae is a unicellular yeast that grows vegetatively by budding and divides every 100 min under optimal conditions (Fig. 1; Herskowitz and Oshima, 1981; Herskowitz, 1988, 1989). Individual cells, if haploid, are of either **a** or **α** mating type (for life cycle, see Fig. 2). A cell of a particular mating type signals its presence to a cell of the opposite mating type by producing mating-type-specific pheromones. The signalling molecule produced by an **a** cell is a peptide (**a** factor) 12 amino acids (aa) in length (Betz *et al.*, 1987) that is extensively modified posttranslationally (Anderegg *et al.*, 1988) and then secreted from the **a** cell via a cell-type-specific secretion system (McGrath and Varshavsky, 1989). An **α** cell has a receptor (the STE3 protein) that specifically binds **a** factor (Hagen *et al.*, 1986). Similarly, **α** cells secrete **α** factor, a 13-aa peptide (Stotzler *et al.*, 1976); **a** cells contain a receptor, the product of the *STE2* gene, specific for **α** factor (Blumer *et al.*, 1988). Binding of the pheromones to their respective receptors on the membrane of the opposite cell type (**a** factor to STE3 product in **α** cells and **α** factor to STE2 product in **a** cells) induces a cascade of events that leads, in each case, to the transient arrest of the cell cycle in the G_1 phase (Bucking-Throm *et al.*, 1973). This arrest insures that only a single copy of the **a** and **α** genome are present at the time of cell fusion. In addition to causing G_1 arrest, binding of the pheromone to its receptor induces the production of cell-surface-bound glycoproteins (McCaffrey *et al.*, 1987; Trueheart *et al.*, 1987) and agglutinins that aid in bringing the cells into stable association (conjugation), followed by the breakdown and reformation of cell membranes and cell walls around what has become a single cell (cytogamy). After cytogamy, nuclear fusion (karyogamy) occurs promptly to form a diploid cell, the form in which *S. cerevisiae* most commonly grows in nature. The diploid cell neither produces nor responds to pheromones and is incapable of mating; it thus represents a unique cell type. Diploids, like haploids, are capable of unlimited vegetative growth. However, upon starvation, the diploid cell stops mitotic growth and proceeds through meiosis. Four haploid ascospores are produced within a single ascus (Fig. 1), two of which are mating-type **a** and two of which are mating-type **α**.

B. Organization and Role of the Mating-Type Locus

The organization and function of the mating-type locus in *Saccharomyces cerevisiae* has been extensively characterized. A great deal is understood about the molecular basis of regulatory decisions during the sexual

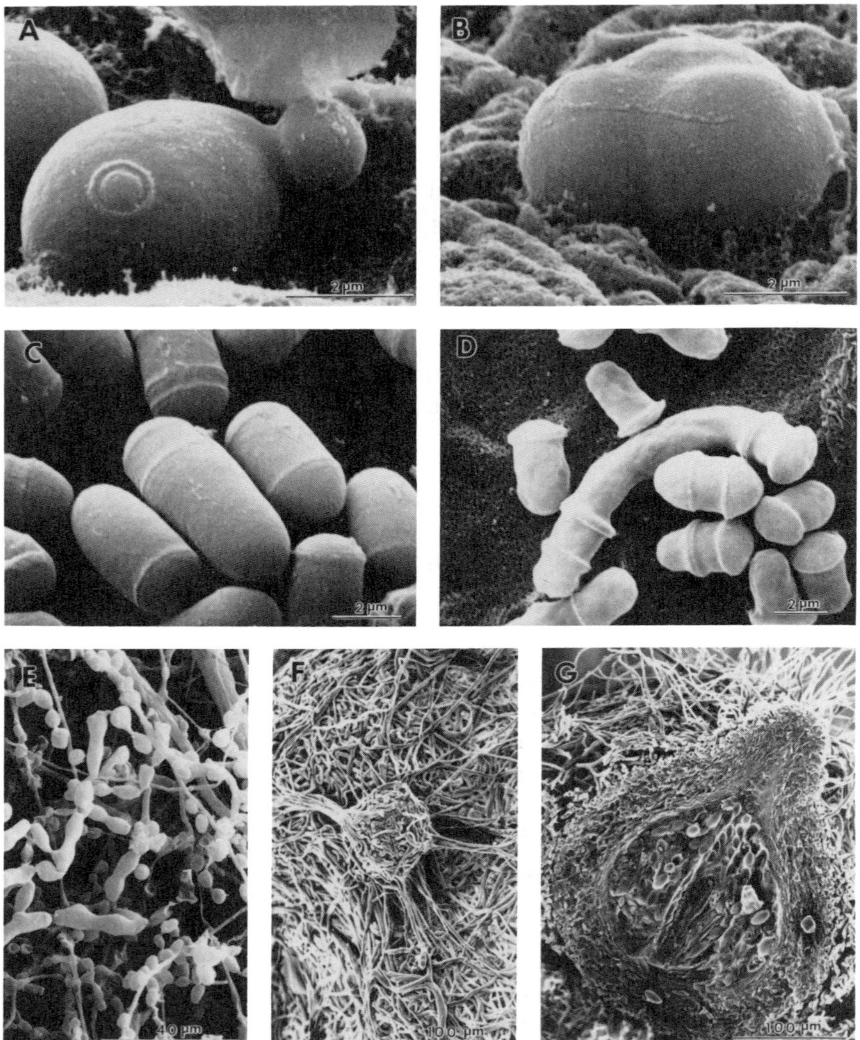

Fig. 1. Scanning electron micrographs of *S. cerevisiae, S. pombe,* and *N. crassa* at different stages of their life cycles. (A) Budding haploid *S. cerevisiae* cell. (B) *S. cerevisiae* ascus. (C) Haploid *S. pombe* cells. (D) *S. pombe* ascus. (E) Vegetative colony of *N. crassa.* (F) *N. crassa* protoperithecium. (G) *N. crassa* perithecium. Micrographs were taken by Lynn Lee.

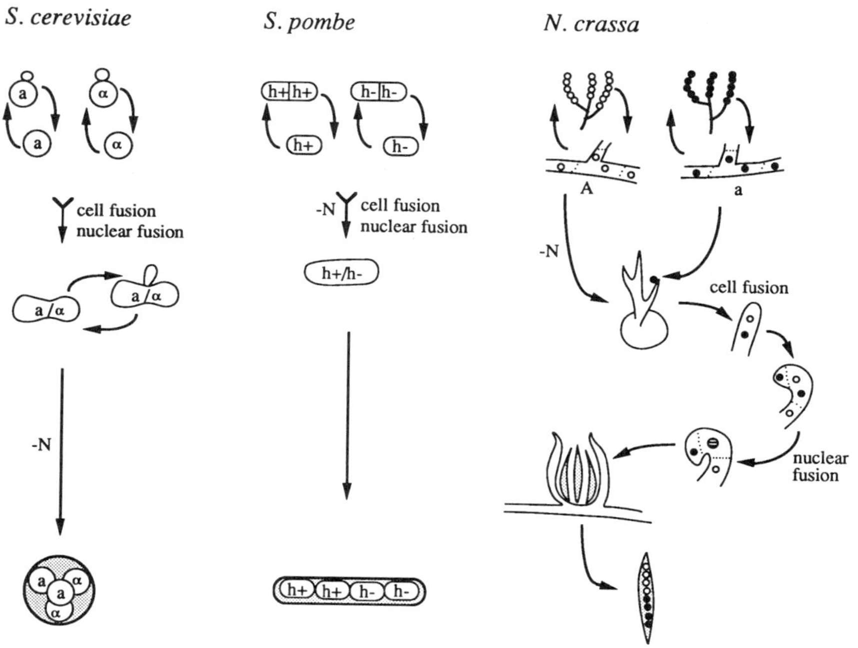

Fig. 2. Schematic representations of the life cycles of *Saccharomyces cerevisiae, Schizosaccharomyces pombe,* and *Neurospora crassa.* Haploid *S. cerevisiae* of either **a** or **α** mating type divide by budding. When two cells of the opposite mating type meet, they fuse to form a diploid **a**/**α** cell. This cell can also replicate vegetatively by budding. Upon starvation for nitrogen (−N), the diploid cell undergoes meiosis and sporulation to form an ascus containing four haploid cells, two of each mating type. In *S. pombe,* haploid cells divide by medial fission. Mating between P and M mating types only occurs upon starvation. The diploid cell does not normally replicate vegetatively, instead it immediately undergoes meiosis and sporulation to form an ascus containing four haploid cells. *N. crassa* is a coenocytic organism (i.e., it contains more than one nucleus per cell). Fusion between cells of opposite mating type during vegetative growth is prevented by the heterokaryon incompatibility response. Upon starvation for nitrogen, either *A* or *a* colonies can form female reproductive structures known as protoperithecia. (In this diagram, *A* cells are shown forming a protoperithecium.) Male cells of the opposite mating type fuse with specialized hyphae on the protoperithecia. The opposite mating type nuclei then go through 8–10 synchronous divisions before fusing and undergoing meiosis. Meiosis is followed by a single mitotic division, to give rise to an ascus containing eight haploid cells, four of each mating type. The asci are contained within a structure called a perithecium.

phase. The following discussion of mating type in *S. cerevisiae* is not meant to be comprehensive, but will cover critical regulatory mechanisms to allow a comparative analysis with mating in *Schizosaccharomyces pombe* and *Neurospora crassa* (Table I).

The mating-type locus (MAT) of *S. cerevisiae* is located near the centromere on chromosome III (Mortimer and Hawthorne, 1969; Fig. 3). The sequences that confer **a** and **α** mating type are completely dissimilar and therefore probably do not have a common evolutionary ancestry (Nasmyth

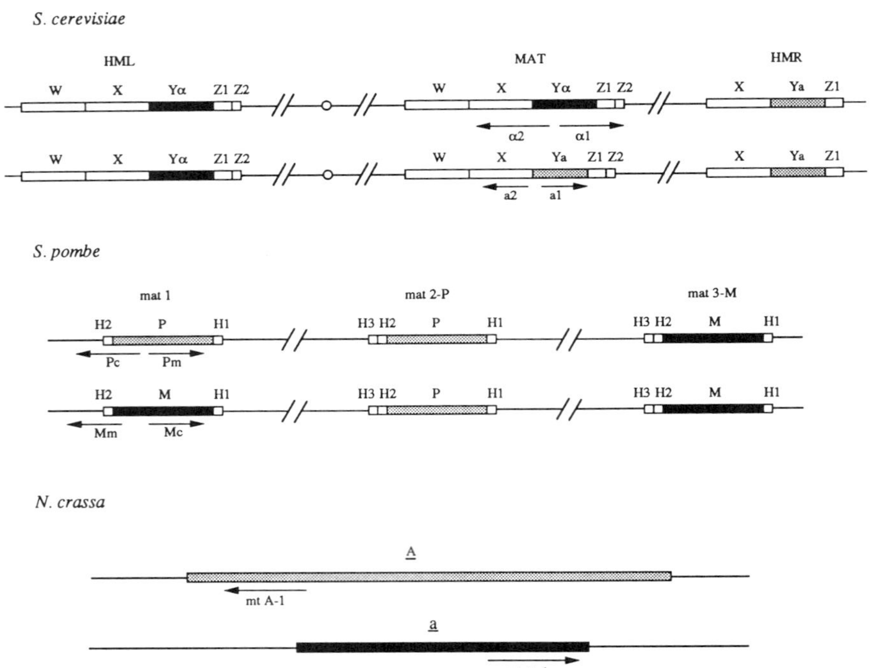

Fig. 3. Structural organization of the mating loci of *S. cerevisiae, S. pombe,* and *N. crassa.* In *S. cerevisiae* **α**-mating-type cells (upper line), the sequence Yα is present at the MAT locus. In *S. cerevisiae* **a**-mating-type cells (lower line), the sequence Y**a** is present at the MAT locus. Both cell types contain a silent copy of the Y**α** sequence at the HML locus and a silent copy of the Y**a** sequence at the HMR locus. In *S. pombe,* P cells (upper line) contain the sequence P at the mat 1 locus, whereas M cells (lower line) contain the sequence M at the mat 1 locus. Both cell types contain silent copies of P and M at the mat 2-P and mat 3-M locus, respectively. *N. crassa* cells contain either an *A* sequence (*A* cells) or an *a* sequence (*a* cells) at the mating locus. Each cell contains only one copy of either *A* or *a* per haploid genome. Identified transcripts encoded by the mating loci are designated by arrows.

and Tatchell, 1980; Strathern *et al.*, 1980; Astell *et al.*, 1981). The term "idiomorph" has been introduced to denote sequences that occupy the same locus on a chromosome but are not related in structure or (probably) descent (Metzenberg and Glass, 1990). The **a** idiomorph is composed of 642 bp of unique sequence (Y**a**); the α idiomorph is composed of 747 bp (Yα). Archival copies of both **a** and α information are also present near the left and right telomeres of chromosome III, at loci designated *HML* and *HMR*, respectively (Fig. 3). Flanking the *MAT* locus are sequences, termed X (704 bp) and Z1 (239 bp), that are also present at *HMR* and *HML*. Two additional identical segments, W (723 bp) and Z2 (88 bp), are present on either side of *MAT* and *HML*. Although *S. cerevisiae* contains copies of both **a** and α idiomorphs within the haploid cell, functional mating type is conferred by the sequence (either **a** or α) that is present at the *MAT* locus; the **a** and α sequences present at *HMR* and *HML* are not transcribed and do not contribute to the mating characteristics of the cell (Klar *et al.*, 1981; Nasmyth *et al.*, 1981). These loci serve as repositories of mating-type information that usually can be activated only by transposition to *MAT*.

Repression of the silent mating-type cassettes in *S. cerevisiae* is accomplished by at least four genes (Ivy *et al.*, 1986; Rine and Herskowitz, 1987). *HML* and *HMR* are maintained in a transcriptionally inactive state by the products of three genes: *SIR2, SIR3,* and *SIR4*. Inactivation of *SIR2, SIR3,* or *SIR4* allows expression of the mating-type genes at *HML* and *HMR* at a level equivalent to that of the genes present at *MAT*. A fourth gene, *SIR1,* is required for full repression of *HML* and *HMR*. The action of the four *SIR* gene products is most likely mediated through *cis*-acting sites (E and I) termed "silencers" that flank *HML* and *HMR* (Abraham *et al.*, 1984; Feldman *et al.*, 1984; Mahoney and Broach, 1989). One model for silencer action postulates the formation of looped DNA that spans the silent cassettes, anchored by attachments to the silencer regions E and I (Nasmyth, 1982; Hofmann *et al.*, 1989). This loop would restrict expression in the region spanning E and I, and could possibly be an integral part of the silencing mechanism. Alternatively, E and I could serve as attachment points for proteins that cause an altered chromatin structure and constrain expression of adjacent sequences.

Two divergent transcripts from *MAT*α have been identified as *MAT*α*1* and *MAT*α*2* (Fig. 3; Tatchell *et al.*, 1981). Both transcripts originate in the unique Yα region, but the *MAT*α*1* transcript terminates in the Z2 region whereas *MAT*α*2* terminates in the X region. Thus, the *MAT*α*1* and *MAT*α*2* transcripts are composed of unique Yα sequences and X, Z1, and Z2 sequences that are common to both *MATa* and *MAT*α. Two transcripts have also been identified from *MATa* (Fig. 3). *MATa1* originates and terminates in the unique Y**a** region. *MATa2* starts and terminates in the conserved X region. (Y**a** may provide part of the promoter sequences for this transcript.) The latter transcript has no known function in conjugation or sporulation.

During vegetative growth, yeast cells of one mating type are programmed to respond to cells of the opposite mating type; fusion only occurs between cells of opposite mating type. The **α** cell type is conferred by a positive function encoded by *MATα1*. MATα1 is a positive activator of **α**-specific genes, such as the genes for **α** factor (*MFα1* and *MFα2*) and for the receptor for **a** factor (*STE3* product; see Fig. 4; Sprague *et al.*, 1983). *matα1* mutants are defective in mating and do not exhibit an **α**-specific phenotype. MATα1 is a basic 175-aa polypeptide that forms a complex with a transcriptional factor MCM1 (also referred to as PRTF and GRM) (Passmore *et al.*, 1988). This complex binds to the upstream regulatory region of **α**-specific genes and activates their transcription (Bender and Sprague, 1987; Ammerer, 1990). *matα1* mutants do not express **α**-specific functions because MCM1 by itself is incapable of activating the transcription of **α**-specific genes, although MCM1 apparently binds to the upstream activating sequences (UASs) of the **α**-specific genes (Tan *et al.*, 1988; Tan and Richmond, 1990). Although MATα1 is required for the **α** cell type and for conjugation, it clearly is not required for meiosis, because *matα1/MATa* diploids sporulate normally (Strathern *et al.*, 1981).

In **a** cells, MCM1 is capable of binding to a site in the upstream regulatory region of **a**-specific genes, such as the genes for **a** factor (*MFA1* and *MFA2*) and **α**-factor receptor (*STE2*), and of stimulating transcription (Tan *et al.*, 1988; Tan and Richmond, 1990). Thus, **a** cells do not require a mating-type-specific factor for activation of the **a**-specific genes (Fig. 4). *MAta1* mutations have no effect on the transcription of **a**-specific genes and *mata1* mutants conjugate normally. In summary, MCM1 is present in all cell types, but its specificity for activating transcription depends on the presence or absence of accessory polypeptides, such as MATα1, that allow it to function at **α**-specific genes.

Since transcription of **a**-specific genes occurs without the need for specific activation, these genes must be repressed in **α** cells. This repression is accomplished by MATα2 (Wilson and Herskowitz, 1984; Johnson and Herskowitz, 1985), which has a region of amino acid sequence similar to the homeodomain of *Drosophila melanogaster Antennapedia* protein and other related proteins (Shepherd *et al.*, 1984). MATα2 interacts directly with a region of dyad symmetry upstream of **a**-specific genes (Sauer *et al.*, 1988). The two symmetrical halves of the operators to which MATα2 binds are on either side of the MCM1 binding site (Kronstad *et al.*, 1987;

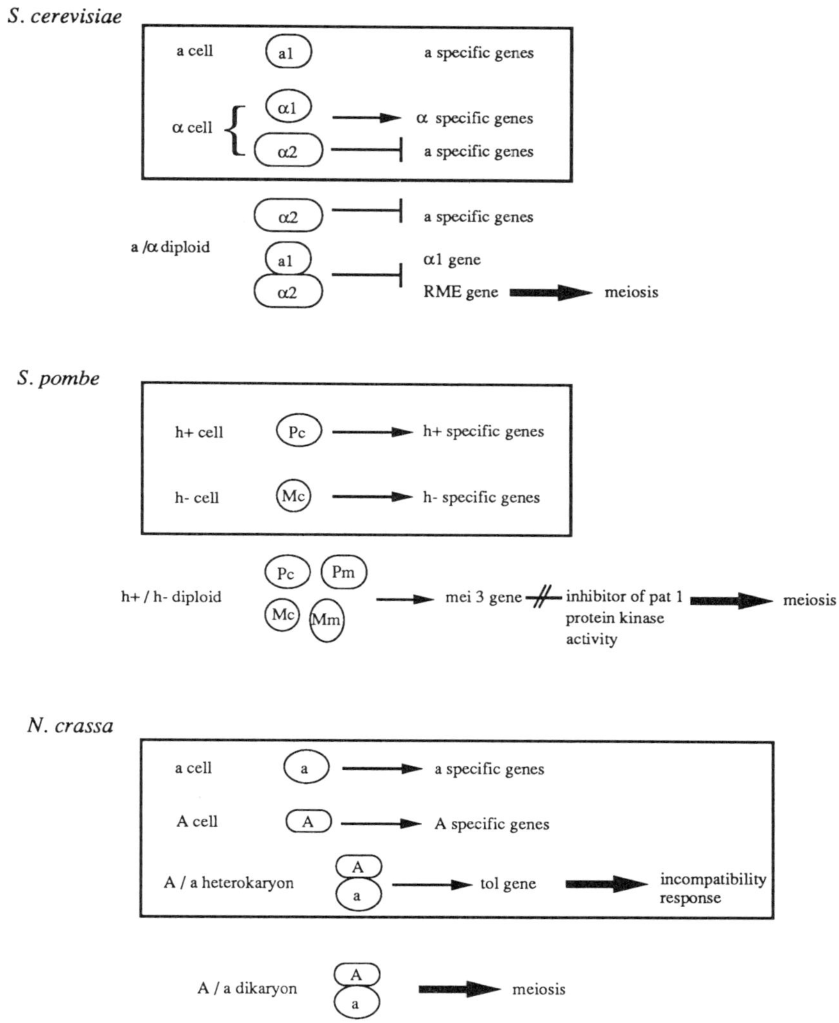

Fig. 4. Models for the mode of action of products of the mating type loci in *S. cerevisiae, S. pombe,* and *N. crassa.* Proteins expressed from the mating type loci in each cell type for each organism are shown, as are their target genes. Activation of gene expression is shown by thin arrows. Inhibition is shown by lines with vertical bars at the end. Thick arrows show the overall biological response initiated by repression or activation of a particular gene. The activities of the products of the *S. pombe* and *N. crassa* mating type loci shown here are hypothetical and have not been directly demonstrated.

Keleher *et al.*, 1988). The binding of MATα2 at these sites prevents activation of transcription of **a**-specific genes by MCM1 (Fig. 4). *matα2* mutants display some characteristics associated with **a** cells and mate, albeit very inefficiently, with $\boldsymbol{\alpha}$ cells (Strathern *et al.*, 1981). Thus, **a** cells exhibit mating-type **a** because they lack MATα2 and can therefore express **a**-specific genes.

Although *mata1* mutants are unaffected in conjugation, *mata1/MATα* cells do not sporulate, indicating that MATa1 is required for meiosis following cell and nuclear fusion (Kassir and Simchen, 1976; Klar *et al.*, 1979). Similarly, *matα2/MATa* strains do not sporulate, which indicates that MATα2 is also required for meiosis. It has been shown that a novel regulatory activity, referred to as **a**1 – $\boldsymbol{\alpha}$2, is formed in a diploid cell, and that it is the "molecular signature" of the unique **a**/$\boldsymbol{\alpha}$ cell type (Goutte and Johnson, 1988). Only this cell type is capable of meiosis and sporulation. The **a**1 and $\boldsymbol{\alpha}$2 polypeptides function as a heterodimer (Miller *et al.*, 1985; Goutte and Johnson, 1988); an association between these two polypeptides has been reported (Herskowitz, 1989; Dranginis, 1990). This **a**1 – $\boldsymbol{\alpha}$2 activity represses the transcription of *MATα1* (thus repressing $\boldsymbol{\alpha}$-specific genes) and a large array of haploid-specific genes, such as *HO, KAR1,* and *RME* (Fig. 4; Miller *et al.*, 1985). The *RME* gene encodes a repressor of meiosis; repression of *RME* transcription by **a**1 – $\boldsymbol{\alpha}$2 causes the onset of meiosis when cells are nutritionally starved (Kassir and Simchen, 1976; Mitchell and Herskowitz, 1986). Thus, α2 has two functions in a diploid cell. As a homodimer it represses **a**-specific genes, and in association with **a**1 it represses *MATα1* and haploid-specific genes.

In summary, *S. cerevisiae* uses associations of the mating-type polypeptides with each other and with MCM1 to generate regulators that ultimately control the diverse cell-type functions associated with **a**, $\boldsymbol{\alpha}$ and **a**/$\boldsymbol{\alpha}$ cells (Fig. 4). In the haploid cell, the mating-type genes of *S. cerevisiae* encode polypeptides that are components of site-specific DNA-binding proteins. These proteins bind to regulatory regions of cell-type-specific target genes. These polypeptides do not act on their own to produce the cell-type-specific pattern of gene expression, but interact with the protein MCM1, which is present in all cell types. In the diploid cell, a novel regulatory factor composed of polypeptides encoded by *both* mating types is responsible for conferring the unique diploid cell type. The association of polypeptides specified by opposite mating types, and the capacity of this protein complex to affect transcription of mating-specific and haploid-specific genes, signals the organism that fusion of opposite mating types has occurred. Finally, meiosis in the diploid cell is triggered by nutritional starvation (reviewed by Malone, 1990).

III. *SCHIZOSACCHAROMYCES POMBE*

A. Life Cycle of *Schizosaccharomyces pombe*

Schizosaccharomyces pombe is a unicellular yeast that is distinguished from *Saccharomyces cerevisiae* by its method of vegetative reproduction. Cells grow by tip extension and divide by medial fission to form two equal-sized daughter cells (reviewed by Egel, 1989; Egel *et al.*, 1990; see Fig. 1). Haploid cells can be of two mating types: P (plus) and M (minus) (Leupold, 1950, 1958). Unlike *S. cerevisiae,* in which conjugation occurs in rich medium and sporulation is induced by starvation, both conjugation and sporulation are induced by starvation in *S. pombe* (for life cycle, see Fig. 2). Under starvation conditions, exposure of cells of opposite mating type to one another induces the secretion of pheromones that cause the elongation of these cells toward each other (Leupold, 1987). The cells become arrested in the G_1 phase of the cell cycle and agglutinate. Following cell fusion, karyogamy and meiosis occur immediately, resulting in the formation of a four-spored ascus (Fig. 1), of which two spores are M mating type and two are P mating type. Thus, unlike *S. cerevisiae,* the diploid stage in *S. pombe* is transient and the organism normally reproduces vegetatively only as a haploid, although the diploid stage can be propagated in the laboratory by manipulating growth conditions (Egel, 1971, 1973).

B. Organization and Role of the Mating Type Locus

The overall organization of the mating-type locus in *S. pombe* is similar to that of *S. cerevisiae* (Fig. 3; Egel and Gutz, 1981; Beach, 1983; Beach and Klar, 1984; Kelly *et al.*, 1988). The *mat1* locus of *S. pombe* is present on the right arm of chromosome II. The mating type of a given cell is determined by the mating-type sequence that is inserted at *mat1* (*mat1-P* or *mat1-M*), the only site at which expression can occur. The *P* allele is composed of 1104 bp; *M* is composed of 1128 bp. As in *S. cerevisiae,* the two mating-type alleles of *S. pombe* are not similar in sequence and, thus, probably reflect separate ancestry; they also are considered idiomorphs. Two additional loci are also present on chromosome II—*mat2* and *mat3*—that contain archival copies of *M* and *P* sequences. A 15-kb segment of DNA separates *mat2* from *mat1* and another 15 kb separates *mat2* from *mat3*. Surrounding the *mat2, mat3,* and *mat1* loci are short regions of homology: H1 (59 bp) and H2 (135 bp). An additional sequence, H3 (57 bp), is present only at *mat2* and *mat3*.

As in *S. cerevisiae,* the archival copies of mating-type information at *mat2* and *mat3* are not expressed in haploid or diploid cells. Chromatin structure may play a role in repressing transcription of the silent *mat* loci because the 15-kb region between *mat2-P* and *mat3-M* is inaccessible to meiotic recombination (Egel, 1984). The existence of a mutant (*rik1*), in which there is increased recombination within the spacer region, indicates that the absence of recombination between *mat2-P* and *mat3-M* is not an intrinsic property of the region but may be imposed by the *rik1*$^+$ protein (Egel, 1989; Egel *et al.*, 1989). Whether mutants derepressed for expression of *mat2-P* and/or *mat3-M* can arise in a *rik1* background has not been tested.

Two transcripts (*Pc* and *Mc*; c = conjugation), one from each idiomorph, are detectable under vegetative growth conditions (Fig. 3; Kelly *et al.*, 1988). When opal mutations generated *in vitro* in *Pc* are reintroduced into *S. pombe* P strains by gene replacement, the resulting strains are unable to conjugate with either P or M strains. Similarly, frameshift mutations generated in *Mc* result in strains unable to mate. Thus, *Pc* and *Mc* appear to encode positive regulators that are responsible for conferring both mating-type and conjugation functions (Fig. 4). Some of the presumed target genes of Pc and Mc include genes that control the production of pheromones and their respective receptors. The predicted polypeptides encoded by *Pc* and *Mc* (118 aa and 181 aa, respectively) are basic, a characteristic of DNA-binding proteins. The function of *Pc* and *Mc* products in the haploid state is proposed to be analogous to that of MATα1, as transcriptional activator of cell-specific genes, although similarities between these regulators are not apparent at the amino acid level and biochemical evidence for Pc and Mc function has not been established.

Although the *Pc* and *Mc* transcripts are detectable under vegetative conditions, an increase in message level is observed when cells are shifted to nitrogen-limited medium. Two additional transcripts, one from *mat1-M* (*mat1-Mm;* m = meiosis) and one from *mat1-P* (mat1-*Pm*), are induced by starvation. The polypeptides corresponding to these two transcripts apparently do not have a role in conjugation, but are required for meiosis and sporulation; *Mc*$^+$*Mm*$^+$ × *Pc*$^+$*Pm*$^-$ and *Mc*$^+$*Mm*$^-$ × *Pc*$^+$*Pm*$^+$ strains conjugate normally, but the diploid zygotes fail to sporulate. The *Mm*-encoded polypeptide is 42 aa in length and bears no resemblance to **a**1, **α**1, or **α**2 of *S. cerevisiae.* However, the carboxy-terminal portion of the 159-aa *Pm* polypeptide has limited, but probably significant, homology with the homeodomain of MATα2, suggesting that is may have similar DNA-binding properties. Whereas MATα2 represses transcription of the **a**-specific genes and **a**1 – **α**2 represses transcription of haploid-specific genes, the *Pm* polypeptide apparently is a transcriptional activator required for expression of a meiosis-specific gene, *mei3*$^{+}$ (see subsequent text; Iino and Yamamoto, 1985; McLeod *et al.*, 1987).

Pc is required for conferring upon P cells their characteristic mating behavior and their ability to conjugate. In addition, *Pc* is also required for meiosis and sporulation. A heterothallic h^{-} (*Mc*$^{+}$*Mm*$^{+}$) strain transformed with only a *Pm*$^{+}$ plasmid fails to enter meiosis (Kelly *et al.*, 1988), whereas the same strain transformed with both *Pc*$^{+}$ and *Pm*$^{+}$ plasmids forms abundant azygotic asci. Evidently the *Mc* product is also required for meiosis and sporulation, because a *Mc*$^{-}$*Mm*$^{+}$ strain transformed with the *Pc*$^{+}$ and *Pm*$^{+}$ plasmids is unable to sporulate. Therefore, unlike MATα1, the *Pc* and *Mc* polypeptides appear to have functions during both the haploid and diploid phases.

In *S. cerevisiae,* the organism recognizes the fusion of opposite mating types by the formation of the novel regulatory factor **a**1$-\alpha$2. **a**1$-\alpha$2 represses the transcription of haploid-specific genes, one of which is *RME* (repressor of meiosis). Removal of RME (under appropriate starvation conditions) allows the cell to enter meiosis. In *S. pombe,* entry into the sexual phase requires a positive function encoded by *Pm.* The RNA transcript for *mei3*$^{+}$ and its translational product are only expressed in *mat1-M*$^{+}$/*mat1-P*$^{+}$ diploids entering meiosis; a null allele of *mei3* is arrested before the initiation of premeiotic DNA synthesis (McLeod *et al.*, 1987). In diploid strains containing a mutation in *Pm, Mc*$^{+}$*Mm*$^{+}$/*Pc*$^{+}$*Pm*$^{-}$ B102, the transcription of *mei3*$^{+}$ is not induced. From this result it is inferred that the *Pm* polypeptide is a transcriptional activator required for the induction of *mei3*$^{+}$ transcription, although this activity has not been shown biochemically. The *mei3*$^{+}$ product binds to a protein kinase encoded by the *pat1*$^{+}$ gene (also referred to as *ran1*$^{+}$), inhibiting its enzymatic activity (McLeod and Beach, 1988). The *pat1*$^{+}$ protein kinase is a negative regulator of meiosis, and its inactivation promptly commits the cell to meiosis. In both organisms, the final commitment to meiosis is under negative control, but in *S. cerevisiae* this control is exerted at the level of transcriptional repression of *RME,* whereas in *S. pombe* it is exerted at the level of modification of protein *pat1*$^{+}$.

In summary, *S. pombe* has two regulatory molecules that determine the haploid mating-type state and four regulatory factors that are required for sporulation, one of which may be a positive regulator of a meiosis-specific gene (Fig. 4). It is possible that unique combinatorial associations between the *S. pombe Pm, Mm, Pc,* and *Mc* polypeptides signal that conjugation has occurred and meiosis can proceed. However, such associations between any *S. pombe* mating-type polypeptides have not been demonstrated at the biochemical level, nor have any of the mating-type products been shown to be involved directly in transcriptional activation or repression.

The elucidation of the molecular mechanism of mating-type regulation in *S. pombe* remains an important aspect of the analysis of mating-type function.

IV. *NEUROSPORA CRASSA*

A. Life Cycle of *Neurospora crassa*

Unlike *Saccharomyces cerevisiae* and *Schizosaccharomyces pombe, Neurospora crassa* is a coenocytic (many nuclei per cell) filamentous fungus that grows as a single individual with genetically identical nuclei. Fusion occurs readily between hyphae within an individual colony. However, productive vegetative fusion between genetically nonidentical colonies is precluded by heteroallelism at any of numerous heterokaryon-incompatibility (*het*) loci. Similar to *S. cerevisiae* and *S. pombe, N. crassa* has two mating types, designated *A* and *a* (reviewed by Glass and Staben, 1990).

Upon nitrogen starvation, either *A* or *a* mating-type cultures form spherical aggregates of hyphae that become the female reproductive structures (protoperithecia; Fig. 1; for life cycle, see Fig. 2). Specialized hyphae (trichogynes) extend from these protoperithecia. Mating occurs between these trichogynes and a target cell of the opposite mating type which is destined to function as the male. This target cell may be a conidium, a microconidium, or a vegetative hypha of the opposite mating type. As in *S. pombe* and *S. cerevisiae,* meiosis is ultimately under nutritional control, although in *N. crassa* that control is at the level of female differentiation, which, in turn, has no genetic or physiological relationship to mating type (Johnson, 1978).

The attraction of trichogynes by target cells of the opposite mating type is apparently mediated by mating-type specific pheromones; trichogynes are neither attracted to, nor fuse with, male cells of identical mating type (Bistis, 1981, 1983). After fusion of a trichogyne and its target, the nucleus that functions as the male element is transported through the trichogyne and into the protoperithecium. Within the protoperithecium, *A* and *a* nuclei divide synchronously to form fertile (ascogenous) hyphae (Fig. 2). After 8 – 10 divisions of these nuclei, karyogamy occurs in the penultimate cell of each ascogenous hypha. Thus, unlike *S. cerevisiae* and *S. pombe,* in *N. crassa* cell fusion and nuclear fusion are substantially separated by both time and space (Fig. 2). In addition to the recognition of opposite mating types required for cell fusion, nuclei of opposite mating type must recognize one another within the protoperithecium during conjugate division and prior to nuclear fusion. Nuclear fusion is followed immediately by two meiotic divisions and a single mitotic division to form the characteristic 8-spored ascus. During the meiotic process, the protoperithecium enlarges, becomes black due to the deposition of pigments, and forms various internal structures (Fig. 1). The hundreds of asci that are formed within each perithecium usually originate from a single fertilization event.

Although *A* and *a* have their most evident role in the sexual phase, they also have a function in the vegetative phase (Table 1). If hyphae of opposite mating types fuse, a reaction called heterokaryon-incompatibility is elicited that results in the death of cell compartments along the line of fusion (Gross, 1952; Pittenger, 1957). This contrasts with vigorous growth of heterokaryons formed between cultures of the same mating type (provided that the alleles at other *het* loci are also identical). Attempts to separate the heterokaryon-incompatibility function and the mating function of *A* and *a* idiomorphs by genetic recombination have been unsuccessful (Newmeyer *et al.*, 1973).

B. Organization and Role of the Mating-Type Locus

The mating type locus in *Neurospora crassa* is on the left arm of chromosome I (Fig. 3). The *A* and *a* mating-type idiomorphs of *N. crassa* are defined as regions of sequence dissimilarity at the mating-type locus in opposite mating-type genomes; this region includes all the sequences necessary for mating-type control of the life cycle (Vollmer and Yanofsky, 1986; Glass *et al.*, 1988). By this criterion, the *A* idiomorph is 5301 bp in length and the *a* idiomorph is 3235 bp in length. Essentially identical sequences flank the mating-type locus in the *A* and *a* haploid genomes (Glass *et al.*, 1990; Staben and Yanofsky, 1990). Unlike *Saccharomyces cerevisiae* and *Schizosaccharomyces pombe, N. crassa* contains only a single copy of either the *A* or *a* idiomorph within a haploid genome. The organism is thus structurally heterothallic and incapable of mating-type switching. This explains the historical observation of mating-type stability in *N. crassa.*

Two transcripts, one from each idiomorph, have been detected under vegetative and mating growth conditions (Glass *et al.*, 1990; Staben and Yanofsky, 1990). These transcripts, *mt A-1* and *mt a-1,* are completely contained within the unique *A* or *a* region. Frameshift mutations identified within *mt A-1* and *mt a-1* result in strains that fail to conduct any of the steps involved in the sexual phase of the life cycle (Griffiths and DeLange, 1978; Griffiths, 1982; Glass *et al.*, 1990; Staben and Yanofsky, 1990). In vegetative growth, the A^{m} and a^{m} mutants are morphologically indistinguishable from wild-type and, in response to nitrogen starvation, form protoperithecia. A^{m} and a^{m} trichogynes, however, fail to respond to the proximity of cells of the opposite mating type, and A^m and a^m conidia fail to attract trichogynes of the opposite mating type (A. J. F. Griffiths and C. Myers, personal communication). The *mt A-1* and *mt a-1* products are thought to encode positive regulators that control target genes conferring cell type, for example, genes for pheromones and their respective receptors (Fig. 4). The *mt A-1* and *mt a-1* products in *N. crassa* may provide functions analogous to those of *Mc* and *Pc* in *S. pombe* and MATα1 in *S. cerevisiae* in establishing mating cell type in the haploid phase. Although putative target genes of *mt A-1* and *mt a-1* have been identified, how the *mt A-1* and *mt a-1* products ultimately control cell type on a molecular level has not been elucidated.

The *mt A-1* transcript contains a 288-aa ORF consisting of two exons (Fig. 3). A search of the databank for sequences similar to the *mt A-1* ORF revealed a region near the amino terminus with similarity to MATα1 of *S. cerevisiae* (Glass *et al.*, 1990). In a region of 15 consecutive amino acids, 10 are identical in *mt A-1* ORF and MATα1, and 3 are conservative substitutions. This region of similarity may indicate that the *mt A-1* polypeptide has functions such as DNA-binding, transcriptional activation, or both in common with MATα1; however, neither of these functions has been demonstrated. Unlike *S. cerevisiae* MATα1, *mt A-1* is required for meiosis as well as for mating. The evidence for this follows. A sterile A^m strain, altered in *mt A-1*, can be combined in a nutritionally forced heterokaryon with a compatible strain that is functional at the mating-type locus (*mt* A^+). Such a heterokaryon is fully fertile when crossed with a normal strain of mating-type *a*, but all the ascospores produced receive their genetic material from the *mt* A^+ component of the heterokaryon. This indicates that the failure of an A^m strain in the sexual phase is not simply a failure to attract *a* mating-type trichogynes and fuse with them, since this does occur with the heterokaryon. An additional deficit prevents A^m-derived nuclei from giving rise to progeny even when they are escorted by a competent partner into ascogenous hyphae.

The functional region of the *a* idiomorph is contained within a 382-aa ORF (*mt a-1*) consisting of three exons (Fig. 3). A search of the databank for sequences related to the *mt a-1* ORF resulted in identification of a region of similarity to the *mat-Mc* polypeptide of *S. pombe* (Staben and Yanofsky, 1990). The similarity is within the amino-terminal half of the *mt a-1* ORF, although it extends over almost the entire length of the shorter *S. pombe Mc* polypeptide. As for the A^m strains, a^m nuclear markers cannot be recovered from a cross of a heterokaryon of the a^m strain and a mating-competent *a* nucleus with an *A* strain (Perkins, 1984), indicating that *mt a-1* function is also required after cell fusion. These observations suggest that both the *mt A-1* and *mt a-1* polypeptides are required after nuclei of opposite mating type are paired off within the ascogenous hyphae. The nuclei in the ascogenous hyphae must be able to recognize each other as potential partners for karyogamy. *N. crassa* asci always contain four *A* and four *a* mating-type ascospores, the progeny from the fusion of opposite mating-type nuclei (Fig. 4).

The A^{m} and a^{m} sterile strains will form vigorous heterokaryons with strains of either mating type if the other *het* loci are homoallelic; indeed, this was the original basis of the selection of A^{m} and a^{m} strains (Griffiths and DeLange, 1978; Griffiths, 1982). Thus, single mutations within *mt A-1* and *mt a-1* usually inactivate both the fertility and heterokaryon-incompatibility function of the mating type idiomorphs. In fungi closely related to *N. crassa,* such as *N. sitophila* and *N. tetrasperma,* opposite mating-type strains are capable of forming vigorous heterokaryons. However, incompatibility is seen when mating-type idiomorphs from *N. tetrasperma* and *N. sitophila* are introgressed into *N. crassa* (Metzenberg and Algren, 1973; Perkins, 1977). Thus, a functional mating-type locus, while necessary for incompatibility, is not sufficient for it. The absence of mating-type incompatibility in related species is due to epistatic differences in unlinked genes that are not closely linked to mating type. In harmony with this hypothesis, a mutation (*tol*) exists that suppresses mating-type heterokaryon-incompatibility in *N. crassa* and is unlinked to mating type (Newmeyer, 1970). A challenge posed by the mating-type system in *N. crassa* is to understand how *mt A-1* and *mt a-1* can control such seemingly unrelated functions in the vegetative and the sexual phases of the life cycle.

It has been suggested that, in *S. cerevisiae* and *S. pombe,* the transcriptional repression of *HMR, HML, mat2,* and *mat3* loci is due to structural features of higher order than DNA sequence, for example, alterations in chromatin structure. Although *N. crassa* does not contain archival copies of mating-type information, higher-order structural features associated with the mating-type locus are again a possibility. The insertion of *A* or *a* mating-type idiomorphs into sterile mutants typically produces transformants capable of forming abundant perithecia, but no ascospores. However, a small minority of fully fertile transformants is also obtained. Examination of the two kinds of transformants shows that, in the major type that forms barren perithecia, the incoming idiomorph is inserted ectopically, whereas in the fertile minor type, the incoming idiomorph is inserted at the original mating-type locus by gene replacement (Glass *et al.*, 1988). It is possible that transformation at an ectopic site affects correct expression of mating-type information, or that, in *Neurospora,* the *A* and *a* sequences must be paired on homologous chromosomes during meiosis and that it is more important that they be at the same place than at the "right" place. An interesting situation exists in the closely related filamentous fungus *Podo-*

spora anserina: the resident mating-type information was found to be dominant to transgenic copies of the opposite mating type not only when they were integrated at ectopic positions, but also when they were integrated by homologous recombination at the mating-type locus (M. Picard, personal communication; for review, see Rossignol and Picard, Chapter 12).

It is possible that the remaining portion of the *A* and *a* idiomorphs encodes functions involved in meiosis and sporulation; initial mutational analysis of this region has resulted in the isolation of mutants that are unable to sporulate (Glass *et al.*, 1990; Staben and Yanofsky, 1990; N. L. Glass, unpublished).

V. CONCLUDING REMARKS

A. Overall Organization of the Mating-Type Loci

Several obvious similarities and differences are seen in the organization of the mating loci of *Saccharomyces cerevisiae, Schizosaccharomyces pombe,* and *Neurospora crassa.* All use a region of DNA that is essentially nonhomologous between mating types to control a set of mating responses. Thus, the gene products from each idiomorph are dissimilar. This feature ultimately controls the mating type of the cell. The mating-type gene products from all three organisms probably encode regulatory factors that control the transcription of mating-specific gene products, although this has been shown only for *S. cerevisiae.* A different regulatory factor is required to specify each mating cell type. This requirement for dissimilarity in regulatory properties presumably led to the evolution of dissimilar sequences at the same locus in haploid cells of different mating type.

The idiomorphs of *N. crassa* are significantly larger than those of the unicellular yeasts. Although the *mt a-1* and *mt A-1* transcripts from the *N. crassa* idiomorphs are contained completely within the idiomorphs, the shorter *S. cerevisiae* and *S. pombe* idiomorphs often supply only the sequences for the 5′ region of the mating-loci transcripts. In each of the three organisms, the regions flanking the idiomorphs are virtually identical for the two mating types of that organism. It is not clear why these sequences are so highly conserved. In *N. crassa,* the mating-type loci are completely functional only when they are situated at the correct chromosomal location, suggesting an important role for the DNA region surrounding the mating-type locus. The transcriptional repression of the *HML, HMR, mat2,* and *mat3* loci and the reduction in meiotic recombination in the intergenic region between *mat2* and *mat3* also appear to be associated with higher-order features of chromatin structure. In both *S.*

cerevisiae and *S. pombe*, silent copies of each sequence are maintained in each haploid cell and a mechanism exists for switching the identity of the sequence present at the expression locus. The absence of these features in *N. crassa* may be a reflection of its adaptation to a particular ecological niche.

B. Products of the Mating-Type Loci

All of the identified open reading frames in the mating loci encode basic polypeptides, a feature frequently seen in DNA-binding proteins. The polypeptides encoded by the *Saccharomyces cerevisiae* mating locus bind to regulatory sequences upstream from target genes. The products of the mating loci of *Schizosaccharomyces pombe* and *Neurospora crassa* show homology to the *S. cerevisiae* proteins. The *N. crassa mt A-1* polypeptide shows appreciable similarity to the *S. cerevisiae MATα1* polypeptide. *S. pombe Pm* product shows homology to *S. cerevisiae* **α**2 and, to a lesser extent, to **a**1. In addition, these three proteins all have a region homologous to the homeodomain in regulatory proteins of higher eukaryotes. The *N. crassa mt a-1* shows some similarity to the *S. pombe Mc* polypeptide. This region also shows a high degree of similarity to an ORF (SRY) identified from the human Y chromosome that is implicated in male sex determination (Sinclair *et al.*, 1990). It seems likely that all these distantly related polypeptides encoded by the mating loci act directly on target DNA sequences to regulate transcription, since this is clearly the function of those proteins from *S. cerevisiae.* There is no evidence that mating loci encode proteins such as protein kinases or proteases that might act by a posttranscriptional mechanism. However, no one particular motif is found in the mating genes of all three organisms, and several of the putative polypeptides do not contain any previously characterized DNA-binding motif. Thus, a particular type of DNA-binding polypeptide encoded by all mating-type idiomorphs is not a requirement for function.

C. Combinatorial Control by Products of the Mating-Type Loci

It appears that mating type in *Neurospora crassa* and *Schizosaccharomyces pombe* is determined by a single polypeptide encoded by each mating-type locus. These proteins presumably act as activators, turning on the transcription of the appropriate set of genes to give a particular mating type (see Fig. 4). Disruption of these mating-type activators abolishes mating. In *Saccharomyces cerevisiae,* two proteins, one acting as a repressor (**α**2) and the other acting as an activator (**α**1), are required to establish the **α** mating type. In **a** cells, **a**-specific genes are expressed spontaneously, so that disruption of *MAT***a***1* results in cells still capable of conjugating as **a** cells. There seems to be no advantage to having this "default" cell type, nor is it known if this mechanism occurs in other organisms. Following cell fusion, the *mt A-1* and *mt a-1* polypeptides and *Pc* and *Mc* polypeptides, respectively, can potentially be inactivated with respect to their haploid functions by forming a combinatorial complex with a new regulatory activity. This hypothesis is suggested by the fact that all the mating-type gene products are required in *N. crassa* and *S. pombe* during meiosis and sporulation. In *S. cerevisiae,* the α2 polypeptide must retain some of its haploid function in the diploid cell to prevent diploids from mating as **a** cells. Moreover, the **a**1–α2 heterodimer formed in the diploid cell must inhibit expression of *MAT*α*1,* since this protein is not inactivated by the formation of a complex with any polypeptide from the **a** idiomorph.

The models we have presented here predict that regulatory proteins encoded by the mating-type loci of *N. crassa* and *S. pombe* confer mating cell type in the haploid phase, and combine upon cell fusion to form new regulatory activities that function in meiosis and sporulation. It also seems likely that these proteins activate mating in concert with other proteins that are not specific to cell type. The elucidation of the similarities and differences in mating-type regulation of *S. cerevisiae, S. pombe,* and *N. crassa* will offer us new insights into the control of cell type and meiosis.

ACKNOWLEDGMENTS

We wish to express our gratitude to Robert L. Metzenberg for his critical reading of this manuscript and for his unending enthusiasm. We are also grateful to Ira Herskowitz and Richard Egel for their comments and suggestions and to Lynn Lee for providing the SEM photographs. I.L. is a recipient of a MRC of Canada Postdoctoral Fellowship.

REFERENCES

Abraham, J., Nasmyth, K. A., Strathern, J. N., Klar, A. J. S., and Hicks, J. B. (1984). Regulation of mating-type information in yeast: Negative control requiring sequences both 5′ and 3′ to the regulated region. *J. Mol. Biol.* **176,** 307–331.

Ammerer, G. (1990). Identification, purification and cloning of a polypeptide (PRTF/GRM) that binds to mating-specific promoter elements in yeast. *Genes Dev.* **4,** 299–312.

Anderegg, R. J., Betz, R., Carr, S. A., Crabb, J. W., and Dutze, W. J. (1988). Structure of *Saccharomyces cerevisiae* mating hormone *a*-factor. *J. Biol. Chem.* **263,** 18236–18240.

Astell, C. R., Ahlstrom-Jonasson, L., Smith, M., Tatchell, K., Nasmyth, K. A., and Hall, B. D. (1981). The sequence of the DNAs coding for the mating-type loci of *Saccharomyces cerevisiae. Cell* **27,** 15–23.

Beach, D. H. (1983). Cell type switching by DNA transposition in fission yeast. *Nature (London)* **305,** 682–688.

Beach, D. H., and Klar, A. J. S. (1984). Rearrangements of the transposable mating-type cassettes in fission yeast. *EMBO J.* **3,** 603–610.

Bender, A., and Sprague, G. F. Jr. (1987). Mat $\alpha 1$ protein, a yeast transcription activator, binds synergistically with a second protein to a set of cell-type-specific genes. *Cell* **50,** 681–691.

Betz, R., Crabb, J. W., Meyer, H. E., Wittig, R., and Duntze, W. (1987). Amino acid sequences of *a*-factor mating peptides from *Saccharomyces cerevisiae. J. Biol. Chem.* **262,** 546–548.

Bistis, G. N. (1981). Chemotrophic interaction between tricogynes and conidia of opposite mating type in *Neurospora crassa. Mycologia* **73,** 959–975.

Bistis, G. N. (1983). Evidence for diffusible, mating-type-specific trichogyne attractants in *Neurospora crassa. Exp. Mycol.* **7,** 292–295.

Blumer, K. J., Reneke, J. E., and Thorner, J. (1988). The STE2 gene product is the ligand binding component of the α-factor receptor of *Saccharomyces cerevisiae. J. Biol. Chem.* **263,** 10836–10842.

Bucking-Throm, E., Duntze, W., Hartwell, L. H. and Manney, T. R. (1973). Reversible arrest of haploid cells at the initiation of DNA synthesis by a diffusible sex factor. *Exp. Cell Res.* **76,** 99–110.

Dranginis, A. M. (1990). Binding of yeast *a*1 and $\alpha 2$ as a heterodimer to the operator DNA of a haploid-specific gene. *Nature (London)* **347,** 682–685.

Egel, R. (1971). Physiological aspects of conjugation in fission yeast. *Planta* **98,** 89–96.

Egel, R. (1973). Commitment to meiosis in fission yeast. *Mol. Gen. Genet.* **121,** 277–284.

Egel, R. (1984). Two tightly linked cassettes in the mating-type region of *Schizosaccharomyces pombe. Curr. Genet.* **8,** 199–203.

Egel, R. (1989). Mating-type genes, meiosis and sporulation. *In* "Molecular Biology of the Fission Yeast" (A. Nasim, P. Young, and B. F. Johnson, eds.), pp. 31–73. Academic Press, San Diego.

Egel, R., and Gutz, H. (1981). Gene activation by copy transposition in mating-type switching of a homothallic fission yeast. *Curr. Genet.* **3,** 5–12.

Egel, R., Willer, M., and Nielsen, O. (1989). Unblocking of meiotic crossing-over between the silent mating-type cassettes of fission yeast, conditioned by the recessive, pleiotropic mutant *rik1. Curr. Genet.* **15,** 407–410.

Egel, R., Nielsen, O., and Weilguny, D. (1990). Sexual differentiation in fission yeast. *Trends Genet.* **6,** 369–373.

Feldman, J. B., Hicks, J. B., and Broach, J. R. (1984). Identification of the sites required for repression of a silent mating type locus in yeast. *J. Mol. Biol.* **178,** 815–834.

Glass, N. L., and Staben, C. (1990). Genetic control of mating in *Neurospora crassa. Sem. Dev. Biol.* **1,** 177–184.

Glass, N. L., Vollmer, S. J., Staben, C., Groteleuschen, J., Metzenberg, R. L., and Yanofsky, C. (1988). DNAs of the two mating-type alleles of *Neurospora crassa* are highly dissimilar. *Science* **241,** 570–573.

Glass, N. L., Groteleuschen, J., and Metzenberg, R. L. (1990). *Neurospora crassa A* mating-type region. *Proc. Natl. Acad. Sci. U.S.A.* **87,** 4912–4916.

Goutte, C., and Johnson, A. D. (1988). *a*1 protein alters the DNA binding specificity of $\alpha 2$ repressor. *Cell* **52,** 875–882.

Griffiths, A. J. F. (1982). Null mutants of the *A* and *a* mating type alleles of *Neurospora crassa. Can. J. Genet. Cytol.* **24,** 167–176.

Griffiths, A. J. F., and DeLange, A. M. (1978). Mutations of the *a* mating-type region in *Neurospora crassa. Genetics* **88,** 239–254.

Gross, S. R. (1952). Heterokaryosis between opposite mating types in *Neurospora crassa. Am. J. Bot.* **39,** 574–577.

Hagen, D. C., McCaffrey, C., and Sprague, G. F., Jr. (1986). Evidence that the yeast STE3 gene encodes a receptor for the peptide pheromone, *a*-factor: Gene sequence and implications for the structure of the presumed receptor. *Proc. Natl. Acad. Sci. U.S.A.* **83,** 1418–1422.

Herskowitz, I. (1988). Life cycle of the budding yeast *Saccharomyces cerevisiae. Microbiol. Rev.* **52,** 526–553.

Herskowitz, I. (1989). A regulatory hierarchy for cell specialization in yeast. *Nature (London)* **342,** 749–757.

Herskowitz, I., and Oshima, Y. (1981). Control of cell type in *Saccharomyces cerevisiae:* Mating type and mating type interconversion. *In* "The Molecular Biology of the Yeast *Saccharomyces:* Life Cycle and Inheritance" (J. N. Strathern, E. W. Jones, and J. R. Broach, eds.), pp. 181–209. Cold Spring Harbor Laboratory Press, Cold Spring Harbor, New York.

Hofmann, J. F.-X., Laroche, T., Brand, A. H., and Gasser, S. M. (1989). RAP-1 factor is necessary for DNA loop formation *in vitro* at the silent mating type locus *HML. Cell* **57,** 725–737.

Iino, Y., and Yamamoto, M. (1985). Negative control for the initiation of meiosis in *Schizosaccharomyces pombe. Proc. Natl. Acad. Sci. U.S.A.* **82,** 2447–2451.

Ivy, J. M., Klar, A. J. S., and Hicks, J. B. (1986). Cloning and characterization of four *SIR* genes of *Saccharomyces cerevisiae. Mol. Cell. Biol.* **6,** 688–702.

Johnson, A. D., and Herskowitz, I. (1985). A repressor (MAT $\alpha 2$ product) and its operator control expression of a set of cell type specific genes in yeast. *Cell* **42,** 237–247.

Johnson, T. E. (1978). Isolation and characterization of perithecial development mutants in *Neurospora. Genetics* **88,** 27–47.

Kassir, Y., and Simchen, G. (1976). Regulation of mating and meiosis in yeast by the mating-type region. *Genetics* **82,** 187–206.

Keleher, C. A., Goutte, C., and Johnson, A. D. (1988). The yeast cell-type-specific repressor $\alpha 2$ acts cooperatively with a non-cell-type-specific protein. *Cell* **53,** 927–936.

Kelly, M., Burke, J., Smith, M., Klar, A., and Beach, D. (1988). Four mating type genes control sexual differentiation in the fission yeast. *EMBO J.* **7,** 1537–1547.

Klar, A. J. S., Fogel, S., and Lusnak, K. (1979). Gene conversion of the mating-type locus in *Saccharomyces cerevisiae. Genetics* **92,** 777–782.

Klar, A. J. S., Strathern, J. N., Broach, J. R., and Hicks, J. B. (1981). Regulation of transcription in expressed and unexpressed mating type cassettes of yeast. *Nature (London)* **289,** 239–244.

Kronstad, J. W., Holly, J. A., and MacKay, V. L. (1987). A yeast operator overlaps an upstream activation site. *Cell* **50,** 369–377.

Leupold, U. (1950). Die Vererbung von Homothallie und Heterothallie bei *Schizosaccharomyces pombe. C.R. Lab. Carlsberg Ser. Physiol.* **24,** 381–480.

Leupold, U. (1958). Studies on recombination in *Schizosaccharomyces pombe. Cold Spring Harbor Symp. Quant. Biol.* **23,** 161–170.

Leupold, U. (1987). Sex appeal in fission yeast. *Curr. Genet.* **12,** 543–545.

McCaffrey, G., Clay, F. J., Kelsay, K. D., and Sprague, G. F., Jr. (1987). Identification and regulation of a gene required for cell fusion during mating of the yeast *Saccharomyces cerevisiae. Mol. Cell. Biol.* **7,** 2680–2690.

McGrath, J. P., and Varshavsky, A. (1989). The yeast STE6 gene encodes a homologue of the mammalian multidrug resistance P-glycoprotein. *Nature (London)* **340,** 400–404.

McLeod, M., and Beach, D. (1988). A specific inhibitor of the *ran1*$^+$ protein kinase regulates entry into meiosis in *Schizosaccharomyces pombe. Nature (London)* **332,** 509–514.

McLeod, M. Stein, M., and Beach, D. (1987). The product of the *mei3*$^{+}$ gene, expressed under control of the mating type locus, induces meiosis and sporulation in fission yeast. *EMBO J.* **6,** 729–736.

Mahoney, D. J., and Broach, J. R. (1989). The HML mating-type cassette of *Saccharomyces cerevisiae* is regulated by two separate but functionally equivalent silencers. *Mol. Cell. Biol.* **9,** 4621–4630.

Malone, R. E. (1990). Dual regulation of meiosis in yeast. *Cell* **61,** 375–378.

Metzenberg, R. L., and Algren, S. K. (1973). Behaviour of *Neurospora tetrasperma* mating-type genes introgressed into *Neurospora crassa. Can. J. Genet. Cytol.* **15,** 571–576.

Metzenberg, R. L., and Glass, N. L. (1990). Mating type and mating strategies in *Neurospora. BioEssays* **12,** 53–60.

Miller, A. M., MacKay, V. L., and Nasmyth, K. A. (1985). Identification and comparison of two sequence elements that confer cell-type specific transcription in yeast. *Nature (London)* **314,** 598–603.

Mitchell, A. P., and Herskowitz, I. (1986). Activation of meiosis and sporulation by repression of the RME1 product in yeast. *Nature (London)* **319,** 738–742.

Mortimer, R. K., and Hawthorne, D. C. (1969). Yeast genetics. *In* "The Yeasts" (A. H. Rose and J. S. Harrison, eds.), pp. 385–460. Academic Press, New York.

Nasmyth, K. A. (1982). The regulation of yeast mating-type chromatin structure by SIR: An action at a distance affecting both transcription and transposition. *Cell* **30,** 567–578.

Nasmyth, K. A., and Tatchell, K. (1980). The structure of the transposable yeast mating type loci. *Cell* **19,** 753–764.

Nasmyth, K. A., Tatchell, K., Hall, B. D., Astell, C. R., and Smith, M. (1981). A position effect in the control of transcription at yeast mating type loci. *Nature (London)* **289,** 244–250.

Newmeyer, D. (1970). A suppressor of the heterokaryon-incompatibility associated with mating type in *Neurospora crassa. Can. J. Genet. Cytol.* **17,** 441–449.

Newmeyer, D., Howe, H. B., Jr., and Galeazzi, D. R. (1973). A search for complexity at the mating-type locus of *Neurospora crassa. Can. J. Genet. Cytol.* **15,** 577–585.

Passmore, S., Maine, G. T., Elble, R., Christ, C., and Tye, B.-K. (1988). *Saccharomyces cerevisiae* protein involved in plasmid maintenance is necessary for mating of MATα cells. *J. Mol. Biol.* **204,** 593–606.

Perkins, D. D. (1977). Behaviour of *Neurospora sitophila* mating-type alleles in heterozygous duplications after introgression into *Neurospora crassa. Exp. Mycol.* **1,** 166–172.

Perkins, D. D. (1984). Advantages of using the inactive-mating-type a^{m1} strain as a helper component in heterokaryons. *Neurospora Newslett.* **31,** 41–42.

Pittenger, T. H. (1957). The mating type alleles and heterokaryon formation in *Neurospora crassa. Microb. Genet. Bull.* **15,** 21–22.

Rine, J. and Herskowitz, I. (1987). Four genes responsible for a position effect on expression from *HML* and *HMR* in *Saccharomyces cerevisiae. Genetics* **116,** 9–22.

Russell, P., and Nurse, P. (1986). *Schizosaccharomyces pombe* and *Saccharomyces cerevisiae:* A look at yeasts divided. *Cell* **45,** 781–782.

Sauer, R. T., Smith, D. L., and Johnson, A. D. (1988). Flexibility of the yeast $\alpha 2$ repressor enables it to occupy the ends of its operator, leaving the center free. *Genes Dev.* **2,** 807–816.

Shepherd, J. C. W., McGinnis, W., Carrasco, A. E., De Robertis, E. M., and Gehring, W. J. (1984). Fly and frog homeodomains show homologies with yeast mating type regulatory proteins. *Nature (London)* **310,** 70–71.

Sinclair, A. H., Berta, P., Palmer, M. S., Hawkins, J. R. Griffiths, B. L., Smith, M. J., Foster,

J. W., Frischauf, A., Lovell-Badge, R., and Goodfellow, P. N. (1990). A gene from the human sex determining region encodes a protein with homology to a conserved DNA-binding motif. *Nature (London)* **346,** 240–244.

Sprague, G. F., Jr., Jensen, R., and Herskowitz, I. (1983). Control of yeast cell type by the mating type locus: Positive regulation of the α-specific STE3 gene by the MATα1 product. *Cell* **32,** 409–415.

Staben, C., and Yanofsky, C. (1990). *Neurospora crassa a* mating-type region. *Proc. Natl. Acad. Sci. U.S.A.* **87,** 4917–4921.

Stotzler, D., Kiltz, H.-H., and Duntze, W. (1976). Primary structure of α-factor peptides from *Saccharomyces cerevisiae. Eur. J. Biochem.* **19,** 397–400.

Strathern, J. N., Spatola, E., McGill, C., and Hicks, J. B. (1980). Structure and organization of transposable mating type cassettes in *Saccharomyces* yeast. *Proc. Natl. Acad. Sci. U.S.A.* **77,** 2839–2843.

Strathern, J., Hicks, J., and Herskowitz, I. (1981). Control of cell type in yeast by the mating type locus. *J. Mol. Biol.* **147,** 357–372.

Tan, S., and Richmond, T. J. (1990). DNA binding induced conformational change of the yeast transcriptional activator PRTF. *Cell* **62,** 367–377.

Tan, S., Ammerer, G., and Richmond, T. J. (1988). Interactions of purified transcription factors: Binding of yeast Matα1 and PRTF to cell-type specific, upstream activating sequences. *EMBO J.* **7,** 4255–4264.

Tatchell, K., Nasmyth, K. A., Hall, B. D., Astell, C., and Smith, M. (1981). *In vitro* mutation analysis of the mating type locus in yeast. *Cell* **27,** 25–35.

Trueheart, J., Boeke, J., and Fink, G. R. (1987). Two genes required for cell fusion during yeast conjugation: Evidence for a pheromone induced protein. *Mol. Cell. Biol.* **7,** 2316–2328.

Wilson, K. L., and Herskowitz, I. (1984). Negative regulation of STE6 gene expression by the α2 product of *Saccharomyces cerevisiae. Mol. Cell. Biol.* **4,** 2420–2427.

Vollmer, S. J., and Yanofsky, C. (1986). Efficient cloning of genes of *Neurospora crassa. Proc. Natl. Acad. Sci. U.S.A.* **83,** 4869–4873.

9

Life Cycle Determinants of the Plant Pathogen *Ustilago maydis*

FLORA BANUETT
Department of Biochemistry and Biophysics
University of California
San Francisco, California

I. INTRODUCTION

Ustilago maydis is a basidiomycetous fungus that causes corn smut disease (huitlacoche). It is a member of the smut fungi (Ustilaginales), a group that causes severe economic losses of cereal grains (barley, sorghum, oats, wheat, millet, corn) and other crops. The disease is characterized by chlorosis, stunting, and tumors (the most distinctive symptom) that can develop on stems, leaves, ears, and tassels (Christensen, 1963). *Ustilago maydis* exhibits two forms in its life cycle. The haploid unicellular form grows by budding, forms compact yeast-like colonies, and is nonpathogenic; the dikaryotic filamentous form requires the plant for growth and is pathogenic (Christensen, 1963; Banuett and Herskowitz, 1988). Growth of the dikaryon in the plant leads to tumor induction. The ability of the

MORE GENE MANIPULATIONS
IN FUNGI

dikaryotic form to undergo karyogamy (nuclear fusion) and to differentiate and produce diploid spores (teliospores) competent to undergo meiosis is strictly dependent on passage through the plant. Thus, the life cycle and pathogenicity of *U. maydis* are intimately intertwined. The developmental program responsible for producing these two forms and for completion of subsequent steps in the life cycle is governed by two master regulatory loci, *a* and *b*—the mating-type or incompatibility loci—known from classical genetic studies (Rowell and DeVay, 1954; Holliday, 1961; Puhalla, 1968, 1970), and by the *fuz*1, *fuz*2, and *rtf*1 genes (Banuett, 1991a). Understanding the mode of action of these different loci will provide insights into the molecular mechanisms underlying the pathogenicity of this organism.

The goal of this chapter is to describe the basic properties of the master regulatory loci, *a* and *b*, and the processes that they govern, which include filamentous growth and pathogenicity. Recent developments have led to cloning and sequencing of these loci. It appears that the *b* locus encodes a regulatory protein that governs expression of target genes. Recent identification and properties of three new genes, *fuz*1, *fuz*2, and *rtf*1 are also described. These genes are involved in filamentous growth and tumor formation and may be the targets for *a* or *b* (Fig. 1). For more extensive reviews on *U. maydis,* smut disease, and other aspects of the life cycle not covered in this chapter, the reader is referred to the reviews by Christensen (1963), Holliday (1961, 1974), and Banuett and Herskowitz (1988).

II. THE MATING TYPE LOCI AND CONTROL OF THE LIFE CYCLE

A. Life Cycle

Initiation of the pathogenic process that leads to disease symptoms and completion of the life cycle requires fusion of two haploid cells that differ at both *a* and *b* (Rowell and DeVay, 1954; Holliday, 1961; Puhalla, 1968;

Fig. 1. Regulatory loci for the life cycle of *U. maydis* and putative target genes. The left panel shows the two master regulatory loci, *a* and *b*, also known as the mating type or incompatibility loci, that govern the life cycle of *Ustilago maydis.* The right panel shows the proposed target genes for the mating type loci. The *fuz* genes (*fuz*1 and *fuz*2) are genes necessary for filament formation and filamentous growth. They are hypothesized to be the targets of *a* or *b* (or both) because the presence of different alleles at both *a* and *b* are necessary for filamentous growth. The *tum* genes are necessary for tumor induction and are hypothesized to be the targets of *b* but not *a* (see text).

Fig. 2). Cell fusion, which normally occurs in the infected plant, results in formation of the filamentous dikaryon. If the strains carry identical *a* or identical *b* alleles, filament formation does not occur. Although the filamentous form requires the plant for its growth, filament formation can be detected using a simple plate assay. When two haploids that differ at *a* and *b* are mixed on charcoal nutrient medium, a white fuzziness develops (the Fuz^+ phenotype) caused by the formation of dikaryotic filaments (Fig. 3). If the haploids carry identical *a* or identical *b* alleles, no fuzziness develops (the Fuz^- phenotype; Fig. 3; Banuett and Herskowitz, 1988, 1989). This plate assay has proved useful for cloning the *a* and *b* loci and for the isolation of mutants defective in filament formation, as will be described in Sections II,B and C.

Growth of the filamentous dikaryon *in planta* results in tumor formation on aerial plant parts (Christensen, 1963; Fig. 2). This symptom is characteristic only of corn smut disease and it is not seen in smut diseases caused by other members of the genus *Ustilago* (see Banuett, 1991b). These tumors are generated by induction of cell division and cell enlargement and presumably require fungal gene products that alter growth control of host cells. Few studies have been done of the infectious process and growth of the fungus *in planta.* Development of antibody probes specific to the fungus will make immunocytological analyses possible and provide a better understanding of these processes in the future.

Fungal differentiation occurs within the tumors. This process entails cell enlargement, gelatinization, karyogamy, and deposition of a thick cell wall, and results in production of diploid spores (teliospores; Fig. 2). It seems likely that initiation of this process involves signals produced by the plant. If so, it would be of great interest to identify the fungal genes that respond to such signals. These are a few of the many unexplored areas concerning the life cycle of this fungus. Once teliospores are formed, they can be removed from the tumors and allowed to undergo meiosis on nutrient medium. Even though meiosis occurs outside the plant, competence to undergo meiosis is acquired by passage through the plant. It is clear that competence for meiosis is not simply the result of karyogamy: diploid strains obtained in the laboratory are incapable of undergoing meiosis unless they infect a plant and yield teliospores. Meiosis is preceded by germination of the teliospore to form a short filament (the promycelium) into which the diploid nucleus migrates for the meiotic divisions (O'Donnell and McLaughlin, 1984a,b; Fig. 2). In field infections, the tumors crack open at maturity, exposing masses of dark spores that can be spread by the wind and carried to a susceptible host. These spores germinate on the plant and start the infection process.

The life cycle can be completed in a period of 2–3 wk in the laboratory

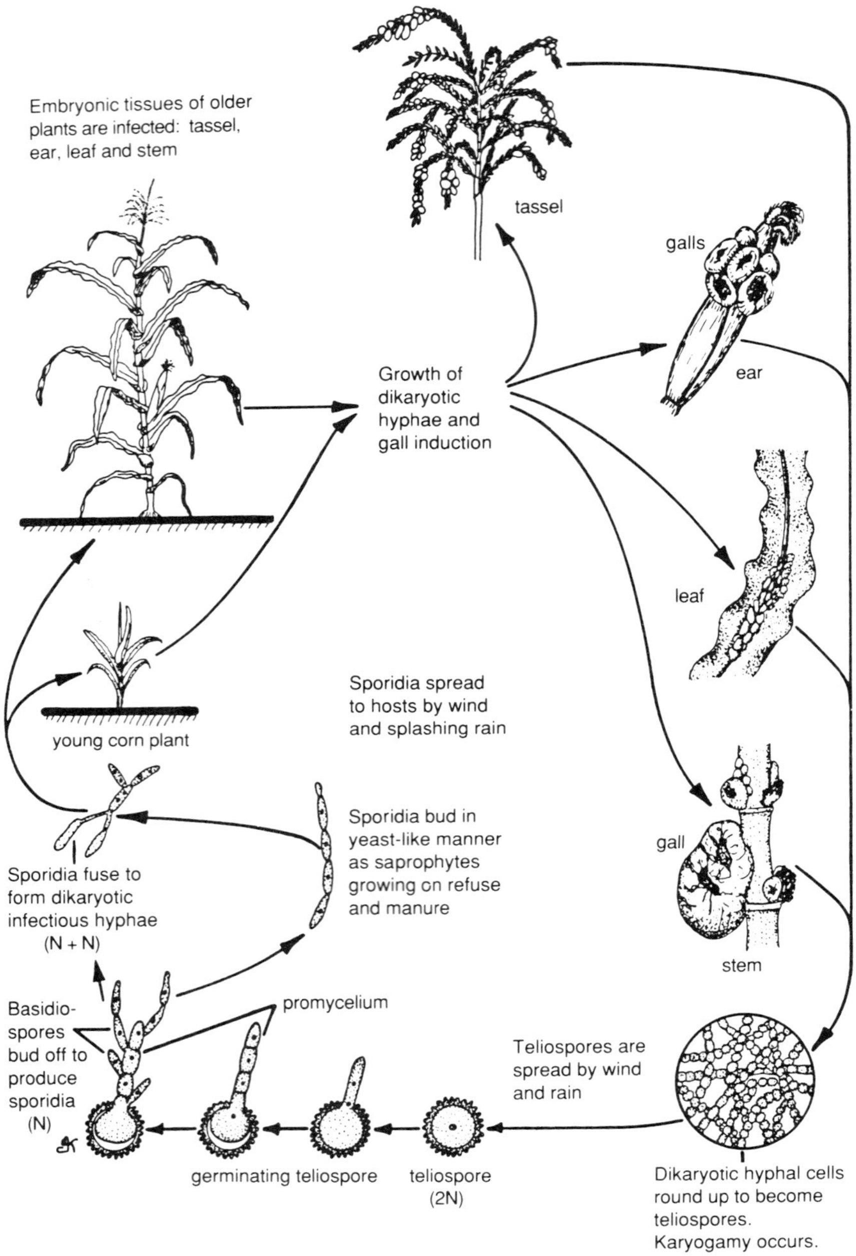

Fig. 2. Lifecycle of *Ustilago maydis.* Modified with permission from Kenaga *et al.* (1971).

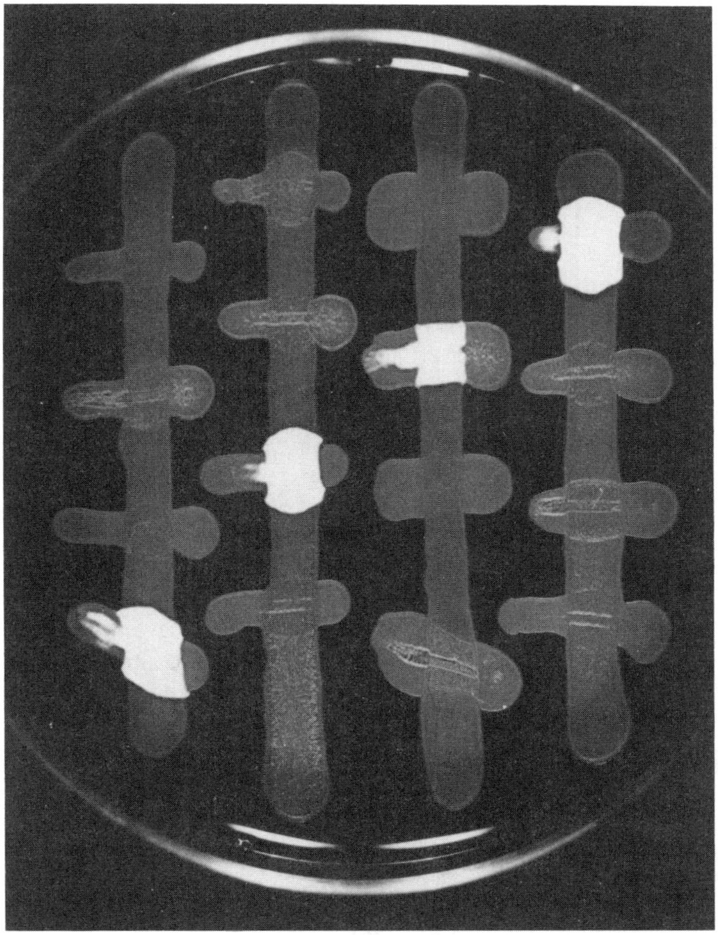

Fig. 3. Mating reaction of haploid strains. Saturated cultures (2×10^8 cells/ml) of haploid strains were cross-streaked against haploid testers *a*1 *b*1, *a*2 *b*2, *a*1 *b*2, and *a*2 *b*1 (*left* to *right*) on charcoal nutrient medium and incubated overnight at room temperature. Strains in the horizontal lines are (*top* to *bottom*) *a*2 *b*2, *a*1 *b*1, *a*2 *b*1, and *a*1 *b* 2. Clear cut reactions such as these are exhibited by a set of isogenic strains (see Banuett and Herskowitz, 1989).

(Banuett and Herskowitz, 1988; see also Holliday, 1974). It is initiated by co-inoculation of the desired strains with a hypodermic syringe into 5-day-old plants (older plants can also be used). The first symptoms of tumor formation are detectable 3–5 days postinoculation. Teliospores become visible as dark brown specks in the tumors 10–12 days postinoculation.

Teliospores germinate in less than 24 hr, and colonies of the segregants can be seen after 2–3 days incubation at 32°C.

The haploid form of *U. maydis* can be grown in the laboratory on a variety of media, whereas the dikaryon is difficult, if not impossible, to propagate under similar conditions. Thus, the transition from one morphological type to the other results in decreased ability to grow on nutrient media in addition to acquisition of pathogenic ability. In contrast, diploids (which can be obtained by mating strains carrying complementing auxotrophic markers; Holliday, 1961, 1974; Day *et al.*, 1971; Banuett and Herskowitz, 1989) exhibit properties of both cell types (if they are heterozygous for both *a* and *b*). They form compact yeast-like colonies and can be grown on a variety of media. On media containing charcoal, their filamentous growth mode is elicited. In addition, diploids are solopathogenic—a pure culture induces tumor formation (Holliday, 1961, 1974; Banuett and Herskowitz, 1989). Vegetative diploids are not a normal part of the life cycle of *U. maydis* or other smut fungi (see Banuett, 1991b).

B. Master Regulators: *a* and *b*

As noted earlier, the *a* and *b* loci govern the developmental switch responsible for the distinct cell types that characterize the *U. maydis* life cycle. This switch entails the acquisition of specialized properties by each cell type and a transition from a haploid, unicellular, nonpathogenic, and saprophytic form to a dikaryotic, multicellular, pathogenic, and parasitic form. Clearly, properties exhibited by these two forms reflect expression of different gene sets. The challenge is to understand how these loci govern expression of these sets of genes (Fig. 1). *U. maydis* is unique among the smut fungi because two master regulatory loci program the life cycle, as opposed to the single regulatory locus (*a*) in the other smut fungi. *U. maydis* closely resembles other basidiomycetes to be described in subsequent sections.

1. a *Locus*

The *a* locus has two naturally occurring alleles, *a*1 and *a*2 (Rowell and DeVay, 1954; Rowell, 1955; Holliday, 1961), and was proposed to govern cell fusion (Rowell and DeVay, 1954; Rowell, 1955), although this function has not been established conclusively. Filamentous growth requires different alleles at both the *a* locus and the *b* locus (Banuett and Herskowitz, 1989). This fact was established by the use of isogenic diploid strains differing at the mating-type loci. As mentioned earlier, stable diploids can easily be constructed in the laboratory by selecting for prototrophy after mating strains that carry complementing auxotrophic markers. Diploids that are heterozyogous for both *a* and *b* (*a*≠ *b*≠) form mycelial or fuzzy colonies and are pathogenic. Diploid strains homozygous at *a* and heterozygous at *b* (*a*= *b*≠), obtained as mitotic recombinants of the *a*≠ *b*≠ diploid, are not capable of mycelial growth (Banuett and Herskowitz, 1989); diploid strains heterozygous at *a* and homozygous at *b* (*a*≠ *b*=) are also incapable of mycelial growth (Puhalla, 1968; Day *et al.*, 1971; Holliday, 1974; Banuett and Herskowitz, 1989). These observations demonstrate that different *a* alleles and different *b* alleles are necessary for filamentous growth.

The two alleles of the *a* locus have recently been cloned and sequenced (M. Bölker and R. Kahmann, unpublished data). The *a* alleles comprise heterologous regions of DNA flanked by homologous regions. Similar gene organization has been observed in the ascomycetes, which have a single mating-type locus with two alleles. In the ascomycetous yeasts, *Saccharomyces cerevisiae* (Astell *et al.*, 1981) and *Schizosaccharomyces pombe* (Kelly *et al.*, 1988), and in the filamentous ascomycetes, *Neurospora crassa* (Glass *et al.*, 1988, 1990; Glass and Staben, 1990; Staben and Yanofsky, 1990; also, Glass and Lorimer, Chapter 8) and *Cochliobolus heterostrophus* (O. Yoder, personal communication), the two alleles comprise heterologous regions of DNA. The open reading frames (ORFs) encoded by the *N. crassa* alleles exhibit some sequence similarity to the regulatory proteins encoded by the mating-type loci of the yeasts (Glass *et al.*, 1990; Staben and Yanofsky, 1990). The *a* locus of *U. maydis* may encode a regulatory protein that governs expression of target genes, some of which may be genes for filamentous growth. It remains possible, however, that *a* might code for a protein kinase, a protease, or a protein of some other function (Banuett and Herskowitz, 1988).

2. b *Locus*

Although diploids that are homozygous at either *a* or *b* exhibit the same phenotype with respect to ability to form filaments—both are nonmycelial—they differ with respect to their ability to induce tumors. *a*= *b*≠ strains are pathogenic (Banuett and Herskowitz, 1989) whereas *a*≠ *b*= strains are nonpathogenic (Puhalla, 1968; Day *et al.*, 1971; Holliday, 1974; Banuett and Herskowitz, 1989). The presence of different *b* alleles is sufficient for tumor induction once cell fusion has taken place. Thus, the *b* locus is the major pathogenicity determinant of *U. maydis.*

The *b* locus is multiallelic and is estimated to have 25 naturally occurring alleles. Any combination of two different *b* alleles leads to development of the pathogenic dikaryon (Rowell and DeVay, 1954; Puhalla, 1968; Holliday, 1961; Day *et al.*, 1971). This observation poses a fascinating molecular challenge: what is the basis of self–nonself recognition of the *b* alleles in the fungal cell? We now know that this recognition occurs intracellularly (Banuett and Herskowitz, 1989) and involves recognition of polypeptides (see subsequent text) rather than, for example, cell-surface recognition. No *b* locus has been found in any of the other smut fungi. However, multiallelic incompatibility loci are a common feature of many basidiomycetes, for example, *Tremella mesenterica* (a jelly fungus; Wong and Wells, 1985), *Schizophyllum commune* (a wood-rotting fungus; Raper, 1983; Stankis *et al.*, 1990; also, Novotny *et al.*, Chapter 10), and *Coprinus cinereus* (Casselton, 1978; also, Pukkila and Casselton, Chapter 6). Completion of the life cycle occurs only when the mating partners carry different alleles at these loci. In *Schizophyllum commune* and in *Coprinus cinereus* there are two factors (*A* and *B*), each consisting of two linked loci α and β, each of which contains multiple alleles. If the mating partners exhibit allelic differences at $A\alpha$ or $A\beta$ and at $B\alpha$ or $B\beta$, then a dikaryon can be established and maintained, and fruiting body formation and meiosis can occur. In *U. maydis* and these other basidiomycetes, self–nonself recognition occurs after cell fusion, and completion of the life cycle occurs only if alleles are different. Although the details of the molecular organization of these loci in *Schizophyllum commune* are more complex than in *U. maydis,* self–nonself recognition appears to involve recognition of polypeptides in both cases.

The fact that the *b* locus is multiallelic and that its function requires the ability of the *b* polypeptides to distinguish self from nonself raises several challenging questions. Understanding the rules that govern interactions between *b* polypeptides has implications for protein structure and for understanding the function of a wide-spread class of regulatory proteins—the homeodomain proteins of eukaryotes—as will be apparent from the description that follows.

a. Molecular Analysis of the *b* Locus. The *b* locus was cloned using a simple functional assay on charcoal nutrient medium. As described earlier, diploid strains homozygous for the *b* locus form nonfuzzy colonies on this medium. The method consisted of screening for Fuz^+ transformants after introduction of a library into a nonfuzzy strain. An *a1/a2 b1/b1* diploid was transformed with a library from an *a2 b2* strain that was constructed in a plasmid vector that confers hygromycin resistance. Hygromycin resistant transformants were selected and screened for the fuzzy phenotype on charcoal medium (Schulz *et al.*, 1990; see also Kronstad and Leong, 1989). A transformant carrying a plasmid with an insert responsible for the fuzzy phenotype and the ability to induce tumors was identified. The region within the DNA insert responsible for the phenotype was sequenced and shown to contain an ORF of 410 amino acids (aa; Schulz *et al.*, 1990). Three other *b* alleles were identified by nucleic acid hybridization using an 8-kb *Bam*H1 fragment containing *b*2 as a probe. All four alleles contained an ORF of 410 aas. Comparison of the amino acid sequences revealed that the ORFs consist of a variable region in the amino-terminal 110 aas and a constant region in the remainder of the ORF (Fig. 4). The constant region contains a homeodomain motif, suggesting that the *b* polypeptides are DNA-binding proteins (see Schulz *et al.*, 1990; Fig. 5). Further support for this proposal came from the finding that the *b* polypeptides exhibit significant similarity with known yeast DNA-binding proteins (see Schulz *et al.*, 1990; Fig. 6). The sequences of additional *b* alleles described recently (*b*5, *b*6, and *b*7; Kronstad and Leong, 1990; Fig. 5) further demonstrate the remarkable conservation of the organization of the *b* locus first described by Schulz and co-workers (1990).

In summary, the biological observations indicate that different *b* alleles are necessary for pathogenic development. The status of *b* alleles in the cell may be monitored by interaction of *b* monomers to form a multimeric regulatory protein that, in turn, governs expression of target genes for filamentous growth (*fuz* genes) and tumor induction (*tum* genes). Thus, the *b* locus appears to be a master regulator that encodes a DNA-binding protein responsible for the switch to pathogenic development (Schulz *et al.*, 1990).

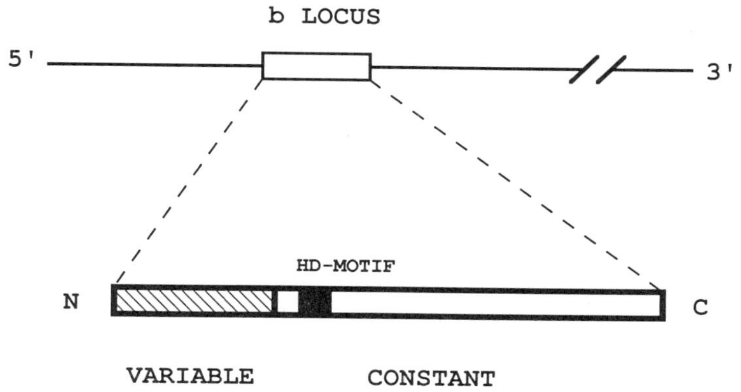

Fig. 4. Organization of the open reading frame encoded by the *b* locus. The *b* locus contains an open reading frame (ORF) of 410 amino acids (or 473 aa after processing of an intron near the 3′ end; see Schulz *et al.*, 1990). The first 110 amino acids are highly variable; the rest are highly conserved and constitute a constant region (see Fig. 5). A homeodomain-related (HD) motif (see Figs. 5 and 6) is found in the highly conserved region as indicated.

```
                    20                  40                  60                  80                 100
b2 MSNYPNFSLTSFVECLNEIEHEFLRDKLENRPVLVRKLQELRRKTPNNVASLSYDPGTIHQIHQTTHRIKVAAKAFIRIDQSFVSLHSDAVEDTSKALKK
b3 --RD-KL--SK-L--------------V-H-----------QQ---KH--K-PHE-EM-Q----AA---DIV-Q----F--K----C-EI-H--T-VMQE
b4 --SD-KI-I---L---S---E------EK-P----S-----QQ-----I-N-DH--K--QK--------N--V----C---T----R------A-R----
b1 --SD-----I--L--------------G--Y-------R--QQ-I--DI-N-PR--E--Q---------RAV-Q----F--K----C-EV-HG---VMQE
b5 --SD--------L--------------E----II----R--QQ----H--D-AH-SK--E-----A------V-T------K----C-EV-HG---VMEE
b6 --RD-KL--SK-L--------------V-H-----------QQ---KH-TN-PN--E--Q----IA--LE--V-V-LH--RK----R--V--------QE
b7 --SD--------L---SQ---------V--P----------QQ---GH----LH--E--Q-----A---E--V-V--H---K-T-GC-EV-HG---VMQE

                   120                 140                 160                 180                 200
                                                                          ** * *
b2 ADASSPVVGCRDLSEDLPAYHMRKHFLHTLDNPYPTQEEKEGLVRLTNESTARVGLSKANRPPLEVHQLTLWFINARRRSGWSHILKKFAREDRSRMKHL
b3 -NVV--GE---N-----------N------E----------------------IRP-N-I--------------------------------------R-
b4 -N---------E------------------E----------T-------------Q-IV-----------------------------------------
b1 FNVV--D----N---------------L-------------T-------------Q-SV--------HV-------------------------------
b5 FNVV--A-V--N---------------L------------QN---------V---S-NPA----------------------------------------
b6 VNVA--A-EY-N-------------------S---------T-------------Q-SV-----------------------------------------
b7 VNVG--A----N---------------L---S-------------------------N-T-------E------------------------------R-

                   220                 240                 260                 280                 300
b2 VRAKLSSSNQSTPPSSTSDSLSNNLDDVLSDNLGRPLTPVDKQQFEDDWASMISWIKYGVKEKVGDWVYDLCAASKKTPKPGMPRPVTTVAKRHPARKTK
b3 ---------------L--EKP-DD--V-----------LA------------------------------------------------------------
b4 ---------------P--EYP------F-----------A-----------------------------------------------------Q------
b1 -----------------Y-------------------------------------------------------------------------N--------
b5 ---------------P--EYP------F-----------A----------------------------------N---------------T--Q------
b6 ---------------L--EKP-DD--V-----------LA-----------------------------------------------------Q------
b7 --------T--S---PMPEYP-----NI-----------A-----------------------------------------------------Q------

                   320                 340                 360                 380                 400
b2 PAAKPKSRTANPRASTTPSIDSTLDSSKLESTPELSMCSTADTSFSTFGSSLSMSHYNPFQDGNDILQSPTVKARGNRKVKALPKRAGKQQPDEVENGKI
b3 -----------------------------------------------------------------------------------------------D----
b4 ---------------------------------------------------------D---Y---------F----------------------------
b1 -----------------------------------------------------------------------------------------------D----
b5 -----------------------------------------------------------------------------------------------D----
b6 ---------------------------------------------------------D---Y---------F----------------------ID----
b7 -----N----------------------------------------------------------------------------------------------

          410
b2 PFLCLSVAFV
b3 ----------
b4 ----------
b1 ----------
b5 ----------
b6 --F---I---
b7 ----------
```

Fig. 5. Comparison of the deduced amino acid sequence of seven different *b* alleles. Dashes denote amino acid identity. The asterisks indicate the four invariant amino acids WF–N–R, found in all higher eukaryotic homeodomain proteins (Scott *et al.*, 1989). Some members of the fungal family of homeodomain-related proteins contain only three of the four invariant amino acids (see Fig. 6; also Novotny *et al.*, Chapter 10). Data are compiled from Schulz *et al.*, (1990) (*b*1–*b*4) and from Kronstad and Leong (1990) (*b*5–*b*7).

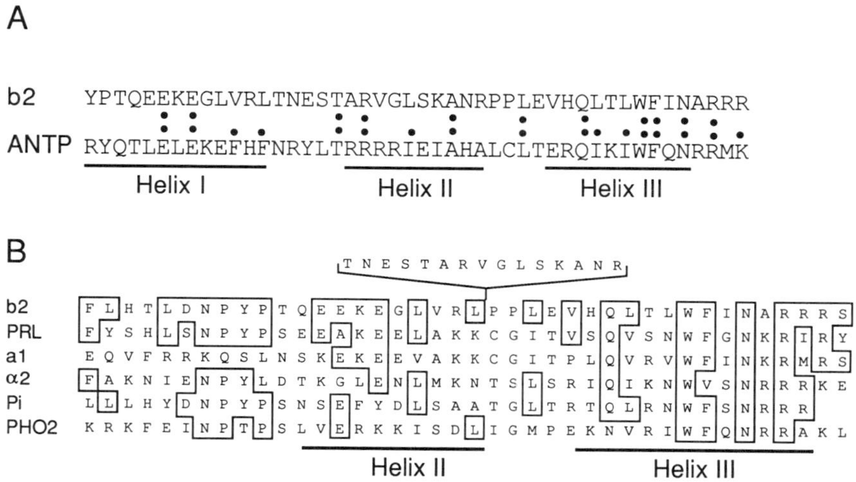

Fig. 6. The homeodomain-related region of *b*. (A) Alignment of the homeodomain-related region of *b*2 with the homeodomain of the *Drosophila* Antennapedia protein (Scott *et al.*, 1989). (B) Alignment with homeodomain-related proteins from humans (*prl;* Kamps *et al.*, 1990; Nourse *et al.*, 1990) and yeast (**a**1, **α**2, matPi, PHO2; Schulz *et al.*, 1990). Helix II and helix III refer to proposed helical regions of the homeodomain (Kissinger *et al.*, 1990). Identical amino acids are indicated with a colon, conservative changes with a single dot. The alignment between *b*2 and the other proteins in B has been maximized by deleting 15 amino acids from *b*2. Amino acids identical in *b*2 and any of the other proteins are boxed.

b. Models. Two broad categories of models have been proposed to explain how the *b* polypeptides might interact to create a functional species that can be a repressor or an activator. Each category encompasses many different possibilities (see Schulz *et al.*, 1990). For simplicity, it was proposed that the monomers interact to form a dimer, but it is, of course, possible that the functional *b* protein is a higher-order multimer (for example, a tetramer). The two categories of models differ in whether the variable or the constant region determines association of the monomers.

In one class of models, the variable region is the determinant of association. If only unlike variable regions are able to associate, then a heteromultimer is formed in the dikaryon. This heteromultimer is the active species and is responsible for expression of genes specific to this cell type. In the haploid cells, this active species would not be present.

In the other class of models, the constant region determines association of monomers. Since all monomers contain the same constant region, it then follows that the variable region is responsible for activity of the species formed. There are many possibilities within this category of models. For example, one could imagine that homomultimers (multimers formed from identical monomers) are the active molecular species, in this case functioning as a repressor of genes for the filamentous cell type in haploid cells. In the dikaryon, nonidentical monomers interact to form an inactive heteromultimer that is unable to repress genes for the filamentous cell type. (In this specific model, it is argued that there is higher affinity for interaction between unlike monomers than like monomers. For higher-order multimers, one does not have to invoke higher affinity between unlike monomers.) An important experimental challenge is the determination of what type of oligomer the *b* polypeptides form and whether the functional species is a repressor or an activator. More complex situations are also possible; for example, homomultimers could modulate expression of certain sets of genes and heteromultimers could regulate other sets. The identification of targets of the *b* protein would facilitate *in vitro* studies aimed at determining the active *b* species. One approach to identify possible target genes for *a* and *b* is described in the next section.

Another goal is to identify the allele-specificity determinants of the *b* polypeptides. The behavior of hybrid alleles indicates that the domain for allele specificity can be roughly localized to a region of 60 aa in the variable region (M. Dahl and R. Kahmann, unpublished observations). Future experiments will make it possible to determine whether new alleles can be constructed by mutating specific amino acid residues.

III. GENES NECESSARY FOR FILAMENTOUS GROWTH AND TUMOR INDUCTION

One way to learn about specific steps regulated by the *a* and *b* loci is to isolate mutants affecting filamentous growth or pathogenicity, identify the genes affected by these mutations, and determine if expression or activity of these genes is governed by *a* or *b*. Identification of such genes will be of interest in itself, since they are expected to be more directly involved in the specific processes that lead to filamentous growth and pathogenicity.

Mutants of an *a*1 *b*1 strain have been isolated that, when mated with an *a*2 *b*2 strain, do not produce filaments (Banuett, 1991a). These mutants are expected to be defective in cell fusion or subsequent steps in development of the pathogenic dikaryotic hyphae. Crosses were performed with some of the mutants to determine if they carry mutation in *a* or *b* or in other genes. Two mutants carry mutations that segregate independently of *a* and *b* and identify two new genes, *fuz*1 and *fuz*2. Two other mutants carry mutations that are linked to *b* and identify a new locus, *rtf*1 (for *r*egulator of *t*umor *f*ormation; Banuett, 1991a).

In addition to being necessary for filament formation, *fuz*1 is also essential for normal size and distribution of tumors and for teliospore formation; *fuz*2 is also necessary for teliospore germination. Mutants in either of these two genes exhibit altered cell morphology—they are longer and thinner than wild-type strains. This morphological alteration suggests that *fuz*1 and *fuz*2 may encode components of the cytoskeleton (Banuett, 1991a). Defects in such components are hypothesized to interfere with localization or deposition of materials needed for tip growth of the filament or bud as well as with cell-wall breakdown. The *A* and *B* factors of both *Schizophyllum commune* and *Coprinus cinereus* control discrete steps in the process of establishment and maintenance of the dikaryotic state (see Novotny *et al.*, Chapter 10; Pukkila and Casselton, Chapter 6). These steps involve nuclear migration, septum dissolution, conjugate division, and formation of clamp connections, which distinguish the dikaryon of these organisms from the monokaryon. Mutations in genes distinct from the mating factors that affect nuclear migration have been identified in *Schizophyllum commune* (Dubovoy, 1976; Raper, 1988). Perhaps some of these genes also encode components of the cytoskeleton (Enos and Morris, 1990; Meluh and Rose, 1990), which is likely to play a role in nuclear migration.

Cloning the *fuz*1 and *fuz*2 genes will be done using the same charcoal plate assay that was used in their isolation. A better understanding of their role in the life cycle of *U. maydis* is expected to be gained from their molecular analysis.

As already mentioned, tumor formation is dependent on the presence of two different *b* alleles. This prerequisite is found to be bypassed by mutations in the *rtf*1 gene. Additional analysis shows that the *rtf*1 mutation is recessive, indicating that the product of the *rtf*1 gene normally inhibits tumor formation. These two observations lead to three specific hypotheses to explain the mode of action of *rtf*1 (Banuett, 1991a). In one model, the *rtf*1 gene product is proposed to be a negative regulator of initiation of tumor formation. The *b* heteromultimer in the dikaryon is proposed to repress *rtf*1. Repression of *rtf*1 would consequently result in tumor induction by the dikaryon (Fig. 7). In the haploid cells, *rtf*1 is expressed, thus inhibiting tumor formation. The regulation of tumor formation by *rtf*1 in this model is analogous to the regulation of meiosis and sporulation by the *RME1* gene of *S. cerevisiae* (Rine *et al.*, 1981; Mitchell and Herskowitz, 1986). The second hypothesis proposes that *rtf*1 is a negative regulator of an ordinarily silent *b* allele. This proposal is similar to the regulation of expression of the silent *MAT* copies in *S. cerevisiae* (Rine and Herskowitz, 1987). It is not known whether *U. maydis* contains cryptic *b* alleles, but precedent for functionally redundant copies of mating-type genes can be

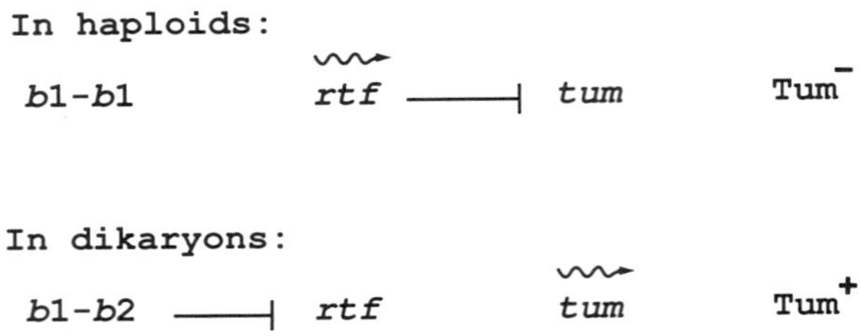

Fig. 7. A possible role for *rtf.* The observation that recessive mutations in the *rtf* gene bypass the requirement for the presence of different *b* alleles for tumor induction leads to three possible hypotheses. In one hypothesis, *rtf* codes for a negative regulator of tumor induction. In haploids, *rtf* is expressed; consequently, tumor formation is inhibited. In dikaryons, *b*1 and *b*2 form a heteromultimer that represses *rtf;* consequently tumors are formed. The other hypotheses are described in the text (see Banuett, 1991a).

found in other basidiomycetes (see Novotny *et al.*, Chapter 10; Pukkila and Casselton, Chapter 6). In the third hypothesis, *rtf*1 is proposed to be a negative regulator of *b* (for details see Banuett, 1991a).

The isolation of mutants defective in filament formation may lead to identification of genes that determine the specific interaction of *U. maydis* with its host. Determining the specific function of these genes will provide a better understanding of filament formation and of tumor induction. If the genes prove to be regulated by *b*, they will also provide a useful molecular reagent in studies of the mode of action of the *b* protein.

IV. CONCLUDING REMARKS

There has been enormous progress in the molecular analysis of mating type of filamentous fungi. The mating-type loci for some filamentous ascomycetes have been cloned and sequenced as have those for some basidiomycetes: the *A*α locus of *Schizophyllum commune* (Giasson *et al.*, 1989; Stankis *et al.*, 1990; also, Novotny *et al.*, Chapter 10), the *A*α and *A*β of *Coprinus cinereus* (cited in Novotny *et al.*, Chapter 10), and the *a* (M. Bölker and R. Kahmann, unpublished data) and *b* (Kronstad and Leong, 1990; Schulz *et al.*, 1990) mating type loci of *Ustilago maydis.* Sequence information of different alleles of the multiallelic *b* locus of *U. maydis* led to a description of the specific organization of these alleles into a variable and constant region and to the identification of a homeodomain-related motif in the ORF encoded by these alleles. The information derived from such analysis led to the proposal that *b* is a DNA-binding protein governing expression of putative target genes (*fuz, tum*) and to specific models of *b* polypeptide interaction to create this functional regulatory species. The implications of this work go beyond *U. maydis* and not only provide a way of thinking about interaction of polypeptides encoded by the multiallelic loci of other basidiomycetes but address fundamental questions of protein structure and function.

The identification of genes distinct from *a* and *b* that affect filamentous growth and pathogenicity using classical genetic approaches was motivated by similar approaches used in budding yeast, *Saccharomyces cerevisiae,* to identify genes distinct from *MAT* that affect the mating process (MacKay and Manney, 1974a,b). Characterization of the genes thus identified in *S. cerevisiae* not only led to the identification of target genes for the regulatory proteins of the mating-type locus (Sprague *et al.*, 1983; Wilson and Herskowitz, 1984; see Herskowitz, 1988, 1989), but many of these genes were found to be part of a signalling pathway for the mating pheromones (Cross *et al.*, 1988; Herskowitz, 1989). These studies have impacted our understanding of the response of mammalian cells to growth factors and of growth control (Chang and Herskowitz, 1990).

It is hoped that we will achieve the same level of understanding with respect to how the mating-type loci of *U. maydis* govern its life cycle as has been achieved in the budding yeast *S. cerevisiae.* Moreover, an understanding of the molecular basis of pathogenicity in this fungus will provide a framework for studying the interactions of other fungal pathogens with their hosts.

ACKNOWLEDGMENTS

I would like to thank Ira Herskowitz for comments on this manuscript, and Regine Kahmann, the members of her laboratory, and Charles Novotny for communicating unpublished results. My work has been supported by the Weingart Program in Developmental Genetics at UCSF and by a research grant (to Ira Herskowitz) from the National Institutes of Health (AI 18738).

REFERENCES

Astell, C. R., Ahlstrom-Jonasson, L., Smith, M., Tatchell, K., Nasmyth, K. A., and Hall, B. D. (1981). The sequence of the DNAs coding for the mating type loci of *Saccharomyces cerevisiae. Cell* **27,** 15–23.

Banuett, F. (1991a). Identification of genes necessary for filamentous growth and tumor induction of the plant pathogen *Ustilago maydis. Proc. Natl. Acad. Sci. U.S.A.* **88,** 3922–3926.

Banuett, F. (1991b). Pathogenicity determinants in the smut fungi of cereals. *In* "Virulence Determinants of Plant Pathogenic Fungi" (T. Böller and F. Meins, eds.) Springer-Verlag, New York.

Banuett, F., and Herskowitz, I. (1988). Smut of maize. In "Genetics of Plant Pathogenic Fungi" (G. S. Sidhu, ed.), Advances in Plant Pathology, Vol. 6, pp. 427–455. Academic Press, New York.

Banuett, F., and Herskowitz, I. (1989). Different *a* alleles of *Ustilago maydis* are necessary for maintenance of filamentous growth but not for meiosis. *Proc. Natl. Acad. Sci. U.S.A.* **68,** 5878–5882.

Casselton, L. A. (1978). Dikaryon formation in higher basidiomycetes, *In* "The Filamentous Fungi" (J. E. Smith and D. R. Berry, eds.), Vol. 3, pp. 275–297. Arnold, London.

Chang, F., and Herskowitz, I. (1990). Identification of a gene necessary for cell cycle arrest by a negative growth factor of yeast: *FAR1* is an inhibitor of a G1 cyclin, *CLN2*. *Cell* **63,** 999–1011.

Christensen, J. J. (1963). "Corn Smut Caused by *Ustilago maydis.*" Monograph No. 2. American Phytopathological Society, St. Paul, Minnesota.

Cross, F., Hartwell, L., Jackson, C., and Konopka, B. (1988). Conjugation in *Saccharomyces cerevisiae. Ann. Rev. Cell Biol.* **4,** 429–457.

Day, P. R., Anagnostakis, S. L., and Puhalla, J. E. (1971). Pathogenicity resulting from mutation at the *b* locus of *Ustilago maydis. Proc. Natl. Acad. Sci. U.S.A.* **68,** 533–535.

Dubovoy, C. (1976). A class of genes affecting *B* factor-regulated development in *Schizophyllum commune. Genetics* **82,** 423–428.

Enos A. P., and Morris, N. R. (1990). Mutation of a gene that encodes a kinesin-like protein blocks nuclear division in *A. nidulans. Cell* **60,** 1019–1027.

Glass, N. L., and Staben C. (1990). Genetic control of mating in *Neurospora crassa.* Seminars in Developmental Biology, **1,** 177–184.

Glass, N. L., Vollmer, S. J., Staben, C., Grotelueschen, J., Metzenberg, R. L., and Yanofsky, C. (1988). DNAs of the two mating-type alleles of *Neurospora crassa* are highly dissimilar. *Science* **241,** 570–573.

Glass, N. L., Grotelueschen, J., and Metzenberg, R. L. (1990). *Neurospora crassa A* mating-type region. *Proc. Natl. Acad. Sci. U.S.A.* **87,** 4912–4916.

Giasson, L., Specht, C. A., Milgrim, C., Novotny, C. P., and Ullrich, R. C. (1989). Cloning and comparison of $A\alpha$ mating-type alleles of the Basdiomycete *Schizophyllum commune. Mol. Gen. Genet.* **218,** 72–77.

Herskowitz, I. (1988). Life cycle of the budding yeast *Saccharomyces cerevisiae. Microbiol. Rev.* **52,** 536–553.

Herskowitz, I. (1989). A regulatory hierarchy for cell specialization in yeast. *Nature (London)* **342,** 749–757.

Holliday, R. (1961). The genetics of *Ustilago maydis. Genet. Res. Camb.* **2,** 204–230.

Holliday, R. (1974). *Ustilago maydis. In* "Handbook of Genetics" (R. C. King, ed.), Vol. 1, pp. 575–595. Plenum Press, New York.

Kamps, M. P., Murre, C., Sun, X-H., and Baltimore, D. (1990). A new homeobox gene contributes the DNA binding domain of the t(1;19) translocation protein in pre-B ALL. *Cell* **60,** 547–555.

Kelly, M., Burke, J. Smith, M., Klar, A., and Beach, D. (1988). Four mating-type genes control sexual differentiation in the fission yeast. *EMBO J.* **7,** 1537–1547.

Kenaga, C. B., Williams, E. B., and Green, R. J. (1971). "Plant Disease Syllabus." Balt, Lafayette, Indiana.

Kissinger, C. R., Beishan, L., Martin-Blanco, E., Kornberg, T. B., and Pabo, C. O. (1990). Crystal structure of an engrailed homeodomain-DNA complex at 2.8Å resolution: a framework for understanding homeodomain-DNA interactions. *Cell* **63,** 579–590.

Kronstad, J. W., and Leong, S. A. (1989). Isolation of two alleles of the *b* locus of *Ustilago maydis. Proc. Natl. Acad. Sci. U.S.A.* **86,** 878–982.

Kronstad, J. W., and Leong, S. A. (1990). The *b* mating type locus of *Ustilago maydis* contains variable and constant regions. *Genes Dev.* **4,** 1384–1395.

MacKay, V. L., and Manney, T. R. (1974a). Mutations affecting sexual conjugation and related processes in *Saccharomyces cerevisiae.* I. Isolation and phenotypic characterization of non-mating mutants. *Genetics* **76,** 255–271.

MacKay, V. L., and Manney, T. R. (1974b). Mutations affecting sexual conjugation and related processes in *Saccharomyces cerevisiae.* II. Genetic analysis of non-mating mutants. *Genetics* **76,** 273–288.

Meluh, P. B., and Rose, M. D. (1990). *KAR3,* a kinesin-related gene required for yeast nuclear fusion. *Cell* **60,** 1029–1041.

Mitchell, A. P., and Herskowitz, I. (1986). Activation of meiosis and sporulation by repression of the *RME1* product in yeast. *Nature (London)* **319,** 738–742.

Nourse, J., Mellentin, J. D., Galili, N., Wilkinson, J., Stanbridge, E., Smith, S.D., and Cleary, M. L. (1990). Chromosomal translocation t(1;19) results in synthesis of a homeobox fusion mRNA that codes for a potential chimeric transcription factor. *Cell* **60,** 535–545.

O'Donnell, K. L., and McLaughlin, D. J. (1984a). Ultrastructure of meiosis in *Ustilago maydis. Mycologia* **76,** 468–485.

O'Donnell, K. L., and McLaughlin, D. J. (1984b). Postmeiotic mitosis, basidiospore development, and septation in *Ustilago maydis. Mycologia* **76,** 486–502.

Puhalla, J. E. (1968). Compatibility reactions on solid medium and interstrain inhibition in *Ustilago maydis. Genetics* **60,** 461–474.

Puhalla, J. E. (1970). Genetic studies of the *b* incompatibility locus of *Ustilago maydis.. Genet. Res. Camb.* **16,** 229–232.

Raper, C. A. (1983). Controls for development and differentiation in the dikaryon in Basidiomycetes. *In* "Secondary Metabolism and Differentiation in Fungi" (J. Bennett and A. Ciegler, eds.), pp. 195–238. Marcel Dekker, New York.

Raper, C. A. (1988). *Schizophyllum commune,* a model for genetic studies of the Basidiomycotina. *In* "Genetics of Plant Pathogenic Fungi" (G. S. Sidhu, ed.), Advances in Plant Pathology, Vol. 6, pp. 511–522. Academic Press, New York.

Rine J., and Herskowitz, I. (1987). Four genes responsible for a position effect on expression from *HML* and *HMR* in *Saccharomyces cerevisiae. Genetics* **116,** 9–22.

Rine, J., Sprague, G. F., Jr., and Herskowitz, I. (1981). *rme1* mutation of *Saccharomyces cerevisiae:* Map position and bypass of mating type locus control of sporulation. *Mol. Cell. Biol.* **1,** 958–968.

Rowell, J. B. (1955). Functional role of compatibility factors and an *in vitro* test for sexual compatibility with haploid lines of *Ustilago zeae. Phytopathology* **45,** 370–374.

Rowell, J. B., and DeVay, J. E. (1954). Genetics of *Ustilago zeae* in relationship to basic problems of its pathogenicity. *Phytopathology* **44,** 356–362.

Schulz, B., Banuett, F., Dahl, M., Schlesinger, R., Schäfer, W., Martin, T., Herskowitz, I., and Kahmann, R. (1990). The *b* alleles of *U. maydis,* whose combinations program pathogenic development, code for polypeptides containing a homeodomain-related motif. *Cell* **60,** 295–306.

Scott, M. P., Tamkun, J. W., and Hartzell, G. W., III. (1989). The structure and function of the homeodomain. *Biochim. Biophys. Acta* **989,** 25–48.

Sprague, G. F., Jr., Jensen, R., and Herskowitz, I. (1983). Control of yeast cell type by the mating type locus: Positive regulation of the α-specific *STE3* gene by the *MAT*α*1* product. *Cell* **32,** 409–412.

Staben, C., and Yanofsky, C. (1990). *Neurospora crassa a* mating-type region. *Proc. Natl. Acad. Sci. U.S.A.* **87,** 4917–4921.

Stankis, M. M., Specht, S. A., and Giasson, L. (1990). Sexual incompatibility in *Schizophyllum commune:* From classical genetics to a molecular view. Seminars in Developmental Biology **1,** 195–206.

Wilson, K., and Herskowitz, I. (1984). Negative regulation of *STE6* gene expression by the $\alpha 2$ product of *Saccharomyces cerevisiae. Mol. Cell. Biol.* **4,** 2420–2427.

Wong, G. J., and Wells, K. (1985). Modified bifactorial incompatibility in *Tremella mesenterica. Trans. Br. Mycol. Soc.* **84,** 834–838.

10

The $A\alpha$ Mating Type Locus of *Schizophyllum commune*

CHARLES P. NOVOTNY
MARY M. STANKIS
Department of Microbiology and Molecular Genetics
University of Vermont
Burlington, Vermont

CHARLES A. SPECHT
HUILING YANG
ROBERT C. ULLRICH
Department of Botany
University of Vermont
Burlington, Vermont

LUC GIASSON
Biotechnology Laboratory
University of British Columbia
Vancouver, British Columbia
Canada

MORE GENE MANIPULATIONS
IN FUNGI

I. INTRODUCTION AND LIFE CYCLE

Filamentous fungi are attractive organisms for studying gene regulation during development. They have fascinating life cycles, are of manageable complexity, and are amenable to most biochemical and genetic techniques. Most importantly, tractable transformation systems now exist for several species, making it possible to take molecular approaches to the study of processes essential for development.

One of these model fungi is the basidiomycete *Schizophyllum commune.* This common wood-rotting mushroom has many features that lend themselves to studies of development: the life cycle can be completed on defined media within 2 wk; cells in two phases of the life cycle (homokaryons and dikaryons) can be maintained in culture indefinitely; and major processes in its development have been well characterized through cytological studies and extensive genetic analysis (Raper 1966, 1983, 1988; Stankis *et al.,* 1990).

Traditional genetic analysis, mostly done by John Raper and his collaborators in the 1960s, showed that sexual development in *S. commune* is governed by four multiallelic loci, called $A\alpha$, $A\beta$, $B\alpha$, and $B\beta$, that regulate two distinct developmental pathways. How these loci serve as "master switches" to regulate other genes has been the subject of much speculation and analysis. That it still remains a mystery should not be surprising; any regulatory circuit with four different switches (loci) and a total of 59 possible alleles should be a challenge to unravel.

Despite this complexity, some progress has been made toward understanding how this system operates at the molecular level. With the use of transformation, we recently succeeded in cloning three alleles of the $A\alpha$ locus. In this chapter we will discuss what we have learned about the structure of this locus and will compare the general organization of $A\alpha$ to the structure of mating-type loci in other filamentous fungi.

A. Life Cycle

The major events in the life cycle of *Schizophyllum commune* are shown in Fig. 1. Haploid spores are produced by the basidial cells of fruiting bodies. Each spore germinates and grows into a mycelium of uninucleate

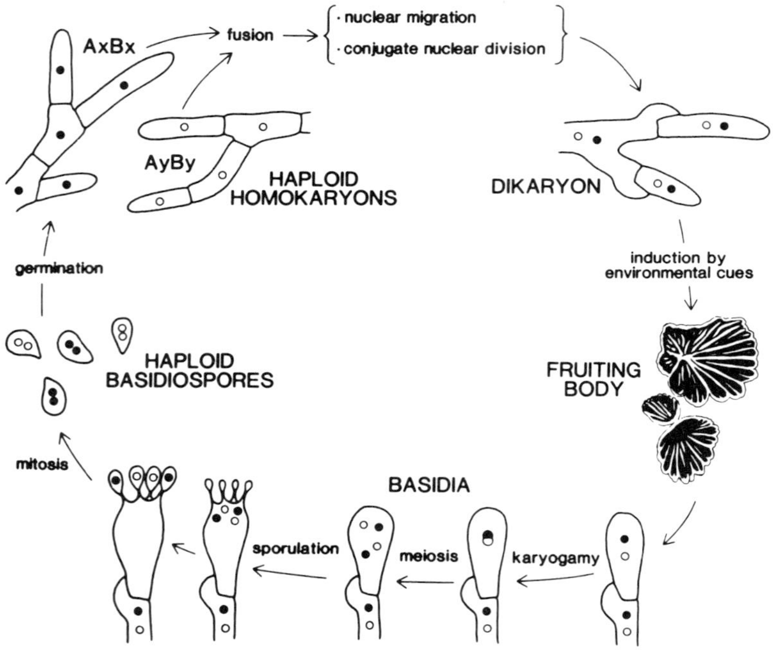

Fig. 1. Life cycle of *S. commune.* White and black circles represent haploid nuclei. *A* and *B* are incompatibility loci that determine mating type. Subscripts x and y refer to different incompatibility factors. (After Stankis *et al.*, 1990.)

cells called the homokaryon. Homokaryons are capable of fusing. Cell fusions occur between all homokaryons, regardless of mating type, but in fusions between cells of the same mating type there is no sexual development and the fused cells vacuolate. If the nuclei of the two mates contain different alleles of both the *A* and *B* mating-type loci, the full cycle of sexual development ensues. If the nuclei have different alleles at either the *A* or *B* loci, but not both, sexual development is incomplete. These hemicompatible matings reveal the existence of two distinct pathways of sexual development; one controlled by the *A* mating-type loci and one controlled by the *B* loci.

In a fully compatible mating, the products of the *A* and *B* alleles trigger a series of events that transform the two homokaryotic mycelia into a single fertile mycelium called the dikaryon. Each cell in the dikaryotic mycelium contains one nucleus from each mate. The dikaryotic mycelium can be propagated indefinitely; in response to appropriate environmental conditions, it will give rise to fruiting bodies. Nuclear fusion and meiosis occur in the basidial cells of the fruiting body, followed by sporulation.

We have a general understanding of events involved in the formation of the dikaryon (Fig. 2). There is a reciprocal exchange of nuclei between the two mates and the exchanged nuclei migrate throughout the mycelium of the mate. The nuclei divide as they travel. In each homokaryotic cell, the septa are degraded to allow the nuclei to pass. These homokaryotic septa are then replaced with different septa characteristic of dikaryotic cells. By this process the two homokaryotic mycelia develop into a single dikaryotic mycelium.

During growth, the dikaryon maintains a 1 : 1 ratio of the two different nuclei in newly formed tip cells. The process called conjugate nuclear division involves the formation of lateral appendages called hook cells, conjugate nuclear division, hook cell septation, and hook cell fusion. A hook cell fused to the subapical cell creates a visible structure called a clamp connection which is often used to distinguish dikaryons from homokaryons in *S. commune.* The two nuclei are closely associated in each cell of the dikaryon.

We have some knowledge of the processes controlled by the *A* and *B* mating-type loci (Fig. 3). Hemicompatible matings show that nuclear migration is controlled by the *B* loci and that nuclear pairing, cellular septation, hook cell formation, conjugate nuclear division, and hook cell septation are controlled by the *A* loci. The fusion of a hook cell to the subapical cell is controlled by the *B* loci.

How the *A* and *B* loci control other genes is, of course, one of the most interesting and puzzling aspects of this developmental system. The first definitive information about *A* and *B* that provided some clues about how the gene products might act was obtained by Dr. John Raper and his colleagues. Using traditional genetic approaches they were able to enumer-

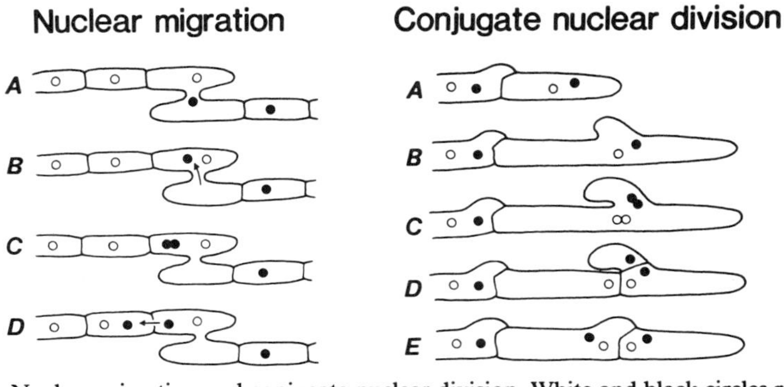

Fig. 2. Nuclear migration and conjugate nuclear division. White and black circles represent haploid nuclei. (After Stankis *et al.*, 1990.)

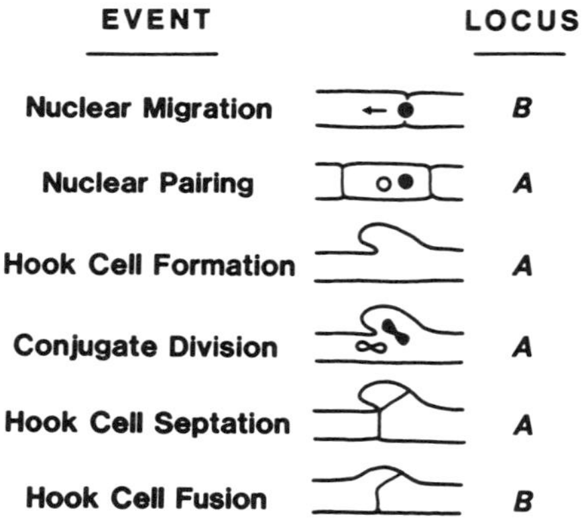

Fig. 3. Processes controlled by the *A* and *B* mating-type loci of *S. commune*. The *B* mating-type loci control nuclear migration and hook cell fusion. The *A* mating-type loci govern nuclear pairing, hook cell formation, conjugate division, and hook cell septation. (After Stankis *et al.*, 1990.)

ate the alleles of each of the four loci and obtain interesting mating-type mutations. Some mutations map to the mating-type loci; others occur in unlinked genes that are presumably controlled by the *A* and *B* loci. A short summary of this work is presented in Section II. The picture of mating type that emerged from these early studies made it possible for us to design molecular approaches and interpret the results; our debt to Dr. John Raper and his colleagues is difficult to overestimate.

II. TRADITIONAL ANALYSIS OF MATING TYPE

A. Mating-Type Loci

Compatibility tests and linkage analysis showed that *Schizophyllum commune* has four mating-type loci—*Aα, Aβ, Bα,* and *Bβ*—that are arranged in pairs on two chromosomes (I and II) as shown in Fig. 4. Statistical analysis of worldwide populations shows that there are 9 *Aα* alleles, 32 *Aβ* alleles, and 9 alleles each for *Bα* and *Bβ* (Raper *et al.*, 1958, 1960; Koltin *et al.*, 1967; Stamberg and Koltin, 1972). Full sexual development ensues from the fusion of two homokaryons only if the two mates have different alleles for *A* and *B*, as shown in Table I.

The mating-type loci appear to be functionally redundant, that is, allelic differences of the mates at either *Aα* or *Aβ* fully activate the *A* pathway and differences at either *Bα* or *Bβ* fully activate the *B* pathway. Why *S. commune* maintains two equivalent sets of mating-type loci (some basidiomy-

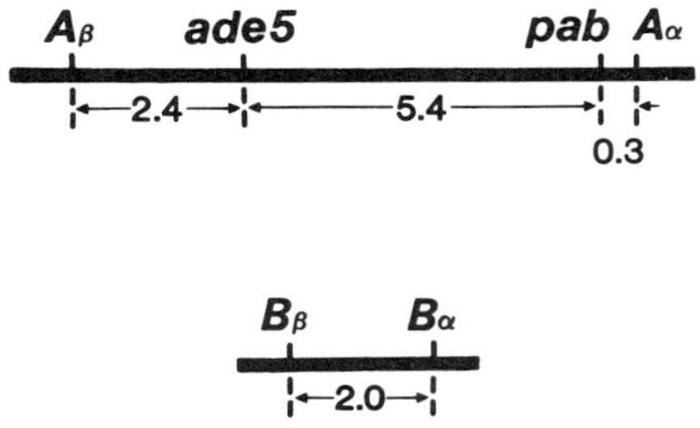

Fig. 4. Map of the mating-type linkage groups. Recombination frequencies are in centiMorgans. *Aα, Aβ, Bα, Bβ,* mating-type loci; *pab1,* synthesis of *p*-aminobenzoic acid; *ade5,* synthesis of adenine (J. R. Raper, 1966).

cetes such as *Ustilago maydis* apparently only have one set; see Banuett, Chapter 9) raises some interesting questions. Are the products of the *Aα* and *Aβ* loci similar, and do the products act on the same targets? Another intriguing aspect of the system is that each allele seems to be equally effective in triggering development.

B. Mutations

Extensive mutational analysis of mating type in *Schizophyllum commune* identified many mutations that influence sexual development. Two types of mutations, constitutive mutations *(con)* and mutations that suppress the constitutive phenotype, may be the key to understanding how the

TABLE I.

Examples of Incompatible, Compatible, and Hemicompatible Matings in *Schizophyllum*

Genotype of mate 1	Genotype of mate 2	Status of regulated events	
		A	*B*
Aα1 Aβ1 Bα1 Bβ2	*Aα1 Aβ1 Bα1 Bβ2*	off	off
Aα1 Aβ1 Bα1 Bβ2	*Aα2 Aβ1 Bα2 Bβ2*	on	on
Aα1 Aβ1 Bα1 Bβ2	*Aα1 Aβ2 Bα1 Bβ1*	on	on
Aα1 Aβ1 Bα1 Bβ2	*Aα2 Aβ2 Bα2 Bβ1*	on	on
Aα1 Aβ1 Bα1 Bβ2	*Aα2 Aβ1 Bα1 Bβ2*	on	off
Aα1 Aβ1 Bα1 Bβ2	*Aα1 Aβ1 Bα2 Bβ2*	off	on

system works. Constitutive mutations map to the mating-type loci and enable a specific allele to activate the developmental pathway that it normally controls in the absence of a compatible allele (Parag, 1962; Raper *et al.,* 1965; Koltin, 1968; Raudaskoski *et al.,* 1976; Koltin *et al.,* 1979). Constitutive mutations may be in structural gene sequences of the *A* and *B* loci, where they affect the nature of the gene products, or they may be in regulatory sequences. Since *con* mutations alleviate the need for a compatible allele, they may reveal how compatible combinations of products do their work. Constitutive mutations have been obtained for $A\beta$, $B\alpha$, and $B\beta$, but not $A\alpha$. Attempts to obtain $A\alpha con$ mutants were not extensive, and it is possible that $A\alpha con$ mutants can also be isolated. In homokaryons, *con* mutations produce recognizable morphological changes. For example, *Bcon* mutants produce a submerged highly branched mycelium, termed "flat," in which nuclear migration and septal dissolution are constantly occurring. *Acon* mutants produce hook cells. Many suppressor mutations of the *Bcon* "flat" phenotype were isolated and analyzed for their effect on sexual development (Raper and Raper, 1964, 1966, 1973; Raper and Raudaskoski, 1968; Koltin, 1970). About 30 loci were identified; most of these involve nuclear migration and map to chromosome II (reviewed by Raper, 1988). It is likely that these loci are targets for the products of the *B* loci. A few suppressors of the *Acon* phenotype were isolated in early studies, but unfortunately the mutants were lost (C. Raper, personal communication). *Acon* suppressor mutations may identify genes controlled by the *A* loci.

III. MOLECULAR STUDIES

A. Genome

Kinetic analysis of DNA reassociations indicates that the size of the *Schizophyllum commune* genome is 3.6×10^7 base pairs (2.5 times the genome of baker's yeast) and that 10% of the DNA is repetitive (Dons *et al.,* 1979; Dons and Wessels, 1980; Ullrich *et al.,* 1980). There is uncertainty about the number of chromosomes. Genetic mapping of over 100 auxotrophic and developmental genes identified seven linkage groups (I–VII) (Raper, 1988), whereas direct observation by electron microscopy indicated the presence of 11 chromosomes (Carmi *et al.,* 1978). Recent attempts to resolve chromosomes by pulsed-field gel electrophoresis have been partially successful. Six bands, two of which are thought to be triplets and one a doublet, have been observed, and bands (chromosomes) corresponding to linkage groups I, II, IV, and V were identified with probes of cloned *S. commune* genes (Horton and Raper, 1991).

B. Development of Transformation

The transformation protocol developed for *Schizophyllum commune* is similar to those used for other filamentous fungi. Protoplasts are prepared with Novozym 234 (Novo Biolabs, Witton, Connecticut) and transformed with linear and circular DNA in the presence of $CaCl_2$ and polyethylene glycol (Muñoz-Rivas *et al.*, 1986a; Specht *et al.*, 1988). Protoplasts are usually prepared from strains deficient in either indole 3-glycerol-phosphate synthetase *(trp1)* or orotidine-5′-phosphate decarboxylase *(ura1)*, and the cloned *S. commune* genes corresponding to these mutations (*TRP1* and *URA1*) are used as selectable markers. The transformation frequency can be as high as 1×10^3 trp^+ or ura^+ transformants per μg of plasmid DNA containing *TRP1* or *URA1* (Specht *et al.*, 1988). This frequency is adequate for complementation of specific mutations by transformation with DNA from *S. commune* gene libraries in plasmids and cosmids (see subsequent text). Transformation of *S. commune* for hygromycin B resistance was also recently reported (Mooibroek *et al.*, 1990).

In *S. commune,* transforming DNA integrates into the genome. Vectors capable of autonomous replication in *S. commune* have not been developed to date. Transforming DNA usually integrates by nonhomologous recombination and is generally stable in mitosis and meiosis. Homologous recombination has also been observed. The frequency of homologous integration increases when the homology between genomic and transforming DNA is extended. Some of these homologous recombination events result in gene replacements. Recently we succeeded in replacing *TRP1* with a mutated copy (*TRP1* with a *URA1* insertion) and found that about 2% of the integration events were replacements (C. Specht, unpublished results). The ability to make gene replacements in *S. commune* will be important for studies of development because loci essential for development can be mutated or deleted and the function of cloned genes can be surmised by gene disruption. Performing gene replacements will add a new and important dimension to our ability to study mating type.

C. Cloning Auxotrophic Genes

One *Schizophyllum* gene, *TRP1,* was isolated by complementation of the *trpC* (indole 3-glycerol-phosphate synthetase and phosphoribosylanthranilate isomerase) mutation in *Escherichia coli* (Muños-Rivas *et al.*, 1986b). Four additional auxotrophic genes of *S. commune*—*URA1, PAB1* (encoding a gene essential for synthesis of *p*-aminobenzoic acid), *ADE2,* and *ADE5* (encoding genes essential for the synthesis of adenine)—were isolated by complementation of the corresponding *S. commune* mutations (*ura1, pab1, ade2,* and *ade5*) with DNA from a plasmid gene library. All four genes were recovered from transformants by plasmid rescue in *E. coli* (Froeliger *et al.*, 1987). This procedure makes it feasible to isolate directly any *S. commune* gene for which mutants can be obtained.

IV. ANALYSIS OF $A\alpha$ LOCI

A. Isolation

Once transformation became tractable, we began the molecular cloning of the mating-type loci. Initially we focused on the $A\alpha$ locus because it seemed feasible to walk the chromosome from the closely linked *PAB1* gene, which had been cloned, to $A\alpha$ (Fig. 4). We made a cosmid gene library in a vector that contained the *TRP1* gene as a selectable marker. The genomic DNA was from an $A\alpha4$ $A\beta6$ strain. Starting at *PAB1,* we walked the chromosome in both directions in the $A\alpha4$ $A\beta6$ cosmid library and transformed an $A\alpha1$ $A\beta6$ *trp1* strain to trp^+ with cosmids obtained in the walk. The rationale was that integration of the $A\alpha4$ locus into the genome of the $A\alpha1$ recipient would yield a homokaryon with two compatible $A\alpha$ alleles; this would activate the *A* pathway of sexual development. These $A\alpha1/A\alpha4$ homokaryons were expected to produce hook cells and mate with both $A\alpha1$ $A\beta6$ and $A\alpha4$ $A\beta6$ tester strains with *B* alleles different from those of the transformant. About 50 kb from *PAB1* we recovered cosmids that transformed cells for these characteristics. An analysis of overlapping cosmids showed that the transforming DNA was within a 9-kb region. DNA fragments containing the $A\alpha4$ locus were then used to probe $A\alpha1$ and $A\alpha3$ cosmid–gene libraries. Cosmids that hybridized to the probes were used to transform a *trp1* $A\alpha4$ $A\beta6$ strain to trp^+. Most of the trp^+ transformants produced hook cells and mated with $A\alpha1$ $A\beta6$ and $A\alpha4$ $A\beta6$ tester strains that had different *B* alleles. Analysis of the overlapping cosmids containing $A\alpha4$ and $A\alpha1$ mapped the putative $A\alpha$ loci to sequences about 7 and 9 kb in size (Fig. 5). The pattern of overlapping cosmids containing $A\alpha3$ mapped the $A\alpha3$ locus to a region about 8 kb in size (data not shown).

To prove that the transforming sequences specified a given allele, we did reciprocal transformation experiments (Table II). The results show that cosmids with a putative $A\alpha$ allele were capable of transforming recipients with a different $A\alpha$ allele but not $A\alpha$ strains with the same allele. This shows that the cloned sequences encode allelic specificity as expected from the analysis of compatible and incompatible matings (Table I).

Southern analysis of representative transformants showed that the transforming $A\alpha$ DNA was integrated at different sites in each transformant and

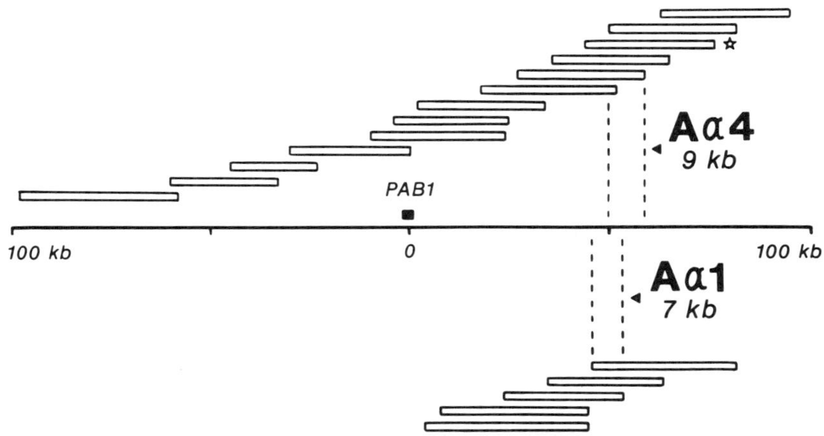

Fig. 5. Map of cosmids with *Aα1* and *Aα4* activity in transformation. The solid line represents chromosome (linkage group) I. Upper bars represent overlapping cosmid inserts obtained by walking the chromosome of an *Aα4* homokaryon from *PAB1* in two directions. Lower bars represent overlapping cosmid inserts obtained by probing a cosmid gene library of an *Aα1* strain with the *Aα4* cosmid marked with a star. Dotted lines intersect inserts with transforming activity for *Aα*. Restriction sites used for mapping are not shown.

that the resident *Aα* locus was intact. Analysis of progeny from some transformants showed that the donor *Aα* was linked to *TRP1* and unlinked to *Aα* of the recipient. These results indicate that the donor *Aα* produces a diffusible product(s) which acts in concert with the *Aα* product(s) of the recipient to activate the *A* pathway of development.

TABLE II

Fraction of *TRP*⁺ Transformants also Transformed for *Aα*[a]

	Aα of the *trp1* recipient			
Aα on the *trp1* cosmid	*Aα1*	*Aα3*	*Aα4*	*Aα5*
Aα1	0/57[b]	—	21/29	9/14
Aα3	—	0/120	28/69	10/30
Aα4	137/210	—	0/29	6/6

[a] *Aα* transformants produce hook cells and mate with strains carrying the same *Aα* and *Aβ* alleles as the recipient. *Aα* nontransformants do not produce hook cells and only mate with strains carrying *Aα* and/or *Aβ* alleles that are different from those of the recipient.

[b] Ratio of *Aα* to trp$^+$ transformants.

B. General Structure

Cosmid sequences that are active in *Aα1, Aα3,* and *Aα4* transformation were subcloned to active fragments 2.8 kb *(Aα1),* 3.2 kb *(Aα3),* and 1.2 kb *(Aα4)* in size. In these first attempts at subcloning we used *Aα1, Aα4,* and *Aα5* strains as recipients. We then examined the homology between the fragments by restriction mapping and Southern hybridizations. Restriction analysis showed extensive restriction site polymorphisms in these sequences and in the flanking regions, indicating that the *Aα* DNA for these three alleles was very heterogeneous (Giasson *et al.,* 1989; M. Stankis, unpublished results). Additional evidence of extensive heterogeneity came from Southern hybridizations. Genomic DNA from *Aα1 – Aα9* strains was restricted and DNA blots were probed with the 1.2-kb, 2.8-kb, and 3.2-kb fragments encoding *Aα4, Aα1,* and *Aα3,* respectively. These probes, under stringent and relaxed stringency conditions, hybridized only to DNA obtained from a strain carrying the same *Aα* allele. This shows that *Aα1, Aα3,* and *Aα4* strains contain only one allele of *Aα*; there are no silent alleles as in baker's yeast (Herskowitz, 1989). In an attempt to define broadly the heterologous *Aα* region for all nine *Aα* alleles, we made a series of small probes from DNA fragments spanning the *Aα4* and *Aα1* regions and probed blots of DNA from *Aα1 – Aα9* strains (C. Specht, unpublished results). The restriction maps made from these results indicated that the heterogeneous regions in all nine *Aα* alleles were about 5 – 8 kb. Thus the *Aα* locus, as pictured from these preliminary studies, consists of a unique region of DNA within which is embedded a shorter sequence that encodes *Aα* specificity in transformation.

C. Analysis of DNA Sequences

As a first approximation to define genes within the *Aα* loci we began to analyze the DNA sequences in the regions that appeared, on the basis of Southern hybridizations, to be unique to each allele. The extent of this analysis to date is summarized in Fig. 6. Except for two small gaps, most of the DNA sequence thought to be unique to *Aα1* and *Aα4* has been determined. Only about 40% of the sequence unique to *Aα3* has been obtained. The nature of the transition from unique to identical DNA is unknown because we have not yet sequenced all the boundaries.

Sequences were analyzed for open reading frames (ORFs) and all the ORFs were compared with respect to DNA sequence and deduced polypeptide sequence. Sequences were then examined for putative introns that had consensus splice sites 3′AG and 5′GT (Mount, 1982) and the conserved branch point sequences common to *Schizophyllum commune*

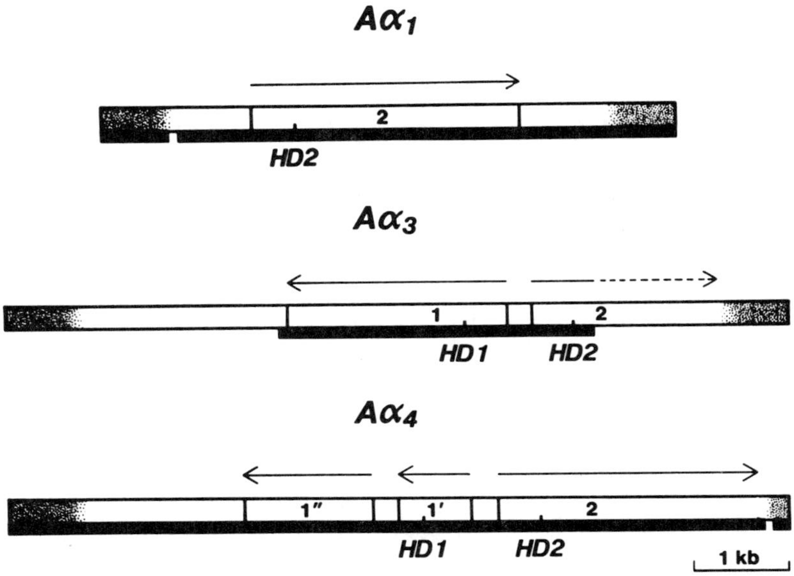

Fig. 6. Sequence analysis of *Aα1, Aα3,* and *Aα4.* Homologous DNA flanking the regions unique to each locus is indicated by stipling. The thick bar delineates regions that have been sequenced. Open reading frames thought to encode proteins are boxed and numbered. Arrows indicate the direction of transcription. *HD1* and *HD2* indicate regions with homeodomain motifs.

(Schuren and Wessels, 1990) and other filamentous fungi (Ballance, 1986; Brown, 1986). Proteins were deduced from the ORFs that remained after the putative introns were deleted and the amino acid sequences of these polypeptides were compared. This analysis was performed to identify regions that may encode similar proteins (possible alleles) and to identify unique motifs in the deduced proteins that might provide clues about how they might function. There are obvious risks in surmising the structure of genes and proteins from sequence data alone: errors may occur in sequencing, some introns may escape detection, and among the putative introns there may be some that are not spliced. Nevertheless it is possible to glean some useful and provocative information from the sequence data. As discussed in subsequent text, several ORFs appear to encode similar proteins and most of the deduced proteins contain a homeodomain motif (Scott *et al.*, 1989).

In *Aα1,* one ORF (*Aα1*ORF2) had significant identity with ORFs in *Aα3* and *Aα4. Aα1*ORF2 contains seven putative introns. The deduced polypeptide of *Aα1*ORF2 is 891 aas and very basic; the predicted isoelectric point is 9.9. This polypeptide shows identity to polypeptides deduced from two partially sequenced ORFs found in *Aα3* (*Aα3*ORF2) and *Aα4* (*Aα4*ORF2). From the existing sequences it is possible to generate partial polypeptides of 141 aas for *Aα3*ORF2 and 882 aas for *Aα4*ORF2. The partial polypeptide deduced from *Aα3*ORF2 shows 55% identity to the first 141 aas of the polypeptides deduced from *Aα1*ORF2 and *Aα4*ORF2. The partial polypeptide from *Aα4*ORF2 shows a 55% identity to the first 882 aas of the deduced polypeptide from *Aα1*ORF2. In general, identity extends through all three polypeptides in small blocks of 1–5 aas. In the N-terminal regions of each polypeptide is a block of 60 aas that is 82% identical in all three deduced proteins. This region, called HD2 (Fig. 6), has a homeodomain motif (Scott *et al.,* 1989). A comparison of the amino acid sequences of HD2 with the amino acid sequences of other homeodomains shows that HD2 is 35% identical to the *Drosophila Antp* homeodomain (Scott *et al.,* 1989) and 26% identical to the homeodomain found in the *b* mating-type polypeptide of *Ustilago maydis* (Schulz *et al.,* 1990). The presence of a homeodomain motif suggests that the deduced polypeptides are able to bind to DNA and thereby regulate transcription of developmental genes. Except for the HD2 regions, the deduced polypeptides of *Aα1*ORF2, *Aα3*ORF2, and *Aα4*ORF2 show no significant identity to proteins in the National Biomedical Research Foundation (NBRF) Protein Database Bank or to other proteins thought to be encoded by *Aα* (see subsequent text).

Aα3 contains a second ORF (*Aα3*ORF1) that is transcribed in the opposite direction from *Aα3*ORF2. There are five putative introns in *Aα3*ORF1. The deduced polypeptide is 660 aas and is very acidic; the predicted isoelectric point is 4.9. This polypeptide shows identity to polypeptides deduced from two ORFs found in *Aα4* (ORF1′ and ORF1″). The peptides deduced from ORF1′ and ORF1″ are 228 and 418 aas, respectively. The 228 aas of the ORF1′ polypeptide show about 45% identity to the first 228 aas of the polypeptide deduced from *Aα3*ORF1. The 418 aas of the polypeptide deduced from ORF1″ show a 45% identity to the last 418 aas of the deduced polypeptide of *Aα3*ORF1. This raises the possibility that *Aα3*ORF1 may actually be two ORFs corresponding to ORF1′ and ORF1″ or the latter could be part of one larger ORF corresponding to ORF1 in *Aα3*. In either case, the degree of identity suggests that these regions encode alleles. In the N-terminal regions of the deduced polypeptides of ORF1 and ORF1′ there is a block of 60 aas with a homeodomain motif (HD1). The two HD1 regions (ORF1HD1 and ORF1′HD1) are 40% identical with regard to their amino acid sequences and show identity to homeodomains found in other fungi (Fig. 7). Two blocks of about 30 aas rich in glutamate and aspartate are found in the C-terminal end of the

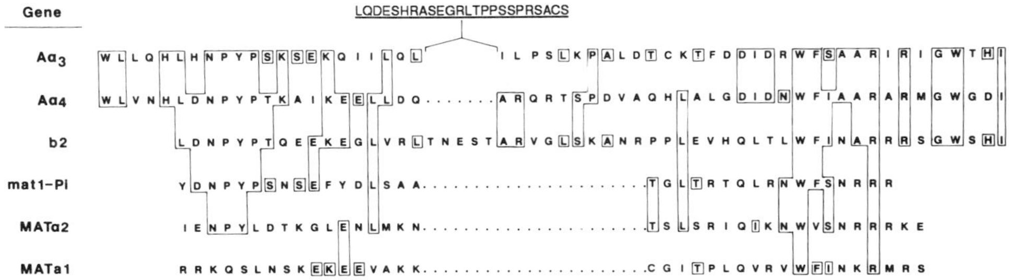

Fig. 7. A comparison of fungal homeodomains. Shown are *Aα3* ORF1 and *Aα4* ORF1′ homeodomains aligned with those of the *b2* mating type gene of *Ustilago maydis; mat1-Pi* of *Schizosaccharomyces pombe;* and *MATα2* and *MATa1* of *Saccharomyces cerevisiae.* (After Schulz *et al.*, 1990.)

deduced protein from *Aα3*ORF1 and the N-terminal end of the deduced protein from *Aα4*ORF1″. These regions resemble the transcriptional activating region of *GAL4* of yeast (Ma and Ptashne, 1987). Except for HD1 and the glutamate–aspartate acidic domains, the deduced polypeptides of ORF1, ORF1′, and ORF1″ show no significant identity to proteins in the NBRF protein Database Bank.

We think it is interesting and significant that all of the putative polypeptides deduced from the related ORFs in *Aα1, Aα3,* and *Aα4* have some characteristics (homeodomains, acidic domains, or both) of known eukaryotic regulatory proteins (Johnson and McKnight, 1989).

Confirmation of the structures of genes and corresponding proteins in *Aα1, Aα3,* and *Aα4* will necessitate the cloning and sequencing of the cDNAs. Sequences from the three cloned loci were used as probes in northern hybridizations with polyA$^+$ RNA from corresponding homokaryons and dikaryons. Several *Aα* specific transcripts, about 3–3.5 kb in size and of low abundance, were detected. We are currently mapping these transcripts and making cDNA libraries for the isolation of *Aα* specific cDNAs.

D. Activity of *Aα* ORFs in Transformation

Restriction fragments from the *Aα1, Aα3,* and *Aα4* loci were tested for their ability to activate *A*-regulated events in transformation (Fig. 8). Four fragments from *Aα1* were transformed into the eight possible compatible *Aα* recipients *(Aα2–Aα9)* and the transformants were tested for activation of *A* events in matings with tester strains. The 2.8-kb fragment containing ORF2 activated *A* events in all eight *Aα* strains. A subfragment, missing the 5′ end of ORF2 and any potential promoter region, was inactive in all

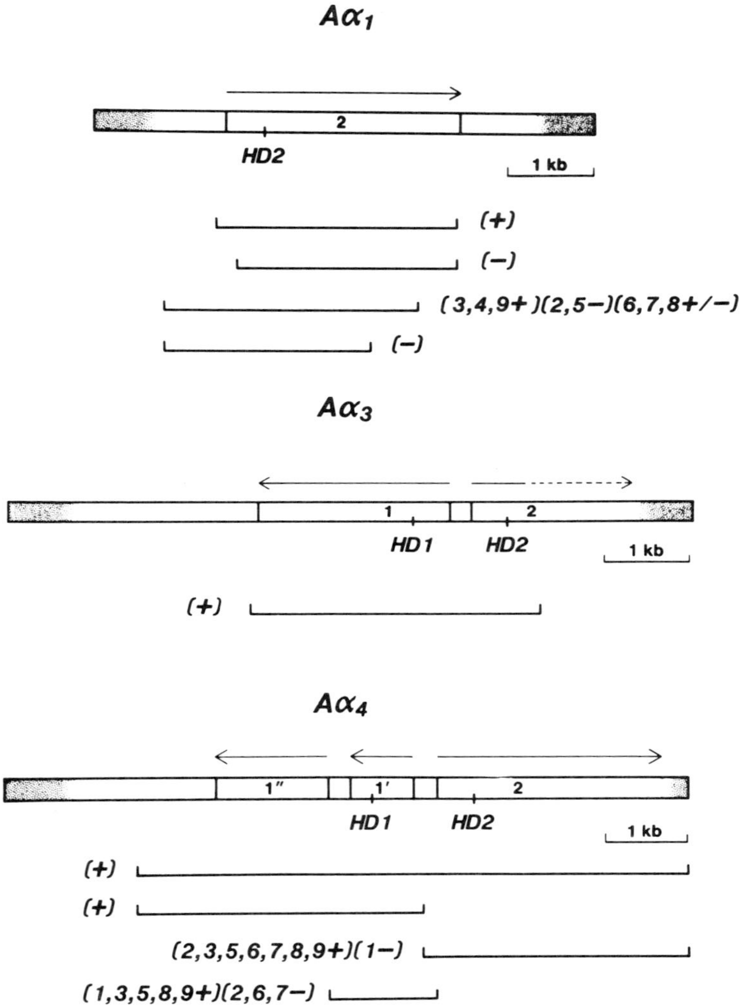

Fig. 8. Transformation analysis of *Aα1, Aβ3,* and *Aα4.* Lines below the map of each locus indicate the subcloned region used for transformation. (+) indicates that the fragment activates *A*-regulated events in all nonself *Aα* strains. (−) indicates that the fragment has no effect in any strain. (+/−) indicates a phenotype intermediate between a positive and negative reaction. Numbers preceding these symbols for activity indicate the *Aα* locus of the recipient.

eight strains, suggesting that ORF2 is one of the components of the *Aα1* locus that is required for activation of *A* events in mating. A fragment missing about 15% of the 3′ end of ORF2 was completely active in three strains (*Aα3, Aα4,* and *Aα9*), partially active in three other strains (*Aα6, Aα7,* and *Aα8*), and inactive in two strains (*Aα2* and *Aα5*). A shorter version of this fragment, missing about 75% of the 3′ end of ORF2 was inactive in all eight strains of *Aα*. This suggests that removal of a small portion (15%) of the 3′ end of the deduced polypeptide does not completely abolish its activity in some recipients. This is consistent with the idea that the deduced *Aα1* polypeptide interacts with one or more multiallelic polypeptides and that the partial *Aα1* polypeptide can interact with some of the polypeptides in this set and not others.

A fragment from *Aα3,* containing ORF 1 and part of ORF2, transformed *Aα1, Aα2,* and *Aα4–Aα9* strains, indicating that it encodes information for the complete range of transforming activity. Smaller fragments, containing only ORF1 and ORF2 have yet to be tested in transformation.

Four fragments from *Aα4* were tested in transformation for ability to activate *A* regulated events. A large fragment containing ORF1′, ORF1″, and ORF2 has a complete range of transforming activity. The left half of this fragment, containing only ORF1′ and ORF1″, is as active in transformation as the larger fragment. This suggests that polypeptides encoded by ORF1′ and ORF1″ are sufficient for the activation of *A* events in all strains except *Aα4*. A smaller fragment containing all of ORF1′ and only a small portion of the 5′ end of ORF1″ is active in transformation with *Aα1, Aα3, Aα5, Aα8,* and *Aα9* strains but not with *Aα2, Aα6,* and *Aα7* strains. A mutant copy of this fragment, with a 450-bp deletion in ORF1′ that removes HD1, is completely inactive in transformation (data not shown). Assuming that ORF1′ and ORF1″ encode two proteins, the data suggest that the deduced protein from ORF1′ is sufficient to activate *A* events in five *Aα* strains but that the deduced polypeptide from ORF1″ is required for transformation of the remaining three strains. If ORF1′ and ORF1″ are really part of a larger open reading frame, then one can argue that a truncated version of the deduced polypeptide is active in some strains and not others.

A fragment from the right half of *Aα4,* containing ORF2, has a full range of activity except that it does not activate *A* events in *Aα1* recipients. Thus, the *Aα4* locus appears to be functionally redundant, that is, two distinct regions have transformation ranges that almost completely overlap. The fact that *Aα1* strains are transformed with *Aα4* DNA containing only ORF1′ but not with *Aα4* DNA containing only ORF2 suggests that activation of *A* events may be mediated by the interaction of two different proteins, in this instance by the deduced polypeptides of *Aα1*ORF2 and *Aα4*ORF1′.

V. DISCUSSION

Previously we reported the molecular cloning of two *Aα* loci (*Aα1* and *Aα4*) and the DNA sequence analysis of two fragments with *Aα* activity in transformation (Giasson *et al.*, 1989; Ullrich *et al.*, 1990). The 2.8-kb fragment from *Aα1* containing ORF2 showed no significant identity to the 1.2-kb fragment from *Aα4* containing ORF1′. We were puzzled by the lack of sequence identity, by the fact that transformation activity seemed to require one large ORF in one case *(Aα1)* and one small ORF in another *(Aα4),* and by the apparent lack of an overall pattern in the organization of both loci. We were also surprised to discover that the 1.2-kb fragment from *Aα4* did not activate *A* events in all *Aα* strains, thus showing that the functional test for *Aα* activity in transformation was dependent on the *Aα* locus of the recipient strain. This suggested that the *Aα* loci may encode more than one polypeptide and that, in a given mating between two compatible *Aα* strains, a specific set of *Aα*-encoded polypeptides interact to activate *A*-regulated events.

A somewhat clearer picture of the similarities and differences among the three *Aα* loci is emerging from an analysis of the *Aα3* locus, which was recently cloned and partially sequenced (M. Stankis, unpublished results). DNA sequence analysis of the 3.2-kb fragment with *Aα3* activity in transformation revealed two ORFs with significant identity to ORFs in *Aα1* and *Aα4*. *Aα3*ORF1 appears to be homologous to ORF1′ and ORF1″ in *Aα4* and *Aα3*ORF2 appears to be homologous to *Aα1*ORF2 and *Aα4*ORF2. The identification of more than one set of ORFs showing similarity among the three *Aα* loci studied supports the notion that each locus consists of one or more genes that encode products that interact with the products of other compatible loci to activate the *A* developmental pathway.

The question of which gene products may be interacting in a given mating is unanswerable at the moment. The identification of homeodomain motifs in several of the deduced polypeptides is consistent with the notion that they bind to DNA. But the importance of homeodomains in the function of these polypeptides has yet to be established.

All the deduced *b* polypeptides of *Ustilago maydis* and the yeast mating-type proteins **a**1 and **α**2 have homeodomain motifs. There is direct evidence that **a**1 and **α**2 bind to DNA (Johnson and Herskowitz, 1985; Goutte and Johnson, 1988). It has been inferred, by analogy, that the *b* polypeptides of *U. maydis* may also bind to DNA (Schulz *et al.*, 1990). A similar argument can be put forth for the putative *Aα* polypeptides of

Schizophyllum commune. For *b* polypeptides it was proposed that two allelic polypeptides form an active complex that binds to DNA, and models for the formation of heterodimers of *b* polypeptides were proposed (Schulz *et al.,* 1990). A general model for how mating-type polypeptides might aggregate into self and nonself tetramers that are inactive and active has also been proposed (Metzenberg, 1990). It seems premature to begin model building in *S. commune,* since the identity of polypeptides involved in compatibility has not been confirmed, and which ones interact in a given mating is unknown. However, given that *Aα4*ORF1 has no counterpart (allele) in *Aα1* but still activates *A*-regulated events when transformed into an *Aα1* strain, one can argue that activation of *A* events is not mediated by heterodimers of allelic polypeptides. Instead, the *Aα4*ORF1′ deduced polypeptide interacts with a different protein, the deduced polypeptide of *Aα1*ORF2. Both of these proteins have a homeodomain motif and the complexes that activate transcription of *A* target genes may be multimers of two homeodomain polypeptides.

A. Other Mating-Type Systems

There are many examples of sexual incompatibility in fungi (Raper, 1966; Fincham *et al.,* 1979) and, when tractable transformation system exist, it has been possible to clone and study the incompatibility loci (mating-type loci). In two ascomycetes, *Saccharomyces cerevisiae* and *Schizosaccharomyces pombe,* mating type is well characterized and the regulatory circuits that control cell type serve as benchmarks for understanding how other mating-type systems work (Kelly *et al.,* 1988; Herskowitz, 1989). In baker's yeast, mating-type is determined by two alleles (**a** and **α**) of a single locus (MAT). Unlike most fungi, yeasts are able to switch mating type. Switching in baker's yeast involves the transfer of genetic information from one of two unexpressed loci to the transcriptionally active *MAT* locus. *MATa* encodes a polypeptide called **a**1. *MATα* encodes two polypeptides called **α**1 and **α**2. The character of **α** cells, **a** cells, and **a**/**α** diploid cells is determined by these three polypeptides. In **α** cells, **α**1 activates transcription of **α**-specific genes and **α**2 inhibits transcription of **a**-specific genes. In **a** cells, **a**-specific genes are expressed constitutively; **a**1 is produced but seems to have no effect on the expression of any cell-type-specific genes. The character of **a**/**α** diploid cells is attributed to the presence of **α**2 and an **a**1–**α**2 heterodimer. **α**2 inhibits the transcription of **a**-specific genes, whereas the **a**1–**α**2 heterodimer blocks transcription of **α**1 and haploid-specific genes.

The molecular analysis of mating type in two filamentous ascomycetes has just begun. The two mating-type loci of *Neurospora crassa* (Glass *et al.,*

1988; Glass and Lorimer, Chapter 8) and the two *MAT* loci of *Cochliobolus heterostrophus* (B. G. Turgeon and O. C. Yoder, personal communication) have been cloned and are being analyzed. Mating type in these species, as in yeast, is determined by either of two mating-type regions at a single locus. There are no silent *MAT* loci; therefore switching does not occur. In *N. crassa* the two mating types, *a* and *A*, are encoded by two nonhomologous regions of DNA. The *a* region is 3235 bp and encodes at least two functions. One segment encodes the polypeptide *a1* which seems to be responsible for heterokaryon incompatibility, designation of mating type, and induction of perithecium. Another segment encodes an uncharacterized gene that is involved in maturation of the perithecium and is activated by *a1* (Glass and Staben, 1990; Staben and Yanofsky, 1990). The *A* region is 5301 bp and encodes at least one polypeptide called *A1*, which is responsible for designation of mating-type fertility and heterokaryon incompatibility. A model for how *A* and *a* regulate vegetative and sexual functions, with features similar to those for control of cell type in yeast, has been proposed (Glass and Staben, 1990). In *C. heterostrophus,* the two *MAT* loci, *MAT1* and *MAT2,* are about 1.5 kb in size and, like *A* and *a* of *N. crassa,* the DNA sequences are dissimilar (B. G. Turgeon and O. C. Yoder, personal communication). It has been proposed that the term "allele" not be used to designate these very different loci because it is unlikely that they evolved from a single ancestral gene. Instead the word "idiomorph" has been suggested for these nonhomologous loci (Metzenberg, 1990; Metzenberg and Glass, 1990).

Mating-type loci from three basidiomycetes, *Schizophyllum commune, Coprinus cinereus,* and *Ustilago maydis,* have been cloned and some similarities and differences are apparent. The mating-type system of *C. cinereus* closely resembles that of *S. commune.* There are four multiallelic loci called *Aα, Aβ, Bα,* and *Bβ* that regulate two developmental pathways as described for the *A* and *B* loci of *Schizophyllum* (Casselton *et al.,* 1989; Pukkila and Casselton, Chapter 6). The system is thought to be functionally redundant, as it is in *S. commune,* because allelic differences at either *Aα* or *Aβ* fully activate the *A* pathway and differences at either *Bα* or *Bβ* fully activate the *B* pathway. Representative *Aα* and *Aβ* loci were cloned by walking the chromosome from the closely linked *pab* gene (Mutasa *et al.,* 1990; L. A. Casselton, personal communication). A fragment with *A* activity was isolated from a different strain of *C. cinereus* by transformation with DNA from a cosmid gene library (May *et al.,* 1991). In *C. cinereus,* the *Aα* and *Aβ* loci are closely linked, so it was possible to clone both loci on a single fragment of DNA. Sequences of these loci do not hybridize to each other (L. A. Casselton, personal communication) or to the cloned *Aα* loci of *S. commune* (C. Specht, unpublished results). Two transcripts that map to the *Aα* and *Aβ* regions have been detected (L. A. Casselton, personal communication). The number of genes in the *C. cinereus Aα* and *Aβ* loci and the differences and similarities between *Aα* and *Aβ* have yet to be determined. A constitutive (self-compatible) mutation of *C. cinereus* was also cloned; a preliminary characterization of this mutation suggests that it is a large deletion that brings the *Aα* and *Aβ* loci together (L. A. Casselton, personal communication).

The mating-type system of *U. maydis* consists of two multiallelic loci, *a* and *b* (see Banuett, Chapter 9). There are two alleles of the *a* locus (*a1* and *a2*) and 25 alleles of the *b* locus. The developmental pathways regulated by *a* and *b* are not known with certainty, but allelic differences at both loci are necessary for development of the dikaryotic filamentous form that is pathogenic. Both *a* alleles have been cloned (Froeliger and Kronstad, 1990; Froeliger and Leong, 1991; R. Kahmann, personal communication). Their structure is still uncertain but restriction analysis and cross-hybridization studies (Froeliger and Kronstad, 1990; Froeliger and Leong, 1991) and DNA sequencing (R. Kahmann, personal communication) show that these loci consist of about 9 kb *(a1)* and 5 kb *(a2)* of nonidentical DNA. These sequences have been subcloned to functionally active fragments about 2.0 kb (Froeliger and Leong, 1991; R. Kahmann, personal communications). Thus, these loci bear some resemblance to the *a* and *A* idiomorphs of *N. crassa.*

Several *b* alleles have been cloned and sequenced (Kronstad and Leong, 1989; Froeliger and Kronstad, 1990; Schulz *et al.,* 1990; Banuett, Chapter 9). The *b* locus appears to consist of a single multiallelic gene. The deduced polypeptides have a homeodomain and the amino acid sequences are very similar. The presence of a homeodomain suggests that the *b* polypeptides are DNA-binding proteins (Schulz *et al.,* 1990). Aside from the presence of homeodomains in *U. maydis b* and in the deduced polypeptides of ORF1′ and ORF2 in *S. commune,* the two mating-type systems bear little resemblance to one another. In *U. maydis* it appears that heterodimers of *b* polypeptides act as transcriptional activators, whereas in *S. commune* there is evidence that different (nonallelic) polypeptides interact to activate transcription.

Important questions remain. For example, how do the mating-type loci of *S. commune* actually do their work? All the data reported in this chapter are consistent with the notion that the nine forms of *Aα* each code for one or two polypeptides that regulate the expression of cell-type-specific genes. In dikaryotic cells the polypeptides encoded by the two compatible *Aα* loci may activate the expression of heterokaryon-specific genes and inhibit the expression of homokaryon-specific genes. How these polypeptides carry out these tasks is one of the most intriguing aspects of sexual development.

We need not stray far from the yeast model for *MAT* regulation to imagine that, after two different $A\alpha$ strains fuse, a characteristic set of $A\alpha$ polypeptides is formed and that these polypeptides act as monomers and heterodimers to activate and repress transcription of cell-type-specific genes. One immediate challenge is to complete the catalog of putative polypeptides encoded by the $A\alpha$ loci and to determine which ones, in any given mating, are necessary for the formation of the dikaryon. More challenging still will be the identification and cloning of the genes that are the direct targets for regulation by the $A\alpha$ polypeptides. With this information in hand it should be possible to characterize the protein–DNA interactions involved in gene regulation. Much of the technology needed to do this work is in place. We think the prospect for solving the mystery of incompatibility in filamentous fungi is good.

ACKNOWLEDGMENTS

We are grateful to Lorna Casselton, Eunice Froeliger, Steve Horton, Georgiana May, Patricia Pukkila, Carlene Raper, Gillian Turgeon, and Olen Yoder for letting us use their unpublished results. We also thank the W.B. Saunders Company for permission to reproduce material in Figures 1, 2, and 3. This research was supported by the Vermont Agricultural Experimental Station and by USPHS grant GM34023.

REFERENCES

Ballance, D. J. (1986). Sequences important for gene expression in filamentous fungi. *Yeast* **2,** 229–236.

Brown, J. W. S. (1986). A catalogue of splice junction and putative branch point sequences from plant introns. *Nucl. Acids Res.* **14,** 9549–9559.

Carmi, P., Holm, P. B., Koltin, Y., Rasmussen, S. W., Sage, J., and Zickler, D. (1978). The pachytene karyotype of *Schizophyllum commune* analyzed by three dimensional reconstruction of synaptonemal complexes. *Carlsberg Res. Commun.* **43,** 117–132.

Casselton, L. A., Mutasa, E. S., Tymon, A., Mellon, F. M., Little, P. F. R., Taylor, S., Bernhagen, J., and Stratmann, R. (1989). The molecular analysis of basidiomycete mating type genes. *In* "Proceedings of the EMBO–Alko Workshop on Molecular Biology of Filamentous Fungi" (H. Navalainen and M. Penttilä, eds.), Vol. 6, pp. 139–148. Foundation for Biotechnical and Industrial Fermentation Research, Helsinki.

Dons, J. J. M., and Wessels, J. G. H. (1980). Sequence organization of the nuclear DNA of *Schizophyllum commune. Biochim. Biophys. Acta* **607,** 385–396.

Dons, J. J. M., de Vries, O. M. H., and Wessels, J. G. H. (1979). Characterization of the genome of the basidiomycete *Schizophyllum commune. Biochim. Biophys. Acta* **563,** 100–112.

Fincham, J. R. S., Day, P. R., and Radford, A. (1979). "Fungal Genetics," 4th Ed. University of California Press, Berkeley.

Froeliger, E. H., and Kronstad, J. W. (1990). Mating and pathogenesis in *Ustilago maydis. Sem. Dev. Biol.* **1,** 185–193.

Froeliger, E. H., and Leong, S. A. (1991). The *a* mating-type alleles of *Ustilago maydis* are idiomorphs. *Gene*. In Press.

Froeliger, E. H., Muñoz-Rivas, A. M., Specht, C. A., Ullrich, R. C., and Novotny, C. P. (1987). The isolation of specific genes from the Basidiomycete *Schizophyllum commune*. *Curr. Genet.* **12,** 547–554.

Giasson, L., Specht, C. A., Milgrim, C., Novotny, C. P., and Ullrich, R. C. (1989). Cloning and comparison of *Aα* mating-type alleles of the Basidiomycete *Schizophyllum commune*. *Mol. Gen. Genet.* **218,** 72–77.

Glass, N. L., and Staben, C. (1990). Genetic control of mating in *Neurospora crassa*. *Sem. Dev. Biol.* **1,** 177–184.

Glass, N. L., Vollmer, S. J., Staben, C., Grotelueschen, J., and Metzenberg, R. L. (1988). DNAs of the two mating-type alleles of *Neurospora crassa* are highly dissimilar. *Science* **241,** 570–573.

Goutte, C., and Johnson, A. D. (1988). *a1* protein alters the DNA binding specificity of **α2** repressor. *Cell* **52,** 875–882.

Herskowitz, I. (1989). A regulatory hierarchy for cell specialization in yeast. *Nature (London)* **342,** 749–757.

Horton, J. S., and Raper, C. A. (1991). Pulsed-field gel electrophoretic analysis of *Schizophyllum commune* chromosomal DNA. *Curr. Genet.* **19,** 77–80.

Johnson, A. D., and Herskowitz, I. (1985). A repressor (*MATα2* product) and its operator control expression of a set of cell type specific genes in yeast. *Cell* **42,** 237–247.

Johnson, P. F., and McKnight, S. L. (1989). Eukaryotic transcriptional regulatory proteins. *Ann. Rev. Biochem.* **58,** 799–839.

Kelly, M., Burke, J., Smith, M., Klar, A., and Beach, D. (1988). Four mating-type genes control sexual differentiation in the fission yeast. *EMBO J.* **7,** 1537–1547.

Koltin, Y. (1968). The genetic structure of the incompatibility factors of *Schizophyllum commune:* Comparative studies of primary mutations in the *B* factor. *Mol. Gen. Genet.* **102,** 196–203.

Koltin, Y. (1970). Studies on mutations disruptive to nuclear migration in *Schizophyllum commune*. *Mol. Gen. Genet.* **106,** 155–161.

Koltin, Y., Raper, J. R., and Simchen, G. (1967). Genetic structure of the incompatibility factors of *Schizophyllum commune:* The B factor. *Proc. Natl. Acad. Sci. U.S.A.* **57,** 55–63.

Koltin, Y., Stamberg, J., Bawnik, N., Tamarkin, A., and Werczberger, R. (1979). Mutational analysis of natural alleles in and affecting the *B* incompatibility factor of *Schizophyllum*. *Genetics* **93,** 383–391.

Kronstad, J. W., and Leong, S. A. (1989). Isolation of two alleles of the *b* locus of *Ustilago maydis*. *Proc. Natl. Acad. Sci. U.S.A.* **86,** 978–982.

Ma, J., and Ptashne, M. (1987). Deletion analysis of GAL4 defines two transcriptional activating segments. *Cell* **48,** 847–853.

May, G., LeChevanton, L., and Pukkila, P. J. (1991). Molecular analysis of the *Coprinus cinereus* mating type *A* factor demonstrates an unexpectedly complex structure. *Genetics* (In press).

Metzenberg, R. L. (1990). The role of similarity and differences in fungal mating. *Genetics* **125,** 457–462.

Metzenberg, R. L., and Glass, N. L. (1990). Mating type and mating strategies in *Neurospora*. *BioEssays* **12,** 53–59.

Mooibroek, H., Kuipers, A. G. J., Sietsma, J. H., Punt, P. J., and Wessels, J. G. H. (1990). Introduction of hygromycin B resistance in *Schizophyllum commune:* Preferential methylation of donor DNA. *Mol. Gen. Genet.* **222,** 41–48.

Mount, S. M. (1982). A catalogue of splice junction sequences. *Nucl. Acids Res.* **10,** 459–472.

Muñoz-Rivas, A. M., Specht, C. A., Drummond, B. J., Froeliger, E., Novotny, C. P., and Ullrich, R. C. (1986a). Transformation of the Basidiomycete *Schizophyllum commune. Mol. Gen. Genet.* **205,** 103–106.

Muñoz-Rivas, A. M., Specht, C. A., Ullrich, R. C., and Novotny, C. P. (1986b). Isolation of the DNA sequence coding indole-3-glycerol phosphate synthetase and phosphoribosyl anthranilate isomerase of *Schizophyllum commune. Curr. Genet.* **10,** 909–913.

Mutasa, E. S., Tymon, A. M., Göttgens, B., Mellon, F. M., Little, P. F. R., and Casselton, L. A. (1990). Molecular organization of an *A* mating type factor of the basidiomycete fungus *Coprinus cinereus. Curr. Genet.* **18,** 223–229.

Parag, Y. (1962). Mutations in the *B* incompatibility factor in *Schizophyllum commune. Proc. Natl. Acad. Sci. U.S.A.* **48,** 743–750.

Raper, C. A. (1983). Controls for development and differentiation in the dikaryon in Basidiomycetes. *In* "Secondary Metabolism and Differentiation in Fungi" (J. Bennett and A. Ciegler, eds.), pp. 195–238. Marcel Dekker, New York.

Raper, C. A. (1988). *Schizophyllum commune,* a model for genetic studies of the Basidiomycotina. *In* "Genetics of Pathogenic Fungi" (G. S. Sidhu, ed.), Advances in Plant Pathology, Vol. 6, pp. 511–522. Academic Press, New York.

Raper, C. A., and Raper, J. R. (1964). Mutations affecting heterokaryons in *Schizophyllum commune. Am. J. Bot.* **511,** 503–513.

Raper, C. A., and Raper, J. R. (1966). Mutations modifying sexual morphogenesis in *Schizophyllum. Genetics* **54,** 1151–1168.

Raper, C. A., and Raper, J. R. (1973). Mutational analysis of a regulatory gene for morphogenesis in *Schizophyllum. Proc. Natl. Acad. Sci. U.S.A.* **70,** 1427–1431.

Raper, J. R. (1966). "Genetics of Sexuality in Higher Fungi." Ronald Press, New York.

Raper, J. R., and Raudaskoski, M. (1968). Secondary mutations at the *Bβ* locus of *Schizophyllum. Heredity* **23,** 109–117.

Raper, J. R., Krongeld, G. S., and Baxter, M. G. (1958). The number and distribution of incompatibility factors in *Schizophyllum commune. Amer. Nat.* **92,** 221–232.

Raper, J. R., Baxter, M. G., and Ellingboe, A. H. (1960). The genetic structure of the incompatibility factors of *Schizophyllum commune:* The *A* factor. *Proc. Natl. Acad. Sci. U.S.A.* **44,** 889–900.

Raper, J. R., Boyd, D. H., and Raper, C. A. (1965). Primary and secondary mutations at the incompatibility loci in *Schizophyllum. Proc. Natl. Acad. Sci. U.S.A.* **53,** 1324–1332.

Raudaskoski, M., Stamberg, J., Bawnik, N., and Koltin, Y. (1976). Mutational analysis of natural alleles at the *B* incompatibility factor of *Schizophyllum commune:* $\alpha 2$ and $\beta 6$. *Genetics* **83,** 507–516.

Schulz, B., Banuett, F., Dahl, M., Schlesinger, R., Schafer, W., Martin, T., Herskowitz, I., and Kahmann, R. (1990). The *b* alleles of *U. maydis,* whose combinations program pathogenic development, code for polypeptides containing a homeodomain-related motif. *Cell* **60,** 295–306.

Schuren, F. H. J., and Wessels, J. G. H. (1990). Two genes specifically expressed in fruiting dikaryons of *Schizophyllum commune:* Homologies with a gene not regulated by mating-type genes. *Gene* **90,** 199–205.

Scott, M. P., Tamkun, J. W., and Hartzell, G. W., III (1989). The structure and function of the homeodomain. *Biochim. Biophys. Acta* **989,** 25–48.

Specht, C. A., Muñoz-Rivas, A. M., Novotny, C. P., and Ullrich, R. C. (1988). Transformation of *Schizophyllum commune:* An analysis of parameters for improving transformation frequencies. *Exp. Mycol.* **12,** 357–366.

Staben, C., and Yanofsky, C. (1990). The *Neurospora crassa a* mating-type region. *Proc. Natl. Acad. Sci. U.S.A.* **87,** 4917–4921.

Stamberg, J., and Koltin, Y. (1972). The organization of the incompatibility factors in higher fungi: The effects of structure and symmetry on breeding. *Heredity* **30,** 15–26.

Stankis, M. M., Specht, C. A., and Giasson, L. (1990). Sexual incompatibility in *Schizophyllum commune:* From classical genetics to a molecular view. *Sem. Dev. Biol.* **1,** 195–206.

Ullrich, R. C., Kohorn, B. D., and Specht, C. A. (1980). Absence of short-period repetitive sequence interspersion in the Basidiomycete *Schizophyllum commune. Chromosoma (Berlin)* **81,** 371–378.

Ullrich, R. C., Giasson, L., Specht, C. A., Stankis, M. M., and Novotny, C. P. (1990). The *A* multiallelic mating type genes of *Schizophyllum commune. In* "UCLA Symposia on Molecular and Cellular Biology New Series" (E. H. Davidson, J. V. Ruderman, and J. W. Pasokony, eds.), Vol. 125, pp. 271–288. Willey-Liss, New York.

NOTE ADDED IN PROOF

We found a sequencing error which, when corrected, fuses $A\alpha4$ORF′ and $A\alpha4$ORF″ (Figs. 6 and 8) together, in frame, to form one larger ORF now called $A\alpha4$ORF1. The similarity between the deduced polypeptides of $A\alpha3$ORF1 and $A\alpha4$ORF1 indicates that these two ORFs are alleles.

11

Repeat-Induced Point Mutation and DNA Methylation

ERIC U. SELKER
Institute of Molecular Biology
University of Oregon
Eugene, Oregon

I. INTRODUCTION

Fungi tend to have small genomes without much obvious "junk" DNA. For example, in the *Neurospora crassa* genome, which is about 10 times larger than that of *Escherichia coli,* the gene family encoding the ribosomal RNAs accounts for the vast majority of repeated sequences (Krumlauf and Marzluf, 1980). Furthermore, cytosine methylation, a hallmark of inactive sequences in many organisms, is uncommon in the *Neurospora* genome. The apparent paucity of extraneous DNA may be partly due to natural selection for individuals that grow rapidly. In addition, at least some fungi have specific mechanisms to inactivate redundant sequences, to reduce their copy number, or to do both. I will summarize in this chapter what is known about one such mechanism discovered in *N. crassa*—repeat-induced point mutation (RIP)—which specifically modifies duplicated sequences. Processes similar or identical to RIP recently have been found

MORE GENE MANIPULATIONS
IN FUNGI

in some other fungi (see Rossignol and Picard, Chapter 12; Selker, 1990b). I will also briefly discuss how RIP can be harnessed to inactivate genes in *Neurospora* and what the process has taught us about DNA methylation in eukaryotes. For more details on RIP, see Selker, 1990b; for its use in mutagenesis, see Selker *et al.,* 1989; and for an overview of DNA methylation, see Selker, 1990a.

II. SPECIFICITY AND CHARACTERISTICS OF RIP

The RIP process mutates DNA sequence duplications between fertilization and nuclear fusion in the *Neurospora* life cycle, a stage consisting of perhaps 10 nuclear divisions (Fig. 1). The process is easily detected in sexual progeny from most strains obtained by DNA-mediated transformation. Tandem repeats of transforming sequences, which commonly result from transformation in *Neurospora,* are most sensitive to RIP (Selker *et al.,* 1987b; Selker and Garrett, 1988). In progeny from crosses involving transformants that carry such linked duplications, the duplicated DNA almost invariably shows the signs of RIP: changes in the arrangement of restriction sites and cytosine methylation. (The acronym RIP was originally introduced as "rearrangement induced premeiotically" to describe the new restriction patterns and the timing of the process.) Unlinked duplications, arising from ectopic integration of gene-size DNA fragments in transformation, are also detected and mutated by RIP, but typically survive a cross unaltered in about 50% of the progeny.

When two copies of a sequence are present in a nucleus, RIP alters either both copies or neither copy (Selker and Garrett, 1988). When three or more copies are present, however, RIP sometimes inactivates a fraction of the copies (Selker and Garrett, 1988; Fincham *et al.,* 1989; Faugeron *et al.,* 1990). These observations are consistent with the idea that RIP involves pairing of homologous sequences present in the haploid nuclei of premeiotic tissue.

It was not initially known whether the restriction site "rearrangements" characteristic of RIP resulted from multiple small classical rearrangement, point mutations, or both (Selker *et al.,* 1987b; Selker and Garrett, 1988). It is now virtually certain, however, that all changes in the primary structure of DNA that result from RIP are polarized transition mutations in which G : C pairs are replaced by A : T pairs; 10% or more of the G : C pairs can be replaced in a single cross (Cambareri *et al.,* 1989). Sequences altered by RIP are sensitive to additional mutation in subsequent crosses, as long as similar partners are provided (Cambareri *et al.,* 1991; Foss *et al.,* 1991). Although the partners in RIP appear to mutate independently, their diver-

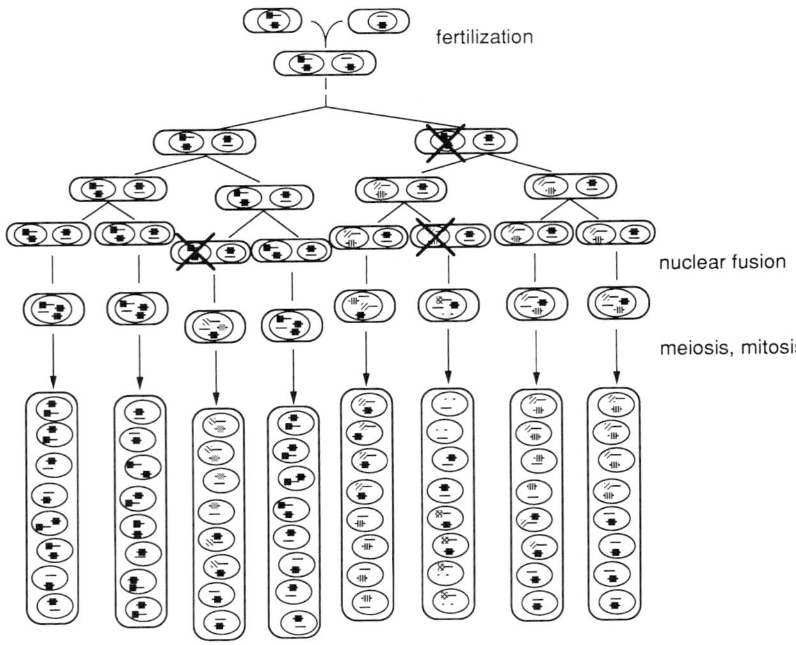

Fig. 1. Timing of RIP. In general, all asci from a given perithecium arise from a common fertilization and the initial dikaryotic cell undergoes more than seven divisions prior to nuclear fusion (see Perkins and Barry, 1977). For illustrative purposes, the lineage is condensed and only two of the seven chromosomes are indicated. DNA sequence alterations by RIP (X) in a gene (filled squares) duplicated in the parent shown on the left (top row) are indicated by different fill patterns. The fact that all derivatives of a duplicated sequence display a common alteration pattern implies that the changes occurred prior to premeiotic DNA synthesis, and therefore prior to karyogamy and meiosis, as illustrated. Depending on segregation and crossing over, four, three, or two meiotic products display alterations. Analysis of progeny from multiple asci of a given perithecium revealed that RIP occurs after fertilization (Selker *et al.*, 1987b).

gence is limited by two levels of site specificity. First, the process is confined to G:C pairs and second, it affects some G:C pairs much more than others. About two-thirds of the cytosines replaced are located 5′ to adenines (Cambareri *et al.*, 1989; Grayburn and Selker, 1989). The most extensive information currently available on site specificity comes from analysis of the $\zeta-\eta$ (zeta–eta) region, a natural duplication that probably went through many cycles of RIP (Selker and Stevens, 1985; Grayburn and Selker, 1989). This information is summarized in Table I.

RIP is triggered by duplications of foreign as well as native DNA. The mutations are generally confined to the duplicated sequences (Selker and Garrett, 1988; Foss *et al.*, 1991). Nevertheless, mutations can extend into

TABLE I.

Site Specificity of GC to AT Mutations in $\zeta-\eta$ Region[a]

Site	Percentage of sites changed for X				Number of mutations classified
	G	A	C	T	
5′XpC	21	41	30	32	236
5′CpX	11	74	2	30	218
5′CpApX	33	49	48	81	120

[a] The sequence context of CG to TA mutations that led to the $\zeta-\eta$ region was examined on the strand showing the C to T change (boldface C). The frequencies do not include the minority of mutations that could not be unambiguously categorized without making assumptions about whether mutations occurred simultaneously or in series (Grayburn and Selker, 1989).

adjacent unique sequences. Apparently the RIP machine is careful in its choice of targets, but sloppy in its application of mutations. Some mutated regions show only C to T or G to A changes on one strand. This asymmetry implies that RIP results from a single type of mutation (i.e., C to T or G to A changes, but not both). It seems most likely that RIP involves enzymatic deamination of cytosines or 5-methylcytosines to give uracils or thymines, respectively (Cambareri *et al.,* 1989; Selker, 1990b).

III. DNA METHYLATION AND RIP

Although sequences altered by RIP typically display DNA methylation, it is not yet known whether methylation is directly involved in the mechanism of RIP. Methylation of a sequence does not seem sufficient to make it sensitive to high frequency mutation, nor does it appear to protect a sequence from further RIP. (Methylated sequences passed through a cross do not show evidence of mutation if not duplicated, but are susceptible to RIP if duplicated; Cambareri *et al.,* 1991.)

The methylation resulting from RIP does not *directly* depend on the duplication. This conclusion comes in part from the observation that methylation persists in altered unique sequences resulting from meiotic segregation of the duplication partners. A tetratype ascus illustrating this point is shown in Fig. 2. Although one could imagine that the methylation of single-copy sequences might reflect an epigenetic system that perpetuates methylation patterns set up earlier (i.e., premeiotically), this is not the

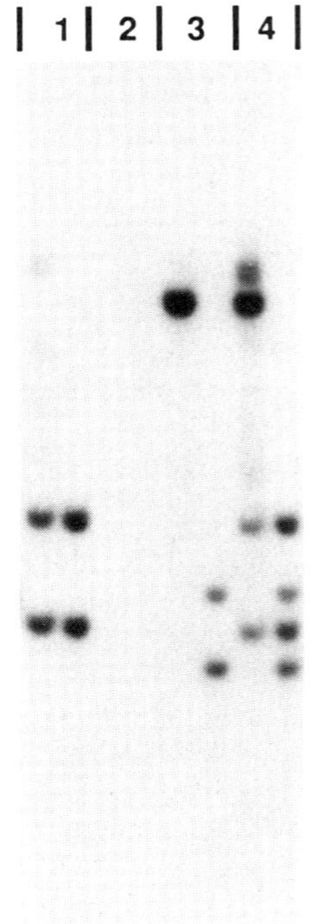

Fig. 2. Segregation of duplicated *am* genes inactivated by RIP. Progeny representing the four meiotic products in an ascus from a cross between a strain having an unlinked duplication of am^+ and a strain deleted for *am* were analyzed using *Sau*3A (left lane of each pair) or *Mbo*I (right lane of each pair). The digests were probed for the *am* region. Isolates 2 and 4 represent the parental types; 1 and 3 represent nonduplication recombinants. The differences detected in *Sau*3A and *Mbo*I digests are indicative of cytosine methylation. The mutant allele that segregated to spores 3 and 4 showed substantial methylation even when it was present as a single copy (isolate 3). The allele that segregated to spores 1 and 4 displayed a low level of methylation in both progeny, although a somewhat higher level of methylation was seen in isolate 4, which has both copies of the gene. Reprinted from Selker and Garrett, 1988, with permission.

correct explanation in this case. We know this because a normally methylated sequence stripped of its methylation by passage through *E. coli,* or by treatment with 5-azacytidine (which prevents cytosine methylation; Selker and Stevens, 1985), becomes methylated *de novo* (Selker *et al.,* 1987a; Cambareri *et al.,* 1991). Thus the G:C to A:T mutations resulting from RIP somehow render the sequences susceptible to DNA methylation.

The conclusion that mutations can induce methylation, coupled with other information on DNA methylation in *Neurospora* (e.g., that foreign sequences introduced into *Neurospora* become methylated in some sequence contexts, but not in others), led to the development of a general model to account for the distribution of methylation on eukaryotic DNA (Selker, 1990a). In this model (Fig. 3), methylation is the default state of chromosomal DNA. Sequence-specific DNA-binding proteins (e.g., transcription factors) are indirectly responsible for preventing DNA methylation. In particular, such nonhistone chromosomal proteins hold the DNA in an open form (Form I chromatin) that is not susceptible to DNA methylation. DNA without bound nonhistone proteins "collapses" into Form II, which is susceptible to methylation upon replication.

In the collapsed-chromatin model, methylation of a region reflects loss of capacity to bind nonhistone chromosomal proteins. This could occur in at least two ways. Binding sites on the DNA could be destroyed by mutation (e.g., by RIP). In addition, proteins that normally bind in the region could be prevented from binding long enough to allow collapse and methylation of the region. For example, proteins might be displaced from the duplicated sequences by the RIP machinery. If such proteins are able to bind only to unmethylated DNA or to sequences in Form I chromatin, this switch from Form I to Form II chromatin could be effectively irreversible.

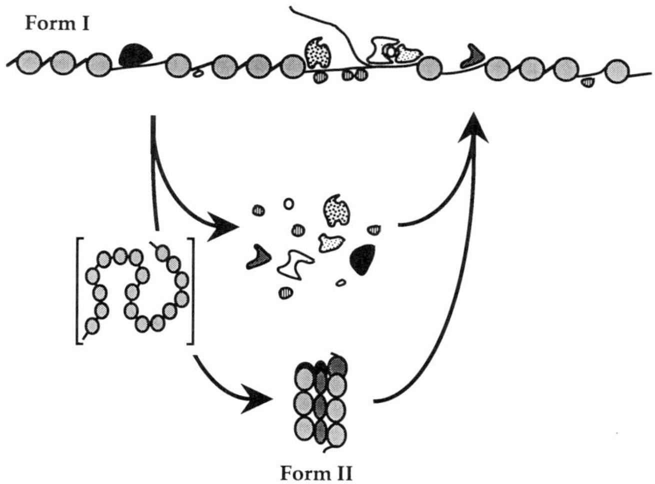

Fig. 3. The collapsed chromatin model. Chromatin exists in the cell in two basic forms. Form I, the extended state, is held open by the presence of various DNA-binding proteins (irregular shapes). DNA lacking bound factors is organized in nucleosomes (shaded circles), and any uninterrupted string of nucleosomes spontaneously condenses into Form II. Only DNA in Form II chromatin is subject to new methylation upon replication. Reproduced from Selker, 1990a, with permission.

Thus, the collapsed-chromatin model provides a possible explanation for epigenetic maintenance of methylation patterns, as observed in a number of systems (see Holliday, 1987; Selker, 1990a). The methylation associated with inactivation of duplicated sequences in *Ascobolus* (which may not involve mutations; see Rossignol and Picard, Chapter 12) may be an example of this effect.

IV. USE OF RIP FOR MUTAGENESIS

RIP can be exploited for *in vivo* mutagenesis of specific chromosomal regions in *Neurospora* (Selker *et al.,* 1987b, 1989; Selker and Garrett, 1988; Marathe *et al.,* 1990). DNA corresponding to the sequence to be mutagenized is introduced into the organism by transformation to render the sequence redundant in the genome. Transformants are then crossed and progeny are scored, phenotypically and at the DNA level, for evidence of mutations. Because RIP can extend into unique sequences adjacent to duplications, one must cautiously interpret phenotypes resulting from RIP. The degree of sequence similarity influences the efficiency of the process, but homologous sequences need not be identical to trigger RIP (Cambareri *et al.,* 1991). Thus RIP can be used to mutagenize genes not yet isolated from *Neurospora,* but available from related organisms (Selker *et al.,* 1989; I. Connerton, personal communication).

ACKNOWLEDGMENTS

I thank Drs. Jette Foss and Vivian Miao for constructive comments on the developing manuscript. The work from my laboratory was supported by National Science Foundation grant DCB 8718163 and Public Health Services grant GM-35690 from the National Institutes of Health. The article was written during the tenure of an Established Investigatorship of the American Heart Association.

REFERENCES

Cambareri, E. B., Jensen, B. C., Schabtach, E., and Selker, E. U. (1989). Repeat-induced G–C to A–T mutations in *Neurospora. Science* **244,** 1571–1575.

Cambareri, E. B., Singer, M. J., and Selker, E. U. (1991). Recurrence of repeat-induced point mutation (RIP) in *Neurospora crassa. Genetics* **127,** 699–710.

Faugeron, G., Rhounim, L., and Rossignol, J.-L. (1990). How does the cell count the number of ectopic copies of a gene in the premeiotic inactivation process acting in *Ascobolus immersus? Genetics* **124,** 585–591.

Fincham, J. R. S., Connerton, I. F., Notarianni, E., and Harrington, K. (1989). Premeiotic disruption of duplicated and triplicated copies of the *Neurospora crassa am* (glutamate dehydrogenase) gene. *Curr. Genet.* **15,** 327–334.

Foss, E. J., Garrett, P. W., Kinsey, J. A., and Selker, E. U. (1991). Specificity of repeat-induced point mutation (RIP) in *Neurospora:* Sensitivity of non-*Neurospora* sequences, a natural diverged tandem duplication, and unique DNA adjacent to a duplicated region. *Genetics* **127,** 711–717.

Grayburn, W. S., and Selker, E. U. (1989). A natural case of RIP: Degeneration of DNA sequence in an ancestral tandem duplication. *Mol. Cell. Biol.* **9,** 4416–4421.

Holliday, R. (1987). The inheritance of epigenetic defects. *Science* **238,** 163–170.

Krumlauf, R., and Marzluf, G. A. (1980). Genome organization and characterization of the repetitive and inverted DNA sequences in *Neurospora crassa. J. Biol. Chem.* **255,** 1138–1145.

Marathe, S., Connerton, I. F., and Fincham, J. R. S. (1990). Duplication-induced mutation of a new *Neurospora* gene required for acetate utilization: Properties of the mutant and predicted amino acid sequence of the protein product. *Mol. Cell. Biol.* **10,** 2638–2644.

Perkins, D. D., and Barry, E. G. (1977). The cytogenetics of *Neurospora. Adv. Genet.* **19,** 133–285.

Selker, E. U., and Garrett, P. W. (1988). DNA sequence duplications trigger gene inactivation in *Neurospora crassa. Proc. Natl. Acad. Sci. U.S.A.* **85,** 6870–6874.

Selker, E. U. (1990a). DNA methylation and chromatin structure: A view from below. *Trends Biochem. Sci.* **15,** 103–107.

Selker, E. U. (1990b). Premeiotic instability of repeated sequences in *Neurospora crassa. Ann. Rev. Genet.* **24,** 579–613.

Selker, E. U., and Stevens, J. N. (1985). DNA methylation at asymmetric sites is associated with numerous transition mutations. *Proc. Natl. Acad. Sci. U.S.A.* **82,** 8114–8118.

Selker, E. U., Jensen, B. C., and Richardson, G. A. (1987a). A portable signal causing faithful DNA methylation *de novo* in *Neurospora crassa. Science* **238,** 48–53.

Selker, E. U., Cambareri, E. B., Jensen, B. C., and Haack, K. R. (1987b). Rearrangement of duplicated DNA in specialized cells of *Neurospora. Cell* **51,** 741–752.

Selker, E. U., Cambareri, E. B., Garrett, P. W., Jensen, B. C., Haack, K. R., Foss, E., Turpen, C., Singer, M., and Kinsey, J. (1989). Use of RIP to inactivate genes in *Neurospora crassa. Fungal Genet. Newslett.* **36,** 76–77.

12

Ascobolus immersus and *Podospora anserina:* Sex, Recombination, Silencing, and Death

JEAN-LUC ROSSIGNOL
MARGUERITE PICARD
Institut de Génétique et Microbiologie
Université Paris-Sud
Orsay, France

I. INTRODUCTION

The two Ascomycetes *Ascobolus immersus* and *Podospora anserina* were introduced as material for genetic studies by Georges Rizet. *A. immersus* is a heterothallic discomycete ("cup fungus"). In crosses between wild-type and ascospore pigmentation mutants, gene conversion events are easily detected as departures from Mendelian segregations in octads (Rizet *et al.,* 1960). This makes *A. immersus* a valuable tool for studying genetic recombination.

P. anserina, another heterothallic species (pyrenomycetes), showed two remarkable properties that were both related to cytoplasmic heredity. The first one was revealed by the existence of "barrages" resulting from mycelium death at the border of two incompatible strains (Rizet, 1952). The second one was a phenomenon of genetically controlled aging, which leads to the death of hyphae after a defined length of mycelial growth (Rizet, 1953a). This senescence phenomenon has been related to changes in the mitochondrial genome. The study of genetic recombination in *A. immersus* (Section III) and of gene incompatibility (Section IV) and cell aging (Section V) in *P. anserina* is well advanced, using both genetic and molecular approaches. Genetic transformation, an indispensable tool for molecular biology studies, has been developed for both organisms. The main features of the transformation systems are described in Section II. The cloning of mating-type genes in *P. anserina* is described in Section VI.

II. TRANSFORMATION SYSTEMS

A. Selectable Markers

Several markers corresponding to both endogenous (homologous) and foreign (heterologous) genes are used in *Podospora anserina* transformation. The *P. anserina ura5* gene encoding orotidylic acid pyrophosphorylase (OMPppase) was cloned by complementation of the *Escherichia coli pyrE* strain (Bégueret *et al.,* 1984; Turcq and Bégueret, 1987). The *ura5* gene is routinely used to transform *ura5*$^-$ recipient strains.

The tRNA mutant genes *su8.1* and *su4.1* encode tRNA isoacceptors for serine that recognize the same UGA stop codon (Picard, 1973; Debuchy and Brygoo, 1985). They have been cloned through their ability to suppress auxotrophic nonsense (UGA) mutations of the yeast *Schizosaccharomyces pombe* (Debuchy and Brygoo, 1985). The *su8.1* tRNA suppressor gene can transform the *leu1.1* nonsense mutant of *P. anserina* to prototrophy; its ability to suppress a spore-color nonsense mutation allows a powerful genetic analysis of the transformants (Brygoo and Debuchy, 1985).

The *leu1* gene was cloned from a cosmid library by complementation of the *leu1.1* mutant strain (Turcq, 1989), using SIB-selection as described in *Neurospora crassa* (Akins and Lambowitz, 1985). This gene, which is homologous to the *leu1* genes from *S. pombe* and *N. crassa* and to the *LEU2* gene of *Saccharomyces cerevisiae* (J. Bégueret, personal communication) efficiently transforms *leu1*⁻ strains.

Heterologous drug resistance markers are also used in *P. anserina*. One system uses the point-mutated β-tubulin gene from *N. crassa* (Orbach *et al.*, 1986) which confers resistance to the fungicide benomyl. The resistance level, which varies between 1 and 12 μg benomyl/ml, is conserved after growth on nonselective medium; however, molecular analysis shows no relationship between the number of transforming gene copies per nucleus and the resistance level (Fernandez-Larrea and Stahl, 1989). Another system is based on phleomycin resistance (Perrière, 1990). The vector carries the bacterial Tn5 *ble* gene under the control of the *P. anserina ura5* promoter as well as the *ura5* gene. Original transformants are weakly resistant to phleomycin, but increased levels of resistance can be obtained by gene amplification.

In *A. immersus*, the homologous *met2* gene encoding homoserine *O*-transacetylase is used to transform the recipient *met2*⁻ strain (Faugeron *et al.*, 1988). The *Ascobolus met2* gene was cloned using heterologous hybridization with the *S. cerevisiae MET2* gene (Goyon *et al.*, 1988).

Foreign genes are also used in *A. immersus*. For example, the wild-type strain does not grow on medium with acetamide as the sole nitrogen source. In transformation experiments using the *amdS* gene from *Aspergillus nidulans* (Hynes *et al.*, 1983; Davis and Hynes, Chapter 7), which encodes acetamidase, *A. immersus* transformant strains are selected for their ability to use acetamide (L. Rhounim and G. Faugeron, unpublished observations).

The *hph* gene from *E. coli* encodes an HMB phosphotransferase which inactivates the antibiotic hygromycin by phosphorylation. This gene has been bordered by the *gpd* (glyceraldehyde-3-phosphate dehydrogenase gene) promoter and the *trpC* terminator, both from *A. nidulans* (Punt *et al.*, 1987). Introduced to *A. immersus*, this construct confers resistance to up to 100 μM hygromycin to the transformants, whereas the wild-type strain is killed by 20 μM of this antibiotic (L. Rhounim and G. Faugeron, unpublished observations).

B. Transformation Procedures

The transformation procedures are those routinely used for filamentous fungi. Protoplasts are formed from young hyphae with Glucanex® or

Novozyme®. The protoplasts are then treated with DNA in the presence of calcium ions and polyethylene glycol. The osmotic agent for stabilizing protoplasts is either sorbitol or sucrose. *Ascobolus immersus* protoplasts can be stored at −80 °C.

In *Podospora anserina,* the efficiencies of transformation were initially 10–20 μg DNA (Brygoo and Debuchy, 1985). Submitting the protoplasts to a 48 °C heat-shock before the addition of DNA increases the efficiency of transformation to 100–200 transformants per μg DNA with plasmid vectors (Bergès and Barreau, 1989) and about 50 transformants per μg DNA with cosmid vectors (Turcq *et al.,* 1990).

In *A. immersus,* the efficiency of transformation is 50–100 per μg DNA with the homologous *met2* marker (Faugeron *et al.,* 1988) and lower (10–20 per μg DNA) with the heterologous markers *amdS* and *hph* (L. Rhounim and G. Faugeron, unpublished).

Co-transformation by separate plasmids, allowing the introduction of a nonselectable marker with a selectable marker brought by a second plasmid, was achieved in *P. anserina* (Razanamparany, 1988). Whatever the selectable marker, the co-transformation efficiency ranged from 10 to 50% (Razanamparany, 1988; Turcq *et al.,* 1990).

C. Fate of Transforming DNA

1. Homologous versus Nonhomologous Integration

Transformation with circular double-stranded plasmids leads to integrative transformation in both *Podospora anserina* and *Ascobolus immersus.* The efficiency of homologous recombination can be compared with the efficiency of nonhomologous (ectopic) integration using resident genes as transformation markers. Homologous recombination may correspond to either homologous integration (leading to a tandem duplication of the target gene, bordering the vector sequence) or gene substitution (substitution of the information of the recipient gene by that of the transforming one). Nonhomologous integration may correspond to three situations: single ectopic integration, multiple ectopic integrations at distinct sites, and multiple integrations at a single ectopic site. Transformation studies using circular double-stranded plasmids with the homologous gene *ura5, su4.1,* and *su8.1* in *P. anserina* and *met2* in *A. immersus* lead to similar observations.

1. Nonhomologous integrations are more frequent than homologous recombinations. They correspond to 70–90% of the transformation events in *P. anserina* (Bégueret *et al.,* 1984; Brygoo and Debuchy, 1985; Sainsard-Chanet and Begel, 1986) and 80–95% in *A. immersus* (L. Rhounim

and G. Faugeron, unpublished). An early study underestimated the frequency of nonhomologous integrations in *A. immersus* because it did not take into account a fraction of the transformants that was highly unstable through meiosis (Faugeron *et al.*, 1988).

2. Multiple nonhomologous integrations at distinct sites are extremely rare. Single and multiple ectopic integrations at a single site occur with similar frequencies (Bégueret *et al.*, 1984; Faugeron *et al.*, 1988; L. Rhounim and G. Faugeron, unpublished).

3. Both single and multiple ectopic integrations at a single site show rearrangement of the integrated plasmids (Bégueret *et al.*, 1984; Sainsard-Chanet and Begel, 1986; Picard *et al.*, 1987; Faugeron *et al.*, 1988; J. Julien, personal communication; L. Rhounim and G. Faugeron, unpublished observations); in some cases, the entire vector sequence has been lost (Faugeron *et al.*, 1988). This differs from nonhomologous integration in mammals, which usually preserves the integrity of most of the plasmid sequence (Roth and Wilson, 1988). Fine molecular analysis of the DNA of three transformants resulting from single ectopic integration events in *P. anserina* showed the presence of deletions at the plasmid integration site extending from 6 to more than 1000 bp (Razanamparany and Bégueret, 1988). The same analysis showed that the plasmid and chromosomal sequences share little homology at the integration site (2–8 bp). In *P. anserina*, co-transformation with two distinct plasmids is accompanied by integration of the two plasmids at the same locus (Razanamparany, 1988); this may reflect either interplasmid recombination prior to the integration event or local chromosomal competence for nonhomologous recombination, creating an integration hotspot.

Nevertheless, the two organisms apparently display a major difference in the nature of the transformants by homologous recombination. In *P. anserina* they result from homologous integration, whereas in *A. immersus* they result from gene substitution. However, we cannot exclude the possibility that *Ascobolus* transformants actually derive from a primary ectopic integration event followed by a transfer of information by recombination from the transformant gene to the recipient one (Faugeron *et al.*, 1988). If so, the only difference between the two organisms would be a somewhat larger (although small) frequency of homologous integration in *P. anserina* compared with *A. immersus*.

2. *Targeting Transforming DNA to a Recipient Gene*

The naturally low frequency of homologous recombination in transformation of *Podospora anserina* and *Ascobolus immersus* can be increased in several ways. In *P. anserina*, the use of cosmids leads to a high proportion of homologous integration (50–95% of the events: Debuchy *et al.,* 1987; Picard *et al.,* 1987; Turcq, 1989; Turcq *et al.,* 1990; V. Contamine and M. Picard, unpublished data). This may be a consequence of the increase in the length of homology between the transforming DNA and the target, which reaches 30–50 kb with cosmids.

In *A. immersus,* the use of single-stranded transforming DNA increases the proportion of gene substitution (up to 65% of the events), but does not lead to homologous integration (Goyon and Faugeron, 1989). A similar result was reported in the yeast *Saccharomyces cerevisiae* (Simon and Moore, 1987). This procedure has not been successful in *P. anserina* (J. Bégueret, personal communication). In *A. immersus,* its efficiency is strongly dependent on the protoplast preparation (G. Faugeron, unpublished observations).

In *A. immersus,* double-strand cutting of the transforming DNA in the homology region increases the proportion of homologous recombination; up to 70% of the transformants result from homologous integration of one or several copies in the recipient gene (Goyon and Faugeron, 1989).

Using these methods, gene replacement was obtained in two steps (homologous integration followed by the excision of the recipient gene and the vector sequence) in both *P. anserina* (Coppin-Raynal *et al.,* 1989; Turcq, 1989) and *A. immersus* (Goyon and Faugeron, 1989). Single-step gene substitution using single-stranded transforming DNA, as described in *A. immersus,* was also used with success in *Aspergillus nidulans* (Diallinas and Scazzocchio, 1989).

3. Nonintegrative Transformation in Podospora anserina

Replicative transformation in filamentous fungi is the exception (Fincham, 1989). This failure to obtain stable self-replicative plasmids, despite active research on ARS (autonomous replicative sequences), could be explained either by the need for a centromere to stabilize the plasmid or by the inability of circular DNAs to replicate in these organisms. *Saccharomyces cerevisiae* centromeres do not work in heterologous systems (Carbon, 1984; Seekrishna *et al.,* 1984). The search for a resident centromere in *Podospora anserina* was unsuccessful (Debuchy, 1987; Debuchy *et al.,* 1987), although linear self-replicating plasmids were obtained by using the telomeric sequences of the linear rDNA plasmid of *Tetrahymena thermophila* (Perrot *et al.,* 1987). Transformants carrying these linear plasmids were unstable on nonselective medium. Moreover, they were transmitted only via the female line in sexual crosses and displayed non-Mendelian segregation of the transformant phenotype in the progeny. Altogether, these genetic properties suggest that the linear plasmids are inefficiently replicated and/or segregated during mitosis (Perrot *et al.,* 1987). Addition of either potential replication origins isolated from fungi and other lower eukaryotes or random cloned *P. anserina* genomic sequences had no effect on the stability of the transformants, although in some cases the transformation efficiency was increased (Bégueret *et al.,* 1989).

D. Premeiotic Instability of the Transformants

1. Instability through Recombinational Excision in Podospora anserina

Transformants obtained with plasmid vectors exhibit high mitotic stability in *Podospora anserina.* For example, ura^- mitotic revertants from ura^+ transformants can be selected by their resistance to 5-fluoroorotate (Razanamparany and Bégueret, 1986). Only transformants resulting from homologous integration revert. Their reversion frequencies are very low, in the range of mutation rates (Bégueret *et al.,* 1984).

Meiotic stability of *su8.1* transformants obtained with plasmids can be estimated by taking advantage of a spore-color mutation that is partly suppressed by the transforming tRNA mutant gene. The presence or absence of the *su8.1* gene can be checked by looking at the ascospore color in meiotic tetrads. A few transformants, probably those arising from homologous integration, exhibit loss of the *su8.1* phenotype (in less than 1% of the progeny; Brygoo and Debuchy, 1985).

Su8.1 transformants obtained with cosmids (which share up to 50 kb homology with the recipient genome) are very stable when they arise from ectopic integration, but are very unstable through sexual reproduction when they result from homologous integration. The instability can be premeiotic or meiotic. Instability is premeiotic when it occurs before premeiotic replication (the last replication before meiosis): in the progeny of a cross between the transformant and a tester strain, the two meiotic products of the same individual tetrad that should exhibit the transformant phenotype have lost it. Instability is meiotic when it occurs after premeiotic replication: only one meiotic product has lost the marker (Picard *et al.,* 1987). The frequency of premeiotic instability increases with the age of the fruiting body, involving 20% of asci from young perithecia and 90% of asci from old ones (E. Vielh and P. Sirand-Pugnet, unpublished). Meiotic instability was observed in about 15% of the asci (Picard *et al.,* 1987). Meiotic and premeiotic instabilities result from the frequent excision of the vector with the selectable marker (Coppin-Raynal *et al.,* 1989).

2. Instability through Cytosine Methylation in Ascobolus immersus

In *Ascobolus immersus,* all the transformants are mitotically stable. Transformants resulting from gene substitution are also meiotically stable (Faugeron *et al.,* 1988). However, transformants resulting from both homologous and nonhomologous integration are highly unstable through sexual reproduction (Faugeron *et al.,* 1988). This loss of the transformant phenotype is due to a premeiotic event. The transforming DNA is usually neither excised nor rearranged (although excision may occur at a low rate). Instead, the restriction sites located within the duplicated DNA sequence (the recipient copy and the newly integrated one) are no longer cut by restriction enzymes sensitive to cytosine methylation (Goyon and Faugeron, 1989). *Sau*3AI and *Mbo*I both recognize the GATC sequence. *Sau*3AI does not cut when the cytosine is methylated, whereas *Mbo*I does. In previously published experiments (Goyon and Faugeron, 1989), both enzymes failed to cut within the duplicated *met2* sequences. Subsequently it has been shown that the *Mbo*I from Pharmacia (France, S.A.) used in this previous study did not cut methylated sequences. All other *Mbo*I stocks from various suppliers used afterward cut GATC restriction sites without exception (L. Rhounim and C. Barry, unpublished). Thus, loss of the transformed phenotype is associated with a heavy methylation of the cytosine residues. More than 500 GATC sites in inactivated sequences have been tested by *Sau*3AI and *Mbo*I: all are cut by *Mbo*I but not by *Sau*3AI. This shows that no point mutation occurred among the 2000 bp tested (L. Rhounim and C. Barry, unpublished). Moreover, all the unstable derivatives studied (several hundreds) are able to revert to the transformant phenotype after vegetative growth (G. Faugeron, L. Rhounim, and V. Haedens, unpublished). This suggests that inactivation is an epigenetic process of gene silencing associated with an extensive cytosine methylation of the sequences that have been duplicated as a consequence of transformation. For this reason, we call this process MIP, which stands for "methylation induced premeiotically."

The efficiency of MIP is close to 100% for tandem repeats. This efficiency is observed in transformants showing tandem repeats obtained via homologous integration as well as in transformants resulting from multiple integrations at a single site of the foreign genes *amdS* or *hph.* This high efficiency of inactivation also accounts for the failure of a consistent fraction of met$^+$ transformants to produce met$^+$ meiotic progeny (Faugeron *et al.,* 1988); most of these transformants actually result from multiple integrations at a single ectopic site (L. Rhounim, unpublished). For dispersed copies (corresponding to either one resident and one ectopic integrated copy or two copies integrated at different sites, obtained by intercrossing two transformants), the efficiency of the process is close to 50% (Faugeron *et al.,* 1988; Faugeron *et al.,* 1990).

A similar process was first discovered in *Neurospora crassa* (Selker *et al.,* 1987; Selker, 1990; Selker, Chapter 11) and was named RIP, repeat-induced point mutations, by the authors. The *A. immersus* and *N. crassa* processes share the following features: they occur premeiotically with similar frequencies, they are triggered by repeats, they simultaneously affect both repeated copies, and they lead to gene inactivation. However, they differ in several ways. Genes silenced in *A. immersus* can recover expression, whereas in *N. crassa* they cannot. In *N. crassa,* the inactivation is presumably due to both cytosine methylation and G:C to A:T mutations (Selker and Garett, 1988); only cytosine methylation is observed in *A. immersus*. Finally, restriction patterns show that the two copies always undergo the same modification in *A. immersus* (cytosine methylation of all the tested sites within the duplicated region; L. Rhounim and C. Barry, unpublished), whereas they usually undergo different modifications in *N. crassa* (i.e., the location of G:C to A:T substitutions and methylated cytosines; Selker and Garrett, 1988).

In summary, three different filamentous fungi are known to have developed efficient ways to eliminate duplicated gene sequences. In *N. crassa* and *A. immersus,* the sequences are inactivated either permanently *(N. crassa)* or transiently *(A. immersus)*. Physical excisions may occur in *N. crassa* and *A. immersus* and, sometimes frequently, in *Podospora anserina*. All these processes occur during the premeiotic stage of the life cycle.

III. GENETIC RECOMBINATION AND SEQUENCE RECOGNITION

A. Heteroduplex Formation and Mismatch Correction

Ascobolus immersus has been used for years as a model organism for studying meiotic recombination. In this species, each ascus is composed of eight spores (four pairs), corresponding to the four meiotic products issued from an individual meiosis. Mutations in the *b2* spore-color gene, which lead to white instead of dark brown (wild-type) ascospores, can be scored under a binocular microscope. In crosses between wild-type and mutant strains, Mendelian segregations yield four wild-type and four mutant spores. Non-Mendelian segregations correspond to gene conversions (six wild-type, two mutant and two wild-type, six mutant) or post-meiotic segregations (one pair of spores corresponding to one meiotic product is composed of one wild-type and one mutant spore). More than 50 mutations have been mapped in *b2*. The study of the segregation patterns and recombination events in crosses heterozygous for one to up to six allelic *b2* mutations gives information about the formation of recombination intermediates (reviewed in Rossignol *et al.,* 1988; Nicolas and Rossignol, 1989). Most, and possibly all, recombination events are accompanied by the formation of heteroduplex DNA, associating one strand of each parental DNA molecule. Three types of heteroduplex that could span the entire *b2* length were observed: (1) asymmetrical heteroduplex, involving only one duplex among the two interacting homologous DNA duplexes; (2) symmetrical heteroduplex, involving both interacting DNA duplexes; and (3) events associating asymmetrical heteroduplex on one side with symmetrical heteroduplex on the other side (Rossignol and Haedens, 1980). These heteroduplex DNAs are initiated close to the left end of the *b2* gene, where they are the most abundant (Paquette and Rossignol, 1978), and are propagated rightwards (Hamza *et al.*, 1981). Asymmetrical heteroduplexes are the most frequent at the left end; the relative fraction of symmetrical heteroduplexes increases from the left to the right of the gene (Rossignol and Paquette, 1979). Two phenomena account for this observation. First, in events associating asymmetrical and symmetrical heteroduplex stretches, the asymmetrical portion is always located on the left side (Rossignol and Haedens, 1980). Second, the symmetrical heteroduplexes always extend to the right end, whereas asymmetrical ones end at various places (Kalogeropoulos and Rossignol, 1989). Hence, genetic studies indicate that heteroduplex DNA appears to be an essential intermediate of meiotic recombination in the *b2* gene of *A. immersus.* Molecular cloning of the *b2* gene should allow more direct assessment of these recombination intermediates.

When mutations are involved in an heteroduplex stretch, a mismatch is formed and may be recognized and corrected by a specific mismatch correction process. Different mutations in *A. immersus* behave in different ways with respect to mismatch correction. Mismatches involving putative single base-pair additions or deletions are very efficiently corrected, with a strong bias to the excision of the shorter strand (thus favoring the conversion to mutant of additions and the conversion to wild-type of deletions; Hastings *et al.,* 1980). The correction of mismatches involving a putative large deletion is also very efficient. This correction favors the installation of the invader strand, whether it bears the wild-type or the deleted sequence (Hamza *et al.,* 1986). Putative base-substitution mutations are corrected with lower efficiency. These observations are not fully accounted for by studies of mismatch correction performed in prokaryotes (Radman, 1988). Cloning *b2* would insure a direct determination of the nature of the mutations, and allow an extensive study of *in vivo* mismatch correction.

B. Recombination and Gene Polymorphism

Laboratory strains used for studying genetic recombination display isogenicity due to inbreeding. Natural isolates display genetic polymorphism. Do natural crosses undergo the same frequencies of recombination as crosses between laboratory strains? To answer this question, *b2* mutations were isolated from two wild-type strains that arose from the same individual natural cross. The two wild-type strains harbored different wild-type haplotypes for the *b2* genomic region (the term haplotype designates the genetic state of a defined small genomic area). When crossed, the *b2* mutants issued from the same parental haplotype gave normal frequencies of allelic recombination, whereas *b2* mutants issued from each of the two different haplotypes failed to give allelic recombination. Moreover, crosses between wild-type and *b2* mutant strains derived from the same haplotype gave normal amounts of gene conversions (usually, between 10 and 20% of total asci), whereas crosses between a wild-type strain of one haplotype and various *b2* mutant strains issued from the other haplotype failed to give gene conversion (their frequency dropped more than 100-fold). This strong effect of gene polymorphism upon genetic recombination at *b2* was observed only when the crosses were heterozygous for the two haplotypes. It was thus a local effect, tightly linked to the polymorphism of this small genomic area. By studying mutations in other genes derived from these same two wild-type strains, it was shown that these two strains displayed gene polymorphism inhibiting genetic recombination in three other genomic areas. The inhibition of recombination in one area was dependent on polymorphism in that area, but did not depend on polymorphism in the three other areas. This showed that the effect of gene polymorphism upon recombination was a local effect (Girard and Rossignol, 1974). This raises interesting questions. In a natural population, for a given genomic area, how many different haplotypes exist that are prone to local suppression of recombination when they are heterozygous? Does gene polymorphism lead to the suppression of genetic recombination by random accumulation of heterozygocities or by accumulation of particular heterozygocities in key sites? If key sites are involved, is the frequency of nonrecombining haplotypes random, or are haplotypes that do not affect recombination favored?

C. The Pairing Puzzle

Genetic recombination requires a previous step of recognition between homologous DNA sequences. A preliminary pairing step between homologous DNA sequences is also probably required in the processes of premeiotic inactivation of repeated genes by excision *(Podospora anserina, Neurospora crassa),* by RIP *(N. crassa),* or by MIP *(Ascobolus immersus).* Indeed, the premeiotic excision of the vector and one copy is likely to result from a recombination process involving homologous pairing between the two copies. The premeiotic cytosine-methylation process in *A. immersus* is triggered only if the haploid parent harbors two or more copies of a DNA segment (a recipient and a transferred gene or two transferred genes) (Faugeron *et al.,* 1990). When two copies are present, either each copy or neither copy is inactivated. When three copies are present, none, two, or three are inactivated. The inactivation of only one copy was never observed. This observation suggests that MIP requires a preliminary pairing step. Paired copies are methylated and one methylated copy is able to undergo a new pairing cycle, allowing the third copy to undergo MIP also (Faugeron *et al.,* 1990). A similar requirement for pairing is seen in *N. crassa* for RIP (Fincham *et al.,* 1989). It is unknown whether the pairing steps involved in meiotic recombination and the postulated pairing steps involved in the different premeiotic excision and inactivation processes utilize common mechanisms.

The meiotic frequencies of recognition between homologous sequences leading to gene conversion in *Saccharomyces cerevisiae* and in *A. immersus* may vary greatly, from 0.1 to 20% for a given genomic area (Nicolas and Rossignol, 1989). These frequencies are much smaller than the very high frequencies (almost 100%) of gene inactivation obtained premeiotically for tandem repeats through RIP in *N. crassa* and through MIP in *A. immersus* and the frequencies of excision (20–90%) observed for large tandem repeats in *P. anserina.* Still more striking is the fact that, in RIP and MIP, sequences dispersed throughout the genome are able to recognize each other with an efficiency close to 50%. This suggests that genomic elements have efficient ways to identify themselves and "talk" to one another. Understanding the mechanisms that allow such communication processes is an exciting challenge for the future.

IV. GENE INCOMPATIBILITY

Incompatibility means the inability for two genomes to coexist (to be expressed) in a cell or for two cells to live together and achieve a specific process. This widespread phenomenon is associated with various syndromes and probably results from various physiological mechanisms. In general, incompatibility can be observed at three levels: transplant failure, fertilization failure, and offspring lethality. Most often, the incompatible partners carry different alleles at one or several of the so-called incompatibility loci. Self-incompatibility in higher eukaryotes is limited to higher plants in which self-fertilization is blocked.

Esser (1959) distinguished two kinds of incompatibility in fungi: homogenic and heterogenic. *Homogenic incompatibility* (or, more logically, heterogenic compatibility) corresponds to the heterothally: mating occurs only between strains that differ at one or more loci. Sexual compatibility in *Podospora anserina* is summarized in Section VI. *Heterogenic incompatibility* was first described by Rizet (1952) in *P. anserina* and subsequently has been observed and studied in a number of other fungi, expecially *Neurospora crassa* (Garnjobst and Wilson, 1956), *Aspergillus nidulans*

(Grindle, 1963), and *Endothia parasitica* (Anagnostakis, 1977). It results in a failure to form heterokaryotic strains by hyphal fusion between different strains. This somatic (also called vegetative or protoplasmic) incompatibility leads to the death of the heterokaryotic cells. When grown on the same Petri dish, the incompatible strains give a "barrage" instead of merging with each other in the zone of contact. The genetic determinism of incompatibility, investigated in detail in *N. crassa* (Perkins, 1975, for a review) and *P. anserina* (Rizet, 1953b; Beisson-Schecroun, 1962; Bernet, 1967; Esser, 1974, for reviews), involves two types of mechanisms: allelic and nonallelic. In the first case, different alleles at *one* locus are responsible for incompatibility whereas in the second case, different alleles at *two* loci are involved.

In *N. crassa,* allelic genes responsible for incompatibility have been identified at 10 loci (Mylyk, 1975). The incompatibility due to these *het* genes is strictly vegetative: it does not reduce fertility. Furthermore, the *A* and *a* mating-type idiomorphs also act as vegetative incompatibility genes (see Glass and Lorimer, Chapter 8).

In *P. anserina,* no vegetative incompatibility is associated with the mating type. Such incompatibility does not occur because ascospores usually contain two nuclei of opposite mating type. Nine genes responsible for allelic and nonallelic incompatibility have been described. Genes *B, Q, S,* and *Z* are involved in allelic incompatibility and genes *C, D, E,* and *R* are responsible for nonallelic incompatibility. Gene *V* is involved in both allelic and nonallelic incompatibility. These genes have been identified in crosses between 16 wild-type races that give rise to lethal self-lysing offspring (Bernet, 1965). Further work using mutagenesis identified several new loci for nonallelic incompatibility (Delettre and Bernet, 1976). At the moment, at least 17 incompatibility genes have been described in *P. anserina.* Both allelic and nonallelic vegetative incompatibility can be associated with sexual incompatibility. In fact, one of the two crosses (for instance, ♀$Z_2 \times$ ♂Z_1 or ♀$C_1 \times$ ♂E_2) is always fertile whereas the reciprocal cross is less fertile (♀$Z_1 \times$ ♂Z_2) or completely sterile (♀$E_2 \times$ ♂C_1) (Bernet, 1965).

The physiological and biochemical processes associated with somatic incompatibility in fungi are not understood at present. The most detailed studies have been performed in *P. anserina* using the temperature-sensitive *R/V* interaction, in which the recombinant *RV* strains display a wild-type phenotype at 32°C and self-lyse at 26°C. Protein synthesis and mRNA are severely reduced. However, two specific proteolytic activities, resulting from *de novo* synthesis of stable mRNA, appear and are responsible for cell disintegration (Bégueret, 1972; Bégueret and Bernet, 1973; Labarère *et al.,* 1974; Boucherie *et al.,* 1981; C. Barreau, personal communication).

Studies carried out on mutations suppressing nonallelic incompatibility reactions, all located in four genes (Belcour and Bernet, 1969, Bernet, 1971; Labarère and Bernet, 1977, 1979), suggest that incompatibility genes are involved in normal *P. anserina* development. Strains carrying these mutations have defects in the formation of protoperithecia, in ascospore germination, or in the renewal of growth from stationary cells (Boucherie and Bernet, 1974; Boucherie *et al.*, 1976; Labarère and Bernet, 1977, 1978; Durrens *et al.*, 1979; Durrens and Bernet, 1982). Considered together, these and more recent data strongly suggest that protoplasmic incompatibility may be the deviant expression of hyphal cell death, which is a condition for production of fertile perithecia (J. Bernet, personal communication).

The *c* gene involved in nonallelic incompatibility and the *S* and *s* alleles of the gene responsible for the most studied allelic incompatibility reaction (Beisson-Schecroun, 1962) have been cloned (Turcq, 1989). The *S/s* incompatibility is triggered by cytoplasmic proteins encoded by the two alleles. Furthermore, in the offspring of the $s \times S$ crosses, most of the strains with the *s* genotype exhibit a new phenotype called s^S. The s^S strains are compatible with both *s* and *S* strains. This result suggests that *s* gene expression is prevented by the early expression of the *S* gene. The s^S phenotype can revert to the *s* phenotype by a protoplasmic contact with an *s* strain, suggesting that *s* gene expression is positively controlled by its own product (Schecroun, 1959). When ♀*s* × ♂*S* crosses are performed at low temperature (18 °C instead of 27 °C), some of the ascospores carrying the *S* gene are abortive. This is explained by the premature expression of the *s* gene in a few asci (Bernet, 1965).

The *s* gene was cloned by SIB-selection through transformation of a recipient strain carrying a neutral allele (s^x) compatible with strains carrying the *s* or *S* alleles. The *S* gene was isolated from a genomic library using the *s* gene as a probe. Outside the coding region, the *s* and *S* loci are highly dissimilar (Turcq *et al.*, 1990): at the 3′ end, the *s* haplotype contains a 2-kb sequence that is missing in the *S* haplotype; at the 5′ end, the *s* strain shows a 349-bp sequence that is similar to the solo-δ, -σ, or -τ elements of *Saccharomyces cerevisiae* (Deleu *et al.*, 1990). The *s* and *S* functions have been correlated with a single open reading frame (ORF), encoding a polypeptide of 289 amino acids. A 1.3-kb mRNA was found in the strains *s, S,* and s^x, as was a 30-kDa protein (B. Turcq and J. Bégueret, personal communication).

Future studies will have to clarify how these genes work in the incompatibility reaction and how they control morphogenetic pathways. Understanding the function of these genes will also be important in terms of evolution, because at least some of the genes favor speciation by limiting outbreeding (Turcq *et al.*, 1990).

V. SENESCENCE, PREMATURE DEATH, AND MITOCHONDRIA

A. Senescence

In *Podospora anserina,* indefinite vegetative growth has not been observed, regardless of the wild-type strain or culture conditions used. This inevitable and irreversible arrest of growth was first described, nearly 40 years ago, by Rizet (1953a), who termed it senescence. Analyses of this phenomenon led to the conclusions that the senescent state is maternally inherited in sexual crosses, and that it is vegetatively contagious (Marcou, 1961). Mitochondria were implicated in the senescence process (Belcour and Begel, 1978; Cummings *et al.,* 1978; Stahl *et al.,* 1978).

The senescent state is correlated with rearrangements of the mitochondrial DNA (mt-DNA) and amplification of specific sequences of mt-DNA as tandemly repeated circular molecules (sen-DNA) (Cummings *et al.,* 1979; Jamet-Vierny *et al.,* 1980; Kück *et al.,* 1981; Wright *et al.,* 1982). The most often amplified sequence, called alpha (α) sen-DNA, or p1-DNA, consists of the first intron (class II) of the mosaic gene CO1 (Osiewacz and Esser, 1984). Molecular and physiological mechanisms responsible for the appearance and amplification of sen-DNA, and for both trigger and contagion of the senescent state, are not completely understood (see Dujon and Belcour, 1989, for review). Nevertheless, several lines of evidence strongly suggest that senescence is primarily due to mt-DNA rearrangements (Vierny *et al.,* 1982; Koll *et al.,* 1984; Belcour and Vierny, 1986). Most strains selected for escaping senescence carry rearrangements of their mt-DNA (see Kück, 1989; Dujon and Belcour, 1989), usually a deletion of the 5′ end of intron α (Belcour and Vierny, 1986; Sainsard-Chanet and Begel, 1990). Amplification of an intronic sequence to multimeric circular molecules seems to be characteristic of *Podospora* and has never been observed elsewhere. The relationship between this amplification and the cell degeneration observed in senescent cultures raises important and intriguing questions.

The *P. anserina* mitochondrial genome was sequenced completely recently (Cummings *et al.,* 1990).

B. Premature Death

The premature death phenotype was found during a systematic study of the effects of protein synthesis alterations on the life span of the fungus. Belcour and Begel (1980) first observed that cycloheximide (an inhibitor of cytosolic protein synthesis) decreased the life span whereas chloramphenicol (an inhibitor of mitochondrial protein synthesis) delayed the onset of senescence. Numerous mutations that alter components of the cytosolic translational apparatus have been described in *Podospora anserina* (Picard, 1973; Picard-Bennoun, 1976; Bégueret *et al.*, 1977; Crouzet and Bégueret, 1980a, b, 1982; Coppin-Raynal and Le Coze, 1982; Dequard-Chablat, 1986; Dequard-Chablat *et al.*, 1986). Some of them, which were selected as suppressors *(su)* or antisuppressors *(AS)*, decrease or increase, respectively, the accuracy in translation (Coppin-Raynal *et al.*, 1988, for a review). Both kinds of mutations can decrease or increase the life span, suggesting that the error rate, *per se,* is not involved in triggering the senescence process (Belcour *et al.*, 1991). Attention was focused on the *AS1-4* mutant. This particular allele of the gene encoding the ribosomal protein *S12* (Dequard-Chablat *et al.*, 1986) confers a very long life span to mycelia of mating type plus (*mat*+), whereas it is responsible for a strikingly short life span (the shortest ever observed) in mycelia expressing the mating type minus (*mat*−). Mitochondrial DNAs from strains containing the *AS1-4* mutation were examined. All mycelia presenting the premature death phenotype bore the same deleted mitochondrial chromosome. The boundaries of the deletion were precisely mapped and the junction region was sequenced from prematurely dying mycelia issued from five independently obtained ascospores. In the five cases, the same 36854-bp deletion was observed. It was ascertained that the deleted molecule did appear *de novo,* as the result of the simultaneous expression of two nuclear genes, *AS1-4* and a gene linked to *mat*−. A remarkable feature of this deletion is that one of the break points is exactly at the 5′ splice site of a peculiar intron called intron α (Belcour *et al.*, 1991). This intronic sequence is found as abundant circular DNA molecules in senescent cultures (see above and Kück *et al.*, 1981; Wright *et al.*, 1982; Osiewacz and Esser, 1984; Vierny *et al.*, 1982). It is a class II intron that may encode a protein similar to retrotransposon reverse transcriptases (Osiewacz and Esser, 1984; Michel and Lang, 1985; Jacquier and Michel, 1987). Another class II intron has recently been shown to transpose in *P. anserina* (L. Belcour and M. Rossignol, personal communication). In yeast, a reverse transcription step is probably involved in transposition of class II introns (Meunier *et al.*, 1990). Taking into account what is known about the mobility of introns in organelle genomes (Dujon and Belcour, 1989, for a review), it is tempting to hypothesize that the premature death deletion even results from mechanisms also involved in intron transposition (Belcour *et al.*, 1991).

These observations support recent observations that strongly suggest that nuclear genes control the stability of mitochondrial genomes in higher and lower eukaryotes (Zeviani *et al.*, 1989; Seidel-Rogol *et al.*, 1989). Cloning the *AS1* gene and the gene linked to *mat*− is in progress. It is of fundamental interest to understand how these genes can control so accurate a rearrangement of the mitochondrial DNA.

VI. MATING TYPES

The processes that take place between fertilization and the beginning of meiosis constitute a "black box" in filamentous fungi. How do nuclei cross-talk from the time of fertilization to the time of karyogamy? Do the *mat*+ and *mat*− (male and female) nuclei recognize each other at the time of dikaryon formation or, rather, at the time of karyogamy? How is this recognition achieved? Y. Brygoo, who initiated the mating-type project in *Podospora anserina,* attempted to open this black box by going directly to the master genes that control sexual reproduction, that is, the mating-type genes. The mating-type region of *P. anserina* also contains one or several genes that are involved in the maintenance of mitochondrial DNA integrity (see previous text), another reason for our interest in studying this locus.

The mating-type locus of *P. anserina* was cloned by using the mating-type *A* DNA from *Neurospora crassa* as a probe (kindly provided to us by L. Glass, C. Staben, R. Metzenberg, and C. Yanofsky). The DNA fragment that hybridized to the *A* probe was obtained from a *mat*− cosmid library. This fragment was, in turn, used as a probe to identify the equivalent region in a *mat*+ cosmid library. The hybridizing cosmids were used to transform *mat*+ and *mat*− strains. Transformants exhibiting a dual *mat*+ *mat*− phenotype were obtained and submitted to molecular and genetic analyses. Since *P. anserina* cosmids integrate preferentially by homologous recombination, the integration event gives rise to a tandem duplication, one copy containing the resident information, the other one containing the copy from the cosmid. This duplicated structure is the target for recombination/excision events before premeiotic replication, that yield gene substitutions. In this way, pure fully functional *mat*− strains were obtained from *mat*+ *mat*− transformants in which the original resident locus was *mat*+. Conversely, *mat*+ strains were obtained from transformants in which the resident locus was *mat*−. Such gene replacement experiments showed that the cloned fragments contained the entire mating-type locus (Picard *et al.,* 1991).

Molecular analyses revealed that the *mat* haplotypes of *P. anserina* are highly dissimilar, as was previously observed for *N. crassa* (Glass *et al.,* 1988; Metzenberg and Glass, 1990). A segment of 4.7 kb is unique to *mat*− whereas a segment of 3.5 kb is unique to *mat*+ (Picard *et al.,* 1991).

Cosmid subcloning and functional analysis of the subcloned fragments through transformation experiments showed that the *mat* haplotypes contain two domains, one sufficient for fertilization, the other one necessary for ascus production. In the case of *mat*−, the fertilization region is bordered by two sporulation domains (E. Coppin and R. Debuchy, unpublished).

Although *mat*+ *mat*− transformants are dual maters (they are able to induce perithecium formation on both *mat*+ and *mat*− strains), the two crosses show dissimilar fertility levels. Ascospore production in a *mat*+ *mat*− × *mat*+ (or *mat*−) cross depends on the nature of the resident and of the transgenic *mat*. The cross is always far more fertile when the tester strain is compatible with the resident *mat* information than when it is compatible with the transgenic one. A *mat*^{0} strain, which bears a deletion of the *mat* locus, was recently obtained (S. Arnaise, unpublished). This strain should permit us to understand how this dominance of the resident information is achieved. Furthermore, cytological analyses are in progress in order to know at which stage fruiting body development is blocked in the poorly fertile crosses (D. Zickler, personal communication).

The *mat*−-specific region has been partly sequenced (R. Debuchy, unpublished). The fertilization domain contains an ORF that shows a high similarity to the *mt A1* gene of *N. crassa* (Glass *et al.*, 1990). This similarity (109 identical amino acids among 195) does not extend to the C-terminal half of the deduced polypeptide sequences. However, it appears that the C-terminal part of the protein is dispensable for fertilization in both fungi (Glass *et al.*, 1990; E. Coppin, unpublished). It is noteworthy that the two proteins share a small domain with the *MATα1* protein of *Saccharomyces cerevisiae* (Astell *et al.*, 1981; Glass *et al.*, 1990; Glass and Lorimer, Chapter 8).

The fertilization domain of the *P. anserina mat*+-specific region has also been sequenced (E. Coppin, unpublished). It contains an ORF, interrupted by two putative introns, that should encode a polypeptide displaying a high similarity to the *mt a1* polypeptide of *N. crassa.* Both proteins show similarity to the *mat-Mc* polypeptide of *Schizosaccharomyces pombe* (Staben and Yanofsky, 1990; Kelly *et al.*, 1988; Glass and Lorimer, Chapter 8). More fascinating is the fact that three proteins required for mating show similarity with the SRY protein thought to be involved in mammalian sex determination (Gubbay *et al.*, 1990; Sinclair *et al.*, 1990). In fact, all three proteins contain a motif (the HMG box) present in the nuclear high mobility group proteins (HMG), which might represent a novel class of DNA-binding domain (Jantzen *et al.*, 1990; R. Debuchy and E. Coppin, unpublished).

Much work remains to be done in order to understand the evolution of mating types and sex in ascomycetes. Furthermore, we must discover how these genes work in filamentous fungi. Genes that are expressed under control of the *mat* genes are useful tools. In *P. anserina,* we will focus our attention on the steps that lie in the black box between fertilization and karyogamy. The use of mutants impaired in crozier formation or in karyogamy, previously isolated by Simonet and Zickler (1978), should facilitate this study.

ACKNOWLEDGMENTS

We thank S. Arnaise, C. Barreau, C. Barry, O. Begel, J. Bégueret, L. Belcour, J. Bernet, Y. Brygoo, V. Contamine, E. Coppin, R. Debuchy, G. Faugeron, V. Haedens, J. Julien, L. Rhounim, M. Rossignol, S. Salama, P. Siran-Pugnet, B. Turcq, E. Vielh, and D. Zickler for communicating unpublished data. We thank S. Arnaise, J. Bégueret, L. Belcour, J. Bernet, E. Coppin, R. Debuchy, G. Faugeron, A. Kalogeropoulos, E. U. Selker, and D. Zickler for critical reading of the manuscript.

REFERENCES

Akins, R. A., and Lambowitz, A. M. (1985). General method for cloning *Neurospora crassa* nuclear genes by complementation of mutants. *Mol. Cell. Biol.* **5,** 2272–2278.

Anagnostakis, S. L. (1977). Vegetative incompatibility in *Endothia parasitica. Exp. Mycol.* **1,** 306–316.

Astell, C. R., Ahlstrom-Jonasson, L., Smith, M., Tatchell, K., Nasmyth, K. A., and Hall, B. D. (1981). The sequence of the DNAs coding for the mating-type loci of *Saccharomyces cerevisiae. Cell* **27,** 15–23.

Bégueret, J. (1972). Protoplasmic incompatibility: Possible involvement of proteolytic enzymes. *Nature (New Biol.)* **235,** 56–58.

Bégueret, J., and Bernet, J. (1973). Proteolytic enzymes and protoplasmic incompatibility in *Podospora anserina. Nature (New Biol.)* **243,** 94–96.

Bégueret, J., Perrot, M., and Crouzet, M. (1977). Ribosomal proteins in the fungus *Podospora anserina:* Evidence for an electrophoretically altered 60S protein in a cycloheximide resistant mutant. *Mol. Gen. Genet.* **156,** 141–144.

Bégueret, J., Razanamparany, V., Perrot, M., and Barreau, C. (1984). Cloning gene *ura5* for the orotidylic acid pyrophosphorylase of the filamentous fungus *Podospora anserina:* Transformation of protoplasts. *Gene* **32,** 487–492.

Bégueret, J., Turcq, B., Razanamparany, V., Perrière, M., Denayrolles, M., Bergès, T., Javerzat, J.-P., and Barreau, C. (1989). Development and use of vectors for *Podospora anserina. In* "Proceedings of the EMBO–Alko Workshop on Molecular Biology of Filamentous Fungi" (H. Nevalainen, and M. Penttilä, eds.), pp. 41–49. Foundation for Biochemical and Industrial Fermentation Research, Helsinki.

Beisson-Schecroun, J. (1962). Incompatibilité cellulaire et interactions nucléocytoplasmiques dans les phénomènes de barrage. *Ann. Genet.* **4,** 1–50.

Belcour, L., and Begel, O. (1978). Lethal mitochondrial genotypes in *Podospora anserina. Mol. Gen. Genet.* **116,** 113–123.

Belcour, L., and Begel, O. (1980). Life-span and senescence in *Podospora anserina:* Effect of mitochondrial genes and functions. *J. Gen. Microbiol.* **119,** 505–515.

Belcour, L., and Bernet, J. (1969). Sur la mise en évidence d'un gène dont la mutation supprime spécifiquement certaines manifestations d'incompatibilité chez le *Podospora anserina. C.R. Acad. Sci. Paris* **269,** 712–714.

Belcour, L., and Vierny, C. (1986). Variable DNA splicing sites of a mitochondrial intron: Relationship to the senescence process in *Podospora. EMBO J.* **5,** 609–614.

Belcour, L., Begel, O., and Picard, M. (1991). A site-specific deletion in mitochondrial DNA of *Podospora* is under the control of nuclear genes. *Proc. Natl. Acad. Sci. USA* **88** (In press).

Bergès, T., and Barreau, C. J. (1989). Heat shock at an elevated temperature improves transformation efficiency of protoplasts from *Podospora anserina. J. Gen. Microbiol.* **135,** 601–604.

Bernet, J. (1965). Mode d'action des gènes de "barrage" et relation entre l'incompatibilité cellulaire et l'incompatibilité sexuelle chez *Podospora anserina. Ann. Sci. Nat. Bot.* **6,** 611–768.

Bernet, J. (1967). Les systèmes d'incompatibilité chez le *Podospora anserina. C.R. Acad. Sci. Paris* **265,** 1330–1333.

Bernet, J. (1971). Sur un cas de suppression de l'incompatibilité cellulaire chez le champignon *Podospora anserina. C.R. Acad. Sci. Paris* **273,** 1120–1122.

Boucherie, H., and Bernet, J. (1974). Protoplasmic incompatibility and female organ formation in *Podospora anserina:* Properties of mutations abolishing both processes. *Mol. Gen. Genet.* **135,** 163–174.

Boucherie, H., Bégueret, J., and Bernet, J. (1976). The molecular mechanism of protoplasmic incompatibility and its relationship to the formation of protoperithecia in *Podospora anserina. J. Gen. Microbiol.* **82,** 59–66.

Boucherie, H., Dupont, C. H., and Bernet, J. (1981). Polypeptide synthesis during protoplasmic incompatibility in the fungus *Podospora anserina. Biochim. Biophys. Acta* **653,** 18–26.

Brygoo, Y., and Debuchy, R. (1985). Transformation by integration in *Podospora anserina.* I. Methodology and phenomenology. *Mol. Gen. Genet.* **200,** 128–131.

Carbon, J. (1984). Yeast centromeres: Structure and function. *Cell* **37,** 351–353.

Coppin-Raynal, E., and Le Coze, D. (1982). Mutations relieving hypersensitivity to paramomycin caused by ribosomal suppressors in *Podospora anserina. Genet. Res.* **40,** 149–164.

Coppin-Raynal, E., Dequard-Chablat, M., and Picard, M. (1988). Genetics of ribosomes and translational accuracy in *Podospora anserina. In* "Genetics of Translation" (M. F. Tuite, M. Picard, and M. Bolotin-Fukuhara, eds.), pp. 431–442. Springer-Verlag, Berlin.

Coppin-Raynal, E., Picard, M., and Arnaise, S. (1989). Transformation by integration in *Podospora anserina.* III. Replacement of a chromosome segment by a two-step process. *Mol. Gen. Genet.* **219,** 270–276.

Crouzet, M., and Bégueret, J. (1980a). A new mutant form of the ribosomal protein L21 in the fungus *Podospora anserina:* Identification of the structural gene for this protein. *Mol. Gen. Genet.* **180,** 177–183.

Crouzet, M., and Bégueret, J. (1980b). Altered ribosomal proteins in mutants of the fungus *Podospora anserina.* J. Biol Chem. **255,** 4996–4999.

Crouzet, M., and Bégueret, J. (1982). Altered ribosomal proteins in emetine resistant mutants of *P. anserina. Curr. Genet.* **6,** 39–45.

Cummings, D. J., Belcour, L., and Grandchamp, C. (1978). Etude au microscope électronique du DNA mitochondrial de *Podospora anserina* et présence d'une série multimérique de molécules circulaires de DNA dans les cultures senescentes. *C.R. Acad. Sci. Paris* **287,** 157–160.

Cummings, D. J., Belcour, L., and Grandchamp, C. (1979). Mitochondrial DNA from

Podospora anserina. II. Properties of mutant DNA and multimeric circular DNA from senescent cultures. *Mol. Gen. Genet.* **171,** 239–250.

Cummings, D. J., McNally, K. L., Domenico, J. M., and Matsuura, E. T. (1990). The complete DNA sequence of the mitochondrial genome of *Podospora anserina. Curr. Genet.* **17,** 375–402.

Debuchy, R. (1987). Mise au point d'un système de transformation du champignon filamenteux *Podospora anserina* et recherche de séquences susceptibles d'assurer l'autonomie de réplication. Ph.D. Thesis. University of Paris VI.

Debuchy, R., and Brygoo, Y. (1985). Cloning of opal suppressor tRNA genes of a filamentous fungus reveals two $tRNA_{UGA}$ genes with unexpected structural differences. *EMBO J.* **4,** 3553–3556.

Debuchy, R., Coppin-Raynal, E., Le Coze, D., and Brygoo, Y. (1987). Chromosome walking towards a centromere in the filamentous fungus *Podospora anserina:* Cloning a sequence lethal at a two-copy state. *Curr. Genet.* **13,** 105–111.

Delettre, Y. M., and Bernet, J. (1976). Regulation of proteolytic enzymes in *Podospora anserina:* Selection and properties of self-lysis mutant strains. *Mol. Gen. Genet.* **144,** 191–197.

Deleu, C., Turcq, B., and Bégueret, J. (1990). Repa, a repetitive and dispersed DNA sequence of the filamentous fungus *Podospora anserina. Nucl. Acids Res.* **18,** 4901–4903.

Dequard-Chablat, M. (1986). Genetics of translational fidelity in *Podospora anserina:* Are all the genes involved in this ribosomal function identified? *Curr. Genet.* **10,** 531–536.

Dequard-Chablat, M., Coppin-Raynal, E., Picard-Bennoun, M., and Madjar, J.-J. (1986). At least seven ribosomal proteins are involved in the control of translational accuracy in a eukaryotic organism. *J. Mol. Biol.* **190,** 167–175.

Diallinas, G., and Scazzocchio, C. (1989). A gene coding for the uric acid–xanthine permease of *Aspergillus nidulans:* Inactivational cloning, characterization, and sequence of a *cis*-acting mutation. *Genetics* **122,** 341–350.

Dujon, B., and Belcour, L. (1989). Mitochondrial DNA instabilities and rearrangements in yeasts and fungi. *In* "Mobile DNA" (D. E. Berg and M. M. Howe, eds.), pp. 861–878. American Society of Microbiology, Washington, D.C.

Durrens, P., and Bernet, J. (1982). *Podospora anserina* mutations inhibiting developmental alternatives and growth renewal. *Curr. Genet.* **5,** 181–186.

Durrens, P., Laigret, F., Labarère, J., and Bernet, J. (1979). *Podospora anserina* mutants defective in protoperithecium formation, ascospore germination, and cell regeneration. *J. Bacteriol.* **140,** 835–842.

Esser, K. (1959). Incompatibilitätsbeziehungen zwischen geographischen Rassen von *Podospora anserina.* III. Untersuchungen zur Genphysiologie der Barragebildung und Semi-incompatibilität. *Z. Vererbungsl.* **90,** 445–456.

Esser, K. (1974). *Podospora anserina. In* "Handbook of Genetics" (R. C. King, ed.), Vol. 1, pp. 531–551. Plenum Press, New York.

Faugeron, G., Goyon, C., and Grégoire, A. (1988). Stable allele replacement and unstable nonhomologous integration events during transformation of *Ascobolus immersus. Gene* **76,** 109–119.

Faugeron, G., Rhounim, L., and Rossignol, J.-L. (1990). How does the cell count the number of ectopic copies of a gene in the premeiotic inactivation process acting in *Ascobolus immersus? Genetics* **124,** 585–591.

Fernandez-Larrea, J., and Stahl, U. (1989). Transformation of *Podospora anserina* with a dominant resistance gene. *Curr. Genet.* **16,** 57–62.

Fincham, J. R. S. (1989). Transformation in fungi. *Microb. Rev.* **53,** 148–170.

Fincham, J. R. S., Connerton, I. F., Notarianni, E., and Harrington, K. (1989). Premeiotic disruption of duplicated and triplicated copies of the *Neurospora crassa am* (glutamate deshydrogenase) gene. *Curr. Genet.* **15,** 327–334.

Garnjobst, L., and Wilson, J. F. (1956). Heterocaryosis and protoplasmic incompatibility in *Neurospora crassa. Proc. Natl. Acad. Sci. U.S.A.* **42,** 613–618.

Girard, J., and Rossignol, J.-L. (1974). The suppression of gene conversion and intragenic crossing-over in *Ascobolus immersus:* Evidence for modifiers acting in the heterozygous state. *Genetics* **76,** 221–243.

Glass, N. L., Wollmer, S. J., Staben, C., Grotelueschen, J., Metzenberg, R. L., and Yanofsky, C. (1988). DNAs of the two mating-type alleles of *Neurospora crassa* are highly dissimilar. *Science* **241,** 470–573.

Glass, N. L., Grotelueschen, J., and Metzenberg, R. L. (1990). The *Neurospora crassa A* mating type region. *Proc. Natl. Acad. Sci. U.S.A.* **87,** 4912–4916.

Goyon, C., and Faugeron, G. (1989). Targeted transformation of *Ascobolus immersus* and *de novo* methylation of the resulting duplicated sequences. *Mol. Cell. Biol.* **9,** 2818–2827.

Goyon, C., Faugeron, G., and Rossignol, J.-L. (1988). Molecular cloning and characterization of the *met2* gene from *Ascobolus immersus. Gene* **63,** 297–308.

Grindle, M. (1963). Heterocaryon compatibility of unrelated strains in the *Aspergillus nidulans* group. *Heredity* **18,** 191–204.

Gubbay, J., Collignon, J., Koopman, P., Capel, B., Economou, A., Münsterberg, A., Vivian, N., Goodfellow, P., and Lovell-Badge, R. (1990). A gene mapping to the sex-determining region of the mouse Y chromosome is a member of a novel family of embryonically expressed genes. *Nature (London)* **346,** 245–250.

Hamza, H., Haedens, V., Mekki-Berrada, A., and Rossignol, J.-L. (1981). Hybrid DNA formation during meiotic recombination. *Proc. Natl. Acad. Sci. U.S.A.* **78,** 7648–7651.

Hamza, H., Nicolas, A., and Rossignol, J.-L. (1986). Two mechanisms for directional gene conversion. *Proc. Natl. Acad. Sci. U.S.A.* **83,** 7386–7390.

Hastings, P., Kalogeropoulos, A., and Rossignol, J.-L. (1980). Restoration to the parental genotype of mismatches formed in recombinant DNA heteroduplex. *Curr. Genet.* **2,** 169–174.

Hynes, M. J., Corrick, C. M., and King, J. A. (1983). Isolation of genomic clones containing the *amdS* gene of *Aspergillus nidulans* and their use in the analysis of structural and regulatory mutations. *Mol. Cell. Biol.* **3,** 1430–1439.

Jacquier, A., and Michel, F. (1987). Multiple exon-binding site in class II self-splicing introns. *Cell* **50,** 17–29.

Jamet-Vierny, C., Begel, O., and Belcour, L. (1980). Senescence in *Podospora anserina:* Amplification of a mitochondrial DNA sequence. *Cell* **21,** 189–194.

Jantzen, H-M., Admon, A., Bell, S. P., and Tjian, R. (1990). Nucleolar transcription factor hUBF contains a DNA-binding motif with homology to HMG proteins. *Nature (London)* **344,** 830–836.

Kalogeropoulos, A., and Rossignol, J.-L. (1989). Non-random transition from asymmetrical to symmetrical hybrid DNA during meiotic recombination. *Genome* **32,** 414–419.

Kelly, M., Burke, J., Smith, M., Klar, A., and Beach, D. (1988). Four mating-type genes control sexual differentiation in the fission yeast. *EMBO J.* **7,** 1537–1547.

Koll, F., Begel, O., Keller, A. M., and Belcour, L. (1984). Ethidium bromide rejuvenation of senescent cultures of *Podospora anserina:* Loss of senescence specific DNA and recovery of normal mitochondrial DNA. *Curr. Genet.* **8,** 127–134.

Kück, U. (1989). Mitochondrial DNA rearrangements in *Podospora anserina. Exp. Mycol.* **13,** 111–120.

Kück, U., Stahl, U., and Esser, K. (1981). Plasmid-like DNA is part of mitochondrial DNA in *Podospora anserina. Curr. Genet.* **3,** 151–156.

Labarère, J., and Bernet, J. (1977). Protoplasmic incompatibility and cell lysis in *Podospora anserina:* Genetic investigation on mutations of a novel modifier gene that suppresses cell destruction. *Genetics* **87,** 249–257.

Labarère, J., and Bernet, J. (1978). A mutation inhibiting protoplasmic incompatibility in *Podospora anserina* that suppresses an extracellular laccase and protoperithecium formation. *J. Gen. Microbiol.* **109,** 187–189.

Labarère, J., and Bernet, J. (1979). Protoplasmic incompatibility in *Podospora anserina:* A possible role for its associated proteolytic activity. *Genetics* **93,** 525–537.

Labarère, J., Bégueret, J., and Bernet, J. (1974). Incompatibility in *Podospora anserina:* Comparative properties of the antagonistic cytoplasmic factors of a nonallelic system. *J. Bacteriol.* **120,** 854–860.

Marcou, D. (1961). Notion de longévité et nature cytoplasmique du déterminant de la senescence. *Ann. Sci. Nat. Bot.* **11,** 653–764.

Metzenberg, R. L., and Glass, N. L. (1990). Mating type and mating strategies in *Neurospora. BioEssays* **12,** 53–59.

Meunier, B., Tian, G. L., Macadre, C., Slonimski, P. P., and Lazowska, J. (1990). Group II introns transpose in yeast mitochondria. *In* "Structure, Function and Biogenesis of Energy Transfer Systems" (E. Quagliariello, S. Papa, F. Palmieri, and C. Saccone, eds.), pp. 169–173. Elsevier Science Publishers, Amsterdam.

Michel, F., and Lang, F. B. (1985). Mitochondrial class II introns encode proteins related to the reverse transcriptases of retroviruses. *Nature (London)* **316,** 641–643.

Mylyk, O. M. (1975). Heterokaryon incompatibility genes in *Neurospora crassa* detected using duplication-producing chromosome rearrangements. *Genetics* **80,** 107–124.

Nicolas, A., and Rossignol, J.-L. (1989). Intermediates in recombination revealed by marker effects in *Ascobolus. Genome* **31,** 528–535.

Orbach, M. J., Porro, E. B., and Yanofsky, C. (1986). Cloning and characterization of the gene for β-tubulin from a benomyl-resistant mutant of *Neurospora crassa* and its use as a dominant selectable marker. *Mol. Cell. Biol.* **6,** 2452–2461.

Osiewacz, H. D., and Esser, K. (1984). The mitochondrial plasmid of *Podospora anserina:* A mobile intron of a mitochondrial gene. *Curr. Genet.* **8,** 299–305.

Paquette, N., and Rossignol, J.-L. (1978). Gene conversion spectrum of 15 mutants giving postmeiotic segregations in the locus *b2* of *Ascobolus immersus. Mol. Gen. Genet.* **163,** 313–326.

Perkins, D. D. (1975). The use of duplication generating rearrangements for studying heterokaryon incompatibility genes in *Neurospora. Genetics* **80,** 87–105.

Perrière, M. (1990). Utilisation d'un gène de résistance à la phléomycine comme marqueur de selection de la transformation de *Podospora anserina.* Ph. D. Thesis. University of Bordeaux II.

Perrot, M., Barreau, C., and Bégueret, J. (1987). Nonintegrative transformation in the filamentous fungus *Podospora anserina:* Stabilization of a linear vector by the chromosomal ends of *Tetrahymena thermophila. Mol. Cell. Biol.* **7,** 1725–1730.

Picard, M. (1973). Genetic evidence for a polycistronic unit of transcription in the complex locus "14" in *Podospora anserina.* II. Genetic analysis of informational suppressors. *Genet. Res.* **21,** 1–15.

Picard-Bennoun, M. (1976). Genetic evidence for ribosomal antisuppressors in *Podospora anserina. Mol. Gen. Genet.* **147,** 199–306.

Picard, M., Debuchy, R., Julien, J., and Brygoo, Y. (1987). Transformation by integration in *Podospora anserina.* II. Targeting to the resident locus with cosmids and instability of the transformants. *Mol. Gen. Genet.* **210,** 129–134.

Picard, M., Debuchy, R., and Coppin, E. (1991). Cloning the mating types of the heterothallic fungus *Podospora anserina:* Developmental features of haploid transformants carrying both mating types. *Genetics* **128** (in press).

Punt, P. J., Oliver, R. P., Dingemanse, M. A., Pauwels, P. H., and van den Hondel, C. A. M. J. J. (1987). Transformation of *Aspergillus* based on the hygromycin B resistance marker from *E. coli. Gene* **56,** 117–124.

Radman, M. (1988). Mismatch repair and genetic recombination. *In* "Genetic recombination" (R. Kuperlachati and G. Smith, eds.), pp. 169–192. American Society for Microbiology, Washington, D.C.

Razanamparany, V. (1988). Mise au point d'un système de transformation chez le champignon filamenteux *Podospora anserina.* Ph.D. Thesis. University of Bordeaux II.

Razanamparany, V., and Bégueret, J. (1986). Positive screening and transformation of *ura5* mutants in the fungus *Podospora anserina:* Characterization of the transformants. *Curr. Genet.* **10,** 811–817.

Razanamparany, V., and Bégueret, J. (1988). Non-homologous integration of transforming vectors in the fungus *Podospora anserina:* Sequences of junction at the integration sites. *Gene* **74,** 399–409.

Rizet, G. (1952). Les phénomènes de barrage chez *Podospora anserina.* I. Analyse génétique des barrages entre souches *S* et *s. Rev. Cytol. Biol. Veg.* **13,** 51–92.

Rizet, G. (1953a). Sur l'impossibilité d'obtenir la multiplication végétative ininterrompue et illimitée de l'Ascomycète *Podospora anserina. C. R. Acad. Sci. Paris* **237,** 838–840.

Rizet, G. (1953b). Sur la multiplicité des mécanismes génétiques conduisant à des barrages chez *Podospora anserina. C. R. Acad. Sci. Paris* **237,** 666–668.

Rizet, G., Engelmann, N., Lefort, C., Lissouba, P., and Mousseau, J. (1960). Sur un ascomycète intéressant pour l'étude de certains aspects de la structure du gène. *C. R. Acad. Sci. Paris.* **250,** 2050–2052.

Rossignol, J.-L., and Haedens, V. (1980). Relationship between asymmetrical and symmetrical hybrid DNA during meiotic recombination. *Curr. Genet.* **1,** 185–191.

Rossignol, J.-L., and Paquette, N. (1979). Disparity of gene conversion in frameshift mutants located in the locus *b2* of *Ascobolus immersus. Proc. Natl. Acad. Sci. U.S.A.* **76,** 2871–2875.

Rossignol, J.-L., Nicolas, A., Hamza, H., and Kalogeropoulos, A. (1988). Recombination and gene conversion in *Ascobolus. In* "The Recombination of Genetic Material" (Low Brook, ed.), pp. 23–72. Academic Press, San Diego.

Roth, D., and Wilson, J. (1988). Illegitimate recombination in mammalian cells. *In* "Genetic Recombination" (R. Kucherlapati and G. Smith, eds.), pp. 621–653. American Society for Microbiology, Washington, D.C.

Sainsard-Chanet, A., and Begel, O. (1986). Transformation of yeast and *Podospora:* Innocuity of senescence-specific DNAs. *Mol. Gen. Genet.* **204,** 443–451.

Sainsard-Chanet, A., and Begel, O. (1990). Insertion of an LrDNA gene fragment and of filler DNA at a mitochondrial exon–intron junction in *Podospora. Nucl. Acids Res.* **18,** 779–783.

Schecroun, J. (1959). Sur la nature de la différence cytoplasmique entre souches *s* et s^S de *Podospora anserina. C. R. Acad. Sci. Paris* **238,** 1394–1397.

Seekrishna, K., Webster, T. D., and Dickson, R. C. (1984). Transformation of *Kluyveromyces lactis* with the kanamycin (G418) resistance gene of Tn903. *Gene* **28,** 73–81.

Seidel-Rogol, B. L., King, J., and Bertrand, H. (1989). Unstable mitochondrial DNA in natural-death nuclear mutants of *Neurospora crassa. Mol. Cell. Biol.* **9,** 4259–4264.

Selker, E. U. (1990). Premeiotic instability of repeated sequences in *Neurospora crassa. Ann. Rev. Genet.* **24,** 579–613.

Selker, E. U., and Garett, P. W. (1988). DNA sequence duplications trigger gene inactivation in *Neurospora crassa*. *Proc. Natl. Acad. Sci. U.S.A.* **85,** 6870–6874.

Selker, E. U., Cambareri, E. B., Jensen, B. C., and Haack, K. R. (1987). Rearrangement of duplicated DNA in specialized cells of *Neurospora*. *Cell* **51,** 741–752.

Simon, J. R., and Moore, P. D. (1987). Homologous recombination between single-stranded DNA and chromosomal genes in *Saccharomyces cerevisiae*. *Mol. Cell. Biol.* **7,** 2329–2334.

Simonet, J.-M., and Zickler, D. (1978). Genes involved in caryogamy and meiosis in *Podospora anserina*. *Mol. Gen. Genet.* **162,** 237–242.

Sinclair, A. H., Berta, P., Palmer, M. S., Hawkings, J. R., Griffiths, B. L., Smith, M. J., Foster, J. W., Frischauf, A.-M., Lovell-Badge, R., and Goodfellow, P. N. (1990). A gene from the human sex-determining region encodes a protein with homology to a conserved DNA-binding motif. *Nature (London)* **346,** 240–244.

Staben, C., and Yanofsky, C. (1990). *Neurospora crassa a* mating-type region. *Proc. Natl. Acad. Sci. U.S.A.* **87,** 4917–4921.

Stahl, U., Lemke, P. A., Tudzynski, P., Kück, U., and Esser, K. (1978). Evidence for plasmid-like DNA in a filamentous fungus, the ascomycete *Podospora anserina*. *Mol. Gen. Genet.* **162,** 341–343.

Turcq, B. (1989). Clonage direct de gènes par complémentation chez le champignon *Podospora anserina:* Application à l'étude de gènes d'incompatibilité. Ph.D. Thesis. University of Bordeaux II.

Turcq, B., and Bégueret, J. (1987). The *ura5* gene of the filamentous fungus *Podospora anserina:* Nucleotide sequence and expression in transformed strains. *Gene* **53,** 201–209.

Turcq, B., Denayrolles, M., and Bégueret, J. (1990). Isolation of the two allelic incompatibility genes *s* and *S* of the fungus *Podospora anserina*. *Curr. Genet.* **17,** 297–303.

Vierny, C., Keller, A. M., Begel, O., and Belcour, L. (1982). A sequence of mitochondrial DNA is associated with the onset of senescence in a fungus. *Nature (London)* **297,** 157.

Wright, R. M., Horrum, M. A., and Cummings, D. J. (1982). Are mitochondrial structural genes selectively amplified during senescence in *Podospora anserina? Cell* **29,** 505–515.

Zeviani, M., Servidei, S., Gellera, C., Bertini, E., Di Mauro, S., and Di Donato, S. (1989). An autosomal dominant disorder with multiple deletions of mitochondrial DNA starting at the D-loop region. *Nature (London)* **339,** 309–311.

13

Sordaria macrospora: The Transition to the Age of Gene Manipulation

LANDRY LE CHEVANTON
DENISE ZICKLER
Institut de Génétique et Microbiologie
Université Paris-Sud
Orsay, France

I. *SORDARIA MACROSPORA:* A MODEL FOR CYTOGENETIC STUDIES OF MEIOTIC RECOMBINATION, PAIRING, AND SYNAPTONEMAL COMPLEX

A. Why *Sordaria macrospora?*

The goal of our work is to understand how meiotic chromosome pairing and recombination are related. We have chosen to focus our studies on the synaptonemal complex (SC). This complex is specifically formed along the

MORE GENE MANIPULATIONS
IN FUNGI

paired homologues in prophase I and is a highly conserved event of meiotic chromosomal differentiation.

Sordaria macrospora is a filamentous ascomycete belonging to the class Pyrenomycetes and the order Sphaeriales. This species was first used by Esser and Straub (1958) and Heslot (1958) for the study of several aspects of recombination by analysis of the segregation patterns of spore-color genes during meiosis. Two other *Sordaria* species, *S. fimicola* and *S. brevicollis,* contributed greatly to our understanding of meiotic recombination. The presence of ordered tetrads allows an easy scoring of meiotic and postmeiotic events by visual analysis of the eight-spored asci. The four nuclei resulting from meiosis are retained in a linear array, their position in which reflects the time of segregation during the two meiotic divisions (pre- and postsegregation of allelic pairs). The four nuclei subsequently undergo mitosis, yielding eight ascospores, which allows the determination of the genetic characteristics of each of the eight DNA strands involved in meiosis.

Postmeiotic segregations in wild-type × mutant crosses were first discovered in *S. fimicola* by Olive (1959). With the same species, Kitani and colleagues (1962) showed another class of postmeiotic segregation, the aberrant 4:4 asci, resulting from reciprocal single-strand exchanges between two homologous DNA molecules. The discovery of these postmeiotic segregations played a key role in all recombination models elaborated so far (review in Holliday, 1964; Whitehouse, 1982; Orr-Weaver and Szostak, 1985).

S. brevicollis was used for gene conversion studies by Fields and Olive (1967). The characterization of two mutations (YS17 and rec1) that change the conversion frequencies within the *Buff* locus gave an interesting test for the mechanisms involved in the recombination-initiation sites (MacDonald and Whitehouse, 1979, 1982). Finally, the studies of Sang and Whitehouse (1979, 1983) provided additional information on the relationship between crossover, gene conversion, and aberrant segregations.

In our laboratory, the use of *S. macrospora* as experimental organism has been prompted by two main attributes. First, all the meiotic stages are easily identified cytologically and a clear image of the SC can be obtained. Relatively few serial sections (30–40) are required for three-dimensional reconstruction of prophase nuclei and SC karyotypes. All seven chromosomes can be recognized with electron microscopy (EM) by centromere index and length, and they can be followed with the light microscope during mitotic and meiotic divisions (Zickler, 1977; Moreau *et al.,* 1985; Zickler *et al.,* 1985). Of special interest is the fact that the recombination nodules associated with the SC can be clearly identified, and that they correlate remarkably well with crossover exchange events (Zickler, 1977; review in Carpenter, 1979; von Wettstein *et al.*, 1984). Although yeasts are of great value for similar projects (review in Esposito and Klapholz, 1981; Giroux, 1988), it is currently difficult to pursue detailed cytological investigations with these organisms.

Second, *S. macrospora* is homothallic and thus able, by self-mating, to form a diploid nucleus homozygous for any induced mutation. Homothallism offers advantages for obtaining recessive mutants affecting the sexual phase of the life cycle, especially meiosis and sporulation. Finally, homothallism guarantees homozygous genomes. For example, Bojko (1988) has shown that, in *Neurospora,* two heterothallic species, *N. crassa* and *N. intermedia,* show abnormal SC but homothallic *N. terricola* and *N. africana* have regular structures. For genetic analyses, selfing can be prevented by using nonallelic self-sterile but cross-fertile mutants.

S. macrospora is closely related to *N. crassa* (Perkins, 1985; Leblon *et al.,* 1987) and has the same life cycle, culture, and cytogenetic advantages already reviewed in Perkins and Barry (1977) and Perkins (1985).

B. Meiotic Mutants

In an effort to identify genes involved in meiotic chromosome pairing and recombination, mutants were isolated by treating protoplasts with ultraviolet (uv) light or ethylmethane sulfonate (EMS). Survivors were recovered on regeneration medium and selfed. Each colony was screened for its ability to produce ascospores. Among the 100 strains that exhibited abnormal sporulation or barren fruiting bodies, 40 meiotic mutants were isolated and characterized. Either they were blocked at specific stages or they exhibited abnormal phenotypes during meiotic development, including alterations in microtubules and spindle-pole bodies. The remaining 60 mutants were more specifically sporulation-deficient.

Mutants with increased nondisjunction frequencies were sought out systematically for use as tools to probe the role of the SC during pairing and recombination. Seven genes (Asy1 – 7) necessary for wild-type levels of pairing were identified (Huynh *et al.,* 1986; Zickler *et al.,* 1988). Three other genes required for normal SC morphogenesis were characterized: mutants *spo76* and *spo77* showed, respectively, abnormal lateral element cohesiveness and structure, and *spo44* exhibited a phenotype suggestive of a failure in central element assembly (Moreau *et al.,* 1985; Zickler *et al.,* 1985). The products of these genes are candidates for structural components of the SC. Their cloning could contribute to the identification of proteins required for SC morphogenesis. Since meiosis is highly conserved, it is likely that results that emerge from studies in fungi will also be applicable to higher eukaryotes.

C. Chromosome Rearrangements

A second topic of interest is the detection and meiotic behavior of chromosome rearrangements. These rearrangements were used to establish the correspondence between genetic and physical maps. Assignment of linkage groups to chromosomes was made by a three-dimensional reconstruction of the SCs. Reciprocal translocation break points were mapped genetically to linkage groups and examined cytologically. Two translocations, with one linkage group in common, were fixed for EM, and the SCs were reconstructed after serial sectioning of their pachytene nuclei. The seven bivalents of *Sordaria macrospora* can be distinguished by centromere index or length, and all reciprocal translocations showed typical quadrivalents in heterozygous crosses. Nine translocations were enough to establish the chromosomal assignment of the seven linkage groups (Zickler *et al.,* 1984). Because break points can be precisely placed on linkage maps by genetic analysis, such SC reconstructions can also facilitate the location of genes within a chromosome arm. In a given translocation quadrivalent, the arm proportions remain similar during prophase, indicating that DNA break points are reflected in SC length. Such a precise relationship between the genetic map and chromosomal location has also been used to compare the frequencies of recombination observed in the gene–centromere interval with the number of SC recombination nodules detected in wild-type and mutant strains (Zickler *et al.,* 1988).

Over 150 uv-induced translocations have been isolated in *S. macrospora.* This fungus is a useful model for studying chromosomal rearrangements and can be used to give reliable quantitative measures of rearrangement frequencies. Induction of rearrangements by uv light is photoreversible (Arnaise *et al.,* 1984). Since *S. macrospora* produces linear octads of spores by meiosis, chromosome rearrangements can be detected and defined by visual observation of asci from heterozygous crosses. Asci that contain aborted ascospores exhibit characteristic patterns that can be used for the diagnosis of the type of rearrangement (Perkins, 1974; Perkins and Barry, 1977). Three types of aberrations were found. According to Perkins' classification (1974), Type I (27% of all rearrangements) corresponds to reciprocal translocations. Type II (8%) represents tip-to-nontip interchanges of arms or transpositions of chromosome segments within or between chromosomes (G. Leblon, personal communication). Finally, the most frequent aberrations, called Type III (65%), are reciprocal translocations with both break points within or adjacent to centromeres (Leblon *et al.,* 1986). Photoreactivation experiments show that reciprocal translocations are related to the formation of pyrimidine dimers (Arnaise *et al.,* 1984). Thus, Type III aberrations may be caused by pyrimidine-rich sequences in the centromere regions (Leblon *et al.,* 1986).

II. TRANSITION TO THE AGE OF GENE MANIPULATION

A. A Transformation System for *Sordaria macrospora*

A transformation system for *Sordaria macrospora* extends the cytological and genetic studies of the meiotic mutants to a molecular level. Such a system facilitates the isolation of meiotic genes by complementation of the corresponding mutants.

When developing a transformation system, we used a selection marker corresponding to a *S. macrospora* gene, ensuring perfect identity between the nucleotide sequences of the donor gene and the chromosomal recipient sequences. This simplifies the transformation system itself and allows the study of phenomena such as targeting or inactivation of duplicated sequences. The marker used was the *Ura5* gene coding for orotate phosphoribosyl transferase (OPRTase). This system of selection had several advantages: first, a mutant was available in our strain collection; second, the growth of the *ura5-1* mutant is completely blocked in the absence of uracil or uridine; and third, all the asci formed after selfing of this strain contain eight underpigmented spores, thus simplifying the genetic analysis of the transformants. This pigmentation characteristic is recessive. However, it is not spore autonomous. Selfing of a *ura5-1* strain gives asci with eight yellow spores. Selfing of a wild-type strain gives asci with eight black spores. In a cross, *ura5-1* × wild-type, only asci with eight black spores are observed, rather than the expected four yellow/four black ratio. The eight spores of an ascus are yellow only if no wild-type nucleus is present in the ascus (for example a 7 *ura5-1*/1 wild-type segregation will appear as an ascus with eight black spores), since uracil is necessary for wild-type sporulation and its production by the wild-type nuclei in the hybrid ascus is enough to allow pigmentation of all eight spores of the ascus.

The *S. macrospora* wild-type gene was isolated from a lambda gene library by heterologous hybridization with the corresponding gene *(Ura5)* from *Podospora anserina* (Turcq and Bégueret, 1987). Identity of the cloned sequence was confirmed by its ability to complement a *pyrE* mutation in *Escherichia coli* (defect in OPRTase). Subclones of the lambda clone were constructed in plasmid vectors (pUC18 or 19 and pBR329) and a 0.9-kb fragment functional in *E. coli* was sequenced. The amino acid sequence deduced from the single reading frame present in this fragment showed similarity with OPRTase from other species and also with other phosphoribosyl transferases (Le Chevanton and Leblon, 1989).

For the transformation experiments, protoplasts obtained after Novozym 234 (Novo Industry, Denmark) treatment of the *ura5-1* hyphae were treated with plasmid DNA containing the *ura5* sequences. The characteristics of *S. macrospora* transformation were established in an intensive genetic study. The frequency of transformation was comparable to the frequencies observed in other filamentous fungi (10–100 transformants per μg DNA) and no significant variations were noted when using different vectors or inserts (Le Chevanton *et al.,* 1989).

Prototrophic *ura*$^+$ transformants (often heterokaryons) were selected on minimal medium. After selfing of these transformants, homokaryotic *ura*$^+$ progeny were isolated from ascospores. Genetic analysis of these progeny prototrophs showed that in 73 of 74 cases, the newly introduced *ura5*$^+$ characteristic was regularly transmitted as a single Mendelian factor through several rounds of crosses, indicating a stable integration of the donor DNA at a unique site in the chromosomes of the recipient strain. Three progeny corresponded to a homologous integration. In one case, two separate points of integration, each of them behaving like a single Mendelian factor, were observed.

Southern blot analyses of DNA samples from transformants indicated, in every case, an integration of the donor sequences at a single location in the genomic DNA. Often, more than one copy of the plasmid was inserted. With multiple-copy integrations, tandem repeats were observed, as were more complicated structures, presumably involving complex rearrangements during or after the integration step. Analysis of the transformants harboring a homologous insertion in the resident *ura5* sequence clearly indicated that the addition had occurred by a mechanism identical to a homologous recombination event (single crossover).

The characteristics of the transformation system developed for *S. macrospora* are similar to those described for the majority of the filamentous ascomycetes (Fincham, 1989), that is, integrative transformation, occurrence of both homologous and ectopic insertion events, presence of complex rearrangements of the donor molecule, as well as possible insertion of multiple copies of the plasmid. These similarities suggest a common mechanism of integration for these organisms.

B. Duplications Are Stable during Meiosis

A number of ascomycetes possess mechanisms for the recognition of duplicated sequences in the genomic DNA, which are then either eliminated or inactivated. In *Neurospora crassa,* linked or unlinked duplicated sequences in premeiotic ascogenous hyphae are subjected to extensive point mutation associated with *de novo* cytosine methylation (RIP; see Selker and Garrett, 1988; Selker, 1990; Selker, Chapter 11). A similar process is observed in *Ascobolus immersus,* in which duplicated sequences are methylated, without occurrence of point mutations (MIP; see Goyon and Faugeron, 1989; Rossignol and Picard, Chapter 12). In both cases, the duplicated sequences are inactivated. Thus, inactivation of duplications resulting from a transformation experiment leads to the loss of the transformed phenotype. In *Podospora anserina,* a somewhat different result is observed (Coppin-Raynal *et al.,* 1989). Only tandem-linked duplications are unstable, whereas ectopic integrations are stable. Tandem-linked duplications are removed by homologous recombination, causing the loss of one copy of the duplication. Moreover, such meiotic instability is high in transformation with cosmids (Coppin-Raynal *et al.,* 1989) and rare in transformation with plasmids (Brygoo and Debuchy, 1985). For all three organisms, instability occurs at a high rate, specifically during the premeiotic dikaryotic stage of sexual reproduction. The biological significance of these inactivations is not known at present. Nevertheless, it seems probable that this process involves a step of "checking" for sequence homology similar to that required during recombination.

Sordaria macrospora does not exhibit meiotic inactivation of duplicated sequences. Such an inactivation would be detected easily with the spore-color defect of the *ura*$^-$ asci. Transformants with one to four additional copies of the *ura5* region, linked or unlinked to the resident locus, were selfed (more than 10^5 ascospores could be checked in a single cross) and no meiotic instability of the transformed phenotype was observed through several generations. This result is surprising, given the evolutionary proximity of *S. macrospora* and *N. crassa* (Perkins, 1985; Leblon *et al.,* 1987) and the high efficiency of the inactivation process (100% for tandem duplications and about 50% for unlinked duplications) observed in *N. crassa* (Fincham, 1989; Selker, Chapter 11). Because *N. crassa* and *A. immersus* have different kinds of processing (methylation and point mutation for the former; only methylation for the latter), we tested whether duplicated sequences could be methylated but not inactivated in *S. macrospora.* In our studies, transformants harboring one to three tandemly repeated copies of the *Ura5* gene, in either ectopic or homologous integration, gave no traces of methylation (Le Chevanton *et al.,* 1989), although such structures are inactivated premeiotically in virtually all the nuclei of *Neurospora* and *Ascobolus.* Thus, in *S. macrospora, ura5* duplicated sequences are neither inactivated nor methylated, and are stably transmitted through meiosis.

C. Transformation with a Heterologous Gene, Co-transformation, and Vegetative Instability of Transformants

The observation that 60% of transformants contain several copies of the transforming plasmid suggests that more than one molecule of DNA is able to enter one nucleus. In order to confirm this hypothesis, a second marker, the *Met2* gene from *Ascobolus immersus* encoding the *O*-acetyl homoserine transferase (Goyon *et al.,* 1988), was used. The corresponding mutant *(met1)* was available in our collection, and we used this marker to extend the observation obtained with an endogenous marker to a second exogenous selectable gene. It was possible to select *met*$^+$ transformants from a *met1*$^-$ strain, demonstrating that the foreign gene was expressed in *Sordaria macrospora,* and confirming the heterologous expression reported for other filamentous ascomycetes (Fincham, 1989).

Co-transformation was conducted with both markers (at equimolar ratio) on two separate plasmids. After selection for one marker, 80% of the transformants also expressed the second selection gene. In all cases, genetic analysis showed that both markers co-segregated, indicating physical linkage between the two genes. This high level of co-transformation has also been observed in other systems (Fincham, 1989), and indicates that at least two independent molecules of DNA can enter a single nucleus. Thus, when duplications are observed in single-marker transformants, they are presumably due to multiple integrations rather than an amplification of a single plasmid.

Homokaryotic prototrophic colonies were obtained after selfing of the co-transformants. Study of the progeny of these homokaryotic colonies after a second selfing revealed an unexpected feature. Among 69 independent second-generation progeny of transformants, 58 showed asci with 8 underpigmented spores, indicating that the transformed phenotype was lost (L. Le Chevanton, unpublished results). Analysis of the asci containing 8 black spores (in hybrid asci with a wild-type allele, the *ura*$^+$ phenotype is expressed by all 8 ascospores, whether or not they carry a *ura5* genotype; see Section II,A) indicated the occurrence of non-Mendelian segregations, that is, asci with 7 *ura*$^+$*met*$^+$ spores and 1 *ura*$^-$*met*$^-$ spore (7 : 1 segregation), as well as 3 : 5, 5 : 3, and 6 : 2 segregations. In each case, the two markers co-segregated. Furthermore, when protoplasts were isolated from homokaryotic co-transformed mycelia (stable through two rounds of selfing), at least half, and sometimes all, the colonies regenerated from these protoplasts had lost the prototrophic phenotype, suggesting that this instability of the transformed phenotype could occur during vegetative divisions.

From Southern blot analysis of the selfed prototrophic progeny of transformants, and their prototrophic and auxotrophic progeny, it is clear that the observed instability is due to the loss of the sequence introduced by transformation in the auxotrophic progenies (Fig. 1).

The cause(s) of this instability is still not clear. It is not due to co-transformation as such: transformations of *met1*$^-$ protoplasts with a plasmid harboring the *Met2* gene alone or with a construct containing both selectable markers on one plasmid (pLLC834) both gave rise to unstable transformants. In the former case, 11 of 26 transformants were unstable and in the latter case, 1 of 9 transformants was unstable. The same pLLC834 plasmid used to transform a *ura5*$^-$ strain gave only stable transformants (32/32).

Mitotic instability of initially stable transformants is very unusual among ascomycetes. In fact, the only previous case described, in which mitotic instability is significantly above the normal background mutation rate, is the transformation of *Podospora anserina* with linear self-replicating plasmids containing the telomeric sequences of the linear plasmid of *Tetrahymena thermophila* (Perrot *et al.,* 1987). In other filamentous fungi, high mitotic instability correlates with autonomous replication of the donor DNA (van Heeswijck, 1986; Tsukuda *et al.,* 1988). The vegetative instability observed here is clearly different from the instability seen in the abortive transformants of *Neurospora crassa* (Fincham, 1989). After selection as prototrophic colonies, these transient transformants abort on the first subcloning to selective medium, whereas all colonies observed in our co-transformation experiments can be maintained and grown after repeated vegetative transfers. We cannot yet exclude extrachromosomal location for the introduced sequences in our unstable transformants. Further studies should contribute to the resolution of several unanswered questions concerning the mechanisms of DNA integration during transformation.

III. FUTURE PROSPECTS

The main goal for developing a transformation system for *Sordaria macrospora* was to obtain a tool for cloning the genes corresponding to the characterized meiotic mutants. The *ura5* marker easily can be used to select transformants, which can then be tested for the ability of the second DNA segment used in the co-transformation to complement the sporulation defect of the mutants. This could be done very efficiently by the same techniques previously used in our laboratory to isolate the revertants of *spo76* (Huynh *et al.,* 1986).

The low transformation frequencies will make the direct isolation of genes, using a gene library, a somewhat laborious task. An alternative method is to utilize the potentially random insertion of the selectable

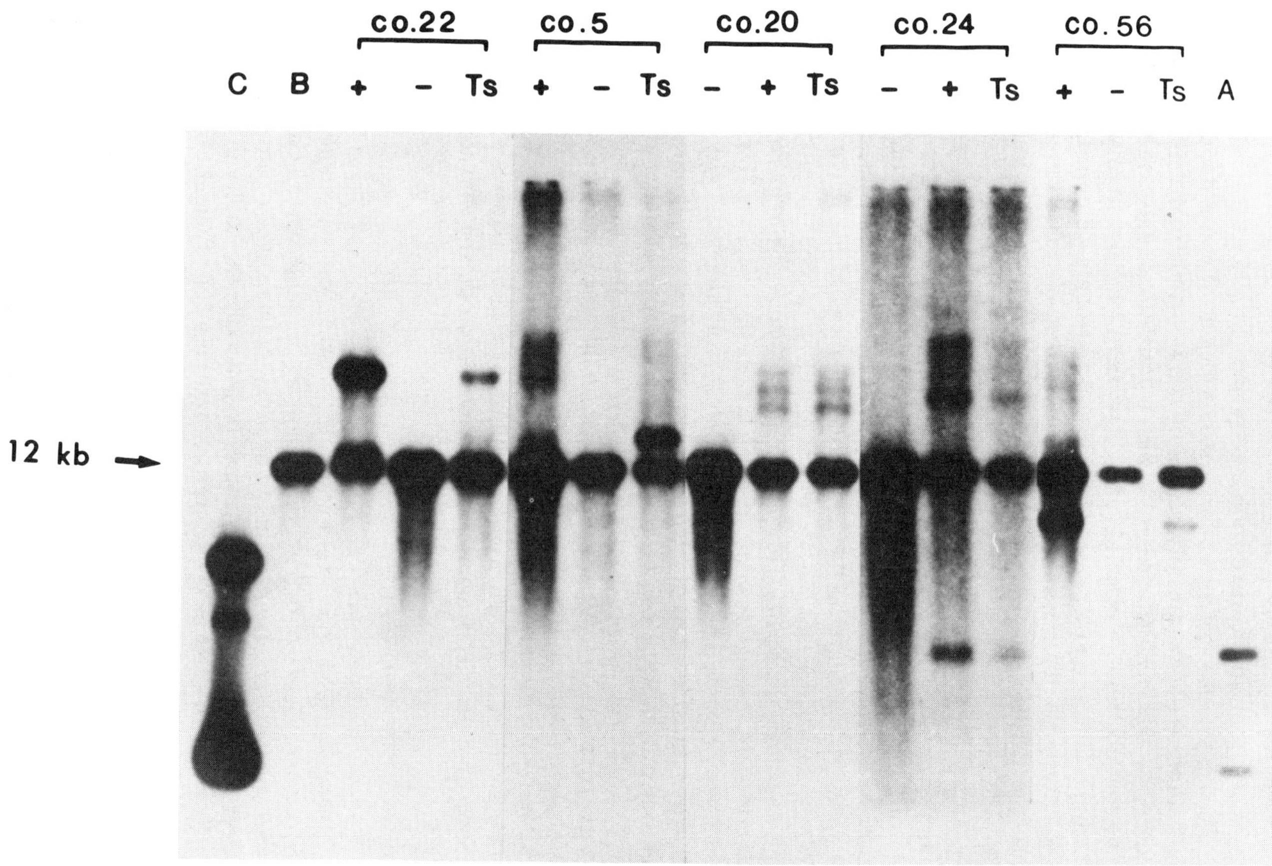
co.22
co.5
co.20
co.24
co.56
C B + − Ts + − Ts − + Ts − + Ts + − Ts A
12 kb

marker, that is, to identify linkage between the selectable marker and the gene of interest. Indeed, the genetic localization of 135 *ura5*$^+$ insertions has revealed four transformants that are strongly linked to *spo76,* another that is tightly linked to an asynaptic mutant *(asy2-17),* and yet another that is linked to a spore-color mutant *(j2).* These insertions can be used to initiate chromosome walks to reach the two meiotic genes, which will be identified by their ability to complement the corresponding mutations after transformation. Moreover, several translocation break points have also been mapped close to the mutations. For example, the mutant strain *spo44,* which is mapped on chromosome 7 and has profound effects on pairing and recombination, results from the translocation of nucleolus-organizer material from chromosome 2 to chromosome 7 (Zickler *et al.,* 1985).

ACKNOWLEDGMENTS

We thank Gérard Leblon and Patrick Moreau for providing information prior to its publication. Thanks also to Jean-Luc Rossignol for ciritcal reading of this manuscript.

REFERENCES

Arnaise, S., Leblon, G., and de Lares, L. (1984). A system for the detection of chromosomal rearrangements using *Sordaria macrospora. Mut. Res.* **25,** 33–42.

Bojko, M. (1988). Presence of abnormal synaptonemal complexes in heterothallic species of *Neurospora. Genome* **30,** 697–709.

Brygoo, Y., and Debuchy, R. (1985). Transformation by integration in *Podospora anserina.* I. Methodology and phenomenology. *Mol. Gen. Genet.* **200,** 128–131.

Fig. 1. After *Xba*I digestion, the DNA samples were submitted to electrophoresis on a 0.6% agarose gel and transferred to a membrane. The membranes were hybridized with a purified fragment containing the *ura5* coding sequence.

The DNA samples are: A, plasmid containing the *met2* gene; B, total DNA of the *S. macrospora ura5*$^-$, *met1*$^-$ strain. *co.22, co.5, co.20, co.24, and co.56* correspond to 5 independent primary tranformants selected on minimal medium after cotransformation of the *ura5*$^-$,*met1*$^-$ strain by two plasmids, one harboring the *ura5* gene and the other the *met2* gene. + and − originate from two spores of the same ascus. + (ura$^+$, met$^+$) and − (ura$^-$, met$^-$) are descendent from the selfing of the Ts (secondary transformant).

The traces of hybridization in A correspond to the contamination of the *ura5* probe by vector (pUC18) sequences. The recipient strain shows a fragment 12 kb in length hybridizing with the probe. This fragment corresponds to the resident *URA5* locus and is present in all strains tested (positive control for hybridization). In the secondary transformants and their ura$^+$, met$^+$ descendents, additional bands are observed corresponding to the introduced DNA. In the ura$^-$, met$^-$ descendents, these bands are not present, indicating the loss of the introduced DNA. Similar results were observed by probing the same membrane with a purified *met2* probe and with a third probe corresponding to the vector sequences from the used plasmids.

Carpenter, A. T. C. (1979). Recombination nodules and synaptonemal complexes in recombination-defective females of *Drosophila melanogaster. Chromosoma (Berlin)* **75,** 259–292.

Coppin-Raynal, E., Picard, M., and Arnaise, S. (1989). Transformation by integration in *Podospora anserina.* III. Replacement of a chromosome segment by a two-step process. *Mol. Gen. Genet.* **219,** 270–276.

Esposito, R. E., and Klapholz, S. (1981). Meiosis and ascospore development. *In* "The Molecular Biology of the Yeast *Saccharomyces:* Life cycle and Inheritance" (J. N. Strathern, E. W. Jones and J. R. Broach, eds.), pp. 109–137. Cold Spring Harbor Laboratory Press, Cold Spring Harbor, New York.

Esser, K., and Straub, J. (1958). Genetische Untersuchungen an *Sordaria macrospora* Auersw., Kompensation und Induktion bei genbedingten Entwicklungsdefekten. *Z. Vererbungsl.* **89,** 729–746.

Fields, W. G., and Olive, L. S. (1967). The genetics of *Sordaria brevicollis.* III. Gene conversion involving a series of hyaline ascospore colour mutants. *Genetics* **57,** 483–499.

Fincham, J. R. S. (1989). Transformation in fungi. *Microbiol. Rev.* **53,** 148–170.

Giroux, C. N. (1988). Chromosome synapsis and meiotic recombination. *In* "Genetic Recombination" (R. Kucherlapati and G. Smith, eds.), pp. 465–496. American Society for Microbiology, Washington, D.C.

Goyon, C., and Faugeron, G. (1989). Targeted transformation of *Ascobolus immersus* and *de novo* methylation of the resulting duplicated DNA sequences. *Mol. Cell. Biol.* **9,** 2818–2827.

Goyon, C., Faugeron, G., and Rossignol, J. L. (1988). Molecular cloning and characterization of the *Met2* gene from *Ascobolus immersus. Gene* **63,** 297–308.

Heslot, H. (1958). Contribution à l'étude cytogénétique et génétique des Sordariacées. *Rev. Cytol. Biol. Végét.* **19** (Suppl. 2), 1–235.

Holliday, R. (1964). A mechanism for gene conversion in fungi. *Genet. Res.* **5,** 282–304.

Huynh, A. D., Leblon, G., and Zickler, D. (1986). Indirect intergenic suppression of a radiosensitive mutant of *Sordaria macrospora* defective in sister-chromatid cohesiveness. *Curr. Genet.* **10,** 545–555.

Kitani, Y., Olive, L. S., and El-Ani, A. S. (1962). Genetics of *Sordaria fimicola.* V. Aberrant segregation at the *g* locus. *Am. J. Bot.* **49,** 697–706.

Leblon, G., Zickler, D., and Le Bilcot, S. (1986). Most U.V. induced reciprocal translocations in *Sordaria macrospora* occur in or near the centromere regions. *Genetics* **112,** 183–204.

Leblon, G., Haedens, V., Huynh, A. D., Le Chevanton, L., Moreau, P. J. F., and Zickler, D. (1987). *Sordaria macrospora* genetic map. *In* "Genetic Maps" (S. J. O'Brien, ed.), 4th Ed., pp. 338–339. Cold Spring Harbor Laboratory Press, Cold Spring Harbor, New York.

Le Chevanton, L., and Leblon, G. (1989). The *Ura5* gene of the ascomycete *Sordaria macrospora:* Molecular cloning, characterization and expression in *E. coli. Gene* **77,** 39–49.

Le Chevanton, L., Leblon, G., and Lebilcot, S. (1989). Duplications created by transformation in *Sordaria macrospora* are not inactivated during meiosis. *Mol. Gen. Genet.* **218,** 390–396.

MacDonald, M. V., and Whitehouse, H. L. K. (1979). A *buff* spore colour mutant in *Sordaria brevicollis* showing high frequency conversion. 1. Characteristics of the mutant. *Genet. Res.* **34,** 87–119.

MacDonald, M. V., and Whitehouse, H. L. K. (1982). A *buff* spore colour mutant in *Sordaria brevicollis* showing high frequency conversion. 2. Loss of the high-frequency conversion. *Genet. Res.* **41,** 155–163.

Moreau, P. J. F., Zickler, D., and Leblon, G. (1985). One class of mutants with disturbed centromere cleavage and chromosome pairing in *Sordaria macrospora. Mol. Gen. Genet.* **198,** 189–197.

Olive, L. S. (1959). Aberrant tetrads in *Sordaria fimicola. Proc. Natl. Acad. Sci. U.S.A.* **45,** 727–732.

Orr-Weaver, T. L., and Szostak, J. W. (1985). Fungal recombination. *Microbiol. Rev.* **49,** 33–58.

Perkins, D. D. (1974). The manifestation of chromosome rearrangements in unordered asci of *Neurospora. Genetics* **77,** 459–489.

Perkins, D. D. (1985). Aspects of the *Neurospora* genome. *In* "Molecular Genetics of Filamentous Fungi" (W. Timberlake, ed.), pp. 277–294. Liss, New York.

Perkins, D. D., and Barry, E. G. (1977). The cytogenetics of *Neurospora. Adv. Genet.* **19,** 133–285.

Perrot, M., Barreau, C., and Bégueret, J. (1987). Nonintegrative transformation in the filamentous fungus *Podospora anserina:* Stabilization of a linear vector by the chromosomal ends of *Tetrahymena thermophila. Mol. Cell. Biol.* **7,** 1725–1730.

Sąng, H., and Whitehouse, H. L. K. (1979). Genetic recombination at the *buff* spore colour locus in *Sordaria brevicollis.* I. Analysis of flanking marker behaviour in crosses between *buff* mutants and wild-type. *Mol. Gen. Genet.* **174,** 327–334.

Sang, H., and Whitehouse, H. L. K. (1983). Genetic recombination at the *buff* spore colour locus in *Sordaria brevicollis.* II. Analysis of flanking marker behaviour in crosses between *buff* mutants. *Genetics* **103,** 161–178.

Selker, E. U. (1990). DNA methylation and chromatin structure: A view from below. *TIBS* **15,** 103–107.

Selker, E. U., and Garrett, P. W. (1988). DNA sequence duplications trigger gene inactivation in *Neurospora crassa. Proc. Natl. Acad. Sci. U.S.A.* **85,** 6870–6874.

Tsukuda, T., Carleton, S., Fotheringham, S., and Holloman, W. K. (1988). Isolation and characterization of an autonomously replicating sequence from *Ustilago maydis. Mol. Cell. Biol.* **8,** 3703–3709.

Turcq, B., and Bégueret, J. (1987). The *ura5* gene of the filamentous fungus *Podospora anserina:* Nucleotide sequence and expression in transformed strains. *Gene* **53,** 201–209.

van Heeswijck, R. (1986). Autonomous replication of plasmids in mucor transformants. *Carlsberg Res. Commun.* **51,** 433–443.

von Wettstein, D., Rasmussen, S. W., and Holm, P. B. (1984). The synaptonemal complex in genetic segregation. *Ann. Rev. Genet.* **18,** 331–413.

Whitehouse, H. L. K. (1982). "Genetic recombination: Understanding the mechanism." Wiley and Sons, New York.

Zickler, D. (1977). Development of the synaptonemal complex and the "Recombination Nodules" during meiotic prophase in the seven bivalents of the fungus *Sordaria macrospora* Auersw. *Chromosoma (Berlin)* **61,** 289–316.

Zickler, D., Leblon, G., Haedens, V., Collard, A., and Thuriaux, P. (1984). Linkage group-chromosome correlations in *Sordaria macrospora:* Chromosome identification by three dimensional reconstruction of their synaptonemal complex. *Curr. Genet.* **8,** 57–67.

Zickler, D., de Lares, L., Moreau, P. J. F., and Leblon, G. (1985). Defective pairing and synaptonemal complex formation in a *Sordaria* mutant *(spo44)* with a translocated segment of the nucleolar organizer. *Chromosoma (Berlin)* **92,** 37–47.

Zickler, D., Huynh, A. D., Moreau, P. J. F., and Leblon, G. (1988). Genetic control of synaptonemal complex formation in the Ascomycete *Sordaria macrospora. In* "Kew Chromosome Conference III" (P. Brandham, ed.), pp. 330–336. HMSO, London.

V

Mycotechnology

14

Fungal Protoplasts

JOHN F. PEBERDY
Department of Botany
University of Nottingham
Nottingham, United Kingdom

Protoplasts have had a major impact on two recent advances in the methodology of fungal genetics: in the mid 1970s through fusion techniques and in the last decade through their use in transformation-mediated recombination. This chapter focuses on the first of these developments; Fincham (1989) has recently reviewed transformation in fungi. Since the subject of fungal protoplasts has been reviewed several times since 1979 (Peberdy, 1979, 1989; Ferenczy, 1981), I shall present only a brief overview of the technology as an introduction to some of the more significant recent examples of its application.

I. METHODOLOGY

Since its inception, protoplast fusion has provided the opportunity to execute genetic recombination in a wide range of fungi. The technique has

MORE GENE MANIPULATIONS
IN FUNGI

been utilized for conventional interstrain crosses as well as for crosses that breach specific and generic boundaries. Intraspecific protoplast fusion enhances the recovery of heterokaryons when compared with conventional hyphal fusion. The generation of interspecies or intergenus hybrids raises interesting questions about the genetic basis of speciation in fungi, and provides novel material for biochemical investigations and possible biotechnological exploitation.

Successful experiments require a capability to isolate adequate numbers of viable protoplasts. Viability is expressed as successful regeneration and reversion to the normal cellular or hyphal form. The most variable component of the "protoplast system" is, not surprisingly, the organism from which the protoplasts are to be prepared. Frequently, little is known about the fungus in question. This lack of physiological data can be a handicap in identifying optimal growth conditions, which are important in determining the susceptibility of the cells or hyphae to mycolytic enzymes. In general, both mycelial and unicellular fungi are most sensitive to these enzymes during early exponential or linear growth phase. It is inferred that, as the cell population ages, the structure of the cell wall changes. In filamentous fungi, the growing hyphal tip is probably the part of the cell most vulnerable to enzymatic lysis. Endogenous lytic enzymes or autolysins, for example, chitinase and β-1,3-glucanase, thought to be associated with *de novo* incorporation of new wall material, may also become involved in the lytic process.

A. Protoplast Isolation

The availability of commercial enzymes that could be used to isolate protoplasts from the mycelial fungi has greatly stimulated research. These enzymes were first reported by Hamlyn and co-workers (1981) and enabled researchers to reap the benefits of higher protoplast yields obtained with minimum effort in a short time. Such benefits had been available to yeast protoplast workers for many years (Ferenczy, 1985).

One of these commercial enzyme preparations, Novozym 234 (Novo Biolabs Ltd. Bagsvaerd, Denmark), is active against a wide range of fungal species (Peberdy and Hocart, 1987; Hocart and Peberdy, 1989). However, a number of oomycetes and basidiomycetes are not susceptible to its action. In the former organisms, this resistance reflects the absence of chitin and presence of cellulose as the skeletal component of the wall. However, it is not clear why certain basidiomycetes that do have chitin as the skeletal component are not susceptible to Novozym 234. Often other commercial complexes, marketed as cellulases, are more effective (Peberdy, 1989). Many of these products have chitinase and β-1,3-glucanase as side activities (Hamlyn *et al.,* 1981; Hocart *et al.,* 1987).

An important feature of these commercial preparations is batch variation. These enzymes are not produced for the purpose of protoplast isolation, so the quality control parameters adopted by the manufacturers do not relate to this particular application. Furthermore, these commercial preparations are complex mixtures of several hydrolytic enzymes and the catalytic activities instrumental in protoplast release are minor components of varying concentrations. The significance of batch variation results in variable protoplast yields with different batches of the enzyme. Properties of the protoplasts, in particular their regenerative potential, are also undoubtedly affected. (To date the data on regeneration are anecdotal, with no published evidence.) If workers were to record the batch numbers of particular enzymes used in their experiments, the problem of variation could be quantified.

The osmotic stabilizer is another important component of the lytic system. The extensive range of inorganic salts, sugars, and sugar alcohols used for this purpose has been described by Davis (1985). The most effective stabilizer is determined empirically and, to date, there has been no rational explanation for one compound's superiority over another.

B. Protoplast Regeneration

Exploitation of protoplasts as genetic systems depends on their capability to regenerate a new cell wall when cultured in a growth medium supplemented with an osmotic stabilizer. Although a large proportion of a protoplast population does produce some cell wall material when cultured, only a small proportion actually goes on to form normal cells or hyphae. These new cells develop as colonies on the regeneration medium. The frequency of regeneration is highly variable among different fungi, ranging from less than 0.1 to 50%. The lower values are found in some of the basidiomycetes and could be associated with a lower growth rate found in many of these organisms. Poor protoplast regeneration is most frequently observed in fungi that have relatively slow growth rates.

The factors responsible for the variable regeneration rates are poorly understood. Not all protoplasts have nuclei and it is presumed that anucleate cells cannot regenerate. More recently it has been shown that the tubulin cytoskeleton is rapidly depolymerized and disorganized in protoplasts isolated from filamentous fungi (J. Jochova and J. F. Peberdy, unpublished data). The proper repolymerization of the cytoskeleton may be a factor in both regeneration and the polarized development of hyphae.

The physiology of protoplast regeneration is affected by environmental factors in the same manner as the growth of normal cells is. A comparative study of different mycolytic enzymes used for protoplast isolation has shown that regeneration may be influenced by the enzyme used. Changes to the plasma membrane, brought about by the enzymes in the lytic complex, may affect the process of wall synthesis at the protoplast surface.

C. Protoplast Fusion

The use of protoplast fusion has had a dramatic effect on recent developments in fungal genetics. For example, one of the many roles of the cell wall is as a determinant for cell-to-cell interaction (Peberdy, 1989), which can be manifested as genetic incompatibility and blocking of cell fusion. In *Aspergillus nidulans,* removal of the wall can overcome incompatibilities (Croft and Dales, 1983). More extreme situations are the interspecies incompatibilities. These effects may arise simply because the recognition molecules at the cell surface of one species do not identify similarly functioning molecules on another organism. In protoplast fusion, no such recognition phenomena are involved.

Other factors may play a part in determining the ability of hyphae or cells of two isogenic strains of a fungus to fuse. The experience of several laboratories working with *Acremonium chrysogenum* (formerly *Cephalosporium acremonium*) was that hyphal fusion to establish heterokaryosis, as a prelude to a parasexual cycle, was very rare compared with the frequency of hyphal fusion in the Aspergilli and Penicillia (Peberdy, 1987). The cause of this phenomenon has not been elucidated. In comparison, the relative ease of establishing crosses through protoplast fusion (Hamlyn *et al.,* 1985) led to the view that the rare hyphal forms may relate to nuclear ploidy and a strict DNA : cytoplasm volume ratio in the vegetative hyphae of this fungus. As a consequence, *A. chrysogenum* is a rare example of a fungus in which protoplast fusion provides the only means of carrying out a classical genetic analysis (Peberdy, 1987).

1. Fusion Procedures

Two major techniques are used to bring about the fusion of protoplasts. Most laboratories use a chemical fusogen, polyethylene glycol, primarily because of cost and convenience. The alternative procedure is electrofusion, which requires carefully established conditions for optimal fusion. Although more demanding, electrofusion yields a greater control of the fusion process. In both systems, fusion events are random, that is, fusion aggregates may be mixtures of protoplasts of different strains or of protoplasts of a single strain.

2. Selection Strategies

The recovery of fusion products derived from crosses between two strains normally requires a selection strategy. For well-studied species such as *Aspergillus nidulans* and *Penicillium chrysogenum,* a large number of genetically marked strains is already available in culture collections such as those housed by the Fungal Genetics Stock Center (University of Kansas Medical Center, Kansas City, Kansas) and the Department of Genetics, University of Glasgow (Glasgow, Scotland). For other species, the generation of mutations that can be subsequently used in selection is usually a straightforward, although tedious, process.

When protoplast fusion is required to generate recombination between strains, for example, in industrially important fungi or plant pathogenic fungi, and when the introduction of conventional auxotrophies may have a detrimental effect on metabolite production or pathogenicity, substrate nonutilization markers are sometimes employed. The better known examples of these markers are the chlorate and selenate resistance mutations that confer changes in nitrate and sulfate utilization, respectively. Mutants resistant to fungicides and other inhibitors, also isolated as a consequence of spontaneous events, have also proved useful (Toyama and Toyama, 1988). Similar selective markers may now be introduced by transformation, for example, resistance to the antibiotics hygromycin and phleomycin (Talbot *et al.,* 1988).

Such physiologically based selection is not always necessary. In some cases, improved recombinants can be isolated by visual screening (Lein, 1986). Protoplasts of *P. chrysogenum* derived from two strains with distinctive morphological features and useful attributes with respect to penicillin production were fused. A few recombinant progeny with the desirable features of both strains were recovered, but many thousands of strains had to be analyzed to find them, a situation that is impracticable for most laboratories.

In comparison with prokaryotes, the approaches adopted to select fusion products in fungi have been rather conservative. In *Streptomyces* and *Bacillus,* for example, strategies based on polarized selection and dead donor selection have been successful (Hopwood, 1981). Protoplasts derived from a wild-type strain are rendered nonviable by heat treatment, bacteriocidal antibiotic, or uv irradiation. They are then fused with viable protoplasts from a suitably marked strain, usually an auxotroph. Fusion products are recovered on minimal medium. An interesting development of this approach is the dramatic increase in recombination frequency observed and the fusion of protoplasts of two uv-treated strains, possibly due to the recombination of highly fragmented DNA in the two protoplast populations (Hopwood, 1981).

3. Characterization of Fusion Products

The ploidy characteristics and nuclear status of protoplast fusion products in fungi are very much a feature of the organism being investigated. For most yeasts, the fusion products are diploid cells that, in some cases, have been induced to undergo a meiotic division to yield recombinant progeny (Ferenczy, 1985). In others, recombinant progeny of undetermined ploidy have been recovered. Heterokaryons are rarely, if ever, found in yeast fusion progeny.

In the filamentous fungi, the presumption is that protoplast fusion will lead to a parasexual cycle that requires the generation of a somatic diploid as a key event (Caten, 1981). Although this has been observed in many cases, for example, Aspergilli (Croft and Dales, 1983; Kevei and Peberdy, 1985) and Penicillia (Anné and Peberdy, 1981; Anné, 1983), there are some notable exceptions in which the ploidy status of fusion products and their derivatives is not as clear, for example *Trichoderma viride* (Hocart and Peberdy, 1989) and related species. In some cases, for example, *Acremonium chrysogenum,* the parasexual events are apparently "telescoped" so the majority of the fusion products are haploid recombinants derived following transient heterokaryon and haploid stages (Peberdy, 1987). Molecular markers have been used increasingly in the characterization of fusion products. These markers include both isoenzyme markers (Anné and Peberdy, 1981; Um *et al.,* 1988; Min *et al.,* 1989) and DNA probes (Hoffmann *et al.,* 1987).

II. APPLICATIONS

A. Fungal Genetics

1. Incompatibility

Early work by Croft and co-workers (Croft, 1985) demonstrated that protoplast fusion provided a means to overcome the incompatibility mechanism operating between some strains of *Aspergillus nidulans,* implicating the cell wall in the incompatibility reaction between these strains. When protoplast fusion failed, it was assumed that the incompatibility mechanism was cytoplasmic or nuclear in nature. Similar incompatibility mechanisms are suggested by results obtained with some plant pathogenic fungi including *Cochliobolus* (Leach and Yoder, 1983) and *Septoria* (Newton and Caten, 1985).

The system controlling mating reactions in basidiomycetes such as *Coprinus* is more complex. Using protoplast fusion, Ohba and colleagues (1988) were able to generate common B dikaryons, some of which produced fruit body primordia.

2. Intra- and Interspecies Hybridization

The protoplast fusion technique has been widely adopted as a tool for intraspecies hybridization. In recent years, the technique has been to establish parasexual crosses in a range of fungi including *Aspergillus sojae,* (Ushijima and Nakadai, 1987; Ushijima *et al.,* 1987), *Penicillium verruculosum* (Chung *et al.,* 1988), *Aspergillus awamori* var. *kawachi* (Ogawa *et al.,* 1988), and *Metarhizium anisopliae* (Silviera and Azevedo, 1987).

The advent of fusion technology also spawned an immediate interest in the possibility of generating novel hybrids in fungi. The availability of novel recombinant strains raised many fundamental questions about speciation, as well as the potential for creating completely new isolates that might have biotechnological importance.

The hybridization of closely related representatives of the *A. nidulans* species group (Croft, 1985; Kevei and Peberdy, 1985) is still the most detailed example, in terms of understanding the underlying genetic events involved. In recent years many other examples of interspecific and intergeneric hybridization have been reported, including *A. niger* × *A. oryzae* (Toyama and Toyama, 1988), *Trichoderma viride* × *Neurospora crassa* (Nutsubidze *et al.,* 1988), *A. awamori* var. *kawachi* × *A. oryzae* (Ogawa *et al.,* 1988), *T. koningii* × *T. reesei* (Park *et al.,* 1986), *Saccharomyces cerevisiae* × *A. niger* (Lee *et al.,* 1989), and *S. cerevisiae* × *Candida tropicalis* (Seu *et al.,* 1989).

Interpretation of the results of these intergeneric fusion crosses also requires caution. For example, the conclusion that fusion between *A. niger* and *T. viride* led to heterokaryon formation was based on the recovery of conidia of the two parental types from the fusion products (Kirimura *et al.,* 1989). The possibility that the heterokaryons were, in fact, mixed colonies of the two fungi growing in close association cannot be ruled out at this stage. Clearly, the characterization of these hybrids must use more rigorous analytical approaches. Molecular probes and pulsed-field gel electrophoresis methods to determine molecular karyotypes can be used to analyze intergeneric hybrids, as has been done with the yeast hybrids of *S. cerevisiae* and *Kluyveromyces marxianus* (Hoffmann *et al.,* 1987). Restriction fragment length polymorphism (RFLP) probes isolated from both parents hybridized against electrophoresed hybrid DNA, providing evidence for the occurrence of recombination events in these intergeneric protoplast fusion products (Witte *et al.,* 1989).

B. Fungal Biotechnology

Recombination has been a valuable approach in strain improvement programs for industrially important fungi for many years; however, the conventional approach to breeding imperfect fungi by the classical parasexual cycle was rarely adopted. The procedures involved were both time consuming and labor intensive. The alternative strategies to select hybrid protoplast fusion products and to isolate recombinant progeny have attracted the interest of many geneticists employed in industrial research laboratories. The newer techniques based on molecular biological methods and transformation are highly relevant when strain improvement can be effected by the manipulation of specific genes. The most relevant strategy for multigene functions, such as growth rate and sporulation, is mass screening of hybrid progeny.

1. Enzyme, Ethanol, and Organic Acid Fermentations

The relevance of intra- and interstrain breeding through protoplast fusion to produce more effective strains for fermentation processes has been discussed in several reviews (Anné, 1983; Ball, 1984; Peberdy and Illing, 1985). An example of this technique is the recovery from fusion crosses in *Aspergillus sojae,* the fungus used in soy-sauce fermentation, of haploid recombinants that produced more protease and glutaminase (Ushijima and Nakadai, 1987; Ushijima *et al.,* 1987.) Diploids from two isolates of *Trichoderma reesei* produced levels of cellulases and xylanase that were intermediate between the levels of the two parents, whereas other enzymes, for example, chitinase, β-glucosidase, and β-1,3-glucanase, were produced at higher levels than in the two parents (Ogawa *et al.,* 1989). Improved citric acid production by diploids of *A. niger,* produced by protoplast fusion, has also been reported (Kirimura *et al.,* 1988).

Some of the more "adventurous" ideas of applying protoplast fusion to generate novel hybrids involve organisms used in various fermentation processes. For several years there has been an interest in developing amylolytic strains of brewing yeasts. Although there are opportunities for exploiting recombinant DNA techniques for this purpose, several groups have tried protoplast fusion instead. A cross between yeast and *A. niger* yielded products that looked like yeast but produced starch-degrading enzymes as well as ethanol. During progressive culture of these hybrids, both ethanol productivity and starch utilization declined, although the fusants maintained prototrophy. A second approach adopted by Seu and colleagues (1989) was to generate hybrids between *Saccharomyces cerevisiae* and *Candida tropicalis.* The fusion products could hydrolyze both α-1,4 and α-1,6 glucosidic linkages of starch. Recombinants between *A. usamii* and *A. niger,* with improved production of amylase and citric acid, have also been described (Ogawa *et al.,* 1989).

2. Biocontrol Fungi

An aspect of fungal biotechnology that has attracted interest in recent years, as part of the movement to reduce the use of chemical control agents in agriculture, is the exploitation of fungi for biocontrol of fungal pathogens and insect pests. *Trichoderma* species have been identified as possible "biocides." It has been suggested that their effectiveness against plant pathogens is the consequence of the production of several factors, including mycolytic enzymes and antifungal antibiotics (Lynch, 1987). Whole-genome recombination offers possibilities for strain development. Several laboratories have reported the use of protoplast fusion to produce strains that have improved performance in the control of various soilborne plant pathogens (Stasz *et al.,* 1989; Pe'er and Chet, 1990). Successful fusion of protoplasts in the insect pathogen *Metarhizium anisopliae* (Silviera and Azevedo, 1987) raises the possibility of breeding more virulent strains as biological pest controls.

3. Edible Fungi

Active programs of strain development in a range of economically important edible mushrooms in Southeast Asia are currently in progress. Protoplast fusion is an important approach to both conventional inter-strain hybridization and attempts to generate hybrids. Examples of the latter include the production of hybrids between *Ganoderma lucidum* and *G. applanantum.* Fusion products were characterized on the basis of mycelial morphology and esterase isozyme patterns (Um *et al.,* 1988). Although most fusion products lacked clamp connections, some grew more vigourously than the parental strains and produced fruit bodies (Park *et al.,* 1988). Intergeneric crosses have also been described, for example, between *Pleurotus ostreatus* and *G. applanantum.* However, the heterokaryons obtained to date have not been fertile (Park *et al.,* 1988; Yoo *et al.,* 1989). Isoenzyme profiles were investigated again to assess hybridity, but the results were not conclusive.

4. Plant Pathology

Despite their importance, until recent years many plant pathogens have been totally ignored from a genetic perspective. Protoplast fusion provides a means of establishing formal genetics in numerous species. In some cases, the fungus has been shown to have a typical parasexual cycle, for example, *Pseudocercosporella herpotrichoides* (Hocart, 1987).

Since the parasexual cycle has been described in an increasing number of species, several deviations from the conventional pattern are revealed (Caten, 1981). In *Fulvia fulva,* for example, heterokaryons are not recovered as fusion products, but as diploids of varying degrees of stability (Talbot *et al.,* 1988). *Phytophthora megasperma* F. sp. *glycinea* provides another interesting variation. In this case the fusants are heterokaryons, some of which produce asexual zoospores carrying recombinant drug resistance markers from the two parents (Layton and Kuhn, 1988).

In *Gibberella zeae,* protoplast fusion between auxotrophic parents yielded heterokaryons that behaved in a manner undescribed in other species. Initially, apical cells of the fusion product are prototrophic and conidia contain only one nuclear type. Subsequently, apical cells become auxotrophic and both types of nuclei can be found in the spores. Fusion between incompatible strains gives a slow growing prototrophic colony, bearing conidia of one nuclear type. The complementing parental types are not recovered, suggesting that the fusion product is heteroploid (Adams *et al.,* 1987).

III. CONCLUSIONS

The technique of protoplast fusion has clearly contributed to our knowledge of the genetics of fungi. As interest in fungal species broadens, protoplast-based systems will provide ready access to a variety of breeding strategies. The ability to produce interspecific and intergeneric hybrids by these methods is technology that offers enormous promise.

REFERENCES

Adams, G., Johnson, N., Leslie, J. F., and Hart, L. P. (1987). Heterokaryons of *Gibberella zeae* formed following hyphal anastomosis or protoplast fusion. *Exp. Mycol.* **11,** 339–353.

Anné, J. (1983). Protoplasts of filamentous fungi in genetics and metabolite production. *In* "Protoplasts 1983" (I. Potrykus, C. T. Harms, A. Hinnen, R. Hutter, P. J. King, and R. D. Shillito, eds.), pp. 167–178. Berkhauser Verlag, Basel.

Anné, J., and Peberdy, J. (1981). Characterisation of interspecific hybrids between *Penicillium chrysogenum* and *P. roqueforti* by isoenzyme-analysis. *Trans. Br. Mycol. Soc.* **77,** 401–408.

Ball, C. (1984). Filamentous fungi. *In* "Genetics and Breeding of Industrial Microorganisms" (C. Ball, ed.), pp. 159–188. CRC Press, Boca Raton, Florida.

Caten, C. E. (1981). Parasexual processes in fungi. *In* "The Fungal Nucleus" (K. Gull and S. G. Oliver, eds.), pp. 191–214. Cambridge University Press, Cambridge.

Chung, K.-C., Park, C.-R., Bai, S., Chun, S.-B., and Kim, K.-C. (1988). Intraspecific protoplast fusion of cellulolytic fungus *Penicillium verruculosum. Korean J. Appl. Microbiol. Bioeng.* **16,** 163–167.

Croft, J. H. (1985). Protoplast fusion and incompatibility in *Aspergillus. In* "Fungal Protoplasts: Applications in Biochemistry and Genetics" (J. F. Peberdy and L. Ferenczy, eds.), pp. 225–440. Marcel Dekker, New York.

Croft, J. H., and Dales, R. B. G. (1983). Interspecific hybridisation in Aspergillus. *In* "Protoplasts 1983" (I. Potrykus, C. T. Harms, A. Hinnen, R. Hutter, P. J. King, and R. D. Shillito, eds.), pp. 179–186. Berkhauser Verlag, Basel.

Davis, B. (1985). Factors affecting protoplast isolation. *In* "Fungal Protoplasts: Applications in Biochemistry and Genetics" (J. F. Peberdy and L. Ferenczy, eds.), pp. 45–72. Marcel Dekker, New York.

Ferenczy, L. (1981). Microbial protoplast fusion. *Symp. Soc. Gen. Microbiol.* **31,** 1–34.

Ferenczy, L. (1985). Protoplast fusion in yeasts. *In* "Fungal Protoplasts: Applications in Biochemistry and Genetics" (J. F. Peberdy and L. Ferenczy, eds.), pp. 279–306. Marcel Dekker, New York.

Fincham, J. R. S. (1989). Transformation in fungi. *Microbiol. Rev.* **53,** 148–170.

Hamlyn, P. F., Bradshaw, R. D., Mellon, F. M., Santiago, C. M., Wilson, J. M., and Peberdy, J. F. (1981). Efficient protoplast isolation from fungi using commercial enzymes. *Enz. Microb. Technol.* **3,** 321–325.

Hamlyn, P. F., Birkett, J. A., Perez, G. and Peberdy, J. F. (1985). Parasexual recombination and genetic analysis in *Cephalosporium acremonium* using protoplast fusion. *J. Gen. Microbiol.* **131,** 2813–2823.

Hocart, M. J. (1987). Fungicide resistance and parasexual recombination in *Pseudocercosporella herpotrichoides.* Ph.D. Thesis. University of Nottingham, England.

Hocart, M. J., and Peberdy, J. F. (1989). Protoplast technology and strain selection. *In* "Biotechnology of Fungi for Improving Plant Growth" (J. M. Whipps and R. D. Lumsden, eds.), pp. 235–258. Cambridge University Press, Cambridge.

Hocart, M. F., Lucas, J. A., and Peberdy, J. F. (1987). Production and regeneration of protoplasts from *Pseudocercosporella herpotrichoides* (Fron) Deighton. *J. Phytopathol.* **119,** 193–205.

Hoffmann, M., Zimmermann, M., and Emeis, C.-C. (1987). Orthoganal field alternation gel electrophoresis (OFAGE) as a means for the analysis of somatic hybrids obtained by protoplast fusion of different *Saccharomyces* strains. *Curr. Genet.* **11,** 599–604.

Hopwood, D. A. (1981). Genetic studies with bacterial protoplasts. *Ann. Rev. Microbiol.* **35,** 237–272.

Kevei, F., and Peberdy, J. F. (1985). Interspecies hybridization after protoplast fusion in *Aspergillus. In* "Fungal Protoplasts: Applications in Biochemistry and Genetics" (J. F. Peberdy and L. Ferenczy, eds.), pp. 241–258. Marcel Dekker, New York.

Kirimura, K., Nakajima, I., Lee, S. P., Kawabe, S., and Usami, S. (1988). Citric acid production by the diploid strains of *Aspergillus niger* obtained by protoplast fusion. *Appl. Microbiol. Biotechnol.* **27,** 504–506.

Kirimura, K., Imura, M., Lee, S. P., Kato, Y., and Usami, S. (1989). Intergenetic hybridization between *Aspergillus niger* and *Trichoderma viride* by protoplast fusion. *Agric. Biol. Chem.* **53,** 1589–1596.

Layton, A. C., and Kuhn, D. N. (1988). Heterokaryon formation by protoplast fusion of drug resistant mutants in *Phytophthora megasperma* F. sp. *glycinea. Exp. Mycol.* **12,** 180–194.

Leach, J., and Yoder, O. (1983). Heterokaryon incompatibility in the plant pathogen fungus *Cochliobolus heterostrophus. J. Hered.* 74, 149–152.

Lee, S. Y., Lee, J. S., and Lee, Y. N. (1989). Ethanol production from starch by protoplast fusion between *Aspergillus oryzae* and *Saccharomyces cerevisiae. Korean J. Microbiol.* **27,** 221–224.

Lein, J. (1986). The Panlab penicillin strain improvement program. *In* "Overproduction of Microbial Metabolites. Strain Improvement and Process Control Strageties" (Z. Vanek and Z. Hostalek, eds.), pp. 105–140. Butterworths, Boston.

Lynch, J. M. (1987). Biological control within microbial communities of the rhizosphere. *Symp. Soc. Gen. Microbiol.* **41,** 55–82.

Min, K.-R., Park, H.-M., and Hah, Y.-C. (1989). Identification of hybrids from intraspecific and interspecific protoplast fusion in *Trichoderma* by electrophoretic patterns of enzymes. *Korean J. Microbiol.* **27,** 27–34.

Newton, A. C., and Caten, C. E. (1985). Heterokaryosis and heterkaryon incompatibility in *Septoria nodorum. In* "Septoria of Cereals" (A. L. Scharen, ed.), p. 13. U.S. Dept. of Agriculture, Publ. No. ARS-12. U.S. Government Printing Office, Washington, D.C.

Nutsubidze, N. N., Prabakaran, K., Dzhafarova, A. N., Sokolovskii, V. Y., Kritskii, M. S., and Klesov, A. A. (1988). Intergeneric protoplast fusion in the fungi *Trichoderma viride* and *Neurospora crassa. Dokl. Aakad. Nauk, SSSR* **300,** 488–490.

Ogawa, K., Ohara, H., and Toyama, N. (1988). Intraspecific hybridization of *Aspergillus awamori* var. *kawachi* by protoplast fusion. *Agric. Biol. Chem.* **52,** 337–342.

Ogawa, K., Ohara, H., Koide, T., and Toyama, N. (1989). Intraspecific hybridization of *Trichoderma reesei* by protoplast fusion. *J. Ferment. Bioeng.* **67,** 207–209.

Ohba, K., Kawasumi, T., and Yanagi, S. O. (1988). Protoplast fusion between common-B mating type monokaryons of *Coprinus cinereus. Trans. Mycol. Soc. Jpn.* **29,** 271–280.

Park, H. M., Jeong, J. M., Hong, S. W., Hah, Y. C., and Seong, N. C. (1986). Interspecific protoplast fusion of *Trichoderma koningii* and *Trichoderma reesei. Korean J. Microbiol.* **24,** 91–97.

Park, Y.-D., Yoo, Y.-B., Shin, P.-G., You, C.-H., Cha, D.-Y., Park, L.-H., and Lee, J.-S. (1988). Interspecific protoplast fusion of *Ganoderma applanatum* and *Ganoderma lucidum* and fruitbody formation of the fusants. *Korean J. Mycol.* **16,** 79–86.

Pe'er, S., and Chet, I. (1990). *Trichoderma* protoplast fusion: A tool for improving biocontrol agents. *Can. J. Microbiol* **36,** 6–9.

Peberdy, J. F. (1979). Fungal protoplasts: Isolation, reversion and fusion. *Ann. Rev. Microbiol.* **33,** 21–40.

Peberdy, J. F. (1987). Genetics of *Acremonium. In* "Biotechnology Handbooks I—*Penicillium* and *Acremonium*" (J. F. Peberdy, ed.), pp. 93–108. Plenum Press, New York.

Peberdy, J. F. (1989). Fungi without coats—Protoplasts as tools for mycological research. *Mycol. Res.* **93,** 1–20.

Peberdy, J. F., and Hocart, M. J. (1987). Protoplasts as tools in the genetics of plant pathogenic fungi. *In* "Genetics and Plant Pathogenesis" (P. R. Day and G. J. Jellis, eds.), pp. 127–142. Blackwell Scientific, Oxford.

Peberdy, J. F., and Illing G. T. (1985). Genetic manipulation of antibiotic producer strains. *Symp. Soc. Gen. Microbiol.* **38,** 283–322.

Seu, J.-H., Jun, D.-Y., and Kim, Y.-H. (1989). Intergeneric protoplast fusion of heterologous transformants of *Saccharomyces cerevisiae* and *Candida tropicalis. Korean J. Appl. Microbiol. Bioeng.* **17,** 1–7.

Silviera, W. D., and Azevedo, J. L. (1987). Protoplast fusion and genetic recombination in *Metarhizium anisopliae. Enz. Microb. Technol.* **9,** 149–152.

Stasz, T. E., Harman, G. E., Gullino, M. L. (1989). Limited vegetative compatibility following intra- and interspecific protoplast fusion in *Trichoderma. Exp. Mycol.* **13,** 364–371.

Talbot, N. J., Coddington, A., Roberts, I. N., and Oliver, R. P. (1988). Diploid construction by protoplast fusion in *Fulvia fulva* synonym *Cladosoporium fulvum:* Genetic analysis of an imperfect fungal plant pathogen. *Curr. Genet.* **14,** 567–572.

Toyama, H., and Toyama, N. (1988). Protoplast fusion between *Aspergillus oryzae* and *Aspergillus kawachii. Hakkokogaku Kaishi* **66,** 377–384.

Um, S.-D., Chae, Y.-A., Park, Y.-H., and Yoo, Y.-B. (1988). Studies on auxotroph induction of *Ganoderma lucidum* and interspecific protoplast fusion between *Ganoderma lucidum* and *Ganoderma applanatum. Korean J. Mycol.* **16,** 16–20.

Ushijima, S., and Nakadai, T. (1987). Breeding by protoplast fusion of koji mold *Aspergillus sojae. Agric. Biol. Chem.* **51,** 1051–1058.

Ushijima, S., Nakadai, T., and Uchida, K. (1987). Improvement of enzyme productivities through mutation of haploidization of heterologous diploids obtained by protoplast fusion of *Aspergillus sojae. Agric. Biol. Chem.* **51,** 2781–2786.

Witte, V., Grossmann, B., and Emeis, C. C. (1989). Molecular probes for the detection of *Kluyveromyces marsianus* chromosomal DNA in electrophoretic karyotypes on intergeneric protoplast fusion products. *Arch. Microbiol.* **152,** 441–446.

Yoo, Y.-B., Song, M.-T., Go, S.-J., You, C.-H., Cha, D.-Y., Park, Y.-H., and Chang, K.-Y. (1989). Interorder protoplast fusion between *Pleurotus ostreatus* and *Ganoderma applanatum. Korean J. Mycol.* **17,** 119–123.

15

Genetics and Molecular Biology of the Lignin-Degrading Basidiomycete *Phanerochaete chrysosporium*

MARGARET ALIC
MICHAEL H. GOLD
Department of Chemical and Biological Sciences
Oregon Graduate Institute of Science and Technology
Beaverton, Oregon

I. INTRODUCTION

White-rot fungi such as *Phanerochaete chrysosporium* are the only known organisms that are capable of completely degrading lignin to CO_2 and H_2O (Crawford, 1981). This heterogeneous, biologically recalcitrant component of plant cell walls is the most abundant aromatic polymer in the biosphere and forms an encrusting matrix around the cellulose fibers of wood (Freudenberg, 1968; Sarkanen and Ludwig, 1971). Much research has been directed at understanding the biochemistry and microbiology of the lignin-degradative system. Under ligninolytic conditions, *P. chrysosporium* secretes two extracellular heme peroxidases, manganese peroxidase (MnP) and lignin peroxidase (LiP), as well as a H_2O_2-generating system. A considerable amount of evidence suggests that these enzymes are responsible for the initial depolymerization of lignin (Higuchi, 1986; Kirk and Farrell, 1987; Gold *et al.,* 1989; Wariishi *et al.,* 1989, 1991; Hammel and Moen, 1991). The structures and the unique nonspecific oxidative mechanisms of these enzymes have been reviewed (Higuchi, 1986; Buswell and Odier, 1987; Kirk and Farrell, 1987; Gold *et al.,* 1989).

Because of the abundance of lignocellulose and its importance as a renewable resource for the production of paper products, chemicals, and fuels, white-rot fungi have many potential applications in industrial bioprocessing schemes. In addition, the lignin-degradative system of *P. chrysosporium* has been utilized successfully for the degradation of a variety of persistent aromatic pullutants (Bumpus and Aust, 1987; Hammel, 1989; Valli and Gold, 1991).

The application of genetic and molecular biological methods to *P. chrysosporium* and the eventual production of fungal strains and enzymes with enhanced degradative capacities are of great interest to both academic and industrial laboratories. The recent literature on the molecular biology and genetics of *P. chrysosporium* is reviewed in this chapter.

II. GENETIC ANALYSIS

A. Auxotrophic Mutants

In initial studies, Gold and co-workers isolated 33 amino acid-, vitamin-, and adenine-requiring auxotrophs of *Phanerochaete chrysosporium* by uv and x-ray mutagenesis, and assigned them to complementation groups by the formation of heterokaryotic mycelia (Gold and Cheng, 1978; Gold *et al.,* 1982a). Additional auxotrophic mutants were isolated using γ irradiation (Liwicki *et al.,* 1985) and UV mutagenesis (Krejčí, 1987; Krejčí and

Homolka, 1991). Although fluorescent staining with mithramycin A has shown that the conidia are predominately multinucleate, the development of a medium that promotes production of uninucleate spores (Gold *et al.*, 1982a) and the use of sonication to "unclump" the conidia (Alic and Gold, 1985) has facilitated the isolation of additional mutants.

P. chrysosporium ura3 mutants, deficient in orotidine-5′-phosphate decarboxylase, and *ura5* mutants, deficient in orotate phosphoribosyl transferase, were selected (M. Alic, C. Letzring, and M. Gold, unpublished) by 5-fluoroorotate resistance (Boeke *et al.*, 1984). Various other auxotrophs have been identified by growth on pathway intermediates; six leucine auxotrophs in three complementation groups were characterized biochemically (Molskness *et al.*, 1986; M. Alic and M. Gold, unpublished).

B. Other Mutant Strains

Some of the earliest studies aimed at elucidating the mechanisms of lignin degradation by *Phanerochaete chrysosporium* involved the isolation of phenol oxidase negative (PO^-) mutants, most of which were pleiotropic for secondary metabolic functions including lignin degradation (Ander and Eriksson, 1976; Gold *et al.*, 1982b; Liwicki *et al.*, 1985). The discovery that several polymeric dyes could serve as substrates for the ligninolytic system suggested a simple new selection, not only for ligninolytic mutants, but also for mutants that were deregulated for nitrogen repression of the lignin-degradative system (Glenn and Gold, 1983; Gold *et al.*, 1988). Utilizing this protocol, Kuwahara and Asada (1987) isolated mutant strains of *P. chrysosporium* that decolorized the dye rhemazol brilliant blue under high nitrogen–low oxygen conditions. Boominathan and coworkers (1990) recently isolated a *lip*$^-$ mutant that had normal levels of MnP and glucose oxidase activities but no detectable LiP. There are also reports of mutant strains lacking cellulases (Ander and Eriksson, 1975; Ander *et al.*, 1980; K.-E. Eriksson *et al.*, 1983; Johnsrud and Eriksson, 1985) or glucose oxidase (Ramasamy *et al.*, 1985; Kelley *et al.*, 1986), and strains with LiP activity that exceeds wild-type levels (Kirk *et al.*, 1986a,b).

C. Genetic Recombination

The demonstration that *Phanerochaete chrysosporium* could be induced to form fruiting bodies and eject viable basidiospores under conditions of nitrogen limitation (Gold and Cheng, 1979) opened the way for detailed examination of genetic mechanisms in this organism. The isolation of prototrophic and double-auxotrophic phenotypes from crosses between various auxotrophs indicated that genetic recombination had occurred between the parent strains (Alic and Gold, 1985; Krejčí, 1987). No evidence of parasexual somatic recombination was found (Alic and Gold, 1985).

The basidiospores of *P. chrysosporium* are binucleate. The frequent occurrence of approximately 1 : 1 : 1 : 1 ratios of parental and recombinant phenotypes among the progeny of crosses suggested that the binucleate basidiospores were homokaryotic, that is, the two nuclei arose from a postmeiotic mitotic event (Alic and Gold, 1985). Prototrophs that were recovered from crosses between auxotrophic strains were fruited; the progeny of most of these self-crosses were shown to be prototrophic. This result indicated that the original prototrophs were wild-type recombinants rather than complementary heterokaryons and that the binucleate basidiospores are indeed homokaryotic. In contrast, the conidia arising from heterokaryotic mycelia were >85% heterokaryotic (Alic and Gold, 1985; Alic *et al.*, 1987).

D. Mating System

Details of the mating system(s) of various *Phanerochaete chrysosporium* strains are still unknown. Although a number of observations suggest that, as indicated by early work on the genus (Boidin, 1971; Gilbertson, 1971; J. Eriksson *et al.*, 1978), at least some strains have a primary homothallic mating system, other evidence suggests that mating type or vegetative incompatibility factors may also exist in some strains.

In most heterothallic basidiomycetes, dikaryons with two mating-type compatible nuclei per cell constitute the predominant vegetative phase (Davis, 1966; Ross, 1979). Cytological studies have demonstrated that various wild-type strains of *P. chrysosporium* are coenocytic, with few septa and no distinguishable secondary (dikaryotic) mycelia (Burdsall, 1985; Alic *et al.*, 1987; Thompson and Broda, 1987). Clamp connections, which are common in heterothallic basidiomycetes, have never been observed in *P. chrysosporium* (Burdsall and Eslyn, 1974; Stalpers, 1984) and the basidia have a variable number of sterigmata (Stalpers, 1984; Johnsrud and Eriksson, 1985; Alic *et al.*, 1987). However, these observations do not, by themselves, constitute proof of homothallism.

Although the genetic mechanisms for self-fertility and homothallism are not well understood in higher fungi (Raper, 1966), the evidence for homothallism in at least some *P. chrysosporium* strains can be summarized as follows.

1. Colonies arising from single conidiospores of various wild-type strains are all capable of forming fruit bodies and producing copious

numbers of viable basidiospores. Likewise, many auxotrophic strains, which presumably arose from a single haploid nucleus, are capable of fruiting (Alic and Gold, 1985; Alic *et al.,* 1987).

2. All monosporous basidiospore isolates of some *P. chrysosporium* strains (ME-416, FP-104297, P-2843f) fruit and produce viable basidiospores. In other strains (OGC101, ME-446, ML-21, BKM-F-1767, P-1271), at least some of monosporous basidiospore isolates fruit and produce viable basidiospores. The basidiospore progeny of these monosporous fruiting strains are, in turn, all capable of fruiting (Fig. 1; Johnsrud and Eriksson, 1985; Alic *et al.,* 1987). Conidia arising from basidiospore isolates always exhibit the parental phenotype. Since the binucleate basidiospores are homokaryotic (the two nuclei arise from a postpackaging mitotic division; Alic *et al.,* 1987), this cannot be a case of secondary homothallism, in which two nuclei of compatible mating types migrate into each basidiospore (Raper, 1966).

3. Auxotrophic mutants derived from ME-446 and OGC101 (a derivative of ME-446) exhibit the same variety of developmental phenotypes as the parental wild-type strains (Alic *et al.,* 1987). However, nutritionally forced heterokaryons can be formed among these auxotrophs. The vast majority of such heterokaryons can undergo karyogamy, demonstrable by genetic recombination (Alic and Gold, 1985), indicating the absence of mating-type incompatibilities (Esser, 1966). In addition, sexually competent auxotrophic strains can be obtained from crosses with fruiting strains (Alic *et al.,* 1989). Finally, in separate experiments, double-auxotrophic strains *(ade2ura5, ade1ura3)* were transformed with each of the complementing biosynthetic genes (Section IV,A,B). The stable single-auxo-

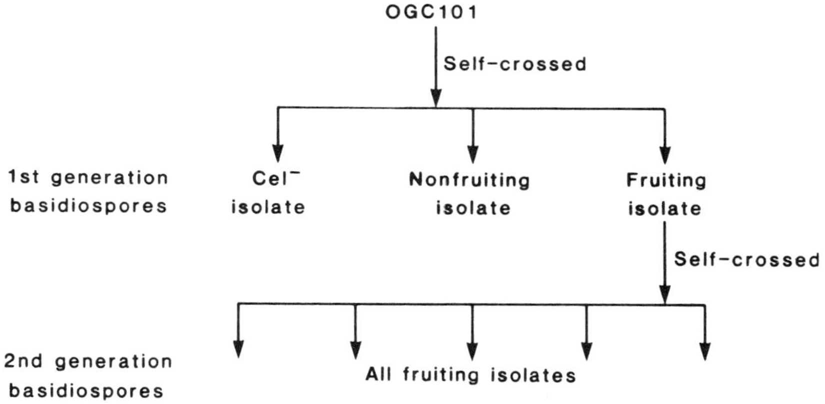

Fig. 1. Fruiting isolates recovered from basidiospores of wild-type strain OGC101 were self-crossed and the progeny phenotypes determined. From Alic *et al.,* 1987.

trophic transformants were crossed and recombinants obtained, demonstrating karyogamy between nuclei that were genetically identical except for the transforming DNA (M. Alic and M. Gold, unpublished).

In contrast to these studies, it has been reported that *P. chrysosporium* strain ME-446 has a bipolar heterothallic mating system with two mating-type alleles. These conclusions were based on observations of vegetative mycelial interactions between basidiospore isolates, without regard to sexual development or subsequent basidiospore formation (Thompson and Broda, 1987; Raeder *et al.,* 1989b). However some nonfruiting basidiospore isolate from strains ME-446 and BKM-F-1767 fruit in pairwise combinations, suggesting the existence of mating type factors (M. Alic and M. Gold, unpublished; D. Cullen, C. A. Reddy, personal communications).

It is conceivable that different laboratory strains of *P. chrysosporium* have different mating systems. Several species of related basidiomycetes are known to contain both homothallic and heterothallic races (Ullrich, 1973; Ullrich and Raper, 1975; Chase *et al.,* 1985). Recently, it has been suggested that fungal mating systems may be more flexible than previously thought and that heterothallism may be easily masked, except under conditions that favor outbreeding (Carlile, 1987).

It is also possible that, in *P. chrysosporium* strains such as OGC101, ME-446, and BKM-F-1767, the fruiting basidiospore isolates are analogous to mutants of *Schizophyllum commune* and *Coprinus lagopus* that are constitutive for those functions controlled by mating-type genes (Parag, 1962; Day, 1963; Raper *et al.,* 1965), whereas the nonfruiting basidiospore isolates are wild-type for mating type. Laboratory strains such as OGC101 may lack compatible mating-type alleles; thus, pairwise combinations of nonfruiting isolates from this strain are not capable of completing the sexual cycle. In this regard, some nonfruiting isolates from OGC101 fruit in combination with some nonfruiting isolates from other strains (Alic *et al.,* 1987).

Unanswered questions about the mating system of *P. chrysosporium* obviously remain. However, the fact that homokaryotic isolates of *P. chrysosporium* are either intrinsically capable of fruiting (Alic *et al.,* 1987), or can acquire this ability via genetic recombination indicates that at least some strains are functionally homothallic. These results, and the absence of apparent barriers to karyogamy among auxotrophic strains (Alic and Gold, 1985), have important implications for strain development and DNA transformation in this organism (Section IV). We are continuing our attempts to clarify the complex mating behavior of this organism.

E. Heterokaryosis

Basidiospore-derived isolates from some *Phanerochaete chrysosporium* strains differ from each other and from the parental wild-type in a variety of characteristics including growth, conidiation, fruiting, and evolution of $^{14}CO_2$ from ^{14}C-labeled lignin, indicating that the parent strains are heterokaryotic (Alic *et al.,* 1987; Fig. 1). Raeder and Broda (1986) confirmed the heterokaryotic nature of the mycelia and conidia and the homokaryotic nature of the basidiospores for strain ME-446 in their analysis of restriction fragment length polymorphisms (RFLPs). Heterokaryosis has important implications both for physiological studies of lignin degradation and for the construction of strains with desirable phenotypes. In addition, it suggests that some of the heterogeneity observed among isolated lignin peroxidase and manganese peroxidase genes (*lip* and *mnp,* respectively) may be due to allelic differences.

F. Genome Content

The estimated haploid genome size of *Phanerochaete chrysosporium* strain ME-446 is 4.4×10^7 base pairs (bp), of which 20–30% is mitochondrial and ribosomal DNA. *P. chrysosporium* strains ME-446 and BKM-F-1767 have chromosomal GC contents of 59% (Raeder and Broda, 1984; 1988). These values are similar to those reported for other basidiomycetes (Kurtzman, 1985).

G. Mapping

Methods for isolating mRNA from growth phase and idiophase cultures of *P. chrysosporium* have been developed (Haylock *et al.,* 1985; Haylock and Broda, 1988). Idiophase-specific cDNA probes were used to identify genomic clones expressed only during secondary metabolism. By probing single basidiospore progeny with these cloned sequences, the segregation of restriction site polymorphisms was demonstrated (Raeder and Broda, 1986). Using additional probes, including *lip* clones and a cellulase gene, a preliminary genetic map was developed for ME-446, with six linkage groups plus four unlinked markers, the latter including a *lip* gene cluster (Raeder *et al.,* 1987, 1989a; Raeder and Broda, 1986; Broda *et al.,* 1989).

Using clamped homogeneous electrical field (CHEF) electrophoresis, Gaskell *et al.,* (1991) resolved seven chromosomal bands from *P. chrysosporium* strain BKM-F-1767 and localized the *lip* gene family to a single dimorphic chromosome.

In a recent report of genetic mapping of *P. chrysosporium,* Krejčí and Homolka (1991) used meiotic segregation analysis to map seven auxotrophic mutations to two linkage groups.

III. GENES ENCODING EXTRACELLULAR PEROXIDASES

After the discovery of extracellular peroxidases involved in the lignin-degradative system of *Phanerochaete chrysosporium* (Gold *et al.,* 1984; Tien and Kirk, 1984; Glenn and Gold, 1985), and the recognition that each of these enzymes is expressed as a series of isozymes (Renganathan *et al.,* 1985; Paszczynski *et al.,* 1986; Leisola *et al.,* 1987), various laboratories began to clone the genes encoding these isozymes. Studies of the structure and regulation of these genes have numerous potential applications including elucidation of physiological and enzymatic mechanisms controlling lignin degradation; site-directed mutagenesis for enzyme modification; expression of these enzymes in other hosts; and manipulation of the lignin-degradative system in *P. chrysosporium.*

A. Lignin Peroxidase

The first cDNA clones for lignin peroxidase were isolated using synthetic oligonucleotide probes based on partial amino acid sequences of the most abundant LiP isozyme from the *Phanerochaete chrysosporium* strain BKM-F-1767 (Zhang *et al.,* 1986; Zhang and Reddy, 1988). In 1987, Tien and Tu reported the first *lip* cDNA sequence for this isozyme, encoding a mature protein product of 345 amino acids with a 28-residue leader sequence. One potential *N*-glycosylation site and residues thought to be essential for peroxidase activity [distal and proximal histidines (His) and the distal arginine (Arg)] were identified. The sequences surrounding these active-center residues are homologous with sequences in other peroxidases (Tien and Tu, 1987).

The sequences of four other *lip* cDNAs from strain BKM-F-1767 have been published. These clones appear to encode additional isozymes of *lip* and allelic variants (de Boer *et al.,* 1987, 1988; Andrawis *et al.,* 1989).

Sequencing of a *lip* cDNA encoding the most abundant LiP isozyme from *P. chrysosporium* strain OGC101 has shown that the mature protein product is composed of 343 amino acids, preceded by a 21-amino-acid signal peptide and a 7-amino-acid propeptide (Fig. 2). The N-terminal, hydrophobic, and C-terminal segments of the signal peptide are typical of those for secreted proteins. Sequencing of a *lip* cDNA *in vitro* translation product cleaved with signal peptidase indicates that the signal peptide is cleaved after the alanine at position 21 (T. Ritch, V. Nipper, L. Akileswaran, A. Smith, D. Pribnow, and M. Gold, unpublished), in agreement with S-factor analysis (von Heijne, 1985). The cleavage site for the propeptide follows the sequence –Lys–Arg–, suggesting that its removal involves an enzyme similar to the yeast Kex2 endopeptidase (Fuller *et al.,* 1988).

A number of genomic sequences of *lip* genes and allelic variants from *P. chrysosporium* strains BKM-F-1767 and ME-446 have also been published (Asada *et al.,* 1988; A. Brown *et al.,* 1988; Smith *et al.,* 1988; Walther *et al.,* 1988; Andrawis *et al.,* 1989; Schalch *et al.,* 1989; Huoponen *et al.,* 1990; Naidu and Reddy, 1990). Most *lip* genes have eight small (49–78 bp) introns interspersed throughout (Fig. 2), with intron–exon junction sequences that conform to those in other filamentous fungi, but internal splice sequences that are not as strictly conserved (Ballance, 1986; Walther *et al.,* 1988). Intron 1 splits the putative signal sequence from the putative propeptide and intron 8 is adjacent to the last proline-rich exon (Schalch *et al.,* 1989; Fig. 2). One *lip* gene reportedly contains nine introns (Gaskell *et al.,* 1991). The 5′ noncoding region includes a CAAT sequence at approximately position −110 and a TATA box at approximately −81 relative to the translation initiation codon (Smith *et al.,* 1988). S1 nuclease mapping has confirmed that transcription initiates downstream of these sequences (Holzbaur and Tien, 1988).

It is apparent that the LiP isozymes are encoded by a family of closely related genes, all of which cross-hybridize to varying extents (A. Brown *et al.,* 1988). *In vitro* translation of poly(A) RNAs from 1–6-day cultures, and immunoprecipitation with polyclonal antibodies to isozyme H8, revealed multiple protein bands with M_r in the range of 36,000–42,000. The first bands appeared on day 2 and increased each day, indicating the presence of LiP isozymes of varying molecular weights, even in the absence of glycosylation (Tien and Tu, 1987). In addition, *lip* gene probes hybridize to multiple sequences in the genomic DNA of other white-rot fungi (Huoponen *et al.,* 1990; Kimura *et al.,* 1990).

At least some of the *lip* genes may be clustered in the *P. chrysosporium* genome. Huoponen and co-workers (1990) and Gaskell *et al.,* (1991) have isolated a λEMBL3 clone carrying two *lip* genes separated by only 1.3 kilobases (kb) of flanking DNA and transcribed in opposite directions. Of the 11 *lip* clones used in RFLP mapping of the *P. chrysosporium* genome, at least 6 mapped to one cluster. Two other *lip* genes, located on a single clone, defined a second cluster mapping to linkage group III (Raeder *et al.,* 1989a). In contrast, Gaskell *et al.,* (1991) localized the *lip* genes to a single dimorphic chromosome using CHEF electrophoresis.

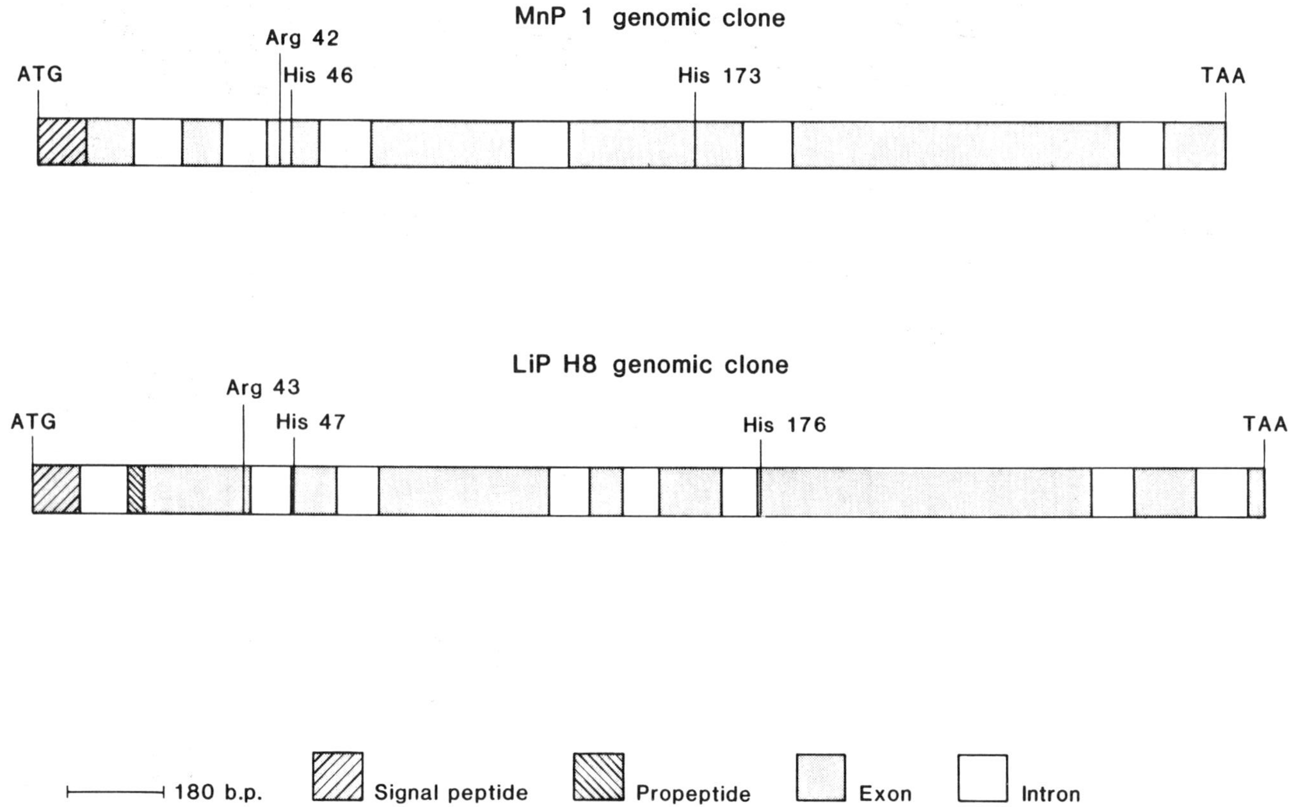

Fig. 2. Comparison of the structures of the *mnp-1* and *lip-H8* genes from the translation initiation codons to the termination codons. His 46/47, Arg 42/43, and His 173/176 are the distal His, distal Arg, and proximal His, respectively, that are involved in the peroxidase mechanism. From Schalch *et al.*, 1989; Godfrey *et al.*, 1990.

B. Manganese Peroxidase

The first *mnp* cDNA was cloned using polyclonal antibodies raised against purified MnP isozyme 1. The coding region of this clone has a 68% GC content, with a high proportion of G–C in degenerate positions (Pribnow *et al.,* 1989). In contrast, the GC content of *lip* genes is similar to that of the organism's genome as a whole (A. Brown *et al.,* 1988; Broda *et al.,* 1989). This bias may reflect the high rate of MnP synthesis; it is similar to that found in highly expressed yeast genes (Ballance, 1986). The deduced mature protein consists of 357 amino acids preceded by a 21-amino-acid leader sequence, with N-terminal, hydrophobic, and C-terminal domains characteristic of signal peptides (von Heijne, 1985; Pribnow *et al.,* 1989; Fig. 2). The deduced mature N-terminal sequence corresponds to the experimentally determined N-terminal sequence of purified MnP isozyme 1. A single *N*-glycosylation site and the proximal His, distal His, and distal Arg (Section III,A) were identified. The sequences flanking these active-center residues are homologous with those of other peroxidases, including LiP. A putative polyadenylation signal, AATACA (Pribnow *et al.,* 1989), matches that in the *lip* cDNA CLG5 (de Boer *et al.,* 1987). The *mnp* and *lip* cDNAs exhibit about 50% identity at the amino-acid level and about 60% identity at the nucleotide level. Both enzymes have an abundance of acidic residues, consistent with their low pIs. Southern blot hybridization analysis suggests the presence of several different *mnp* genes. A cDNA clone for MnP isozyme 2 has been isolated and sequenced (M. Mayfield, D. Pribnow, and M. Gold, unpublished). A *mnp* cDNA from strain BKM-F-1767 has also been sequenced. It encodes a deduced protein of 358 amino acids with a 24-amino-acid leader sequence (Pease *et al.,* 1989).

Recently, a genomic clone encoding MnP-1 was isolated and sequenced (Godfrey *et al.,* 1990). The gene contains six introns ranging in size from 57 to 72 bp; intron splice-junction sequences are conserved. However, there is little similarity to the intron positions in *lip* genes (Fig. 2). The distal His and distal Arg are located on the same exon in the *mnp* gene, whereas they are split by an intron in the *lip* genes, although the two residues are separated by only three amino acids in both proteins. The *mnp-1* gene lacks an obvious propeptide ending in a dibasic amino-acid pair, as well as the intron found in the *lip* leader sequences. The *mnp* gene also lacks the short C-terminal proline-rich sequence of *lip* (Fig. 2). In contrast to the extreme GC bias of the *mnp* coding regions, the *mnp* introns have a GC content of 56%, with 59% G–C in the 5′ upstream region and 41% in the 3′ downstream region (Godfrey *et al.,* 1990).

The 5′ untranslated region of the *mnp* gene contains a TATAA element at position −81 and three inverted CCAAT elements at −181, −195, and

−304 relative to the translation initiation codon. It also contains putative recognition sites for the general transcription factor SP-1 (GGGCGG) with the preferred flanking bases (Dynan and Tjian, 1985) at −446, and factor AP-2 (TGGGGA) (Wingender, 1988) at −285 relative to the initiation codon (Fig. 3). In addition, the *mnp-1* gene contains three putative heat-shock elements (HSEs) similar to the consensus C--GAA--TTC--G sequence (Lindquist, 1986) within 275-bp upstream of the translation initiation codon, and at least five putative consensus metal response elements (MREs; Culotta and Hamer, 1989; Godfrey *et al.,* 1990). The proximal MREs are arranged in two pairs with identical overlaps, suggesting that this arrangement may have physiological significance (Fig. 3).

C. Regulation of LiP and MnP

Tien and Tu (1987) used *in vitro* translation of poly(A) RNAs and Northern blot analysis to demonstrate that LiP is regulated by nutrient nitrogen at the level of mRNA transcription. Northern blot analysis indicates that MnP is also regulated by nutrient nitrogen at the level of gene transcription (Pribnow *et al.,* 1989).

In addition, accumulation of MnP activity in the extracellular medium of nitrogen-limited cultures depends on the presence of manganese (Mn) the substrate for the enzyme (Glenn, 1986; Bonnarme and Jeffries, 1990; J. A. Brown *et al.,* 1990; Gold *et al.,* 1991). Western blots of intracellular and extracellular *Phanerochaete chrysosporium* protein demonstrate that MnP protein is present under nitrogen-limiting conditions only in the presence of Mn (J. A. Brown *et al.,* 1990).

No *mnp* mRNA is detected on Northern blots of RNA isolated from Mn-deficient cultures. The *mnp* mRNA is first detectable in 4-day-old cultures and reaches a maximum in 5-day-old nitrogen-limited cultures grown in the presence of 180 μM Mn (J. A. Brown *et al.,* 1990). Only Mn-supplemented cultures yield MnP protein upon *in vitro* translation of poly(A)RNAs (Brown *et al.,* 1991). Addition of Mn to 5-day-old nitrogen-limited Mn-deficient cultures results in detectable *mnp* mRNA within 1 hr (Gold *et al.,* 1991); the amount of *mnp* mRNA detected is a function of the amount of Mn added. Addition of actinomycin D and cycloheximide indicates that Mn acts at the level of gene transcription, even in the absence of ongoing protein synthesis (Brown, *et al.,* 1991). In contrast, Mn has no significant effect on mycelial dry weights, rates of carbon and nitrogen consumption, amount of extracellular protein, or on synthesis of the secondary metabolite, veratryl alcohol (Bonnarme and Jeffries, 1990; J. A. Brown *et al.,* 1990). It is possible that the putative MREs of the *mnp*-1 promoter region (Fig. 3) are involved in the Mn regulation of *mnp* gene transcription.

```
-526                              GGTACCGGCACATTAACCTCGCCCTTTCCCT   -496
           MRE                                          SP-1
-495  CGAAGGCTGCACTCGCCCATTTGTGAGGCACAGCCGTGCACGTTGGGCGGTATTC   -441

-440  CACAACAACCACTCCCACACCTGAGCTGGGGGTGCGGGATGCATCTGGCAACATA   -386

-385  TGGAAGGACGCGCGCCGATGGCGGTGTCCCTGCCACCGACAGCACTCGCTCTCTC   -331
          MRE                                        AP-2
-330  ACCAAGCGTGCACACCGAGGGCATTGGCCGACACAGTGGGTGGGGACCCGCATCG   -276
         MRE
-275  CTGGATTATTCCCGAGCGATGGGATTCGCCCACGATCCCGCCTTGAGCGACCCAC   -221

-220  ACTCCTGTGAGTTACGGAATTGGCCGCTCATTATTGGACACTTACGCGAGTCGCT   -166

-165  GGCAGCAGCTCCTGATTATTCGTTGGAAACTTGGACCGCTGCTGTTGTCGAGCTC   -111
                 MRE
-110  GAATAGGGCGTGTGCACGCGCAGTATAAAACCTAGGCGGCCTCGACAGAGGTTGG   -56
               MRE
-55   CCAGGACATTCCAGTTCTACTTCCTCCTCCTCCGGTCAACGGCTTGGTATTCCAG   -1
```

Fig. 3. The sequence of the promoter region of the *mnp-1* gene. The TATAAA sequence is boxed. The three putative HSE sequences are underlined. Nucleotide residues in the HSEs that vary from the consensus C--GAA--TTC--G are indicated with dots. Five putative MREs are indicated. The three inverted CCAAT elements (ATTGG) are overlined. The putative SP-1 and AP-2 sites are also indicated. From Godfrey *et al.*, 1990.

Northern blot analysis demonstrates that a 45 °C heat shock for one hour results in the accumulation of *mnp*-1 mRNA even in cells grown in the absence of Mn (Godfrey *et al.*, 1990). This suggests that the putative heat shock elements (HSEs) found in the promoter region of the *mnp*-1 gene (Fig. 3) are physiologically functional. The regulation of *mnp*-1 mRNA by heat shock mechanisms is being examined in our laboratory.

Indirect evidence suggests that intracellular cAMP may be involved in the regulation of the lignin degradation system of *P. chrysosporium* (MacDonald *et al.*, 1984; Broda *et al.*, 1989). The adenylate cyclase of *P. chrysosporium* apparently requires Mn for maximal activity (MacDonald *et al.*, 1985).

D. Other *P. chrysosporium* Genes

A cellulase gene from *P. chrysosporium* was cloned by hybridization of a λEMBL3 genomic library with a fragment of the exo-cellobiohydrolase I (CBHI) gene from *Trichoderma reesei*. The *P. chrysosporium* gene encodes a deduced protein of 516 residues and contains two introns. The transcript is induced by cellulose but not by glucose (Sims *et al.*, 1988a). The *P. chrysosporium* gene encoding phosphoribosylanthranilate isomerase (PRAI) was isolated by complementation of an *E. coli trpC* mutant with a

P. chrysosporium λEMBL3 genomic library (Sims *et al.*, 1988b). The *adel* gene encoding phosphoribosyl aminoimidazole synthetase (Section IV,A) from *P. chrysosporium* and the glyceraldehyde-3-phosphate dehydrogenase *(gpd)* gene have recently been cloned from a λEMBL3 library of strain OGC101 (M. Mayfield, M. Alic, and M. Gold, unpublished), using heterologous probes from the fungi *Schizophyllum commune* and *Aspergillus nidulans* (Punt *et al.*, 1988), respectively. Several other *P. chrysosporium* genes are currently being cloned in our laboratory.

IV. TRANSFORMATION OF *PHANEROCHAETE CHRYSOSPORIUM*

DNA transformation of *Phanerochaete chrysosporium* has a number of potential applications. It will facilitate analysis of *lip* and *mnp* regulatory sequences and genomic position effects, as well as the study of enzyme structure–function relationships via site-directed mutagenesis. It is also possible that high-level expression of these enzymes can be achieved by increasing gene copy number or modifying promoter sequences. In addition, transformation-mediated gene disruption experiments may become feasible, facilitating analysis of the role of LiP and MnP isozymes in lignin degradation.

A. Heterologous Transformation

In the first transformation of *Phanerochaete chrysosporium* (Alic *et al.*, 1989), a plasmid containing an adenine biosynthetic gene from *Schizophyllum commune* (pADE2) (Froeliger *et al.*, 1987) was used to complement an adenine-requiring auxotroph, *ade2.* Swollen basidiospores were protoplasted with Novozym 234 (Novo Industries, Denmark) and Cellulase CP (Sturge Enzymes, United Kingdom) (Gold *et al.*, 1983) and transformed in a protocol similar to that developed for the basidiomycete *S. commune* (Muñoz-Rivas *et al.*, 1986; Specht *et al.*, 1988). In these initial transformations, frequencies of approximately 100 transformants per μg DNA were obtained. The transforming character was mitotically and meiotically stable on both selective and nonselective media. Southern blot analysis demonstrated that the plasmid DNA was integrated into the chromosomal DNA in multiple tandem copies. Genetic crosses between double mutants transformed for adenine prototrophy and other auxotrophic strains yielded *ade*⁻ progeny, indicating that integration occurred at a site or sites other than the resident adenine biosynthetic gene (Alic *et al.*, 1989). These results are similar to those reported with transformations of other filamentous fungi (Fincham, 1989).

A second *P. chrysosporium* adenine auxotroph, *adel,* was transformed to prototrophy (Alic *et al.*, 1990) with a plasmid containing a different adenine biosynthetic gene from *S. commune, pADE5* (Froeliger *et al.,* 1987). Fragments containing the *ade2* and *ade5* genes from *S. commune* were subcloned into pUC18. Between 0.05 and 0.15% of viable *P. chrysosporium* protoplasts were transformed with these subclones. Again, the transforming DNA was mitotically and meiotically stable. Southern analysis of transformants showed that the plasmids were chromosomally integrated in multiple copies at various sites in the *P. chrysosporium* genome. There were also indications of plasmid rearrangements or deletions.

The subclones, pADE2-3b and pADE5-2g, were mapped for restriction sites. The approximate locations of the complementing genes within the inserts were determined by transformation experiments with plasmid digests. In many cases, such digests reduced but did not eliminate transformation, suggesting that the fragments were incorporated into the genome in tandem, or that endogenous DNA ligase activity caused the fragments to religate.

pADE5 and pADE5-2g were used to transform the *Neurospora crassa ade2* strain to prototrophy, tentatively identifying the product of the *S. commune ade5* biosynthetic gene as phosphoribosyl aminoimidazole synthetase. These experiments also demonstrated the expression of a basidiomycete gene in an ascomycetous fungus (Hynes, 1989; Alic *et al.,* 1990).

In addition to the adenine transformations, *P. chrysosporium ura3* and *ura5* auxotrophs have been successfully transformed with plasmids containing the complementary pyrimidine nucleotide biosynthetic enzymes from *S. commune* (Froeliger *et al.,* 1987) and the ascomycete *Podospora anserina* (Bégueret *et al.,* 1984), respectively (M. Alic, L. Akileswaran, and M. Gold, unpublished).

B. Homologous Transformation

A 6.0-kb *Kpn*I fragment of a genomic clone containing the *Phanerochaete chrysosporium ade1* gene (Section III, D) has been subcloned in pUC18 (pADE1) and mapped for restriction sites. Although this plasmid transforms the *P. chrysosporium ade1* auxotroph at approximately the same frequency as pADE5-2g, transformants obtained with pADE1 appear significantly earlier than those obtained with pADE5-2g. Transformants obtained with pADE1 are being examined for homologous integration of the transforming *P. chrysosporium* DNA (M. Alic, M. Mayfield, L. Akileswaran, and M. Gold, unpublished).

C. Transformation to Drug Resistance

In fungi such as *Phanerochaete chrysosporium,* in which auxotrophic genetic markers are available, transformation systems based on complementation can take advantage of the variety of metabolic genes that have been cloned in other organisms (Alic *et al.,* 1989, 1990), as well as provide a method for cloning homologous genes from a genomic library (Froeliger *et al.,* 1987). However, dominant selective markers are useful for the transformation of wild-type strains. Randall and colleagues (1989) reported low-frequency transformation of *P. chrysosporium* to G418 resistance with the bacterial kanamycin-resistant determinant of Tn903 (Grindley and Joyce, 1980) and a *P. chrysosporium* sequence that apparently supports autonomous replication in yeast (Rao and Reddy, 1984; Turner and Ballance, 1985; Katayose *et al.,* 1986) and in *P. chrysosporium* (Randall *et al.,* 1991).

D. Utilization of DNA Transformation

The plasmid pADE5-2g is being used in the construction of expression vectors for *Phanerochaete chrysosporium* peroxidase genes and in reporter gene constructions to study promoter sequences. In addition, the plasmids containing adenine and uridylate biosynthetic genes can be used in co-transformation experiments with plasmids containing the cloned peroxidase genes for which there is no efficient selection. Ultimately, transformation of *P. chrysosporium* should facilitate the analysis of structure–function relationships via site-directed mutagenesis of LiP and MnP, the study of peroxidase function via gene disruptions and replacements, and elucidation of regulatory mechanisms for lignin biodegradation.

V. CONCLUSIONS AND FUTURE DIRECTIONS

The application of classical and molecular genetics to the lignin-degradative system of *Phanerochaete chrysosporium* has progressed rapidly in the last few years and can be used not only to elucidate the mechanisms of lignin degradation by white-rot basidiomycetes, but to learn more about the genetic system of *P. chrysosporium.* More research is needed to elucidate the mating system(s) of this organism. Genetic recombination of homokaryotic strains could be used to construct strains with specific lignin-degrading phenotypes. RFLP analysis could be used to map further the *P. chrysosporium* genome. Expression vectors containing the *lip* and *mnp* genes in conjunction with a complementary biosynthetic gene could be used to analyze expression of introduced peroxidase genes and to examine integration sites within the *P. chrysosporium* genome. It is possible that expression and/or regulation of these genes may be site specific. Depending on the frequency of homologous recombination in transformations of *P. chrysosporium,* experiments involving gene disruptions and replacements (Paietta and Marzluf, 1985; Frederick *et al.,* 1989; Kronstad *et al.,* 1989) could prove very useful, both for creating auxotrophic mutants by reverse genetics and for creating specific mutations in the lignin-degrading system. It may also be possible to increase production of lignin-degrading peroxidases by multiple-copy transformations or by placing the genes under the regulation of a strong promoter such as that of *gpd* (Punt *et al.,* 1988). X-ray crystallographic studies of LiP and MnP are currently underway (T. Poulos, S. Edwards and M. Gold, unpublished). These studies should provide information for the design of experiments involving site-directed mutagenesis (Smith, 1985; Gerlt, 1987) of *lip* and *mnp* genes. Ultimately, such studies could lead to the rational design of peroxidases with improved stability and/or catalytic characteristics.

ACKNOWLEDGMENTS

The authors wish to thank J. A. Brown, T. G. Ritch, B. J. Godfrey, and M. B. Mayfield from our laboratory, and R. C. Ullrich, University of Vermont, for useful discussions. The work in the authors' laboratory was funded by grants DE-FG07-87ER-13715 from the U.S. Department of Energy, Office of Basic Energy Sciences, and DMB-8904358 from the National Science Foundation, to M.H.G.

REFERENCES

Alic, M., and Gold, M. H. (1985). Genetic recombination in the lignin-degrading basidiomycete *Phanerochaete chrysosporium. Appl. Environ. Microbiol.* **50,** 27–30.

Alic, M., Letzring, C., and Gold, M. H. (1987). Mating system and basidiospore formation in the lignin-degrading basidiomycete *Phanerochaete chrysosporium. Appl. Environ. Microbiol.* **53,** 1464–1469.

Alic, M., Kornegay, J. R., Pribnow, D., and Gold, M. H. (1989). Transformation by complementation of an adenine auxotroph of the lignin-degrading basidiomycete *Phanerochaete chrysosporium. Appl. Environ. Microbiol.* **55,** 406–411.

Alic, M., Clark, E. K., Kornegay, J. R., and Gold, M. H. (1990). Transformation of *Phanerochaete chrysosporium* and *Neurospora crassa* with adenine biosynthetic genes from *Schizophyllum commune. Curr. Genet.* **17,** 305–311.

Ander, P., and Eriksson, K.-E. (1975). Influence of carbohydrates on lignin degradation by the white-rot fungus *Sporotrichum pulverulentum. Sven. Papperstid.* **18,** 643–652.

Ander, P., and Eriksson, K.-E. (1976). The importance of phenol oxidase activity in lignin degradation by the white-rot fungus *Sporotrichum pulverulentum. Arch. Microbiol.* **109,** 1–8.

Ander, P., Hatakka, A., and Eriksson, K.-E. (1980). Vanillic acid metabolism by the white-rot fungus *Sporotrichum pulverulentum. Arch Microbiol.* **125,** 189–202.

Andrawis, A., Pease, E. A., Kuan, I., Holzbaur, E., and Tien, M. (1989). Characterization of two lignin peroxidase clones from *Phanerochaete chrysosporium. Biochem. Biophys. Res. Commun.* **162,** 673–680.

Asada, Y., Kimura, Y., Kuwahara, M., Tsukamoto, A., Koide, K., Oka, A., and Takanami, M. (1988). Cloning and sequencing of a ligninase gene from a lignin-degrading basidiomycete, *Phanerochaete chrysosporium. Appl. Microbiol. Biotechnol.* **29,** 469–473.

Ballance, D. J. (1986). Sequences important for gene expression in filamentous fungi. *Yeast* **2,** 229–236.

Bégueret, J., Razanamparany, V., Perrot, M., and Barreau, C. (1984). Cloning gene *ura5* for the orotidylic acid pyrophosphorylase of the filamentous fungus *Podospora anserina:* Transformation of protoplasts. *Gene* **32,** 487–492.

Boeke, J. D., LaCroute, F., and Fink, G. R. (1984). A positive selection for mutants lacking orotidine-5′-phosphate decarboxylase activity in yeast: 5-Fluoro-orotic acid resistance. *Mol. Gen. Genet.* **197,** 345–346.

Boidin, J. (1971). Nuclear behavior in the mycelium and the evolution of the basidiomycetes. *In* "Evolution in the Higher Basidiomycetes" (R. H. Petersen, ed.), pp. 129–148. University of Tennessee Press, Knoxville.

Bonnarme, P., and Jeffries, T. W. (1990). Mn(II) regulation of lignin peroxidases and manganese-dependent peroxidases from lignin-degrading white rot fungi. *Appl. Environ. Microbiol.* **56,** 210–217.

Boominathan, K., Balachandra-Dass, S., Randall, T. A., Kelley, R. L., and Reddy, C. A. (1990). Lignin peroxidase-negative mutant of the white-rot basidiomycete *Phanerochaete chrysosporium. J. Bacteriol.* **172,** 260–265.

Broda, P., Sims, P. F. G., and Mason, J. C. (1989). Lignin biodegradation: A molecular biological approach. *Essays Biochem.* **24,** 82–114.

Brown, A., Sims, P. F. G., Raeder, U., and Broda, P. (1988). Multiple ligninase-related genes from *Phanerochaete chrysosporium. Gene* **73,** 77–85.

Brown, J. A., Glenn, J. K., and Gold, M. H. (1990). Manganese regulates the expression of manganese peroxidase by *Phanerochaete chrysosporium. J. Bacteriol.* **172,** 3125–3130.

Brown, J. A., Alic, M., and Gold, M. H. (1991). Manganese peroxidase gene transcription in *Phanerochaete chrysosporium:* Activation by manganese. *J. Bacteriol.* In press.

Bumpus, J. A., and Aust, S. D. (1987). Biodegradation of environmental pollutants by the white rot fungus *Phanerochaete chrysosporium:* Involvement of the lignin-degrading system. *BioEssays* **6,** 166–170.

Burdsall, H. H., Jr. (1985). "A Contribution to the Taxonomy of the Genus *Phanerochaete* (Corticiaceae, Aphyllophorales)." Mycologia Memoir No. 10. J. Cramer, Braunschweig, Germany.

Burdsall, H. H., Jr., and Eslyn, W. E. (1974). A new *Phanerochaete* with a *chrysosporium* imperfect state. *Mycotaxon* **1,** 123–133.

Buswell, J. A., and Odier, E. (1987). Lignin degradation. *CRC Crit. Rev. Biotechnol.* **6,** 1–60.

Carlile, M. J. (1987). Genetic exchange and gene flow: Their promotion and prevention. *In* "Evolutionary Biology of the Fungi" (A. D. M. Rayner, C. M. Brasier, and D. Moore, eds.), pp. 203–214. Cambridge University Press, Cambridge.

Chase, T. E., Ullrich, R. C., and Korhonen, K. (1985). Homothallic isolates of *Heterobasidion annosum. Mycologia* **77,** 975–977.

Crawford, R. L. (1981). "Lignin Biodegradation and Transformation." Wiley and Sons, New York.

Culotta, V. C., and Hamer, D. H. (1989). Fine mapping of a mouse metallothionein gene metal response element. *Mol. Cell. Biol.* **9,** 1376–1380.

Davis, R. H. (1966). Heterokaryosis. *In* "The Fungi: An Advanced Treatise" (G. C. Ainsworth and A. S. Sussman, eds.), Vol. II, pp. 567–588. Academic Press, New York.

Day, P. R. (1963). Mutations of the A mating-type factor in *Coprinus lagopus. Genet. Res.* **4,** 55–64.

de Boer, H. A., Zhang, Y. Z., Collins, C., and Reddy, C. A. (1987). Analysis of nucleotide sequences of two ligninase cDNAs from a white-rot filamentous fungus, *Phanerochaete chrysosporium. Gene* **60,** 93–102.

de Boer, H. A., Zhang, Y. Z., Collins, C., and Reddy, C. A. (1988). Corrigendum: Analysis of nucleotide sequences of two ligninase cDNAs from a white-rot filamentous fungus, *Phanerochaete chrysosporium*. *Gene* **69,** 369.

Dynan, W. S., and Tjian, R. (1985). Control of eukaryotic messenger RNA synthesis by sequence-specific DNA-binding proteins. *Nature (London)* **316,** 774–778.

Eriksson, J., Hjortstam, K., and Ryvarden, L. (1978). *In* "The Conticiaceae of North Europe" (J. Eriksson and L. Ryvarden, eds.), Vol. 5, (p. 989.) Fungiflora, Oslo.

Eriksson, K.-E., Johnsrud, S. C., and Vallander, L. (1983). Degradation of lignin and lignin model compounds by various mutants of the white-rot fungus *Sporotrichum pulverulentum*. *Arch Microbiol.* **135,** 161–168.

Esser, K. (1966). Incompatibility. *In* "The Fungi: An Advanced Treatise" (G. C. Ainsworth and A. S. Sussman, eds.), Vol. II, pp. 661–676. Academic Press, New York.

Fincham, J. R. S. (1989). Transformation in fungi. *Microbiol. Rev.* **53,** 148–170.

Frederick, G. D., Asch, D. K., and Kinsey, J. A. (1989). Use of transformation to make targeted sequence alterations at the *am* (GDH) locus of *Neurospora*. *Mol. Gen. Genet.* **217,** 294–300.

Freudenberg, K. (1968). The constitution and biosynthesis of lignin. *In* "Constitution and Biosynthesis of Lignin" (A. C. Neish and K. Freudenberg, eds.), pp. 47–122. Springer-Verlag, New York.

Froeliger, E. H., Muñoz-Rivas, A. M., Specht, C. A., Ullrich, R. C., and Novotny, C. P. (1987). The isolation of specific genes from the basidiomycete *Schizophyllum commune*. *Curr. Genet.* **12,** 547–554.

Fuller, R. S., Sterne, R. E., and Thorner, J. (1988). Enzymes required for yeast prohormone processing. *Ann. Rev. Physiol.* **50,** 345–362.

Gaskell, J., Dieperink, E., and Cullen, D. (1991). Genomic organization of lignin peroxidase genes of *Phanerochaete chrysosporium*. *Nucleic Acids Res.* **19,** 599–603.

Gerlt, J. A. (1987). Relationships between enzymatic catalysis and active site structure revealed by applications of site-directed mutagenesis. *Chem. Rev.* **87,** 1079–1105.

Gilbertson, R. L. (1971). Phylogenetic relationships of hymenomycetes with resupinate, hydnaceous basidiocarps. *In* "Evolution in the Higher Basidiomycetes" (R. H. Petersen, ed.), pp. 275–307. University of Tennessee Press, Knoxville.

Glenn, J. K. (1986). "Purification and Characterization of an Extracellular Manganese Oxidizing Peroxidase from the White Rot Fungus *Phanerochaete chrysosporium*." Ph.D. Thesis. Oregon Graduate Center.

Glenn, J. K., and Gold, M. H. (1983). Decolorization of several polymeric dyes by the lignin-degrading basidiomycete *Phanerochaete chrysosporium*. *Appl. Environ. Microbiol.* **45,** 1741–1747.

Glenn, J. K., and Gold, M. H. (1985). Purification and characterization of an extracellular Mn(II)-dependent peroxidase from the lignin-degrading basidiomycete, *Phanerochaete chrysosporium*. *Arch. Biochem. Biophys.* **242,** 329–341.

Godfrey, B. J., Mayfield, M. B., Brown, J. A., and Gold, M. H. (1990). Characterization of a gene encoding a manganese peroxidase from *Phanerochaete chrysosporium*. *Gene* **93,** 119–124.

Gold, M. H., and Cheng, T. M. (1978). Induction of colonial growth and replica plating of the white rot basidiomycete *Phanerochaete chrysosporium*. *Appl. Environ. Microbiol.* **35,** 1223–1225.

Gold, M. H., and Cheng, T. M. (1979). Conditions for fruit body formation in the white rot basidiomycete *Phanerochaete chrysosporium*. *Arch. Microbiol.* **121,** 37–41.

Gold, M. H., Cheng, T. M., and Mayfield, M. B. (1982a). Isolation and complementation studies of auxotrophic mutants of the lignin-degrading basidiomycete *Phanerochaete chrysosporium*. *Appl. Environ. Microbiol.* **44,** 996–1000.

Gold, M. H., Mayfield, M. B., Cheng, T. M., Krisnangkura, K., Shimada, M., Enoki, A., and Glenn, J. K. (1982b). A *Phanerochaete chrysosporium* mutant defective in lignin degradation as well as several other secondary metabolic functions. *Arch. Microbiol.* **132,** 115-122.

Gold, M. H., Cheng, T. M., and Alic, M. (1983). Formation, fusion and regeneration of protoplasts from wild type and auxotrophic mutant strains of the white rot basidiomycete *Phanerochaete chrysosporium. Appl. Environ. Microbiol.* **46,** 260-263.

Gold, M. H., Kuwahara, M., Chiu, A. A., and Glenn, J. K. (1984). Purification and characterization of an extracellular H_2O_2-requiring diarylpropane oxygenase from the white rot basidiomycete *Phanerochaete chrysosporium. Arch. Biochem. Biophys.* **234,** 353-362.

Gold, M. H., Glenn, J. K., and Alic, M. (1988). Use of polymeric dyes in lignin biodegradation assays. *Methods Enzymol.* **161,** 74-78.

Gold, M. H., Wariishi, H., and Valli, K. (1989). Extracellular peroxidases involved in lignin degradation by the white rot basidiomycete *Phanerochaete chrysosporium. In* "Biocatalysis in Agricultural Biotechnology" (J. R. Whitaker and P. E. Sonnet, eds.), pp. 127-140. American Chemical Society, Washington, D.C.

Gold, M. H., Brown, J. A., Godfrey, B. J., Mayfield, M. B., Wariishi, H., and Valli, K. (1991). Structure and regulation of a manganese peroxidase gene from *Phanerochaete chrysosporium. In* "Enzymes in Biomass Conversion." (G. F. Loatham and M. E. Himmel, eds), pp. 188-199, American Chemical Society, Washington, D.C.

Grindley, N. D. F., and Joyce, C. M. (1980). Genetic and DNA sequence analysis of the kanamycin resistance transposon Tn903. *Proc. Natl. Acad. Sci. U.S.A.* **77,** 7176-7180.

Hammel, K. E. (1989). Organopollutant degradation by ligninolytic fungi. *Enz. Microb. Technol.* **11,** 776-777.

Hammel, K. E., and Moen, M. A. (1991). Depolymerization of a synthetic lignin *in vitro* by lignin peroxidase. *Enzyme Microb. Technol.* **13,** 15-18.

Haylock, R., and Broda, P. (1988). Preparation and characterization of mRNA from ligninolytic fungi. *Methods Enzymol.* **161,** 221-227.

Haylock, R., Liwicki, R., and Broda, P. (1985). The isolation of mRNA from the basidiomycete fungi *Phanerochaete chrysosporium* and *Coprinus cinereus* and its *in vitro* translation. *J. Microbiol. Methods* **4,** 155-162.

Higuchi, T. (1986). Catabolic pathways and role of ligninases for the degradation of lignin substructure models by white-rot fungi. *Wood Res.* **73,** 58-81.

Holzbaur, E. L. F., and Tien, M. (1988). Structure and regulation of a lignin peroxidase gene from *Phanerochaete chrysosporium. Biochem. Biophys. Res. Commun.* **55,** 626-633.

Huoponen, K., Ollikka, P., Kälin, M., Walther, I., Mäntsälä, P., and Reiser, J. (1990). Characterization of lignin peroxidase-encoding genes from lignin-degrading basidiomycetes. *Gene.* **89,** 145-150.

Hynes, M. J. (1989). Complementation of an *Aspergillus nidulans* mutation by a gene from the basidiomycete *Coprinus cinereus. Exp. Mycol.* **13,** 196-198.

Johnsrud, S. C., and Eriksson, K.-E. (1985). Cross-breeding of selected and mutated homokaryotic strains of *Phanerochaete chrysosporium* K-3: New cellulase deficient strains with increased ability to degrade lignin. *Appl. Microb. Biotechnol.* **21,** 320-327.

Katayose, Y., Shishido, K., and Ohmasa, M. (1986). Cloning of *Lentinus edodes* mitochondrial DNA fragment capable of autonomous replication in *Saccharomyces cerevisiae. Biochem. Biophys. Res. Commun.* **138,** 1110-1115.

Kelley, R. L., Ramasamy, K., and Reddy, C. A. (1986). Characterization of glucose oxidase-negative mutants of a lignin degrading basidiomycete *Phanerochaete chrysosporium. Arch. Microbiol.* **144,** 254-257.

Kimura, Y., Asada, Y., and Kuwahara, M. (1990). Screening of basidiomycetes for lignin peroxidase genes using a DNA probe. *Appl. Microbiol. Biotechnol.* **32,** 436-442.

Kirk, T. K., and Farrell, R. L. (1987). Enzymatic "combustion": The microbial degradation of lignin. *Ann. Rev. Microbiol.* **41,** 465–505.

Kirk, T. K., Croan, S., Tien, M., Murtagh, K. E., and Farrell, R. L. (1986a). Production of multiple ligninases by *Phanerochaete chrysosporium:* Effect of selected growth conditions and use of a mutant strain. *Enz. Microb. Technol.* **8,** 7–32.

Kirk, T. K., Tien, M., Johnsrud, S. C., and Eriksson, K.-E. (1986b). Lignin degrading activity of *Phanerochaete chrysosporium* Burds.: Comparison of cellulase-negative and other strains. *Enz. Microb. Technol.* **8,** 75–80.

Krejčí, R. (1987). Genetic recombination in auxotrophic strains of *Phanerochaete chrysosporium. Folia Microbiol.* **32,** 177–180.

Krejčí, R., and Homolka, L. (1991). Genetic mapping in the lignin-degrading basidiomycete *Phanerochaete chrysosporium. Appl. Environ. Microbiol.* **57,** 151–156.

Kronstad, J. W., Wang, J., Covert, S. F., Holden, D. W., McKnight, G. L., and Leong, S. A. (1989). Isolation of metabolic genes and demonstration of gene disruption. *Gene* **79,** 97–106.

Kurtzman, C. P. (1985). Molecular taxonomy of the fungi. *In* "Gene Manipulations in Fungi" (J. W. Bennett and L. L. Lasure, eds.), pp. 35–63. Academic Press, Orlando, Florida.

Kuwahara, M., and Asada, Y. (1987). Production of ligninases, peroxidases and alcohol oxidase by *Phanerochaete chrysosporium* and its mutant. *In* "Lignin Enzymic and Microbial Degradation" (E. Odier, ed.), pp. 171–176. Institut National de la Recherche Agronomique, Paris.

Leisola, M. S. A., Kozulic, B., Meussdoerffer, F., and Fiechter, A. (1987). Homology among multiple extracellular peroxidases from *Phanerochaete chrysosporium. J. Biol. Chem.* **262,** 419–424.

Lindquist, S. (1986). The heat-shock response. *Ann. Rev. Biochem.* **55,** 1151–1191.

Liwicki, R., Paterson, A., MacDonald, M. J., and Broda, P. (1985). Phenotypic classes of phenoloxidase-negative mutants of the lignin-degrading fungus *Phanerochaete chrysosporium. J. Bacteriol.* **162,** 641–644.

MacDonald, M. J., Paterson, A., and Broda, P. (1984). Possible relationship between cyclic AMP and idiophasic metabolism in the white rot fungus *Phanerochaete chrysosporium. J. Bacteriol.* **160,** 470–472.

MacDonald, M. J., Ambler, R., and Broda, P. (1985). Regulation of intracellular cyclic AMP levels in the white-rot fungus *Phanerochaete chrysosporium* during the onset of idiophasic metabolism. *Arch. Microbiol.* **142,** 152–156.

Molskness, T. A., Alic, M., and Gold, M. H. (1986). Characterization of leucine auxotrophs of the white rot basidiomycete *Phanerochaete chrysosporium. Appl. Environ. Microbiol.* **51,** 1170–1173.

Muñoz-Rivas, A., Specht, C. A., Drummond, B. J., Froeliger, E., Novotny, C. P., and Ullrich, R. C. (1986). Transformation of the basidiomycete, *Schizophyllum commune. Mol. Gen. Genet.* **205,** 103–106.

Naidu, P. S., and Reddy, C. A. (1990). Nucleotide sequence of a new lignin peroxidase gene *GLG3* from the white-rot fungus, *Phanerochaete chrysosporium. Nucl. Acid. Res.* **18,** 7173.

Paietta, J. V., and Marzluf, G. A. (1985). Gene disruption by transformation in *Neurospora crassa. Mol. Cell. Biol.* **5,** 1554–1559.

Parag, Y. (1962). Mutations in the B incompatibility factor of *Schizophyllum commune. Proc. Natl. Acad. Sci. U.S.A.* **48,** 743–750.

Paszczyński, A., Huynh, V.-B., and Crawford, R. (1986). Comparison of ligninase-I and peroxidase-M2 from the white-rot fungus *Phanerochaete chrysosporium. Arch. Biochem. Biophys.* **244,** 750–765.

Pease, E. A., Andrawis, A., and Tien, M. (1989). Manganese-dependent peroxidase from *Phanerochaete chrysosporium:* Primary structure deduced from cDNA sequence. *J. Biol. Chem.* **264,** 13531–13535.

Pribnow, D., Mayfield, M. B., Nipper, V. J., Brown, J. A., and Gold, M. H. (1989). Characterization of a cDNA encoding a manganese peroxidase, from the lignin-degrading basidiomycete *Phanerochaete chrysosporium. J. Biol. Chem.* **264,** 5036–5040.

Punt, P. J., Dingemanse, M. A., Jacobs-Meijsing, B. J. M., Pouwels, P. H., and van den Hondel, C. A. M. J. J. (1988). Isolation and characterization of the glyceraldehyde-3-phosphate dehydrogenase gene of *Aspergillus nidulans. Gene* **69,** 49–57.

Raeder, U., and Broda, P. (1984). Comparison of the lignin-degrading white rot fungi *Phanerochaete chrysosporium* and *Sporotrichum pulverulentum* at the DNA level. *Curr. Genet.* **8,** 499–506.

Raeder, U., and Broda, P. (1986). Meiotic segregation analysis of restriction site polymorphisms allows genetic mapping. *EMBO J.* **51,** 1125–1127.

Raeder, U., and Broda, P. (1988). Preparation and characterization of DNA from lignin-degrading fungi. *Methods Enzymol.* **161,** 211–220.

Raeder, U., Thompson, W., and Broda, P. (1987). Establishing molecular genetics for *Phanerochaete chrysosporium. Phil. Trans. R. Soc. Lond., Series A* **321,** 475–483.

Raeder, U., Thompson, W., and Broda, P. (1989a). RFLP-based genetic map of *Phanerochaete chrysosporium* ME446: Lignin peroxidase genes occur in clusters. *Mol. Microbiol.* **3,** 911–918.

Raeder, U., Thompson, W., and Broda, P. (1989b). Genetic factors influencing lignin peroxidase activity in *Phanerochaete chrysosporium* ME446. *Mol. Microbiol.* **3,** 919–924.

Ramasamy, K., Kelley, R. L., and Reddy, C. A. (1985). Lack of lignin degradation by glucose oxidase-negative mutants of *Phanerochaete chrysosporium. Biochem. Biophys. Res. Commun.* **131,** 436–441.

Randall, T., Reddy, C. A., and Boominathan, K. (1991). A novel extrachromosomally maintained transformation vector for the lignin-degrading basidiomycete *Phanerochaete chrysosporium. J. Bacteriol.* **173,** 776–782.

Randall, T., Rao, T. R., and Reddy, C. A. (1989). Use of a shuttle vector for the transformation of the white rot basidiomycete, *Phanerochaete chrysosporium. Biochem. Biophys. Res. Commun.* **161,** 720–725.

Rao, T. R., and Reddy, C. A. (1984). DNA sequences from a ligninolytic filamentous fungus *Phanerochaete chrysosporium* capable of autonomous replication in yeast. *Biochem. Biophys. Res. Commun.* **118,** 821–827.

Raper, J. R. (1966). "Genetics of Sexuality in Higher Fungi." Ronald Press, New York.

Raper, J. R., Boyd, D. H., and Raper, C. A. (1965). Primary and secondary mutations at the incompatibility loci in *Schizophyllum. Proc. Natl. Acad. Sci. U.S.A.* **53,** 1324–1332.

Renganathan, V., Miki, K., and Gold, M. H. (1985). Multiple molecular forms of diarylpropane oxygenase, an H_2O_2-requiring, lignin-degrading enzyme from *Phanerochaete chrysosporium. Arch. Biochem. Biophys.* **241,** 304–314.

Ross, I. K. (1979). "Biology of the Fungi." McGraw-Hill, New York.

Sarkanen, K. V., and Ludwig, C. H. (1971). "Lignins. Occurrence, Formation, Structure and Reactions." Wiley-Interscience, New York.

Schalch, H., Gaskell, J., Smith, T. L., and Cullen, D. (1989). Molecular cloning and sequences of lignin peroxidase genes of *Phanerochaete chrysosporium. Mol. Cell. Biol.* **9,** 2743–2747.

Sims, P., James, C., and Broda, P. (1988a). The identification, molecular cloning and characterisation of a gene from *Phanerochaete chrysosporium* that shows strong homology to the exo-cellobiohydrolase I gene from *Trichoderma reesei. Gene* **74,** 411–422.

Sims, P., Brown, A., James, C., Raeder, U., Schrank, A., and Broda, P. (1988b). Molecular genetics of lignocellulose degradation in *Phanerochaete chrysosporium*. *In* "Biochemistry and Genetics of Cellulose Degradation" (J.-P. Aubert, P. Beguin, and J. Millet, eds.), pp. 365–373. Academic Press, London.

Smith, M. (1985). *In vitro* mutagenesis. *Ann. Rev. Genet.* **19,** 423–462.

Smith, T. L., Schalch, H., Gaskell, J., Covert, S., and Cullen, D. (1988). Nucleotide sequence of a ligninase gene from *Phanerochaete chrysosporium*. *Nucl. Acids Res.* **16,** 1219.

Specht, C. A., Muñoz-Rivas, A., Novotny, C. P., and Ullrich, R. C. (1988). Transformation of *Schizophyllum commune:* An analysis of parameters for improving transformation frequencies. *Exp. Mycol.* **12,** 357–366.

Stalpers, J. A. (1984). A revision of the genus *Sporotrichum*. *Stud. Mycol.* **24,** 1–105.

Thompson, W., and Broda, P. (1987). Mating behaviour in an isolate of *Phanerochaete chrysosporium*. *Trans. Br. Mycol. Soc.* **89,** 285–294.

Tien, M., and Kirk, T. K. (1984). Lignin-degrading enzyme from *Phanerochaete chrysosporium:* Purification, characterization, and catalytic properties of a unique H_2O_2-requiring oxygenase. *Proc. Natl. Acad. Sci U.S.A.* **81,** 2280–2284.

Tien, M., and Tu, C.-P. D. (1987). Cloning and sequencing of a cDNA for a ligninase from *Phanerochaete chrysosporium*. *Nature (London)* **326,** 520–523.

Turner, G., and Ballance, D. J. (1985). Cloning and transformation in *Aspergillus*. *In* "Gene Manipulations in Fungi" (J. W. Bennett and L. L. Lasure, eds.), pp. 259–278. Academic Press, Orlando.

Ullrich, R. C. (1973). Sexuality, incompatibility, and intersterility in the biology of the *Sistotrema brinkmannii* aggregate. *Mycologia* **65,** 1234–1249.

Ullrich, R. C., and Raper, J. R. (1975). Primary homothallism—Relation to heterothallism in the regulation of sexual morphogenesis in *Sistotrema*. *Genetics* **80,** 311–321.

Valli, K., and Gold, M. H. (1991). Degradation of 2,4-Dichlorophenol by the lignin-degrading fungus *Phanerochaete chrysosporium*. *J. Bacteriol.* **173,** 345–352.

von Heijne, G. (1985). Signal sequences: The limits of variation. *J. Mol. Biol.* **184,** 99–105.

Walther, I., Kalin, M., Reiser, J., Suter, F., Fritsche, B., Salohemio, M., Leisola, M., Teeri, T., Knowles, J., and Fiechter, A. (1988). Molecular analysis of a *Phanerochaete chrysosporium* lignin peroxidase gene. *Gene* **70,** 127–137.

Wariishi, H., Valli, K., and Gold, M. H. (1989). Oxidative cleavage of a phenolic diarylpropane lignin model dimer by manganese peroxidase from *Phanerochaete chrysosporium*. *Biochemistry* **28,** 6017–6023.

Wariishi, H., Valli, K., and Gold, M. H. (1991). *In vitro* depolymorization of lignin by manganese poroxidase of *Phorenochaete* chrysosporium. *Biochem. Biophys. Res. Common.* **176,** 269–275.

Wingender, E. (1988). Compilation of transcription regulating proteins. *Nucl. Acids. Res.* **16,** 1879–1902.

Zhang, Y. Z., and Reddy, C. A. (1988). Use of synthetic oligonucleotide probes for identifying ligninase cDNA clones. *Methods Enzymol.* **161,** 228–237.

Zhang, Y. Z., Zylstra, G. J., Olsen, R. H., and Reddy, C. A. (1986). Identification of cDNA clones for ligninase from *Phanerochaete chrysosporium* using synthetic oligonucleotide probes. *Biochem. Biophys. Res. Commun.* **137,** 649–656.

16

Molecular Genetics of Human Fungal Pathogens

MYRA B. KURTZ
Infectious Disease Research
Merck Sharp & Dohme Research Laboratories
Rahway, New Jersey

STEWART SCHERER
Department of Microbiology
University of Minnesota Medical School
Minneapolis, Minnesota

I. INTRODUCTION

With a rise in the incidence of human systemic fungal disease, there have been increased efforts to develop the means to manipulate the important human pathogens genetically. The emphasis in this chapter has been

MORE GENE MANIPULATIONS
IN FUNGI

placed on *Candida albicans,* not only because it is responsible for over 70% of fungal disease requiring hospitalization, but because the genetic groundwork for studying this organism is most advanced. A wide variety of classical and modern molecular techniques is now available for genetic manipulation of *C. albicans.* Many of these topics have been covered in recent reviews (Magee *et al.,* 1988b; Kurtz *et al.,* 1990; Scherer and Magee, 1990), which should be consulted for greater depth.

The known sexual cycle of several other human pathogenic fungi should, in theory, make progress in genetic analysis more rapid than in the imperfect fungi such as *C. albicans.* The reluctance to work with these more virulent organisms and the lower incidence of severe infection from these agents have contributed to the paucity of studies on these species. We will concentrate on *Cryptococcus neoformans* and *Histoplasma capsulatum* in this review.

II. *CANDIDA ALBICANS*

A. Introduction

Any review of the molecular genetics of *Candida* species must, by necessity, be restricted primarily to *C. albicans,* the most important human pathogen and the most studied of this genus. Therefore, this review will concentrate on the genetics of *C. albicans.* Despite the medical importance of this fungus, genetic manipulation has been a relatively new approach for the study of *C. albicans* because of its diploid genome and lack of a sexual cycle. Even the most basic studies have been done only in the past 10–15 years. Now that the fundamental techniques are available, rapid progress is expected.

B. Parasexual Genetic System

Due to the lack of a known sexual cycle for *Candida albicans,* it has been necessary to develop a series of specialized methods for conducting basic genetic analyses. A parasexual genetic system has been established that, in many ways, resembles the analysis of mammalian somatic cells in culture. In this section, we consider the central features of the *Candida* parasexual system. These approaches can be applied to a variety of asexual fungi.

Many of the initial genetic experiments were aimed at establishing the ploidy of clinical isolates. Early workers noted that death as a result of DNA damage more closely resembled diploid than haploid *Saccharomyces cerevisiae* (Olaiya and Sogin, 1979). Also, it was found that many clinical isolates of *C. albicans* gave a highly biased spectrum of auxotrophs; particular strains yielded almost exclusively a single type of mutant (Whelan *et al.*, 1980). A simple explanation of such results would be a natural heterozygosity for a particular mutation in those strains. The most revealing experiments involved the analysis of sectored colonies after treatment with uv light to induce mitotic recombination. Beginning with a putative heterozygote, one found certain colonies in which, after uv treatment, half of the cells became mutant. The other half of the colony differed from the parent because it no longer gave rise to the biased auxotrophic spectrum, since both the alleles were wild-type due to the elimination of the mutation by mitotic recombination. One need not start with a natural heterozygote. After chemical induction of a mutation and subsequent reversion to heterozygosity, one can treat with uv light and easily recover the same mutation again. Analysis of such experiments at many loci led to the conclusion that *C. albicans* must be diploid over much of its genome (Poulter, 1987). In summary, different isolates of *C. albicans* are naturally heterozygous for a great variety of mutations, including drug resistance (Whelan *et al.*, 1981) and recessive lethality (Whelan and Soll, 1982). Although the extent of natural heterozygosity might suggest a sexual phase for the species, the recessive lethals and lack of conservation in the karyotype would predict a high degree of inviability in the putative meiotic products.

Several mutagenesis protocols are effective with *C. albicans*. This is somewhat surprising, since its diploid nature would suggest the need for severe mutagenesis to recover mutations. In fact, many mutagens induce mitotic recombination and mutants can be detected as sectored colonies. One can also induce recombination with uv light after mutagenesis or permit a period of nonselective growth after mutagenesis, during which homozygotes readily arise. *C. albicans*, like *S. cerevisiae*, yields red adenine auxotrophs at two loci (Poulter and Rikkerink, 1983). These can be used to monitor the effectiveness of mutagenic treatments.

Spheroplast fusion can be used for complementation testing of mutants (Poulter *et al.*, 1981; Sarachek *et al.*, 1981; Kakar and Magee, 1982). The red adenine auxotrophs map in two complementation groups among the red adenine auxotrophs. The physiological characterization of these mutants was important for the development of a DNA transformation system for *C. albicans*. Although many auxotrophs have been isolated, the biochemical defects in only a few have been determined (Scherer and Magee, 1990). Recently, it has become possible to use genes, cloned by their ability to complement mutations in other organisms, to identify mutants derived in conventional mutageneses (Goshorn and Scherer, 1989).

The parasexual genetic system has been used extensively for linkage analysis (Poulter *et al.*, 1982; Hilton *et al.*, 1985). Because of the small number of chromosomes in the species ($N = 8$), it is not surprising that among the first auxotrophic markers studied, several were found to be linked to one another. Two methods for mitotic mapping have been used in *C. albicans.* Induction of mitotic recombination with uv light gives gene order, but will not detect linkage across a centromere. Heat-shock-induced chromosome loss can map across centromeres but will not give gene order. At least five linkage groups have been defined using these techniques. However, the collective data have not been adequate to predict electrophoretic karyotype chromosome number or relative chromosome size. Although linkage relationships are reasonably conserved within the *C. albicans* group, it is clear that there is also a significant degree of variability. The parasexual genetic system has been used to demonstrate that cloned genes that hybridize to the same sized band on electrophoretically separated chromosomes are genetically linked, ruling out the possibility that they are located on two distinct linkage groups that happen to be the same size. An alternative to heat-shock-induced chromosome loss is selection for reduction of ploidy using recessive 5-fluorocytosine resistance mutations (Whelan *et al.,* 1985). This latter method has not been used to date for linkage analysis.

Recently, genetic mapping data and the physical karyotype have been aligned for certain cloned genes (Scherer and Magee, 1990). The *ADE1* linkage group is on one of the largest chromosomes and the *ADE2* gene is on chromosome 3. Using genes cloned by direct complementation of *C. albicans* mutants, it has been possible to show that the *LYS1* linkage group is chromosome 4 and that an arginine auxotroph known to be unlinked to all other markers is on the smallest chromosome. These approaches, combined with the use of restriction fragment length polymorphisms (RFLPs), should produce a complete combined physical and genetic map in the near future.

The parasexual genetic system can be extended to clinical isolates by specialized methods. Strong selective methods are needed to distinguish rare spheroplast fusion products from the parental strains. As described earlier, auxotrophies have been widely used for this purpose. This procedure necessitates the isolation of auxotrophic mutants in each new isolate before genetic analysis can be initiated. There are several drawbacks to this approach: some strains are surprisingly resistant to mutagenesis; mutagenesis can be accompanied by chromosome rearrangements; and auxotrophs frequently have reduced virulence in animal models. However, by starting with a strain that contains both a dominant selectable marker and a recessive marker such as an auxotrophy, it is possible to select fusion products with any strain of interest. Mycophenolic acid resistance is an excellent dominant marker for this type of selection because the range of natural variation in resistance is small, spontaneous mutation to resistance is extremely infrequent, and the heterokaryons that arise during spheroplast fusion are sensitive to the drug (Goshorn and Scherer, 1989).

Potential virulence determinants, drug resistance markers, and genes that affect the colony morphology of *C. albicans* have been analyzed using the parasexual cycle. The *C. albicans* secreted proteinase has been the subject of both molecular and genetic analyses. Mutants that fail to make the protease are avirulent whereas revertants are virulent (Kwon-Chung *et al.,* 1985). Some strains may be naturally heterozygous for proteinase production (Crandall and Edwards, 1987). It is not certain if any of these mutations are in the structural gene. Resistance to 5-fluorocytosine has also been studied extensively, and natural heterozygosity for resistance has been noted. Most recessive mutants fall into two complementation groups: cytosine deaminase and UMP-pyrophosphorylase (Whelan *et al.,* 1986). Recently, dominant resistant mutants have also been identified (Goshorn and Scherer, 1989).

Like many medically important fungi, *C. albicans* is dimorphic, that is, it can grow in either a yeast or mycelial form in response to environmental conditions. A number of mutants have been isolated with defects in the dimorphic transition. These mutants also appear to be avirulent, but they have not been as extensively analyzed as the proteinase mutants. The ability of *Candida* to vary its colony morphology has also been the subject of recent genetic analysis. In culture, some isolates will produce a wrinkled phenotype from a smooth parent; one strain will undergo a reversible transition from white to opaque colonies. The abilities to undergo the "white–opaque" transition and to produce wrinkled colonies are both recessive to wild-type. The white–opaque switching strain has a second dominant genetic change in addition to the recessive mutation (W.-S. Chu, personal communication). The mutations that lead to wrinkled colonies are known to fall into at least two complementation groups (Gil *et al.,* 1988).

Parasexual fusion products vary in their genetic composition. A number of laboratories noted variability in the ability of individual spheroplast fusion products to segregate parental markers. A systematic way to examine this process is to identify directly fusants that display recessive parental markers (termed "exceptions") (Goshorn and Scherer, 1989). Exceptions generally arise when both chromosomes with the wild-type copies from one parent are absent, leaving only the corresponding pair from the other parent for that linkage group. Most fusion products are heterozygous with respect to parental chromosomes for any given unselected linkage group; however, 5–10% lack both homologues of at least one chromosome from one or the other parent. Extensive genetic analysis indicates that the exceptions arise by unidirectional transfer of a subset of chromosomes from one nucleus to the other during fusion. The exceptional fusion products have ploidies ranging from near diploid to near tetraploid. Different linkage groups are absent at comparable frequencies. In a single exceptional fusant, there is a high probability that both homologues of more than one linkage group from one of the parents will be absent. Exceptions lacking genetic information from both parents are extremely rare. It should be possible to use the exceptional fusion products to detect linkage in one step rather than through spheroplast fusion and heat-shock and uv analysis. The exceptions detect linkage about 10 times more easily than the other two approaches (Goshorn and Scherer, 1989). The physical basis of the exceptions can be demonstrated using the extensive chromosome length polymorphism seen when different *Candida* isolates are compared.

Similar approaches have proven effective with other *Candida* species. Natural heterozygosity has been demonstrated in both *C. parapsilosis* and *C. krusei* (Whelan and Kwon-Chung, 1988). *C. tropicalis* has been the subject of genetic investigations (Corner and Poulter, 1988). It has been possible to enrich for auxotrophs in *C. tropicalis* using inositol-less death in a manner similar to that used with *S. cerevisiae.* Interspecific complementation in fused protoplasts between *C. tropicalis* and *C. albicans* has been useful for assignment of biochemical defects in mutants (Corner and Poulter, 1989). The closest relative to *C. albicans* is *C. stellatoidea* (Kwon-Chung *et al.,* 1989). Whereas some *C. stellatoidea* strains appear to be trivial variants of *C. albicans* (the former cannot utilize sucrose as a carbon source), others display numerous genetic differences. With the high degree of sequence conservation between the species, it is not surprising that interspecific complementation analysis is quite straightforward. *C. stellatoidea* has several small chromosomes in addition to the standard-sized set found in *C. albicans,* as well as rearrangements in the large chromosomes (Kwon-Chung *et al.,* 1989). The extensive chromosomal changes include duplications of genes; thus linkage analysis is greatly complicated. Even with these difficulties, it has been possible to demonstrate that at least one of the genes duplicated in the small *C. stellatoidea* chromosomes is functional (Rikkerink *et al.,* 1990).

C. Physical Characterization of the Genome

1. Electrophoretic Karyotype

In higher eukaryotes, karyotypes were traditionally determined by cytology. The inability to visualize the chromosomes of yeasts led to the determination of chromosome number by dense genetic analysis. The lack of a sexual cycle and the limited number of mutants in *Candida albicans* delayed determination of chromosomal number until parasexual genetic techniques and molecular technology improved. When pulsed-field gel electrophoresis was developed, *C. albicans* was one of the first organisms studied, since its large chromosomes (1200–3000 kb) were a useful model system for developing separation technology for the much larger mammalian chromosomes (Snell and Wilkins, 1986; Snell *et al.*, 1987; Lott *et al.*, 1987; Magee and Magee, 1987; Vollrath and Davis, 1987; Merz *et al.*, 1988). Two recent articles discuss the number and variability of *Candida* chromosomes (Magee *et al.*, 1988a,b; Lasker *et al.*, 1989). Between 7 and 8 pairs of chromosomes can be resolved by OFAGE (orthogonal-field agarose gel electrophoresis), FIGE (field-inversion gel electrophoresis), and CHEF (clamped homogeneous electric field) techniques. (See Skinner *et al.*, Chapter 4, for a more detailed discussion of pulsed-field gel electrophoresis). Different investigators have reported different haploid chromosome numbers for *C. albicans.* These discrepancies are caused by differences in the electrophoretic separation technology used and by the natural heterogeneity of *C. albicans* strains. Two bands of similar size on these gels can either be two different chromosomes or two separated homologues of the same chromosome. The larger chromosomes are the most difficult to resolve. Using molecular probes, chromosome length polymorphisms, and parasexual analysis to rule out coincidentally sized chromosomes, the most likely result is that *C. albicans* strains have eight chromosomes (Lasker *et al.*, 1989; Scherer and Magee, 1990). In addition to chromosome length polymorphism, strains maintained in the lab vary in ploidy and undergo chromosome rearrangements that appear to be related to morphological alterations (Suzuki *et al.*, 1989; Rustchenko-Bulgac *et al.*, 1990). Thus, for any natural isolate, the physical map may vary from the better studied reference strains. The locations of known markers by DNA hybridization in strain FC18 are shown in Table I.

2. *Mitochondrial Genome*

The mitochondrial genome of *Candida albicans* has been studied in detail by Riggsby and co-workers (1990). Electron microscopy and restriction mapping has indicated a size of 40 kb, which is about half the size of the *Saccharomyces cerevisiae* mitochondrial genome. Based on buoyant density analysis, the base composition is 38.2% GC. The genome contains a 5-kb inverted repeat; restriction analysis indicates a pattern consistent with recombination between the inverted repeats to yield a heterogeneous population containing the nonrepeated regions in both orientations with respect to one another. The locations of sequences encoding the large and

TABLE I.

Assignment of Genes to the Karyotype of *Candida albicans*[a]

Group	Size (kb)	Genes
R	3000	ADE1 CDC10 MGL1 RDN1 (ARG4 INO MET URA)
1	3000	ACT1 CDC3 GAL LYS2 TRP1 TUB2 (SER)
2	2600	ERG7 HEM3 HIS3 HPT1 PRA1 RAS1 TMP1
3	2200	ADE2 ILV2 SOR2 URA3 (MET)
4	1700	LYS1 (PRO HIS ILV)
5	1200	CAG1 ERG16
6	1070	ARS1 BEN1
7	1020	ARG57 DFR1 LEU2

[a] The chromosome sizes vary with individual strains. Group R (for rDNA-containing) is not readily resolved from group 1 in strain 1006 on CHEF gels. When different strains are examined or different electrophoretic separation methods are used, group R may be larger or smaller than group 1. Underlined genes indicate loci for which there are corresponding mutant strains (spontaneous, mutagen-induced, or disrupted by transformation) that can be complemented by DNA transformation. Additional auxotrophs with biochemical defects listed have been mapped by parasexual methods to the cloned markers. Other gene designations conform where possible to that used for *S. cerevisiae.* Data from reviews by Kurtz *et al.,* 1990; Poulter, 1990; Scherer and Magee, 1990, and personal communications from B. Wickes, K. J. Kwon-Chung, E. Rustchenko-Bulgac, and B. B. Magee.

small ribosomal genes as well as those specifying products corresponding to known mitochondrial proteins have been assigned by mapping with homologous probes from *S. cerevisiae.* A detailed map is provided in a recent review (Riggsby, 1990).

3. Telomeres and Centromeres

Putative telomeric sequences have been isolated as middle repetitive elements. They are found on every chromosome as judged by hybridization to electrophoretically separated *Candida* chromosomes (J. Hicks, personal communication). DNA sequencing has confirmed that the telomere has the basic structure of telomeres found in other organisms (J. Hicks, personal communication). Like their counterparts in *Saccharomyces cerevisiae* an autonomously replicating sequence (ARS) is found in close association with the telomeric element. To date, there have been no reports of isolation or characterization of centromeric sequences.

D. Molecular Methods

1. Gene Isolation

Gene isolation from *Candida* species relies on the same variety of techniques used for other yeasts, fungi, and higher eukaryotes. Each method has distinct advantages and clear limitations. Cloning of *C. albicans* genes by expression in a heterologous fungus has been successful for a variety of soluble and membrane-bound biosynthetic enzymes and structural components. At this time, cloning *C. albicans* genes by expression has been reported in the heterologous hosts *Saccharomyces cerevisiae, Schizosaccharomyces pombe,* and *Escherichia coli* (Kurtz *et al.,* 1988; Scherer and Magee, 1990; J. K. Bhattacharajee, personal communication). Expression in *E. coli* in simple vectors implies fortuitous bacterial promoters located in or adjacent to the cloned gene region and a lack of introns. *C. albicans* genes that have been expressed in *E. coli* are *URA3* (Gillum *et al.,* 1984), *LEU2* (Kelly *et al.,* 1988), and *DFR1* (Kurtz *et al.,* 1987). Although there have been no reports of cloning *C. albicans* genes by expression of a cDNA library in a lambda cloning vector such as lambda GT11, it should be possible to detect desired clones by the production of antigen recognizable by the appropriate antibodies.

The high degree of success for expression of *C. albicans* genes in *S. cerevisiae* suggests similarities between the two organisms with respect to transcription, translation, and splicing signals. There have been no reports of success (or failure) of expression of *Candida* genes in filamentous fungi such as *Aspergillus nidulans* or *Neurospora crassa.* Cloning by expression can be achieved by complementation of known mutants for which a phenotype or assayable function exists. Restoration of wild-type phenotype to auxotrophs or temperature-sensitive mutants is clearly the most straightforward approach. Overexpression of some genes may lead to dominant resistance to inhibitors of the gene product and has been used to clone such genes as the HMG Co A reductase in *S. cerevisiae* (Rine *et al.,* 1983). This method has not yet been reported for *C. albicans,* but the fact that overexpression of the dihydrofolate reductase gene in *Candida* produces resistance to the combination of methotrexate and sulfanilamide indicates that the method could be applied to homologous cloning (Kurtz *et al.,* 1987). If the gene product functions analogously in a heterologous system, and the transcription and translation signals are efficiently used, resistance cloning may also be possible in *S. cerevisiae* and other fungi. With improved cloning vehicles and transformation frequencies available, we may expect to see more examples of cloning by complementation directly in *C. albicans.* Recently, clones complementing the lysine auxotrophy of strain 1006 have been isolated by directly selecting lysine prototrophs after transformation (Goshorn *et al.*, 1990). Genetic and biochemical data show that the cloned gene and the original mutant are from the *LYS1* locus, using the *S. cerevisiae* nomenclature. A clone complementing an arginine auxotrophy was also isolated by transformation of this strain. It should be noted that isolation of some genes may require a selectable marker for the DNA library in the recipient strain in addition to the mutation or phenotype to be cloned. Parent strains for mutant isolation should, therefore, be carefully selected to permit future manipulation, either by the parasexual cycle or by DNA transformation.

Cloning by cross-hybridization and related methods provides alternative approaches to isolating genes of interest in *C. albicans*. In cases in which no mutants for cloning via complementation are available or if the *Candida* gene is not expressed in the recipient (overexpression may be lethal and single-copy expression may not complement the mutation), cross-hybridization has proven very useful for isolating the desired genes. The successful examples in the literature—actin (Mason *et al.*, 1987), tubulin (Smith *et al.*, 1988), rDNA (Magee *et al.*, 1987; Whelan, cited in Kurtz *et al.*, 1990), EF-1α (Sundstrom *et al.*, 1990), and lanosterol demethylase (Kirsch *et al.*, 1988)—are primarily genes for highly conserved structural components. Since many *Candida* genes do not demonstrate a high degree of homology to other fungi at the DNA level, it is likely that this approach will be successful only for genes that encode products that are structural elements, require close interaction with such elements, or have strong pressure for evolutionary conservation. Supporting this view is the recent isolation of the genes for a number of membrane-associated proteins, as well as members of the G protein class, by cross-hybridization (J. Hicks, personal communication). The examples described have used heterologous probes from conserved regions of the genes involved and subsequent isolation from either plasmid or bacteriophage lambda libraries. Polymerase chain reaction (PCR) methodology could also be used to retrieve homologous DNA directly from genomic DNA preparations. This approach should be used with caution, however, since the PCR chemistry can introduce sequence errors at a rate of almost 1/1000 bp. An alternative to the sequence homology approach could be used to clone the gene specifying a purified protein if part of the amino acid sequence is known. A set of oligonucleotide sequences deduced from the amino acid sequence can be generated for use as hybridization probes to isolate the gene.

Cloning via expression and cross-hybridization has resulted in the isolation of a number of genes that are of interest for developing gene manipulation techniques, understanding the physiology of the organism, and exploring avenues for medical treatment. A recent review discusses the details for isolation of specific genes. Also, all cloning techniques except the PCR methods require genomic or cDNA libraries, and the characteristics of these libraries have been reviewed (Kurtz *et al.*, 1990). To date, there are several bacteriophage libraries that are particularly useful for cross-hybridization experiments and expression in *E. coli*. Current plasmid libraries for heterologous expression are *E. coli–S. cerevisiae* shuttle vectors. These vectors are either high-copy-number vectors in yeast by virtue of the 2 μ replicon or single copy in yeast due to a centromere-containing construction. Libraries prepared in vectors that replicate autonomously in *C. albicans* are now available and should be useful for cloning by complementation directly in *Candida* (Kelly *et al.*, 1988; Goshorn *et al.*, 1990).

2. *Transformation Methods*

The best-studied method for introducing DNA into *Candida albicans* is the procedure that has been developed for *Saccharomyces cerevisiae* (Kurtz *et al.*, 1986). DNA is introduced into enzymatically produced spheroplasts in the presence of calcium and the fusogen polyethylene glycol. Treated spheroplasts are regenerated in osmotically supported selective medium. Unlike *S. cerevisiae* spheroplasts, which regenerate cell walls best within an osmotically supported agar suspension, *C. albicans* spheroplasts can regenerate cell walls efficiently on the surface of osmotically supplemented media. Like other yeasts, *C. albicans* strains vary in their transformability, but at this time all strains have been capable of transformation. Some of the host, vector, and environmental factors influencing transformation frequency have been reviewed recently (Kurtz *et al.*, 1990). Alternative transformation methods that eliminate the need for spheroplasting and the use of PEG may be especially valuable for *Candida,* to avoid cell fusion and the resulting increase in ploidy (Harashima *et al.*, 1984). The lack of a sexual cycle makes it difficult to resolve effects of polyploidy (or aneuploidy) and transformation. The only alternative to spheroplasting reported to date has been the use of cation-treated whole cells (J. D. Aitchison, personal communication; M. B. Kurtz, unpublished observations). The method, essentially the same as the *S. cerevisiae* lithium acetate procedure (Ito *et al.*, 1983), requires vigorously growing cells, removal of PEG after treatment, and a period of recovery (30 min) in fully supplemented medium before plating for increased yield. This method reduces the transformation frequency greater than 100-fold over the spheroplasting method, but avoids wall removal and cell fusion. Refrigerated treated cells can be transformed as many as 3 days after preparation without loss of efficiency. The utility of electroporation for transforming *Candida* species has not been reported.

3. Integrative Transformation

Integrative transformation in *Candida albicans,* as in most fungi, is a low-frequency event (0.5 – 5 transformants/μg), and therefore requires a good selective marker for detection. Currently, a number of host – vector systems suitable for detecting integrative transformation are available. The utility of each system has been reviewed in detail (Kurtz *et al.,* 1990; Scherer and Magee, 1990), but it is worth remarking on the unusual features of several of the systems. The *ADE2* system offers a convenient color marker that has been exploited both for *C. albicans* and *Saccharomyces cerevisiae.* The *ade2* adenine auxotrophs are red, whereas the prototrophs are white. Although the auxotrophs cannot be selected directly, they can be detected easily in a lawn of prototrophs in appropriate media. Unfortunately, no nonreverting *ade2* mutants have been described. The *URA3* system, on the other hand, offers a nonreverting mutant (Kelly *et al.,* 1987) and a recently sequenced selectable gene (Losberger and Ernst, 1989). However, *ura3* mutants tend to grow slowly.

In the absence of an ARS on transformation vectors, DNA introduced into *C. albicans* is generally incorporated into the genome by homologous recombination (Kurtz *et al.,* 1986). The integration generates either a simple gene replacement (indistinguishable from a reversion) or a direct repeat of the homologous region separated by vector sequences. Using incorporation of vector sequences as a measure, the frequency of integration of the entire plasmid is at least 50%. The frequency of integration at the homologous site by plasmid DNA can be enhanced by using restriction enzymes to introduce recombinogenic ends within the region of chromosomal homology. This feature of transformation has been exploited to generate specific gene disruptions (discussed subsequently). Nonhomologous recombination can occur and has been detected when little or no homologous DNA was present on the transforming DNA (Kurtz *et al.,* 1987). The foreign DNA introduced by integrative transformation is quite stable and recovery of vector sequence by selective restriction of DNA from transformants can be used for chromosome walking.

4. Autonomously Replicating Sequences

ARSs have been sought for *Candida albicans* to overcome the low transformation frequency and low copy number of integrative transformation vectors. The first described ARS was isolated based its instability, high transformation frequency (10^3 transformants/μg DNA), and high copy number (> 10/genome) in *Candida* (Kurtz *et al.,* 1987). This ARS, designated CARS1, was isolated as a 0.35-kb *Rsa*I fragment and is present in a single copy in the *C. albicans* genome. It is unable to promote autonomous replication in *Saccharomyces cerevisiae* and has the unexpected property of propagating as a head-to-tail multimeric plasmid. Several convenient *C. albicans–Escherichia coli* shuttle cloning vectors have been engineered with this ARS fragment, and libraries have been constructed that are useful for cloning by complementation in *Candida* (Kelly *et al.,* 1988; Goshorn *et al.,* 1990). Recently, a fragment (CaARS) was isolated from *C. albicans* that replicates autonomously in both *C. albicans* and *S. cerevisiae,* the first report of an element capable of replication in both organisms (Cannon *et al.,* 1990). A plasmid constructed with the ARS and both the *C. albicans URA3* and *LEU2* genes (pRC3920) transforms a *C. albicans* strain *(ura3)* to Ura^+ at high frequency (2.1×10^3 transformants/μg) using the *URA3* marker. *S. cerevisiae* can be transformed to Leu^+ at a similar frequency (1.91×10^3 transformants/μg DNA). Sequence analysis has shown that the fragment contains two 11 bp ARS core consensus sequences (5′TTTTATGTTTT3′) that match the consensus of ARS cores from other yeast species. The sequence is located in the nontranscribed region of the putative 5 S rRNA. *Candida* transformations with CaARS yield large and small colony types. Southern blot analysis suggests that the large colonies, which have plasmid sequences in high-molecular-weight DNA, are integrative transformants, whereas the small colonies, which have low-molecular-weight sequences, have extrachromosomally replicating DNA. A convenient cloning vector has been constructed with this *Candida* ARS (pRC2312). Initial observations with this plasmid in *S. cerevisiae* estimate 11–18 copies per haploid genome but only 2–3 copies per *Candida* genome (R. D. Cannon, personal communication). A discussion of other sources of autonomously replicating sequences is found in a recent publication (Kurtz *et al.,* 1990).

The first ARSs described for *C. albicans* provide vectors with increased transformation efficiency by several orders of magnitude over integrative vectors, and at least one with increased copy number. The resulting increased gene expression by virtue of increased gene dosage may allow isolation and characterization of rare gene products. Genes can now be isolated more easily from *C. albicans* because intact plasmids can be recovered directly from transformants. Finally, these vectors can be used in co-transformation experiments, which have been extremely useful for gene disruption strategies (Kelly *et al.,* 1988). Co-transformation methods rely on the observation that the relatively rare spheroplasts capable of taking up exogenous DNA can frequently incorporate more than one molecule.

Selection for transformants with the ARS marker results in a high proportion of cells with the unselected second DNA. The second DNA, if it is a restriction fragment, can integrate by homologous recombination or, if it is a plasmid, can recombine with the ARS vector to give a useful recoverable plasmid. In the former case, "curing" the transformants of the ARS (by growth on nonselective medium) leads to restoration of the marker for later transformation, but retains the integrated DNA (see Section II,D,5 for examples).

5. Gene Disruption

A number of strategies can now be employed to disrupt genes in *Candida albicans.* All strategies require that the gene has been cloned and that a suitably marked host strain is available. Unlike the methods used for *Saccharomyces cerevisiae,* all the strategies must take into consideration the diploid state of *C. albicans.* Both homologues of the gene must be mutated to detect the alteration (unless dominant) and the mutants must have a viable phenotype under at least one set of recovery conditions. Lethal gene disruptions in *S. cerevisiae* can be detected as inviable spores from a heterozygous diploid, an impossibility for *C. albicans.* However, conditional lethal mutations in essential genes could be obtained from engineered heterozygous diploids by conventional mutagenesis.

A simple "one-step" gene replacement technique adapted from *S. cerevisiae* methodology can be used to isolate mutants if the phenotype of the mutant is known in advance. For example, a heterozygous gene insertion of the *ADE2* gene at the *URA3* locus was constructed and the resulting strain was treated with uv to induce mitotic recombination. A screen of uracil-requiring survivors produced uracil auxotrophs at a high frequency (1/300; Kelly *et al.,* 1987). Although this approach should be useful for many gene disruptions in which the loss-of-function mutant is viable (by supplementation, as in this case), theoretical considerations indicate that the strategy will fail, or be difficult, if the gene is closely linked to the centromere or to a recessive lethal mutation. This explanation has been invoked to explain the failure to obtain homozygous disruptants of the *LEU2* gene by a similar strategy. Instead, a presumptive point mutant was found at low frequency (1/10,000), presumably induced by the uv treatment (Kelly *et al.,* 1988).

When the phenotype of a loss-of-function mutant is not known in advance, a direct selection for a disruption can be made by a second round of disruption using another selectable marker. This technique requires a recipient with two markers for which cloned genes are available. A number of such strains have been constructed and their properties reviewed (Kurtz *et al.,* 1990). Another variation of the sequential gene disruption technique preserves the original selectable marker in the strain while introducing mutations in the desire gene (Kelly *et al.,* 1988). This method takes advantage of the phenomenon of co-transformation. Instead of disrupting the gene of interest with the selectable marker, a piece of "foreign" DNA such as bacteriophage lambda DNA, which can be detected in a colony blot, is the disrupting sequence. Linear recombinogenic fragments are introduced along with an ARS plasmid containing the selectable marker (*URA3* in this example). The Ura$^+$ transformants are scored for the presence of the nonselected foreign DNA and the transformants are grown in uracil-containing medium to allow loss of the ARS plasmid. The DNA of the resulting Ura$^-$ colonies is analyzed by Southern blots to select those colonies with the appropriate disruption. The process can be now be repeated to retain the *ura3* marker, or the *URA3* gene can be inserted into the gene of interest and the Ura$^+$ disruptants selected directly. Other permutations of this approach are also possible.

III. *CANDIDA TROPICALIS*

The information for genetic manipulation of other *Candida* species is limited and has been recently reviewed (Rachubinski, 1990). *Candida tropicalis,* the next most prevalent pathogen after *C. albicans,* has been a model organism for the study of the biogenesis of peroxisomes, organelles responsible for the breakdown of fatty acids and the metabolism of H_2O_2. Like *C. albicans,* it is also a diploid. Although a number of *C. tropicalis* genes encoding peroxisomal proteins and two cytochrome P-450 genes have been cloned, no transformation system has been developed yet (Rachubinski, 1990). Two classes of red adenine-requiring mutants of *C. tropicalis* have been isolated and correspond to the *ADE1* and *ADE2* loci of *C. albicans* and *Saccharomyces cerevisiae* (Corner and Poulter, 1989). The complementation of these mutants by *C. albicans* in interspecific hybrids suggests that markers from the better studied organism may be useful in developing transformation systems for *C. tropicalis* (Corner and Poulter, 1989).

IV. *CRYPTOCOCCUS NEOFORMANS*

Sexual cycles are known for a number of the major human fungal pathogens, *Cryptococcus neoformans, Histoplasma capsulatum,* and many of the dermatophytic fungi. Most importantly, haploid phases are known to exist, greatly facilitating both classical and molecular genetic analysis in these systems.

Cryptococcus species have been the subject of significant molecular and genetic analyses. Of particular interest is the relationship between the serotypes, the varieties *neoformans* and *gattii*, mating type, and pathogenicity. Analysis of RFLPs in the mitochondrial DNA supports the existence of the two distinct varieties, these markers should be useful for epidemiologic studies (Varma and Kwon-Chung, 1989). This possibility is of particular interest because of the geographical distribution of the two varieties and the radically different distribution of types seen in AIDS patients.

Electrophoretic karyotypes of isolates of the two varieties of *C. neoformans* are available (Polacheck and Lebens, 1989). Initial studies suggested a relatively high degree of conservation in the karyotype in var. *gattii*, with at least 8 chromosomes larger than 580 kb. More variability was observed in variety *neoformans*, with at least 9 chromosomes in the same size range on OFAGE gels. More recent experiments with the CHEF system suggest 10–12 chromosomes, containing rDNA on one or two of the larger bands (Perfect *et al.*, 1988). The total haploid genome size is in the same range as that of *C. albicans*.

A major recent development with *C. neoformans* has been the establishment of a DNA transformation system. As described earlier for *C. albicans*, the key tools were the isolation of a well-defined auxotroph and isolation of the corresponding gene. In this case, 5-fluoroorotic acid was used to select mutants in the *URA5* gene (OMP pyrophosphorylase; Worsham and Goldman, 1989). A cDNA clone of the *URA5* gene was isolated by expression in bacteria and was subsequently used to obtain a genomic clone. The DNA was introduced into *C. neoformans* by electroporation. Both unstable autonomously replicating transformants and more stable integrated transformants were obtained (Edman and Kwon-Chung, 1990). The DNA transformation system, combined with the ability to do classical genetic analysis, makes known virulence factors susceptible to molecular genetics analysis. Significant genetic studies of virulence have been conducted on the role of the polysaccharide capsule (Fromtling *et al.*, 1983; Still and Jacobson, 1983) and melanin production (Kwon-Chung and Rhodes, 1986). The genetics of resistance to 5-fluorocytosine and canavanine have also been examined in *C. neoformans* (Polacheck and Kwon-Chung, 1986; Whelan, 1987).

V. *HISTOPLASMA CAPSULATUM*

Histoplasma capsulatum has also been examined in some detail at the molecular level (Maresca and Kobayashi, 1989). Molecular probes for epidemiologic investigations have been developed. Although histoplasmosis is initially acquired from environmental sources, there is considerable interest in using DNA probes to explore possible reservoirs for subsequent disease from calcified foci left in the lung. Some of the earliest studies in this area involved the use of mitochondrial DNA sequences as probes; however, only a small number of subtypes could be detected (Vincent *et al.,* 1986). Ribosomal DNA probes have also been used in combination with the mitochondrial sequences (Spitzer *et al.,* 1989). Significantly, strains from North America could be distinguished from South American isolates. Originally, the Downs strain, widely used in biochemical studies, was thought to be unique in its DNA type. Recently, additional strains have been identified as similar to the Downs strain.

Examination of the *H. capsulatum* electrophoretic karyotype indicates that at least seven chromosomes are present, and polymorphism is observed within the species. Using cloned DNA segments of known or unknown function, specific probes have been obtained for all of the currently resolved bands in the electrophoretic karyotype (Steele *et al.,* 1989).

H. capsulatum has one of the most extensively studied developmental systems among the fungal pathogens. Although there have been extensive biochemical studies of thermal dimorphism in *Histoplasma,* only recently have molecular genetic studies been initiated with this group. Like *Saccharomyces cerevisiae, H. capsulatum* has one gene each for α and β tubulin. Differences in transcription and posttranslational modification have been reported for tubulin when the yeast and mycelial phases are compared (Harris *et al.,* 1989). Keath and co-workers have sought genes with expression specific for the yeast phase and have studied one such gene, *yps-3,* in detail (Keath *et al.,* 1989). Although the transition to the yeast phase required elevated temperatures, expression was not simply temperature dependent. This gene does not appear to be required for conversion to the yeast phase form since it is not expressed at detectable levels in the Downs strain. Although the Downs strain is sensitive to elevated temperatures and is attentuated in virulence, it still undergoes the yeast–mycelial transition. Because of the role of temperature in the dimorphic transition, *Histoplasma* heat-shock genes have also been studied (Caruso *et al.,* 1987). The heat-shock 70 gene was differentially expressed in *H. capsulatum* strains that had different levels of thermotolerance and pathogenicity.

Although the *H. capsulatum* sexual cycle has not yet been extensively used for genetic analyses, its existence adds considerably to the overall potential of the system. For example, strain construction will be considerably more straightforward than with asexual species such as *Candida albicans.* Interpretation of the electrophoretic karyotype will also be much easier, given the prospect of genetic linkage data to complement the physical mapping data.

VI. FUTURE PROSPECTS

We expect that the techniques of molecular biology will be applied in the near future to many other human pathogens. A recent example is the development of a DNA transformation system for one of the dermatophyte fungi. Sexual cycles are known for many of these fungi. The most extensively studied member of this group is *Trichophyton mentagrophytes* (perfect form *Arthroderma benhamiae*). This species gives eight-spored unordered asci that show 4:4 segregation for mating type. *T. mentagrophytes* has been the subject of only limited classical genetic studies. Recently, a DNA transformation system has been reported for this fungus (Gonzalez *et al.,* 1989). A bacterial hygromycin resistance gene fused to a promoter from *Cochliobolus heterostrophus* was used as the selectable marker. Stable transformants were obtained with variable sites of integration for the transforming DNA. This is consistent with the results obtained in DNA transformation of other filamentous fungi. Significantly, the transformants retained virulence in animal models.

Although the important human fungal pathogens are few in number, they span the biological diversity of the fungi and are largely unrelated to one another evolutionarily: *Candida albicans* is an imperfect fungus; *C. neoformans,* a basidiomycete; and many of the dermatophytes are ascomycetes. As a result, specialized techniques have had to be developed for each of them. Future progress in the molecular genetics of human fungal pathogens depends on several factors. First are the intrinsic biological properties of each organism, for example, ploidy and knowledge of its sexual cycle. Second is the degree of relatedness of the pathogen to the intensively studied fungal model systems such as *Saccharomyces cerevisiae* and *Neurospora crassa.* Third is our existing knowledge base for each species, which ranges from our considerable data on *C. albicans* to essentially no molecular biology on important primary pathogens such as *Coccidioides immitis.* Studies on the organisms we have discussed in this chapter make use of different combinations of these factors and considerable advances have been made on each in the past few years. In the future, we can expect the methodologies described here to be applied to diverse problems in fungal pathogenesis, for example, the growth of *H. capsulatum* in macrophages, the role of morphologic variation in disease, and the role of mating type in cryptococcal infections.

ACKNOWLEDGMENTS

We would like to thank J. D. Aitchison, J. K. Bhattacharajee, R. D. Cannon, W.-S. Chu, J. B. Hicks, and E. Rustchenko-Bulgac for sharing unpublished results. Our research has been supported by USPHS grant A123850 (to S. S.) and by Merck Sharp & Dohme Research Laboratories (M. B. K.).

REFERENCES

Cannon, R. D., Jenkinson, H. F., and Shepherd, M. G. (1990). Isolation and nucleotide sequence of an autonomously replicating sequence (ARS) element functional in *Candida albicans* and *Saccharomyces cerevisiae*. *Mol. Gen. Genet.* **221,** 210–218.

Caruso, M., Sacco, M., Medoff, G., and Maresca, B. (1987). Heat shock 70 gene is differentially expressed in *Histoplasma capsulatum* strains with different levels of thermotolerance and pathogenicity. *Mol. Microbiol.* **1,** 151–158.

Corner, B. E., and Poulter, R. T. (1989). Interspecific complementation analysis by protoplast fusion *Candida tropicalis* and *Candida albicans* adenine auxotrophs. *J. Bacteriol.* **171,** 3586–3589.

Crandall, M., and Edwards, J. E. (1987). Segregation of proteinase-negative mutants from heterozygous *Candida albicans. J. Gen. Microbiol.* **133,** 2817–2824.

Edman, J. C., and Kwon-Chung, K. J. (1990). Isolation of the *URA5* gene from *Cryptococcus neoformans* var. *neoformans* and its use as a selective marker for transformation. *Mol. Cell Biol.* In press.

Fromtling, R. A., Kaplan, A. M., and Shadomy, H. J. (1983). Immunization of mice with stable, acapsular, yeast-like mutants of encapsulation and melanin formation as indicators of virulence in *Cryptococcus neoformans. Sabouraudia* **21,** 113–119.

Gil, C., Pomes, R., and Nombela, C. (1988). A complementation analysis by parasexual recombination of *Candida albicans* morphological mutants. *J. Gen. Microbiol.* **134,** 1587–1595.

Gillum, A. M., Tsay, E. Y. H., and Kirsch, D. R. (1984). Isolation of the *Candida albicans* gene for orotidine-5′-phosphate decarboxylase by complementation of *S. cerevisiae ura3* and *E. coli pyr F* mutations. *Mol. Gen. Genet.* **198,** 179–182.

Gonzalez, R., Ferrer, S., Buesa, J., and Ramon, D. (1989). Transformation of the dermatophyte *Trichophyton mentagrophytes* to hygromycin B resistance. *Infect. Immun.* **57,** 2923–2925.

Goshorn, A. K., and Scherer, S. (1989). Genetic analysis of prototrophic natural variants of *Candida albicans. Genetics* **123,** 667–673.

Goshorn, A. K., Grindle, S. M., and Scherer, S. (1990). Gene isolation by complementation in *Candida albicans. Mol. Cell. Biol.* Submitted.

Harashima, S., Takagi, A., and Oshima, Y. (1984). Transformation of protoplasted yeast cells is directly associated with cell fusion. *Mol. Cell. Biol.* **4,** 771–778.

Harris, G. S., Keath, E. J., and Medoff, J. (1989). Expression of alpha- and beta-tubulin genes during dimorphic-phase transitions of *Histoplasma capsulatum. Mol. Cell. Biol.* **99,** 2042–2049.

Hilton, C., Markie, D., Corner, B., Rikkerink, E., and Poulter, R. T. (1985). Heat shock induces chromosome loss in the yeast *Candida albicans. Mol. Gen. Genet.* **200,** 162–168.

Ito, H., Fukuda, Y., Murata, K., and Kimura, A. (1983). Transformation of intact yeast cells treated with alkali cations. *J. Bacteriol.* **153,** 163–168.

Kakar, S. N., and Magee, P. T. (1982). Genetic analysis of *Candida albicans:* Identification of different isoleucine-valine, methionine and arginine alleles by complementation. *J. Bacteriol.* **151,** 1247–1252.

Keath, E. J., Painter, A. A., Kobayashi, G. S., and Medoff, G. (1989). Variable expression of a yeast-phase-specific gene in *Histoplasma capsulatum* strains differing in thermotolerance and virulence. *Infect. Immun.* **57,** 1384–1390.

Kelly, R., Miller, S. M., Kurtz, M. B., and Kirsch, D. R. (1987). Directed mutagenesis in *Candida albicans:* One-step gene disruption to isolate *ura3* mutants. *Mol. Cell. Biol.* **7,** 199–208.

Kelly, R., Miller, S. M., and Kurtz, M. B. (1988). One-step gene disruption by cotransformation to isolate double auxotrophs in *Candida albicans. Mol. Gen. Genet.* **214,** 24–31.

Kirsch, D. R., Lai, M. H., and O'Sullivan, J. (1988). Isolation of the gene for cytochrome P450 L1A1 (lanosterol 14 α-demethylase) from *Candida albicans. Gene* **68,** 229–237.

Kurtz, M. B., Cortelyou, M. W., and Kirsch, D. R. (1986). Integrative transformation of *Candida albicans,* using a cloned *Candida albicans ADE2* gene. *Mol. Cell. Biol.* **6,** 142–149.

Kurtz, M. B., Cortelyou, M. W., Miller, S. M., Lai, M., and Kirsch, D. R. (1987). Development of autonomously replicating plasmids for *Candida albicans. Mol. Cell. Biol.* **7,** 209–217.

Kurtz, M. B., Kirsch, D. R., and Kelly R. (1988). The molecular genetics of *Candida albicans. Microbiol. Sci.* **5,** 58–63.

Kurtz, M. B., Kelly, R., and Kirsch, D. R. (1990). Molecular genetics of *Candida albicans. In* "Genetics of *Candida*" (D. R. Kirsch, R. Kelly, and M. B. Kurtz, eds.), pp. 21–73. CRC Press, Boca Raton, Florida.

Kwon-Chung, K. J., and Rhodes, J. C. (1986). Encapsulation and melanin formation as indicators of virulence in *Cryptococcus neoformans. Infect. Immun.* **51,** 218–223.

Kwon-Chung, K. J., Lehman, D., Good, C., and Magee, P. T. (1985). Genetic evidence for role of extracellular proteinase in virulence of *Candida albicans. Infect Immun.* **49,** 571–578.

Kwon-Chung, K. J., Riggsby, W. S., Uphoff, R. A., Hicks, J. B., Whelan, W. L., Reiss, E., Magee, B. B., and Wickes, B. L. (1989). Genetic differences between type I and II *Candida stellatoidea. Infect. Immun.* **57,** 527–532.

Lasker, B. A., Carle, G. F., Kobayashi, G. S., and Medoff, G. (1989). Comparison of the separation of *Candida albicans* chromosome-sized DNA by pulsed-filed gel electrophoresis techniques. *Nucl. Acids. Res.* **17,** 3783–3793.

Losberger, C., and Ernst, J. F. (1989). Sequence and transcript analysis of the *Candida albicans URA3* gene encoding orotidine-5′-phosphate decarboxylase. *Curr. Genet.* **16,** 153–157.

Lott, T. J., Boiron, P., and Reiss, E. (1987). An electrophoretic karyotype for *Candida albicans* reveals large chromosomes in multiples. *Mol. Gen. Genet.* **209,** 170–174.

Magee, B. B., and Magee, P. T. (1987). Electrophoretic karyotypes and chromosome numbers in *Candida* species. *J. Gen. Microbiol.* **133,** 425–430.

Magee, B. B., D'Souza, T. D., and Magee, P. T. (1987). Ribosomal DNA restriction fragment length polymorphism can be used to identify species and biotypes of medically important yeasts. *J. Bacteriol.* **169,** 1639–1643.

Magee, B. B., Koltin, Y., Gorman, J., and Magee, P. T. (1988a). Assignment of cloned *Candida albicans* genes to bands on the electrophoretic karyotype. *Mol. Cell. Biol.* **8,** 4721–4726.

Magee, P. T., Rikkerink, E. H. A., and Magee, B. B. (1988b). Methods for the genetics and molecular biology of *Candida albicans. Anal. Biochem.* **175,** 361–372.

Maresca, B., and Kobayashi, G. S. (1989). Dimorphism in *Histoplasma capsulatum:* A model for the study of cell differentiation in pathogenic fungi. *Microbiol. Rev.* **53,** 186–209.

Mason, M. M., Lasker, B. A., and Riggsby, W. S. (1987). Molecular probe for identification of medically important *Candida* species and *Torulopsis glarata. J. Clin. Microbiol.* **25,** 563–566.

Merz, W. G., Connelly, C., and Hieter, P. (1988). Variation of electrophoretic karyotype among clinical isolates of *Candida albicans. J. Clin. Microbiol.* **26,** 842–845.

Olaiya, A. F., and Sogin, S. J. (1979). Ploidy determination of *Candida albicans. J. Bacteriol.* **140,** 1043–1049.

Perfect, J. R., Magee, B. B., and Magee, P. T. (1989). Separation of chromosomes of *Cryptococcus neoformans* by pulsed field gel electrophoresis. *Infect. Immun.* **57,** 2624–2627.

Polacheck, I., and Kwon-Chung, K. J. (1986). Canavanine-resistance in *Cryptococcus neoformans. Antimicrob. Agents Chemother.* **29,** 468–473.

Polacheck, I., and Lebens, G. A. (1989). Electrophoretic karyotype of the pathogenic yeast *Cryptococcus neoformans. J. Gen. Microbiol.* **135,** 65–71.

Poulter, R. T. (1987). Natural auxotrophic heterozygosity in *Candida albicans. CRC Crit. Rev. Microbiol.* **15,** 97–101.

Poulter, R. T., and Rikkerink, E. H. (1983). Genetic analysis of red, adenine-requiring mutants of *Candida albicans. J. Bacteriol.* **156,** 1066–1077.

Poulter, R. T. (1990). Classical methods for genetic analysis of *Candida albicans. In* "Genetics of *Candida*" (D. R. Kirsch, R. Kelly, and M. B. Kurtz, eds.) pp. 75–123. CRC Press, Boca Raton, Florida.

Poulter, R. T., Jeffrey, K., Hubbard, M. J., Shepherd, M. G., and Sullivan, P. A. (1981). Parasexual genetic analysis of *Candida albicans* by spheroplast fusion. *J. Bacteriol.* **146,** 833–840.

Poulter, R. T., Hanrahan, V., Jeffrey, K., Markie, D., Shepherd, M. G., and Sullivan, P. T. (1982). Recombination analysis of naturally occurring diploid *Candida albicans. J. Bacteriol.* **152,** 969–975.

Rachubinski, R. A. (1990). Genetic methods for and gene structure in other *Candida* species. *In* "Genetics of *Candida*" (D. R. Kirsch, R. Kelly, and M. B. Kurtz, eds.), pp. 177–196. CRC Press, Boca Raton, Florida.

Riggsby, S. (1990). Physical characterization of the *Candida albicans* genome. *In* "Genetics of *Candida*" (D. R. Kirsch, R. Kelly, and M. B. Kurtz, eds.), pp. 125–146. CRC Press, Boca Raton, Florida.

Rikkerink, E. H., Magee, B. B., and Magee, P. T. (1990). Extra chromosomes and gene duplication in *Candida stellatoidea. Infect. Immun.* **58,** 949–954.

Rine, J., Hansen, W., Hardeman, E., and Davis, R. W. (1983). Targeted selection of recombinant clones through gene dosage effects. *Proc. Natl. Acad. Sci. U.S.A.* **80,** 6750–6754.

Rustchenko-Bulgac, E. P., Sherman, F., and Hicks, J. B. (1990). Chromosomal rearrangements associated with morphological mutants provide a means for genetic variation of *Candida albicans. J. Bacteriol.* **172,** 1276–1283.

Sarachek, A., Rhoads, D. D., and Schwartzhoff, R. H. (1981). Effects of growth temperatures on plating efficiencies and stabilities of heterokaryons of *Candida albicans. Mycopathologia* **83,** 1–8.

Scherer, S., and Magee, P. T. (1990). Genetics of *Candida. Microbiol. Rev.* **54,** 226–241.

Smith, H. A., Allaudeen, H. S., Whitman, M. H., Koltin, Y., and Gorman, J. A. (1988). Isolation and characterization of a β-tubulin gene from *Candida albicans. Gene* **63,** 53–63.

Snell, R. G., and Wilkins, R. J. (1986). Separation of chromosomal DNA molecules from *C. albicans* by pulsed field gel electrophoresis. *Nucl. Acids Res.* **14,** 4401–4406.

Snell, R. G., Hermans, I. F., Wilkins, R. J., and Corner, B. E. (1987). Chromosomal variations in *Candida albicans. Nucl. Acids Res.* **15,** 3625.

Spitzer, E. D., Lasker, B. A., Travis, S. J., Kobayashi, G. S., and Medoff, G. (1989). Use of mitochondrial and ribosomal DNA polymorphisms to classify clinical and soil isolates of *Histoplasma capsulatum. Infect. Immun.* **57,** 1409–1412.

Steele, P. E., Carle, G. F., Kobayashi, G. S., and Medoff, G. (1989). Electrophoretic analysis of *Histoplasma capsulatum* chromosomal DNA. *Mol. Cell. Biol.* **9,** 983–987.

Still, C. N., and Jacobson, E. S. (1983). Recombinational mapping of capsule mutations in *Cryptococcus neoformans. J. Bacteriol.* **156,** 460–462.

Sundstrom, P., Smith, D., and Sypherd, P. S. (1990). Sequence analysis and expression of the two genes for elongation factor-1α from the dimorphic yeast, *Candida albicans. J. Bacteriol.* **135,** 2036–2045.

Suzuki, T., Kobayashi, T., Kanbe, T., and Tanaka, K. (1989). High frequency variation of colony morphology and chromosome reorganization in the pathogenic yeast *Candida albicans. J. Bacteriol.* **152,** 893–896.

Varma, A., and Kwon-Chung, K. J. (1989). Restriction fragment polymorphism in mitochondrial DNA of *Cryptococcus neoformans. J. Gen. Microbiol.* **135,** 3353–3362.

Vincent, R. D., Goewert, R., Goldman, W. E., Kobayashi, G. S., Lambowitz, A. M., and Medoff, G. (1986). Classification of *Histoplasma capsulatum* isolates by restriction fragment polymorphisms. *J. Bacteriol.* **165,** 813–818.

Vollrath, D., and Davis, R. W. (1987). Resolution of DNA molecules greater than 5 megabases by contour-clamped homogeneous electric fields. *Nucl. Acids Res.* **15,** 7865–7876.

Whelan, W. L. (1987). The genetics of medically important fungi. *Crit. Rev. Microbiol.* **14,** 99–170.

Whelan, W. L., and Kwon-Chung, K. J. (1988). Auxotrophic heterozygosities and the ploidy of *Candida parapsilosis* and *Candida krusei. J. Med. Vet. Mycol.* **26,** 163–171.

Whelan, W. L., and Soll, D. R. (1982). Mitotic recombination in *Candida albicans:* Recessive lethal alleles linked to a gene required for methionine biosynthesis. *Mol. Gen. Genet.* **187,** 477–485.

Whelan, W. L., Partridge, R. M., and Magee, P. T. (1980). Heterozygosity and segregation in *Candida albicans. Mol. Gen. Genet.* **180,** 107–113.

Whelan, W. L., Beneke, E. S., Rogers, A. L., and Soll, D. R. (1981). Segregation of 5-fluorocytosine resistant variants by *Candida albicans. Antimicrob. Agents Chemother.* **29,** 726–729.

Whelan, W. L., Markie, D., Simpkin, K. G., and Poulter, R. M. (1985). Instability of *Candida albicans* hybrids. *J. Bacteriol.* **161,** 1131–1136.

Whelan, W. L., Markie, D., and Kwon-Chung, K. J. (1986). Complementation analysis of resistance to 5-fluorocytosine in *Candida albicans. Antimicrob. Agents Chemother.* **29,** 726–729.

Worsham, P. L., and Goldman, W. E. (1988). Selection and characterization of *ura5* mutants of *Histoplasma capsulatum. Mol. Gen. Genet.* **214,** 348–352.

17

Molecular Biology of the Beta-Lactam-Producing Fungi

PAUL L. SKATRUD
Department of Molecular Genetics
Lilly Research Laboratories
Indianapolis, Indiana

I. INTRODUCTION

In the past 6 years, the application of molecular biology to the filamentous fungi *Cephalosporium acremonium* (syn. *Acremonium chrysogenum*) and *Penicillium chrysogenum* has progressed at an incredibly rapid pace. This advance has provided new insight into the biosynthesis of beta-lactam antibiotics and the overall genetic architecture of these industrially important microorganisms. The unique tools of molecular biology have allowed a rational approach to strain improvement that has resulted in improved

MORE GENE MANIPULATIONS
IN FUNGI

productivity, advances toward altered biosynthetic pathways, and formation of novel beta-lactam compounds. An evolutionary hypothesis has been developed that suggests that beta-lactam genes were originally present in prokaryotes and were transferred horizontally to eukaryotes. This chapter will provide a brief historical review of beta-lactams, followed by a focus on significant events in molecular genetic studies of these organisms.

A. Comments on the History of Beta-Lactam Antibiotics

Of all the numerous significant advances made in the history of the practice of clinical medicine, the use of beta-lactam antibiotics in the treatment of infectious disease may represent the single most important advance. Weiner (1986) stated a similar sentiment: "There is perhaps no better example of the accomplishments of human intelligence in the control of man's natural enemies than that of the development of antibacterial agents in general and the beta-lactams in particular." McDermott and Rogers (1982) noted that the introduction of antibiotics to medicinal practice has added about 10 years to the life expectancy of the average American. In comparison, if all forms of cancer could be cured, this would add approximately 2 years to our life span. The role that beta-lactams have played in the elimination of human suffering and death is difficult to exaggerate.

Alexander Fleming was credited with discovering penicillin. He observed growth inhibition of *Staphylococcus aureus* surrounding a contaminant in a Petri plate. The contaminant was identified as *Penicillium notatum* (Fleming, 1929). However, as Queener pointed out (Queener, 1986), the conceptual framework that allowed Fleming and others to understand the significance of this antagonistic response was developed during several decades prior to the now-famous "contaminated experiment." As early as 1877, investigators had recognized potential therapeutic value of such growth inhibition (Pasteur and Joubert, 1877), yet it was more than a decade after Fleming's observation that the clinical efficacy of penicillin was demonstrated (Abraham *et al.,* 1941). The initial clinical trial of penicillin failed because the available supply of partially purified penicillin was inadequate to complete therapy. In contrast, today enough highly purified penicillin and semi-synthetic penicillins are produced worldwide every year to provide at least one round of therapy for every individual on this planet (Lowe, 1986). Manufacture of beta-lactam antibiotics is now a multinational, multibillion dollar industry. Despite an ever-growing array of antibiotics and the incidence of penicillin-resistant clinical isolates, the beta-lactam antibiotics are still by far the most frequently utilized.

Another chapter in the history of beta-lactam antibiotics was opened when Abraham and Newton described the structure of cephalosporin C, a novel beta-lactam antibiotic produced by *Cephalosporium acremonium* (Abraham and Newton, 1961). The discovery of cephalosporin C generated a whole new group of clinically significant beta-lactams. These beta-lactams have significantly extended the utility of this family of antibiotics in terms of antimicrobial spectrum and resistance to penicillinases produced by many penicillin-resistant pathogens. The family of beta-lactam compounds has continued to expand with the discovery of the carbapenems, cephamycins, and monobactams.

Another important chapter in the history of the beta-lactams is currently unfolding. Molecular biology has provided tools, heretofore unavailable in this field, that will allow issues of a broader scope to be addressed with a greater degree of specificity. The use of beta-lactam compounds may expand to activities other than antibacterial as the growing awareness of their therapeutic value leads scientists down new avenues in research.

B. Structure, Activity, and Biosynthesis

Beta-lactam antibiotics can be divided into two general catagories, penicillins and cephalosporins. The basic structural difference between penicillin and cephalosporin compounds is in their respective bicyclic ring systems. All beta-lactam compounds contain a four-membered azetidinone ring (beta-lactam ring), illustrated in the inset of Fig. 1. The four-membered beta-lactam ring is fused with a five-membered thiazolidine ring in penicillin and a six-membered dihydrothiazine ring in cephalosporin C and cephamycin C. All beta-lactams kill bacteria by the same process: inhibition of enzymes involved in cell wall biosynthesis. Beta-lactams are relatively nontoxic compounds because mammalian cells do not possess the same cell wall synthesizing enzymes. This lack of toxicity, coupled with their extremely potent antibacterial activity, results in a high therapeutic index compared with other clinically used antibiotics.

The branched biosynthetic pathway that leads to the production of penicillin G, cephalosporin C, and cephamycin C has been described (Fig. 1) (Queener and Neuss, 1982; Jensen, 1985). Initial enzymatic reactions are the same for the synthesis of all naturally occurring beta-lactams, whereas later steps are specific for penicillins, cephalosporin C, or cephamycin C. The nomenclature describing beta-lactam biosynthetic genes reflects the branched nature of the pathway (Ingolia and Queener, 1989). The first two genes involved in penicillin and cephalosporin biosynthesis were referred to as *pcb* genes. The abbreviation *pen* described the genes which were involved only in penicillin biosynthesis. Those genes involved in cephalosporin biosynthesis were designated *cef*. The original nomenclature

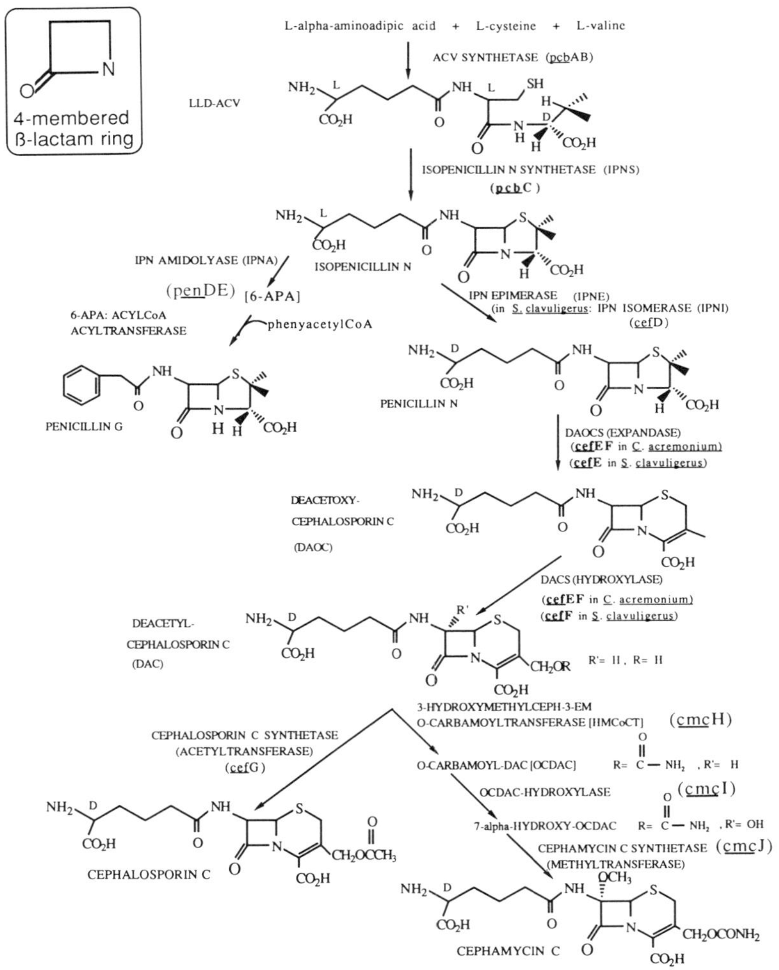

Fig. 1. Branched beta-lactam biosynthetic pathway leading to the production of penicillin G, cephalosporin C, and cephamycin C.

was recently modified to accommodate more fully the rationale used to establish this nomenclature (Skatrud *et al.*, 1991). Those genes involved in cephamycin C synthesis had initially been included with the *cef* genes. Now, *cef*H, *cef*I, and *cef*J are renamed *cmc*H, *cmc*I, and *cmc*J, consistent with the suggestion of Chen and associates (1988). This revised nomenclature recognizes all branches of the pathway that produce significantly different end-products.

C. Beta-Lactam Production by Filamentous Fungi

Many microorganisms produce beta-lactams, including representatives of the lower eukaryotes and prokaryotes. Studies related to penicillin and cephalosporin C production in the filamentous fungus *Aspergillus nidulans* and the gram positive bacterium *Streptomyces clavuligerus* have added much to our knowledge of the biosynthesis of beta-lactams and the evolution of the pathway. However, only *Penicillium chrysogenum* and *Cephalospurium acremonium* are used in large scale industrial fermentations to produce penicillin and cephalosporin C, respectively.

P. chrysogenum and *C. acremonium* are both representatives of the fungi imperfecti (deuteromycetes), that is, no sexual phase has been observed in these organisms. Due to this lack of meiosis, classical Mendelian hybridization studies are impossible. Consequently, genetic manipulation of these organisms has been difficult and unpredictable.

In the past, genetic studies of *P. chrysogenum* and *C. acremonium* were directed mainly at increasing the yield of antibiotic. The initial isolates from nature produced industrially insignificant quantities of antibiotic. During the last five decades, extensive mutagenesis, natural selections, and more recently protoplast fusion (Hamlyn and Ball, 1979), coupled with massive screening programs, have produced new strains with tremendously improved capacities for production of penicillin and cephalosporin C. In more recent years, the success of such programs has been less dramatic, perhaps because mutations that result in improved productivity are less frequent in these already heavily mutagenized strains. In addition to declining success, these methods are limited in scope, and they do not promise production of new or modified antibiotics. Further advances in strain improvement or strain alteration to produce new products in fermentation required us to move away from the empirical approaches of the past and embrace rational approaches based on our growing knowledge of biochemistry and genetics. The first step in the application of recombinant DNA technology to the study of beta-lactam producing organisms was the development of transformation systems.

II. APPLICATION OF RECOMBINANT DNA TECHNOLOGY

A. Genetic Transformation Systems

The transformation system described for *Aspergillus nidulans* (Yelton *et al.*, 1984) was the first for a beta-lactam producing filamentous fungus. The first industrially significant beta-lactam producing filamentous fungus to be transformed was *Cephalosporium acremonium* (Queener *et al.*, 1984,

1985; Penalva *et al.,* 1985). This event was followed approximately 3 years later with the transformation of *Penicillium chrysogenum* (Beri and Turner, 1987; Cantoral *et al.,* 1987; Sanchez *et al.,* 1987; Skatrud *et al.,* 1987a).

The method adopted for transformation of *C. acremonium* and *P. chrysogenum* was initially devised for the yeast *Saccharomyces cerevisiae* (Hinnen *et al.,* 1978), and typically involves the following steps: (1) formation of protoplasts by enzymatic removal of the fungal cell wall in the presence of an osmotic stabilizer; (2) exposure of osmotically stabilized washed protoplasts to transforming DNA in the presence of polyethylene glycol and calcium ions; and (3) regeneration of protoplasts in the presence of osmotically stabilized medium containing a selective agent permitting survival of transformants only.

1. Transformation of **Cephalosporium acremonium**

The use of dominant selectable markers is a key issue in the transformation of industrial organisms. Such markers allow one to transform an organism without first inducing a specific mutation. Resistance to antibiotics not used in human medicine is a particularly attractive choice because strains produced by transformation may some day be used in large scale industrial fermentations. The aminocyclitol antibiotic hygromycin B, which is not used in humans, has become the most frequently used antibiotic resistance marker in fungal transformation systems. The initial use of hygromycin B in a filamentous fungus was for transformation of *Cephalosporium acremonium* (described subsequently). Other dominant selectable markers used for transformation of *C. acremonium* include G418 resistance (Penalva *et al.,* 1985) and phleomycin resistance (P. L. Skatrud, unpublished results).

In the initial transformation system described for *C. acremonium,* less than one transformant per μg of transforming DNA was recovered (Queener *et al.,* 1984). A bacterial gene encoding hygromycin B phosphotransferase (HPT) fused in frame with the phosphoglycerate kinase (PGK) promoter of yeast was used to confer hygromycin resistance to these transformants (Queener *et al.,* 1985). However, the extremely low frequency of transformation indicated that the yeast promoter was nonfunctional in *C. acremonium.* Successful transformation probably occurred by integration of the transforming DNA adjacent to endogenous expression signals (Skatrud *et al.,* 1987b).

The first *C. acremonium* promoter became available when the *pcbC* gene from *C. acremonium* was cloned and characterized (Samson *et al.,* 1985). The *pcbC* promoter region was attached in-frame to the HPT gene in vectors used to transform *C. acremonium.* The frequency of transformation was improved 100-fold with the new hybrid HPT gene (Skatrud *et al.*, 1986, 1987b; Chapman *et al.*, 1987). Transformation frequencies obtained with vectors containing this hybrid gene can be raised an additional 15-fold by exposure of transformed protoplasts to low-level uv irradiation just after transformation (J. Wood, personal communication). One other promoter from *C. acremonium* has been fused with the HPT gene. This promoter, from the beta-isopropyl malate dehydrogenase encoding gene *(Leu2)* of *C. acremonium,* was apparently not as active as the *pcb*C promoter. The frequency of transformation obtained with vectors containing the *LEU2* promoter/HPT hybrid gene was much less than that obtained with vectors with the *pcb*C promoter/HPT hybrid gene (P. L. Skatrud, unpublished results).

Enzymatic removal of *C. acremonium* cell walls produced protoplasts of various sizes. Harvesting a fraction enriched for larger protoplasts also enhanced transformation frequency, particularly when industrial strains of *C. acremonium* were transformed (Skatrud *et al.*, 1987b). Careful titration of the amount of hygromycin B used during selection of transformants also proved critical for transformation of industrial strains.

Other less successful attempts were made to improve the transformation frequency of *C. acremonium.* For example, a fragment of mitochondrial DNA was cloned from *C. acremonium* that conferred autonomous replication (ARS function) to plasmids in yeast (Skatrud and Queener, 1984). In yeast, vectors containing an autonomously replicating sequence (ARS) transform at a much higher frequency than integrating vectors. Unfortunately, when this ARS-containing fragment of mitochondrial DNA was included on vectors used to transform *C. acremonium,* no ARS function was detected, nor did it increase transformation frequency. In an attempt to increase the integration frequency, the ribosomal RNA genes from *C. acremonium* were cloned (Carr *et al.*, 1987) and included in transforming vectors. It was anticipated that this might increase integration frequency by providing a larger target for homologous recombination. Once again, no improvement in transformation frequency was observed (Skatrud *et al.*, 1987b).

Currently, using the HPT gene driven by the *C. acremonium pcb*C promoter as a selectable marker, approximately 10–100 rapidly growing stable transformants can be obtained per μg of transforming DNA. In addition to these transformants, another group of transformants typically appears somewhat later. This group is characterized by a slower growth rate and reduced stability. Analysis of these transformants suggests that free plasmid exists in at least some of these "abortive" transformants (P. L. Skatrud, unpublished results). A similar phenomenon has been observed in *Penicillium chrysogenum* and *Aspergillus nidulans* (Yelton *et al.*, 1984;

Sanchez *et al.*, 1987). Stable transformants of *C. acremonium* arise by integration of transforming DNA into chromosomal DNA. Usually multiple heterologous integration events are observed in each transformant, with tandem reiteration of the transforming DNA (Queener *et al.*, 1985). Approximately 3% homologous integration has been observed in gene disruption studies in *C. acremonium* (Skatrud *et al.*, 1991). Transformants are mitotically stable in the absence of selection (Skatrud *et al.*, 1986). This is an extremely important characteristic when considering application of this technology to industrial strains of *C. acremonium*.

2. *Transformation of* **Penicillium chrysogenum**

In *Penicillium chrysogenum,* similar transformation frequencies, mitotic stability of transformants, and multiple heterologous integration events have been observed (Beri and Turner, 1987; Sanchez *et al.*, 1987; Skatrud *et al.*, 1987a). Reports of homologous integration differ. Sanchez and co-workers (1987) reported 10% homologous integration at the *trpC* locus. In comparison, Picknett and colleagues (1987) reported that the majority of integration events were homologous when the *trpC* locus was used for selection; however the number of transformants examined was small in this case. Gene disruption has not been reported for *P. chrysogenum* to date.

Two types of selectable marker have been employed for transformation of *P. chrysogenum.* Auxotrophic complementation has been useful in generating transformants that complement *trpC* (Picknett *et al.*, 1987; Sanchez *et al.*, 1987) and *pyr4* (Cantoral *et al.*, 1987). However, previous studies demonstrated that mutagenic induction of auxotrophic markers frequently lowered beta-lactam productivity in *P. chrysogenum* (MacDonald *et al.*, 1963). Therefore, auxotrophic complementation is undesirable for transformation of industrial strains. In response to this situation, several dominant selectable markers have been developed for *P. chrysogenum,* including use of the acetamidase *(amdS)* gene (Beri and Turner, 1987; Skatrud *et al.*, 1987a) isolated from *Aspergillus nidulans* by Hynes and co-workers (1983), phleomycin resistance (Kolar *et al.*, 1988), sulfonamide resistance (Carramolino *et al.*, 1989), benomyl resistance (Picknett and Saunders, 1989), oligomycin resistance (Bull *et al.*, 1988), and hygromycin B resistance (Durand *et al.*, 1990).

The ability to transform *P. chrysogenum* and *Cephalosporium acremonium* is firmly established. Because genes involved in beta-lactam biosynthesis have been cloned, it is now possible to modify existing biosynthetic pathways by gene dosage, express foreign genes that modify the end product, and remove undesirable functions by gene disruption.

B. Cloning of Beta-Lactam Biosynthetic Genes

Representatives of nearly all genes involved in penicillin G and cephalosporin C production, as illustrated in Fig. 1, have been cloned and characterized (*cef*G must still be cloned). This section describes these studies of the enzymes of the biosynthetic pathway.

1. pcb*AB Encoding ACV Synthetase*

The first enzymatic reaction involved in biosynthesis of all naturally occurring penicillins and cephalosporins is formation of the tripeptide δ-(L-α-aminoadipyl)-L-cysteine-D-valine (ACV). The gene encoding this enzymatic activity is designated *pcb*AB. ACV is formed in an ATP-dependent reaction by the sequential condensation of L-α-aminoadipic acid, L-cysteine, and L-valine. The valine residue is epimerized to the D form during incorporation. Formation of ACV is catalyzed by a large multifunctional enzyme (ACV synthetase) via a nonribosomal mechanism, most likely a thiotemplate mechanism similar to that of fatty acid synthetase (Wakil, 1989). ACV synthetase has been purified and characterized from *Aspergillus nidulans* (van Liempt *et al.*, 1989), *Cephalosporium acremonium* (Baldwin *et al.*, 1990), and *Streptomyces clavuligerus* (Jensen *et al.*, 1990). Size estimates for ACV synthetase range from 220 kDa in *A. nidulans* to 300 kDa in *C. acremonium.* Jensen and associates (1990) suggested that the active form of ACV synthetase in the bacterium *S. clavuligerus* is a homodimer of approximately 560 kDa. Investigators working with the eukaryotic ACV synthetases have also observed aggregation of the enzyme during the purification process.

Cloning of *pcb*AB from *C. acremonium* has centered around two approaches in this laboratory. The first approach used gene disruption to locate the *pcb*AB gene. The region immediately upstream of *pcb*C was selected for disruption. A restriction fragment from within this upstream region was radiolabeled and hybridized with DNA from *Penicillium chrysogenum* and *A. nidulans.* A distinct response was observed in both cases, yielding restriction fragments of a size consistent with the region upstream of *pcb*C in these organisms. Therefore, a vector (pPS96) was designed to disrupt this region by homologous integration. Approximately 5% of transformants recovered from transformation with pPS96 lacked the ability to produce a beta-lactam yet retained isopenicillin N synthetase activity. When the intact region upstream of *pcb*C was transformed back into these beta-lactam$^-$ transformants, the ability to produce a beta-lactam was restored (Skatrud *et al.*, 1990). These results were consistent with a location for *pcb*AB just upstream of *pcb*C in *C. acremonium.*

The second approach used to find *pcb*AB was reverse genetics. A collaborative effort with O'Callaghan of Oxford University led to the purification of the ACV synthetase protein (Baldwin *et al.,* 1991). An amino acid sequence was obtained from several internal peptides isolated from an enzymatic digestion of purified ACV synthetase. This amino acid sequence was used to confirm results of gene disruption studies just described. DNA sequence analysis of the region upstream of *pcb*C revealed a high degree of similarity with *grs*A and *tyc*A of the gramicidin S and tyrocidine biosynthetic pathways, respectively. Furthermore, there appear to be three related domains in this enzyme (Skatrud *et al.,* 1991). Baldwin and co-workers (1990) demonstrated that phosphopantetheine is associated with purified ACV synthetase from both *C. acremonium* and *S. clavuligerus* in a 1:1 molar ratio. These data strongly suggest that ACV synthetase operates via a thiotemplate mechanism similar to that observed for gramicidin S and tyrocidine synthetases.

The location of the *P. chrysogenum pcb*AB gene can be inferred from studies of D. J. Smith and colleagues (1990a). These investigators transferred a 38-kb fragment of *P. chrysogenum* DNA into *Aspergillus niger* and *Neurospora crassa,* which endowed these organisms with the ability to produce a beta-lactam. Previous studies had demonstrated that *pcb*C and *pen*DE are linked in *P. chrysogenum* (Veenstra *et al.,* 1989) and *A. nidulans* (Tobin *et al.,* 1990). The *pen*DE gene lies just downstream of *pcb*C. D. J. Smith and co-workers (1989) also demonstrated that the approximately 38-kb fragment was amplified in strains of *P. chrysogenum* that produced high titers of penicillin. Amplification of the beta-lactam biosynthetic genes was also observed in a somewhat distantly related strain in our laboratories (M. B. Tobin, personal communication). Thus, it is probable that all genes required for penicillin production are clustered in *P. chrysogenum;* the *pcb*AB gene occupies a position analogous to that observed in *C. acremonium.* D. J. Smith and associates (1990b) extended their earlier studies and confirmed that the *pcb*AB gene was physically located near the *pcb*C gene in *P. chrysogenum.*

2. **pcb***C Encoding Isopenicillin N Synthetase*

Isopenicillin N synthetase (IPNS), a nonheme Fe(II)-dependent enzyme, is responsible for the conversion of ACV to the first bioactive compound in the beta-lactam biosynthetic pathway [i.e., isopenicillin N (IPN)]. This enzyme has also been referred to as "cyclase" or "ACV cyclase." IPNS, which has a calculated molecular weight of 38,416 in *Cephalosporium acremonium* (Samson *et al.,* 1985), catalyzes the removal of four hydrogen atoms and uses one oxygen molecule to form the beta-lactam and thiazolidine rings of IPN (Ingolia and Queener, 1989).

The gene which encodes IPNS is designated *pcbC* (Fig. 1) and was first cloned from *C. acremonium* through reverse genetics. The IPNS protein was purified to homogeneity by Sir Edward Abraham and colleagues at Oxford University. The N-terminal amino acid sequence was obtained at Lilly Research Laboratories. Synthetic oligonucleotide probes were synthesized and used to screen a genomic DNA library from *C. acremonium.* The *pcbC* gene was isolated, sequenced, and expressed in *Escherichia coli* (Samson *et al.,* 1985). Other *pcbC* genes have been cloned by heterologous hybridization or reverse genetics from *Penecillium chrysogenum* (Carr *et al.,* 1986), *Aspergillus nidulans* (Ramon *et al.,* 1987), *Streptomyces lipmanii* (Weigel *et al.,* 1988), *S. clavuligerus* (Leskiw *et al.,* 1988), *S. jumonjinensis* (Shiffman *et al.,* 1988), and *Flavobacterium* (Shiffman *et al.,* 1990). It was originally anticipated that cloning several *pcbC* genes might provide some insight into structure–function relationships in this unique enzyme. However, the unusually high degree of similarity exhibited by *pcbC* genes from divergent sources has limited such insight to only a few amino acids in IPNS.

Site-directed *in vitro* mutagenesis is a directed method to probe structure–activity relationships of certain amino acid residues in the *C. acremonium* IPNS protein (Samson *et al.,* 1987a). Biochemical experiments indicated that sulfhydryl-reactive alkylating agents inactivated IPNS (Perry *et al.,* 1988), suggesting that cysteine residues may be required for activity. All seven cloned *pcbC* genes encode conserved cysteines corresponding to the Cys-106 and Cys-255 residues of the *C. acremonium* IPNS protein. These cysteines, which are the only cysteine residues in the *C. acremonium* IPNS, were changed to serine residues individually (IPNS-Ser-106 and IPNS-Ser-255) and in combination (IPNS-Ser-106, 255). The mutant genes were expressed in *E. coli* and extracts were analyzed for activity. The specific activity of the IPNS-Ser-106 mutant enzyme was substantially reduced and the K_m significantly increased. The IPNS-Ser-255 mutation had little effect on activity. The double mutant, serine inserted at both positions, reflected the properties of the IPNS-Ser-106 mutant enzyme. Thus, a free sulfhydryl is important but it is not essential for activity; even the double mutant retained easily detectable enzymatic activity.

The cloning of the *pcbC* gene has permitted efficient production of IPNS protein from *E. coli.* This has greatly facilitated the study of this unique enzyme. For example, the mechanism of action for IPNS has been examined by using a variety of substrate analogs (Baldwin and Abraham, 1988). Postulated mechanisms evolved from these studies suggested a catalytic role for iron in association with ACV, oxygen, or both. Chen and coworkers (1989) demonstrated through Mossbauer, EPR, and optical spectroscopies that the iron environments of the enzyme and its nitrosyl adduct were changed in the presence of substrate, suggesting that ACV binds directly to the active-site iron via the cysteinyl thiol of ACV. Further analysis of IPNS by NMR suggested a six-coordinate metal center at the active site, with three endogenous histidine ligands and sites for the thiolate of ACV, oxygen, and solvent (Ming *et al.,* 1990). These data were consistent with the *in vitro* mutagenesis studies just described.

The N-2 mutant of *C. acremonium* (Shirafugi *et al.,* 1979) produces no beta-lactam antibiotic. Ramos and colleagues (1986) characterized this mutant strain. They found an accumulation of ACV but no discernible IPNS activity, epimerase activity, or expandase/hydroxylase activity. The mutation in N-2 may be in an effector gene (a *trans*-acting positive regulatory gene), because a single mutation in a structural gene encoding the biosynthetic enzymes would not explain the removal of all three enzymatic activities. The N-2 mutation was complemented by transformation with plasmid pPS26 containing the *pcb*C gene, approximately 3.5 kb of 5′ sequence, and 0.3 kb of 3′ sequence, suggesting that the N-2 mutation was in the *pcb*C gene (Chapman *et al.,* 1987). Ramsden and associates (1989) further characterized the N-2 mutant at the molecular level. They discovered a single basepair change in *pcb*C that changed proline to leucine in a highly conserved region and caused inactivation of IPNS. They also demonstrated restoration of antibiotic production by transforming N-2 with a functional *pcb*C gene (i.e., a DNA fragment containing *pcb*C with 1.2 kb 5′ and 1.0 kb 3′ sequences). Furthermore, in another similar experiment in this laboratory, we complemented the N-2 mutation with *pcb*C plus only 0.7 kb 5′ and 0.3 kb 3′ DNA sequences. There is no evidence for an additional open reading frame (ORF) on the latter clone. These results confirmed that the mutation responsible for the phenotype of N-2 was the single base substitution in *pcb*C. Perhaps isopenicillin N, the product of IPNS, serves as a positive stimulus for production of epimerase and expandase/hydroxylase.

Overall, the regulation of expression of beta-lactam biosynthetic genes is a complex issue that has not been addressed at the molecular level to a large extent. A. W. Smith and co-workers (1990) have initiated such studies by analyzing expression of IPNS activity in *C. acremonium.* They observed rapid accumulation of IPNS mRNA between days 2–4 in a 7-day fermentation. After day 4 of the fermentation, there was virtually no detectable message produced, yet IPNS activity persisted, indicating a certain degree of intracellular stability of the protein. These data suggested that *pcb*C was under transcriptional control in *C. acremonium.* S-1 endonuclease mapping studies indicated two major and two minor transcription initiation sites (−74 and −64 relative to *pcb*C) located on either side of a pyrimidine-rich region. In contrast, the transcription initiation site for the *P. chrysogenum pcbC* gene is located only 11 bp upstream from the translation initiation site (Barredo *et al.,* 1989). The region surrounding the translation initiation site for *pcbC* in *C. acremonium* contained the sequence 5′-TCACCATGG, which is frequently found associated with fungal genes (Ballance, 1986). The corresponding region in *P. chrysogenum* matched this sequence with exception of the initial T. The *pcbC* genes from these two organisms contained other sequence motifs important for the expression of genes in other organisms; however, the function of these motifs is undefined for *C. acremonium* and *P. chrysogenum.*

3. pen*DE Encoding Acyl CoA:Isopenicillin N Acyltransferase*

The final step in beta-lactam biosynthesis in *Penicillium chrysogenum* and *Aspergillus nidulans* is the formation of penicillin G or V from isopenicillin N (Fig. 1). Queener and Neuss (1982) suggested a model for the conversion of isopenicillin N to penicillins involving a two-step enzymatic process in which the L-α-aminoadipyl side chain is removed from isopenicillin N by an isopenicillin N amidolyase to form 6-aminopenicilllanic acid (6-APA). 6-APA is converted to penicillin G by the addition of the CoA derivative of phenylacetic acid by acylCoA:6-APA acyltransferase. (Penicillin V is produced if the CoA derivative of phenoxyacetic acid is used.) A ligase activity that forms the CoA derivative of phenylacetic acid has also been observed in *P. chrysogenum* but is not formally considered a function specific to beta-lactam production.

The initial report of acyltransferase (AT) purification from *P. chrysogenum* indicated that a 29-kDa protein was responsible for formation of penicillin G (Alvarez *et al.,* 1987). Veenstra and co-workers (1989), using an N-terminal amino acid sequence obtained from the purified 29-kDa protein, cloned the corresponding gene (*pen*DE) from *P. chrysogenum.* The *pen*DE ORF was located immediately downstream of *pcbC* and encoded a protein of approximately 40 kDa. The N terminus of the protein encoded by *pen*DE did not correspond to the amino acid sequence used to isolate the gene. The amino acid sequence used to isolate the gene was found within the protein. The *P. chrysogenum pen*DE gene contained three introns. This was the first example of a beta-lactam gene containing introns.

Investigators at Oxford University independently purified AT from both *P. chrysogenum* and *A. nidulans* (Whiteman *et al.,* 1990). These investigators isolated a 29-kDa and an 11-kDa protein in purified preparations of AT. Furthermore, they demonstrated that active AT exists as a heterodimer with subunits totalling 40 kDa. Tobin and associates (1990) cloned and characterized the *A. nidulans* and *P. chrysogenum pen*DE genes. The N-terminal amino acid sequence from the 11-kDa protein corresponded to the N terminus of the protein encoded by *pen*DE. Introns were observed in the *A. nidulans* gene in the same positions as those observed for *P. chrysogenum.* These two *pen*DE genes share 73% identity at the nucleotide level and 76% identity at the amino acid level. The *P. chrysogenum pen*DE gene, cloned from a cDNA library, was modified to allow expression in *Escherichia coli.* The recombinant AT produced in *E. coli* converted both 6-APA and IPN to penicillin G. The ORF in *E. coli* encoded the full 40-kDa protein. The specific activity of this protein was not compared with that of the protein produced in *P. chrysogenum.* If processing to the heterodimer state is necessary for full activity, then it is unlikely that it occurred in *E. coli,* unless it is a process induced by AT itself. Tobin and co-workers (1990) proposed that conversion of IPN to penicillin G in *P. chrysogenum* and *A. nidulans* is catalyzed by a single enzyme of 40 kDa, consisting of two subunits of approximately 11 kDa and 29 kDa, both derived from the same 40-kDa preprotein in a posttranslational processing event.

4. cef*D Encoding Isopenicillin N Epimerase*

The initial step in that branch of the biosynthetic pathway committed to cephalosporin production is the conversion of isopenicillin N to penicillin N (Fig. 1). Purification of the enzyme that catalyzes this reaction from *Cephalosporium acremonium* proved very difficult. Efforts were successfully redirected toward purification of an analogous protein from a prokaryotic source *(Streptomyces clavuligerus).* Although this book is devoted to fungal molecular biology, it is appropriate to discuss the prokaryotic *cef*D gene here because information about it will be used to help establish the basis for discussions later in the text.

The epimerase from *S. clavuligerus* behaves as an isomerase; it establishes an equilibrium between isopenicillin N and penicillin N in extracellular reactions (Usui and Yu, 1989). Using the *S. clavuligerus* epimerase N-terminal amino acid sequence provided by Usui and Yu, investigators at Lilly Research Laboratories cloned the *cef*D gene from *S. clavuligerus* (Kovacevic *et al.,* 1990) and from *S. lipmanii* (P. L. Skatrud, unpublished data). Approximately 78% identity at the nucleotide level and 70.5% identity at the amino acid level was exhibited by *cef*D from these two prokaryotes. The genes involved in beta-lactam biosynthesis in the streptomycetes are linked. The *cef*D and *cef*E genes are expressed on the same transcript, and *pcb*AB and *pcb*C have been isolated on the same cosmid as the *cef*D and *cef*E genes.

5. *cefEF, cefE, and cefF Encoding Expandase/Hydroxylase, Expandase, and Hydroxylase*

The product of the *cef*D gene catalyzes the formation of penicillin N which, in turn, is transformed to deacetoxycephalosporin C (DAOC) by deacetoxycephalosporin C synthase (DAOCS, also referred to as expandase; Fig. 1). DAOC is the first cephalosporin compound in the beta-lactam biosynthetic pathway. (DAOC contains a 6-membered dihydrothiazine ring rather than the 5-membered thiazolidine ring of penicillin.) DAOC is hydroxylated to form deacetylcephalosporin C (DAC) by deacetylcephalosporin C synthase (DACS) or hydroxylase. In the streptomycetes, these two activities are present in separate proteins encoded by *cef*E and *cef*F, respectively. In the filamentous fungus *Cephalosporium acremonium,* these activities cannot be separated biochemically. Both activities are present in a single polypeptide frequently referred to as expandase/hydroxylase and encoded by the *cef*FE gene.

The bifunctional expandase/hydroxylase protein purified from *C. acremonium* (Dotzlaf and Yeh, 1986) was used to prepare synthetic oligonucletide probes based on internal amino acid sequence. The *cef*EF gene from *C. acremonium* was cloned, the DNA sequence was determined, and the ORF, which encodes a protein of 36,462 Da, was expressed in *Escherichia coli* (Samson *et al.,* 1987b). Dotzlaf and Yeh (1989) also purified the monofunctional expandase protein from *Streptomyces clavuligerus.* Kovacevic and associates (1989) cloned and expressed the corresponding gene (*cef*E) in *E. coli.* The *cef*E gene encoded a protein that was 21 amino acids shorter than *cef*EF (34,519 Da). The amino acid sequence of *cef*E was 57% identical to that of *cef*EF, indicating that the bifunctional expandase/hydroxylase and the monofunctional expandase genes were highly related.

The ability to produce large quantities of expandase and expandase/hydroxylase in *E. coli* has facilitated enzymatic studies. Interestingly, the expandase protein originally from *S. clavuligerus* does have a low-level hydroxylase activity and catalyzes a novel hydroxylation of 3-exomethylenecephalosporin C to DAC (Dotzlaf and Yeh, 1989). The high degree of homology of the fungal *cef*EF gene to the bacterial *cef*E gene suggested to Ingolia and Queener (1989) that the bifunctional nature of the *cef*EF gene may have arisen as a result of a fusion of the *cef*E and *cef*F genes in a prokaryote. An alternative possibility is that the bifunctional gene existed first. This gene was transferred to the eukaryotic progenitor of modern *C. acremonium.* The presence of separate but related expandase and hydroxylase enzymes in *S. clavuligerus* might be explained by a gene duplication of the bifunctional gene followed by loss of expandase activity in one and loss of hydroxylase activity in the other.

Recent results lend support to the latter hypothesis. The monofunctional hydroxylase enzyme was purified from *S. clavuligerus* (J. E. Dotzlaf and W.-K. Yeh, personal communication). The *cef*F gene encoding hydroxylase was cloned via reverse genetics and characterized at the molecular level (Kovacevic and Miller, 1990). This gene was found closely linked to *cef*D and *cef*E, and encoded a protein with a molecular weight of 34,584. The amino acid sequence of hydroxylase was 59% identical to expandase. The DNA sequences of these two genes were 71% identical. Careful analysis of enzyme activity revealed that the hydroxylase protein from *S. clavuligerus* possesses a low but reproducible expandase activity and the expandase protein has detectable hydroxylase activity (J. E. Dotzlaf and W.-K. Yeh, personal communication).

6. Other Genes Involved in Beta-Lactam Biosynthesis

The *cef*G gene from *Cephalosporium acremonium,* the last gene involved in cephalosporin C biosynthesis (Fig. 1), has not been studied at the molecular level to date.

Another gene important to beta-lactam biosynthesis in the bacterium *Streptomyces clavuligerus* is the *LAT* gene, which encodes lysine-ϵ-aminotransferase. This enzyme is instrumental in the formation of α-aminoadipate in *S. clavuligerus,* a required precursor for ACV synthetase. In this bacterial genus, two routes of lysine catabolism are present. Cadaverine and δ-aminovalerate are breakdown intermediates in one route, which is present in all species of *Streptomyces* tested by Madduri and co-workers (1989). The other route involves formation of α-aminoadipate from lysine by the action of lysine-ϵ-aminotransferase and 1-piperidine dehydrogenase. Lysine-ϵ-aminotransferase is found only in those species of *Streptomyces* that produce beta-lactam antibiotics. Lysine-ϵ-aminotransferase is expressed in a pattern similar to isopenicillin N synthetase, suggesting that this catabolic pathway may be specific for secondary metabolism (Kern *et al.,* 1980). Furthermore, Madduri and colleagues (1990) recently reported that the *LAT* gene is located between *pcb*C and *cef*E in the beta-lactam biosynthetic gene cluster of *S. clavuligerus.*

In the filamentous fungi *Penicillium chrysogenum* and *C. acremonium,* the anabolic pathway leading to lysine includes the production of α-aminoadipate. However, this is a primary metabolic pathway. Adequate production of α-aminoadipate during secondary metabolism from a primary metabolic pathway may not be possible. One might anticipate the presence of lysine-ϵ-aminotransferase activity in these fungi (see Section III).

The genes that produce those enzymes specific for cephamycin C production (cmcH, I, and J) have not been studied at a detailed molecular level. A 29.3-kb DNA fragment from *S. catteleya* was transferred into *S.*

lividans, a species that does not produce beta-lactams. Cephamycin C production was detected in this previously naive host (Chen *et al.,* 1988). When the same fragment was transferred into a cephamycin C-producing recipient, the total amount of cephamycin C produced was increased. Such studies emphasize the fact that the beta-lactam biosynthetic genes are linked in *Streptomyces* species.

C. Practical Applications of Molecular Biology to the Beta-Lactam-Producing Fungi

The availability of adequate transformation systems and cloned antibiotic biosynthetic genes led immediately to useful applications in this new arena of applied biotechnology. The beta-lactam biosynthetic genes in *Cephalosporium acremonium* are present in single copy (M. B. Tobin, personal communication; Skatrud *et al.,* 1989). Therefore, gene dosage of a beta-lactam biosynthetic gene responsible for a rate-limiting step in the pathway should produce a yield improvement.

1. Gene Dosage for the cef*EF gene in* Cephalosporium acremonium

Analyses of large scale *Cephalosporium acremonium* fermentations revealed an accumulation of penicillin N in the broth. Penicillin N is the substrate for expandase/hydroxylase in *C. acremonium.* This observation suggested that the rate of cephalosporin C biosynthesis was limited by the amount of expandase/hydroxylase present in this strain of *C. acremonium.* Therefore a vector (pPS56) was designed to deliver extra copies of the *cef*EF gene to *C. acremonium.* Analysis of transformants obtained with pPS56 revealed approximately 25% that were improved in cephalosporin C production. Transformant LU4-79-6, which produced 47% more cephalosporin C than the untransformed recipient strain in laboratory scale fermentations, was studied in detail (Skatrud *et al.,* 1989).

A single copy of the transforming DNA had integrated by heterologous recombination into chromosome III of strain LU4-79-6. The endogenous *cef*EF gene is located on chromosome II of *C. acremonium* (Skatrud and Queener, 1989). The transforming DNA was mitotically stable in the absence of selection. Approximately twice as much expandase/hydroxylase activity was detected in this strain as in the recipient. In contrast to the recipient strain, LU4-79-6 did not accumulate penicillin N in fermentation broths. Thus, it would appear that the rate-limiting step in cephalosporin C biosynthesis in this industrial strain had been modified successfully by the addition of a single copy of *cef*EF.

The increase in cephalosporin C production in pilot plant fermentations (150 liters) was less than in laboratory scale fermentations (0.1 liter)

(Skatrud *et al.,* 1989). Nevertheless, the 15% yield improvement observed in these larger fermentation vessels was a dramatic improvement, considering the fact that this strain was already producing high levels of cephalosporin C. Strain LU4-79-6 has been approved for use on an industrial scale under physical conditions less than BL1-LS by the National Institutes of Health (Wyngaarden, 1988).

2. **pcb***C Gene Dosage Studies*

In an earlier study, additional copies of the *pcb*C gene were added to *Cephalosporium acremonium* simply because *pcb*C was the first available cloned gene from the pathway. Several hundred transformants of both wild-type *C. acremonium* and industrial strains were examined for increased cephalosporin C production. No gene dosage effect on overall antibiotic production was observed for *pcb*C (Skatrud *et al.,* 1986; Chapman *et al.,* 1987). In retrospect, it is obvious that the wrong gene was being added in that set of experiments.

The *pcb*C gene of *Penicillium chrysogenum* also was added in multiple copies to an industrial strain of *P. chrysogenum* (Skatrud *et al.,* 1987a). Despite the fact that increased IPNS levels were observed in selected transformants, no improvement in overall productivity was observed. Subsequently it was discovered that the beta-lactam biosynthetic genes were already amplified in certain high-producing strains of *P. chrysogenum,* including the strain used in that study (M. B. Tobin, personal communication; Barredo *et al.,* 1989; Smith *et al.,* 1989). Thus, it is unlikely that gene dosage for any particular structural gene in the formal biosynthetic pathway would result in yield increase. However, increasing the concentration of required primary metabolites (such as α-aminoadipate) during secondary metabolism could have a beneficial effect on productivity in *P. chrysogenum.*

3. Enzymatic Formation of Novel Beta-Lactams

Thousands of unnatural penicillins have been made chemically by changing the side chain on penicillins produced in fermentation (Wolfe *et al.,* 1984). In all these changes, the bicyclic beta-lactam nucleus remained the same. If one could start with a different beta-lactam nucleus, one might produce many new structural variations. New beta-lactam nuclei might be produced enzymatically if the substrate specificity of isopenicillin N synthetase would permit cyclization of tripeptides other than ACV.

The availability of large quantities of recombinant *Cephalosporium acremonium* isopenicillin N synthetase stimulated investigation of its substrate requirements (Huffman and co-workers at Lilly Research Laboratories, unpublished data; Baldwin *et al.,* 1986a,b,c). Investigators at Oxford

University and Lilly Research Laboratories analyzed approximately 150 chemically synthesized unnatural tripeptides for conversion to a beta-lactam product by IPNS in *in vitro* reactions. About 80 substrates were converted to detectable quantities of a beta-lactam. Of these, roughly 20 exhibited antibacterial activity. Unfortunately, compounds produced in this manner were not examined for activities other than antibacterial. However, these studies did point out that it was possible to form novel beta-lactams using isopenicillin N synthetase.

4. *Biosynthetic Pathway Engineering*

Biosynthetic pathway engineering implies for our purposes, altering a biosynthetic pathway to form a product previously unseen in that organism. In the production of oral cephalosporins, many chemical modifications are made to the starting beta-lactam compound in order to achieve stable final products. Currently, the process begins with a penicillin nucleus that is ring-expanded chemically. A major advantage to beginning with a penicillin is the nature of the side chain attached at position 7. Cephalosporin C contains a D-α-aminoadipyl side chain that is not easily removed chemically. In comparison, the side chain of penicillin G is readily removed. Fig. 2a illustrates the current chemical process for obtaining 7-aminodeacetoxycephalosporanic acid (7-ADCA), an obligate intermediate in many of the chemical routes of oral cephalosporin synthesis. Fig. 2b illustrates a proposed route that would involve an *in vivo* synthesis of deacetoxycephalosporin V, which could then be enzymatically deacylated to form 7-ADCA. Such a route would eliminate the need for blocking and deblocking steps and significantly reduce the use of organic solvents. Solvents are costly to recover and, if not fully recovered, will pollute the environment. In order to implement this alternative route, one must modify the *Streptomyces clavuligerus* expandase to accept penicillin V as a substrate, perhaps through protein engineering. It will also be necessary to express foreign genes efficiently in *Penicillium chrysogenum.*

One aspect of this proposal has been completed. Cantwell and associates (1990) reported expression of a hybrid *S. clavuligerus cef*E gene in *P. chrysogenum.* The monofunctional bacterial *cef*E gene was fused with the *P. chrysogenum pcb*C promoter and transformed into *P. chrysogenum.* Although expandase activity was detectable, the amount produced was very low. Recently the *cef*E gene was further modified by addition of the appropriate 3′ expression signals from the *P. chrysogenum pcb*C gene. Using this new hybrid gene, the amount of expandase produced in *P. chrysogenum* transformants is similar to that seen in production strains of *Cephalosporium acremonium.* This breakthrough represents an initial step toward production of 7-ADCA in the absence of organic solvents.

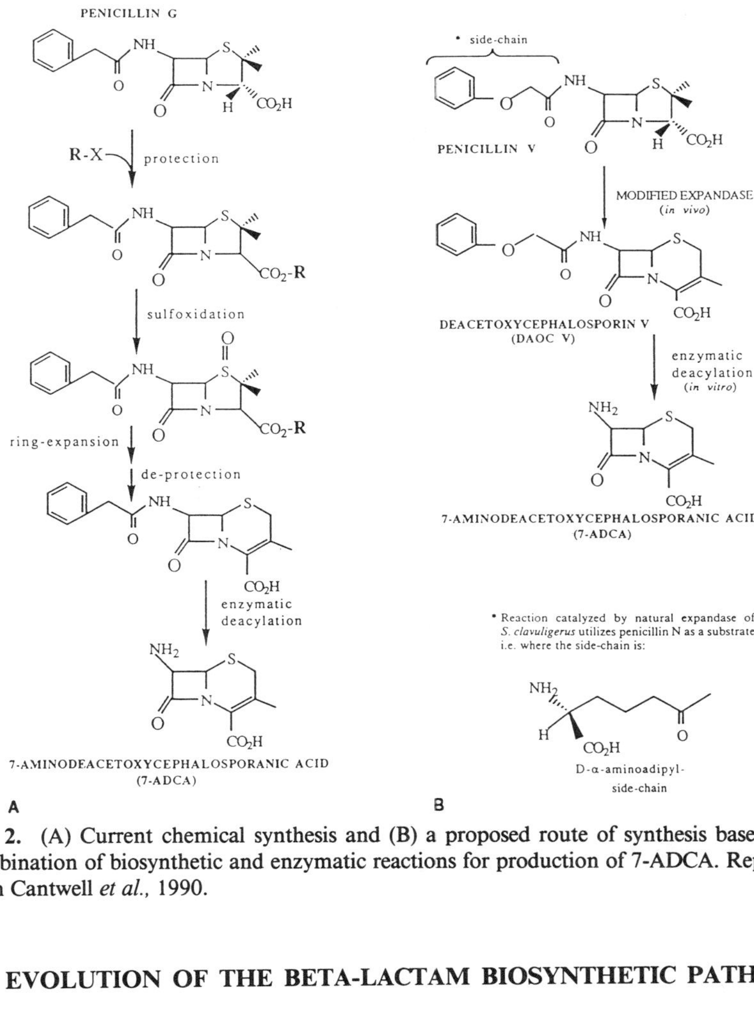

Fig. 2. (A) Current chemical synthesis and (B) a proposed route of synthesis based on a combination of biosynthetic and enzymatic reactions for production of 7-ADCA. Reprinted from Cantwell *et al.*, 1990.

III. EVOLUTION OF THE BETA-LACTAM BIOSYNTHETIC PATHWAY

An evolutionary hypothesis first suggested by Ingolia (Carr *et al.*, 1986; Weigel *et al.*, 1988) was based on a comparison of DNA sequence information obtained from the *pcb*C genes of fungi and gram positive bacteria (Fig. 3). This hypothesis proposed horizontal transfer of the beta-lactam biosynthetic genes from a prokaryote to a eukaryote approximately 370 million years ago (Fig. 4). The *pcb*C gene from a gram negative bacterium, *Flavobacterium* sp. SC 12,154, was recently cloned and sequenced (Shiffman *et al.*, 1990). It also shared 63% identity with fungal *pcb*C genes. All other genes from the beta-lactam biosynthetic pathway that have been

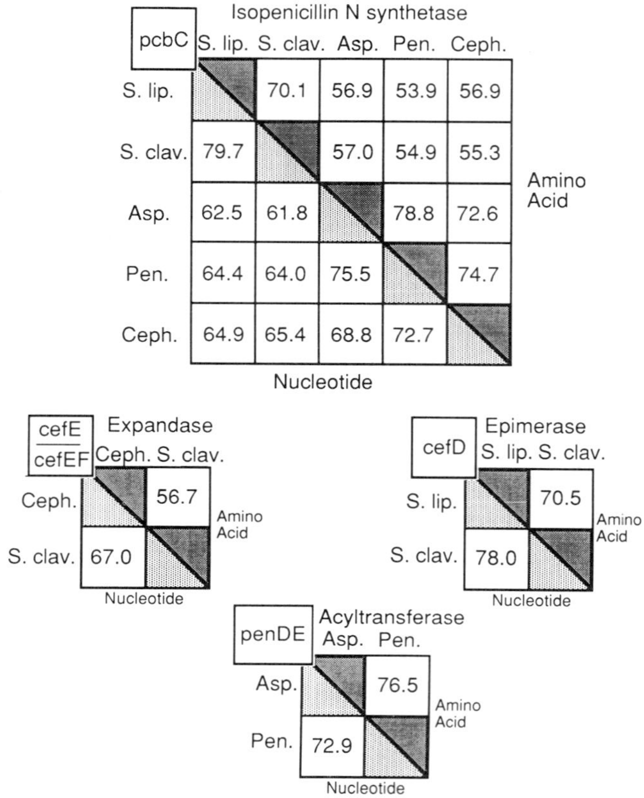

Fig. 3. Comparison of the DNA identities *(lower left)* and amino acid identities *(upper right)* of cloned beta-lactam biosynthetic genes. Numbers in the boxes indicate the percentage identity for a particular gene or protein in the organisms compared. S. lip., *S. lipmanii;* S. clav., *Streptomyces clavuligerus;* Asp., *Aspergillus nidulans;* Pen., *Penicillium chrysogenum;* Ceph., *Cephalosporium acremonium.* Adapted from Weigel *et al.,* 1988.

cloned and compared at the molecular level (i.e., *pen*DE, *cef*D, and *cef*E/*cef*EF) show similar DNA and amino acid sequence conservation, supporting the horizontal transfer theory (Fig. 3).

In addition to nucleic acid sequence information, the genomic arrangement of the beta-lactam biosynthetic genes has been informative with respect to the evolutionary process. In the prokaryote *Streptomyces clavuligerus,* the beta-lactam biosynthetic genes are physically linked within approximately 20–30 kb (Miller and Ingolia, 1989). Genes involved in

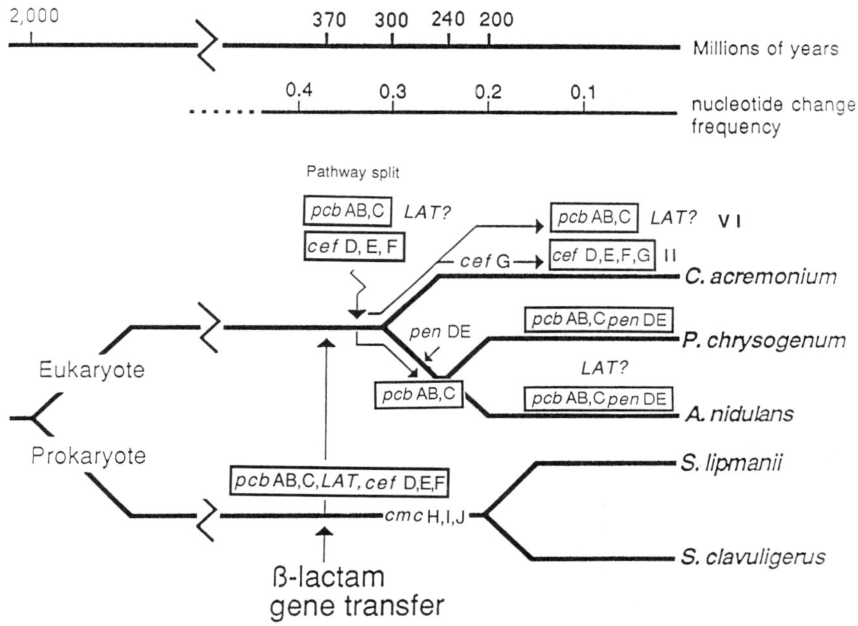

Fig. 4. Phylogenetic tree illustrating the evolution of the beta-lactam biosynthetic pathway based on DNA sequence identities of cloned *pcbC* genes. Adapted from Weigel *et al.*, 1988.

beta-lactam production in the filamentous fungi *Penicillium chrysogenum* and *Aspergillus nidulans* are also clustered (Skatrud *et al.*, 1991; A. W. Smith *et al.*, 1990; Tobin *et al.*, 1990). This similar physical arrangement of beta-lactam biosynthetic genes in these two fungi may reflect their prokaryotic origin. Clustering of the biosynthetic genes in the eukaryotes suggests that all genes required for beta-lactam biosynthesis were transferred to a eukaryotic ancestor in a single event.

In contrast to the gene clustering in *P. chrysogenum* and *A. nidulans,* two genes (*pcb*C and *cef*EF) involved in cephalosporin C biosynthesis in the filamentous fungus *Cephalosporium acremonium* are not physically linked (Skatrud and Queener, 1989). Molecular karyotype analysis indicated that the *pcb*C gene resides on chromosome VI; the *cef*EF gene is located on chromosome II. In light of these results, and considering the horizontal transfer hypothesis, it was speculated that the *pcb*C gene and other genes active prior to the first branch in the biosynthetic pathway might be clustered on chromosome VI in *C. acremonium.* Furthermore, the *cef*EF gene might be associated with another cluster of genes specific for cephalosporin C biosynthesis (i.e., *cef*D and *cef*EF) on chromosome II (Kleinkauf and von Dohren, 1990). The observation that *pcb*AB lies just upstream of *pcb*C in *C. acremonium* supports the notion of two clusters of biosynthetic genes in this fungus (Skatrud *et al.,* 1991).

The horizontal transfer hypothesis, coupled with the observed physical arrangement of genes, suggests at least one explanation for the lack of cephalosporin-specific genes in *P. chrysogenum* and *A. nidulans.* The beta-lactam biosynthetic genes were probably transferred together in a single block from a prokaryote, perhaps an ancestor of one of the streptomycetes, to a common ancestor of *C. acremonium, A. nidulans,* and *P. chrysogenum.* In that common eukaryotic ancestor, the block of beta-lactam biosynthetic genes was split onto two chromosomes, separating the early part of the pathway from the branch that leads to cephalosporin C synthesis. Consistent with this suggestion is the presence of *pcb*C and *cef*EF on separate chromosomes in a modern strain of *C. acremonium.*

Approximately 70 million years after the horizontal transfer (Fig. 4), an ancestor common to *P. chrysogenum* and *A. nidulans* diverged from the *C. acremonium* lineage. If a cluster of genes encoding *pcb*C and *pcb*AB resided on a chromosome separate from the cephalosporin-specific genes in the ancient common ancestor, then it is possible to envision how the more recent ancestor common to *P. chrysogenum* and *A. nidulans* received only the early biosynthetic genes (*pcb*AB and *pcb*C) during divergence.

The cloning and sequence analysis of the *pcb*AB gene suggests that this gene is evolutionarily diverged from a large family of peptide synthetases. The high degree of homology of *pcb*AB with the peptide synthetases involved in gramicidin S and tyrocidine biosyntheses suggests that these enzymes may share a similar mechanism of action (thiotemplate based). Furthermore, it suggests that peptide synthetases as a general class may have a common origin. One would predict that, as more data become available, a definite relationship will emerge between *pcb*AB and other peptide synthetases using a nonribosomal mechanism of synthesis, for example, peptide-based siderophores; echinocandin B; cyclosporin; polymixin; actinomycin; glycopeptides of the vancomycin group; depsipeptide antibiotics; a variety of peptide toxins of mammalian, plant, and bacterial origin; bleomycin; and protease inhibitors important in the immune response and metabolic homeostasis.

The origin of the last gene required for production of penicillin G is not obvious (*pen*DE; Fig. 4). Moreover, there is no known prokaryotic gene analogous to *pen*DE. The only beta-lactam biosynthetic gene characterized to date that contains introns is the *pen*DE gene (Veenstra *et al.,* 1989). The introns in the *pen*DE gene from both *A. nidulans* and *P. chrysogenum* occur in precisely the same position (Tobin *et al.*, 1990). These data are consistent with the formation of a functional *pen*DE gene in the eukaryotic ancestor common to both *P. chrysogenum* and *A. nidulans,* which presumably existed between 240 and 300 million years ago, according to the phylogenetic tree (Fig. 4). However, origin of at least a portion of that gene must predate that time.

Linkage of *pen*DE to the *pcb*C gene in fungi suggests that a sequence functionally related to *pen*DE was present in the prokaryotic ancestor, was transferred along with *pcb*C, and later was modified to its current functional form. One of the functions of the bifunctional protein produced by *pen*DE is an amidolyase activity that removes the side chain of isopenicillin N. A tightly regulated gene encoding isopenicillin N amidolyase, expressed in a pattern reflecting beta-lactam concentration, may have been present in the prokaryotic ancestor at a position adjacent to *pcb*C. When the level of beta-lactam became toxic to the producing organism, the amidolyase gene was transcriptionally activated. The amidolyase removed the L-α-aminoadipyl side chain from isopenicillin N to form 6-APA, which is less toxic to the producing organism, and prevented further accumulation of the toxic beta-lactam. Thus, the postulated ancient amidolyase gene could be considered a resistance gene in its primordial host. Genetic linkage of a resistance gene to antibiotic biosynthetic genes is quite common in *Streptomyces* species (Seno and Baltz, 1989). The existence of a gene in *Streptomyces* that is related to *pen*DE will be verified or refuted as more DNA sequence information becomes available for the region surrounding *pcb*C. One might anticipate finding a region that is approximately 63% identical with at least some portion of *pen*DE, based on the similarity of the *pcb*C genes (Fig. 2). There may also be a region downstream of *pcb*AB in *C. acremonium* that shares some degree of identity with the *pen*DE gene of *P. chrysogenum* and *A. nidulans.* This possibility was recently examined by Southern blot analysis. The *pen*DE gene from *P. chrysogenum* hybridized with a single restriction fragment of genomic DNA from *C. acremonium* under nonstringent conditions (M. B. Tobin, personal communication).

Since the fungi are not sensitive to beta-lactam antibiotics, there would have been no apparent selective advantage in having an amidolyase activity in the fungi. However, if the amidolyase were coupled with the capacity to replace the L-α-aminoadipyl side chain with a side chain associated with greater antibacterial activity, then a selective advantage may have been operational. The presence of introns in *pen*DE suggests eukaryotic exon shuffling, resulting in the modern gene. All introns in the *pen*DE gene are located within the first half of the ORF. The first half of the *pen*DE gene is

58% GC in the third position of the reading frame. In contrast, the second half of the gene, which is devoid of introns, is 71% GC in the third position of the reading frame. This shift in third position bias might suggest fusion of two genes, perhaps from different species.

The last enzymatic reaction involved in the biosynthesis of cephalosporin C in *C. acremonium* is the transfer of an acetyl moiety from acetyl-CoA to the –OH moiety of deacetylcephalosporin C on the 3′C. The gene encoding this enzyme, *cef*G, has not been studied at the molecular level. As for *pen*DE, there is no analogous gene in the prokaryotic species producing beta-lactams. However, there are enzymes that modify the 3′C of cephalosporins in the prokaryotes (Ingolia and Queener, 1989). Perhaps some degree of relatedness will be found between the acetyltransferase which produces cephalosporin C in *C. acremonium* and the 3-hydroxymethylceph-3-em-*O*-carbamoyltransferase which catalyzes the carbamylation of deacetylcephalosporin C in *S. clavuligerus.* If *cef*G is discovered in one of the beta-lactam biosynthetic gene clusters of *C. acremonium,* then one might anticipate a relationship to a prokaryotic gene. If *cef*G is found elsewhere in the genome, then an independent eukaryotic origin is likely. If the evidence supports independent eukaryotic origin, then it would be reasonable to suggest that the cephamycin C specific genes (*cmc*H, *cmc*I, and *cmc*J) arose in the prokaryotes after the horizontal transfer (Fig. 4) or were simply not part of the transfer.

The fate of the lysine-ε-aminotransferase encoding gene *(LAT)* during the horizontal transfer is uncertain. *LAT* is physically linked to other beta-lactam biosynthetic genes in *S. clavuligerus,* suggesting that it could have been transferred to the eukaryotes along with the other beta-lactam biosynthetic genes. The fungi produce L-α-aminoadipic acid as an intermediate to lysine during primary metabolism, in contrast to the prokaryotes, which produce beta-lactams. Limited supply of L-α-aminoadipic acid would be inhibitory to beta-lactam production. Thus it would be advantageous to the beta-lactam-producing fungi to utilize a catabolic reaction producing L-α-aminoadipic acid during secondary metabolism to increase the pool concentration of this essential precursor of penicillins and cephalosporins. Is it possible that the *LAT* gene resides in the beta-lactam producing fungi?

The horizontal transfer hypothesis has led to several predictions regarding the genes involved in beta-lactam biosynthesis—some proven, some unproven. The time estimates of evolutionary divergence are less important than the ability to predict relatedness and location of genes. This hypothesis has, in fact, aided our studies of the molecular genetics of beta-lactam biosynthesis.

IV. FUTURE IMPLICATIONS

All the genes formally recognized in the beta-lactam antibiotic biosynthetic pathway (Fig. 1) will probably be cloned and characterized at the molecular level in the very near future. Genes exerting regulatory effects on antibiotic production are also likely to be discovered (perhaps related to transcriptional regulation or carbon-source control). We have seen one success in the area of industrial strain yield improvement (Skatrud *et al.*, 1989), and it will undoubtedly be followed by many other improvements. New beta-lactam fermentation products will be produced that may prove useful as intermediates in multistep production processes. All these advances could have a profound impact on how beta-lactams are manufactured.

A certain degree of substrate promiscuity has been observed for both isopenicillin N synthetase and ACV synthetase. Efforts have been directed at altering the substrate specificity of expandase. Understanding the evolutionary process that has led to the formation of these highly related biosynthetic enzymes may be informative with respect to structure–activity relationships and may lead to rational decisions in protein engineering studies.

Studies devoted to efficient expression of foreign genes in beta-lactam producing organisms, for example *Penicillium chrysogenum* (Cantwell *et al.*, 1990), will force scientists to learn which factors govern expression of the beta-lactam biosynthetic genes. These genes are expressed during idiophase; therefore, what is learned about the control of beta-lactam biosynthesis may have implications for secondary metabolism in general, as well as for sporulation and other developmental and morphological changes.

For a long time, the beta-lactam compounds have been viewed strictly as antibiotics. During development of clinically useful beta-lactam antibiotics, side effects were considered undesirable and were carefully groomed out of the compounds. Perhaps it is time to reconsider this narrow view, study those side effects, and reassess this class of compounds more broadly as a group of potent serine protease inhibitors. Serine proteases are involved in a number of normal physiological processes such as digestion, blood clotting, and immunity. In many instances, naturally occurring enzymes can lead to a disease state, for example, elastase as a factor in emphysema. Doherty and associates (1986) demonstrated that modified cephalosporins were potent inhibitors of human leukocyte elastase, which is a serine protease. These results suggest application of beta-lactams as a new therapeutic class. Other degenerative processes may also be amenable to treatment with beta-lactams.

Beta-lactams and their breakdown products are inhibitors of DNA replication in mammalian cells and certain viruses (mechanism uncharacterized; Neftel and Hubscher 1987). Observations such as this stimulate consideration of beta-lactams in diverse areas such as antiviral, antitumor, and antifungal therapy. Queener (1990) observed that a great deal had been accomplished in this new era of beta-lactam research, but much remains to be done. Modifications of biosynthetic enzymes through molecular biology may some day lead to compounds effective against other targets of medicinal importance, further extending the utility of this class of compounds that has already had such an unparalleled impact on our health.

REFERENCES

Abraham, E. P., and Newton, G. G. F. (1961). The structure of cephalosporin C. *Biochem. J.* **79,** 377–393.

Abraham, E. P., Chain, E., Fletcher, C. M., Gardner, A. D., Heatley, N. G., Jennings, M. A., and Florey, H. W. (1941). Further observations on penicillin. *Lancet* **241,** (2) 177–188.

Alvarez, E., Cantoral, J. M., Barredo, J. L., Diez, B., and Martin, J. F. (1987). Purification to homogeneity and characterization of acylcoenzyme A:6-aminopenicillanic acid acyltransferase of *P. chrysogenum. Antimicrob. Agents Chemother.* **31,** 1675–1682.

Baldwin, J. E., and Abraham, E. P. (1988). The biosynthesis of penicillins and cephalosporins. *Natural Product Reports* **5,** 129–145.

Baldwin, J. E., Adlington, R. M., Basak, A., Flitsch, S. L., Forrest, A. K., and Ting, H.-H. (1986a). Penicillin biosynthesis structure–reactivity profile of unsaturated substrates for isopenicillin *N* synthetase. *J. Chem. Soc. Chem. Commun.* **3,** 273–275.

Baldwin, J. E., Adlington, R. M., Basak, A., Flitsch, S. L., Petursson, S., and Turner, N. J. (1986b). Enzymatic synthesis of a new type of penicillin. *J. Chem. Soc. Chem. Commun.* **12,** 975–976.

Baldwin, J. E., Adlington, R. M., Basak, A., and Ting, H.-H. (1986c). Penicillin biosynthesis structure–reactivity profile of allenic substrates for isopenicillin *N* synthetase. *J. Chem. Soc. Chem. Commun.* **16,** 1280–1281.

Baldwin, J. E., Bird, J. W., Field, R. A., O'Callaghan, N. M., and Schofield, C. J. (1990). Isolation and partial characterization of ACV synthetase from *C. acremonium* and *S. clavuligerus. J. Antibiot.* Submitted.

Ballance, J. D. (1986). Sequences important for gene expression in filamentous fungi. *Yeast* **2,** 229–236.

Barredo, J. L., Diez, B., Alvarez, E., and Martin, J. F. (1989). Large amplification of a 35-kb DNA fragment carrying two penicillin biosynthetic genes in high penicillin producing strains of *P. chrysogenum. Curr. Genet.* **16,** 453–459.

Beri, R. K., and Turner, G. (1987). Transformation of *P. chrysogenum* using the *A. nidulans amdS* gene as a dominant selective marker. *Curr. Genet.* **11,** 639–641.

Bull, J. H., Smith, D. J., and Turner, G. (1988). Transformation of *P. chrysogenum* with a dominant selectable marker. *Curr. Genet.* **13,** 377–382.

Cantoral, J. M., Diez, B., Barredo, J. L., Alvarez, E., and Martin, J. E. (1987). High-frequency transformation of *Penicillium chrysogenum. Bio/Technology* **5,** 494–497.

Cantwell, C. A., Beckmann, R. J., Dotzlaf, J. E., Fisher, D. L., Skatrud, P. L., Yeh, W.-K., and Queener, S. W. (1990). Cloning and expression of a hybrid *S. clavuligerus cef*E gene in *P. chrysogenum. Curr. Genet.* **17,** 213–221.

Carr, L. G., Skatrud, P. L., Scheetz, M. E., Queener, S. W., and Ingolia, T. D. (1986). Cloning and expression of the isopenicillin *N* synthetase gene from *P. chrysogenum*. *Gene* **48,** 257–266.

Carr, L. G., Skatrud, P. L., Ingolia, T. D., and Queener, S. W. (1987). Organization of the 5.8S, 16–18S, and 23–28S ribosomal RNA genes of *C. acremonium*. *Curr. Genet.* **12,** 209–214.

Carramolino, L., Lozano, M., Perez-Aranda, A., Rubio, V., and Sanchez, F. (1989). Transformation of *P. chrysogenum* to sulfonamide resistance. *Gene* **77,** 31–38.

Chapman, J. L., Skatrud, P. L., Ingolia, T. D., Samson, S. M., Kaster, K. R., and Queener, S. W. (1987). Recombinant DNA studies in *C. acremonium*. *In* "Developments in Industrial Microbiology" (G. Pierce, ed.), Vol. **27,** pp. 165–174. Society for Industrial Microbiology, Arlington, Virginia.

Chen, C. W., Lin, H.-F., Kuo, C. L., Tsai, H.-L., and Tsai, F.-Y. (1988). Cloning and expression of a DNA sequence conferring cephamycin C production. *Bio/Technology* **6,** 1222–1224.

Chen, V. J., Orville, A. M., Harpel, M. R., Frolik, C. A., Surerus, K. K., Munck, E., and Lipscomb, J. D. (1989). Spectroscopic studies of isopenicillin *N* synthase. *J. Biol. Chem.* **264,** (36)21677–21681.

Doherty, J., Ashe, B. M., Argenbright, L., Barker, P., Bonney, R. J., Chandler, G. O., Dahlgren, M. E., Dorn, Jr., C. P., Finke, P., Firestone, R. A., Pisano, J. M., Shah, S. K., Thompson, K. R., and Zimmerman, M. (1986). Modified cephalosporins as acylating inhibitors of human leukocyte elastase. *Nature (London)* **322,** 192.

Dotzlaf, J. E., and Yeh, W.-K. (1986). Copurification and characterization of deacetoxycephalosporin C synthase/hydroxylase from *C. acremonium*. *J. Bacteriol.* **16,** 1611–1618.

Dotzlaf, J. E., and Yeh, W.-K. (1989). Purification and properties of deacetoxycephalosporin C synthase from recombinant *E. coli* and its comparison with the native enzyme purified from *S. clavuligerus*. *J. Biol. Chem.* **264,** 10219–10227.

Durand, N., Reymond, P., and Fevre, M. (1990). Transformations of *Penicillium* to hygromycin B and phleomycin resistance. British Mycological Society, 8th General Meeting—April 9–12, 1990. Nottingham, England, Abstr. 51.

Fleming, A. (1929). On the antibacterial action of cultures of a *Penicillium,* with special reference to their use in the isolation of B. influenzae. *Br. J. Exp. Pathol.* **10,** 226–236.

Hamlyn, P. F., and Ball, C. (1979). Recombination studies with *C. acremonium*. *In* "Genetics of Industrial Microorganisms" (O. K. Sebek and A. I. Laskin, eds.), pp. 185–191. American Society for Microbiology, Washington, D.C.

Hinnen, A., Hicks, J. B., and Fink, J. R. (1978). Transformation of yeast. *Proc. Natl. Acad. Sci. U.S.A.* **75,** 1929–1933.

Hynes, M. J., Corrick, C. M., and King, J. A. (1983). Isolation of genomic clones containing the *amdS* gene of *A. nidulans* and their use in the analysis of structural and regulatory mutations. *Mol. Cell. Biol.* **3,** 1430–1439.

Ingolia, T. D., and Queener, S. W. (1989). Beta-lactam biosynthetic genes. *Med. Res. Rev.* **9,** (2)245–264.

Jensen, S. E. (1985). Biosynthesis of cephalosporins. *Crit. Rev. Biotechnol.* **3,** 277–301.

Jensen, S. E., Wong, A., Rollins, M. J., and Westlake, D. W. S. (1990). Purification and characterization of δ-(L-α-aminoadipyl)-L-cysteinyl-D-valine synthetase from *S. clavuligerus*. *J. Bacteriol.* **172,** (12) 7269–7271.

Kern, B. A., Hendlin, D., and Inamine, E. (1980). L-Lysine-ϵ-aminotransferase involved in cephamycin C synthesis in *S. lactamdurans*. *Antimicrob. Agents Chemother.* **17,** 679–685.

Kleinkauf, H., and von Dohren, H. (1990). Bioactive peptides—Recent advances and trends. *In* "Biochemistry of Peptide Antibiotics" (H. Kleinkauf and H. von Dohren, eds.), pp. 2–26. de Gruyter, Berlin.

Kolar, M., Punt, P. J., van den Hondel, C. A. M. J. J., and Schwab, H. (1988). Transformation of *P. chrysogenum* using dominant selection markers and expression of an *E. coli lacZ* fusion gene. *Gene* **62**, 127–134.

Kovacevic, S., and Miller, J. R. (1990). Cloning and sequencing of the beta-lactam hydroxylase gene (*cef*F) from *Streptomyces clavuligerus.* Gene duplication may have led to separate hydroxylase and expandase activities in the actinomycetes. *J. Bacteriol.* Submitted.

Kovacevic, S., Weigel, B. J., Tobin, M. B., Ingolia, T. D., and Miller, J. R. (1989). Cloning, characterization, and expression in *E. coli* of the *S. clavuligerus* gene encoding deacet oxycephalosporin C synthetase. *J. Bacteriol.* **171,** (2)754–760.

Kovacevic, S., Tobin, M. B., and Miller, J. R. (1990). The beta-lactam biosynthesis genes for isopenicillin *N* epimerase and deacetoxycephalosporin C synthetase are expressed from a single transcript in *Streptomyces clavuligerus. J. Bacteriol.* **172,** (7) 3952–3958.

Leskiw, B. K., Aharonowitz, Y., Mevarech, M., Wolfe, S., Vining, L. C., Westlake, D. W. S., and Jensen, S. E. (1988). Cloning and nucleotide sequence determination of the isopenicillin *N* synthetase gene from *S. clavuligerus. Gene* **57,** 187–196.

Lowe, D. (1986). Manufacture of penicillins. *In* "Beta-lactam Antibiotics for Clinical Use" (S. F. Queener, J. A. Webber, and S. W. Queener, eds.), pp. 117–161. Marcel Dekker, New York.

McDermott, W., and Rogers, D. E. (1982). Social ramifications of control of microbial disease. *Johns Hopkins Med. J.* **151,** 302–312.

MacDonald, K. D., Hutchinson, J. M., and Gillet, W. A. (1963). Isolation of auxotrophs of *P. chrysogenum* and their penicillin yields. *J. Gen. Microbiol.* **33,** 365–374.

Madduri, K., Stuttard, C., and Vining, L. C. (1989). Lysine catabolism in *Streptomyces* spp. is primarily through cadaverine: Beta-lactam producers also make alpha-aminoadipate. *J. Bacteriol.* **171,** (1)299–302.

Madduri, K., Stuttard, C., and Vining, L. C. (1990). Cloning of a gene governing lysine-ϵ-aminotransferase, the first enzyme in beta-lactam biosynthesis in *Streptomyces. In* "Journal of Cellular Biochemistry. UCLA Symposia on Molecular & Cellular Biology—Abstracts," Suppl. 14A, pp. 108. Wiley-Liss, New York.

Miller, J. R., and Ingolia, T. D. (1989). Cloning and characterization of beta-lactam biosynthetic genes. *Mol. Microbiol.* **3,** (5)689–695.

Ming, L.-J., Que, Jr., L., Kriauciunas, A., Frolik, C. A., and Chen, V. A. (1990). Coordination chemistry of the metal binding site of isopenicillin *N* synthase. *Inorganic Chem.* **29,** (6)1111–1112.

Neftel, K. A., and Hubscher, U. (1987). Effects of beta-lactam antibiotics on proliferating eukaryotic cells. *Antimicrob. Agents Chemother.* **31,** (11)1657–1661.

Pasteur, L., and Joubert, J. (1877). Charbon et septicemie. *C. R. Acad. Sci. Paris* **85,** 101–115.

Penalva, M. A., Tourino, A., Patino, C., Sanchez, F., Fernandez Sousa, J. M., and Rubio, V. (1985). Studies of transformation of *C. acremonium. In* "Molecular Genetics of Filamentous Fungi" (W. E. Timberlake, ed.), pp. 59–68. Liss, New York.

Perry, D., Abraham, E. P., and Baldwin, J. E. (1988). Factors affecting the isopenicillin *N* synthetase reaction. *Biochem. J.* **255,** 345–351.

Picknett, T. M., and Saunders, G. (1989). Transformation of *P. chrysogenum* with selection for increased resistance to benomyl. *FEMS Microbiol. Lett.* **60,** 165–168.

Picknett, T. M., Saunders, G., Ford, P., and Holt, G. (1987). Development of a gene transfer system for *Penicillium chrysogenum*. *Curr. Genet.* **12,** 449–455.

Queener, S. F. (1986). History and origins of beta-lactam antibiotics. *In* "Beta-lactam Antibiotics for Clinical Use" (S. F. Queener, J. A. Webber, and S. W. Queener, eds.), pp. 3–15. Marcel Dekker, New York.

Queener, S. W. (1990). Molecular biology of penicillin and cephalosporin biosynthesis. *Antimicrob. Agents Chemother.* **34,** (6) 943–948.

Queener, S. W., and Neuss, N. (1982). The biosynthesis of β-lactam antibiotics. *In* "The Chemistry and Biology of β-lactam Antibiotics" (R. B. Morin and M. Morgan, eds.), pp. 1–81. Academic Press, London.

Queener, S. W., Ingolia, T. D., Skatrud, P. L., Chapman, J. L., and Kaster, K. R. (1984). A system for genetic transformation of *Cephalosporium acremonium*. ASM Conference on Genetics and Molecular Biology of Industrial Microorganisms. Sept. 30–Oct. 3, 1984. Bloomington, Indiana.

Queener, S. W., Ingolia, T. D., Skatrud, P. L., Chapman, J. L., and Kaster, K. R. (1985). A system for genetic transformation of *Cephalosporium acremonium*. *In* "Microbiology—1985" (L. Lieve, ed.), pp. 468–472. American Society for Microbiology, Washington, D.C.

Ramon, D., Caramolino, L., Patino, C., Sanchez, F., and Penalva, M. A. (1987). Cloning and characterization of the isopenicillin *N* synthetase gene mediating the formation of the β-lactam ring in *Aspergillus nidulans*. *Gene* **57,** 171–181.

Ramos, F. R., Lopez-Nieto, M. J., and Martin, J. F. (1986). Coordinate increase of isopenicillin *N* synthetase, isopenicillin *N* epimerase, and deacetoxycephalosporin C synthetase in a high cephalosporin-producing mutant of *Acremonium chrysogenum* and simultaneous loss of the three enzymes in a non-producing mutant. *FEMS Microbiol. Lett.* **35,** 123–127.

Ramsden, M., McQuade, B. A., Saunders, K., Turner, M. K., and Harford, S. (1989). Characterization of a loss-of-function mutation in the isopenicillin *N* synthetase gene of *Acremonium chrysogenum*. *Gene* **85,** 267–273.

Samson, S. M., Belagaje, R., Blankenship, D. T., Chapman, J. L., Perry, D., Skatrud, P. L., VanFrank, R. M., Abraham, E. P., Baldwin, J. E., Queener, S. W., and Ingolia, T. D. (1985). Isolation, sequence determination and expression in *E. coli* of the isopenicillin *N* synthetase gene from *C. acremonium*. *Nature (London)* **318,** 191–194.

Samson, S. M., Chapman, J. L., Belagaje, R., Queener, S. W., and Ingolia, T. D. (1987a). Analysis of the role of cysteine residues in isopenicillin *N* synthetase activity by site-directed mutagenesis. *Proc. Natl. Acad. Sci. U.S.A.* **84,** 5705–5709.

Samson, S. M., Dotzlaf, J. E., Slisz, M. L., Becker, G. W., Van Frank, R. M., Veal, L. E., Yeh, W.-K., Miller, J. R., Queener, S. W., and Ingolia, T. D. (1987b). Cloning and expression of the fungal expandase/hydroxylase gene involved in cephalosporin biosynthesis. *Bio/Technology* **5,** 1207–1214.

Sanchez, F., Lozano, M., Rubio, V., and Penalva, M. A. (1987). Transformation of *Penicillium chrysogenum*. *Gene* **51,** 97–102.

Seno, E. T., and Baltz, R. H. (1989). Structural organization and regulation of antibiotic biosynthesis and resistance genes in *Actinomycetes*. *In* "Regulation of Secondary Metabolism in *Actinomycetes*" (S. Shapiro, ed.), pp. 1–73. CRC Press, Boca Raton, Florida.

Shiffman, D., Mevarech, M., Jensen, S. E., Cohen, G., and Aharonowitz, Y. (1988). Cloning and comparitive sequence analysis of the gene coding for isopenicillin *N* synthase in *Streptomyces*. *Mol. Gen. Genet.* **214,** 562–569.

Shiffman, D., Cohen, G., Aharonowitz, Y., Palissa, H., von Dohren, H., Kleinkauf, H., and

Mevarech, M. (1990). Nucleotide sequence of the isopenicillin *N* synthase gene *(pcbC)* of the gram negative *Flavobacterium* sp. SC 12,154. *Nucl. Acids Res.* **18,** 660.

Shirafuji, H., Fujisawa, Y., Kida, M., Kanzaki, T., and Yoneda, M. (1979). Accumulation of tripeptide derivatives by mutants of *Cephalosporium acremonium. Agric. Biol. Chem.* **43,** 155-160.

Skatrud, P. L., and Queener, S. W. (1984). Cloning of a DNA fragment from *C. acremonium* which functions as an autonomous replication sequence in yeast. *Curr. Genet.* **8,** 155-163.

Skatrud, P. L., and Queener, S. W. (1989). An electrophoretic molecular karyotype for an industrial strain of *C. acremonium. Gene* **78,** 331-338.

Skatrud, P. L., Fisher, D. L., Ingolia, T. D., and Queener, S. W. (1986). Improved transformation of *C. acremonium. In* "Fifth International Symposium on the Genetics of Industrial Microorganisms" (M. Alacevic, D. Hranueli, and Z. Toman, eds.), pp. 111-119. American Society for Microbiology, Washington, D.C.

Skatrud, P. L., Fisher, D. L., Chapman, J. L., Cantwell, C. A., and Queener, S. W. (1987a). Strain improvement studies in *Penicillium chrysogenum* using the cloned *P. chrysogenum* isopenicillin *N* synthetase gene and the *amdS* gene of *Aspergillus nidulans. SIM News* **37,** (4)77.

Skatrud, P. L., Queener, S. W., Carr, L. G., and Fisher, D. L. (1987b). Efficient integrative transformation of *C. acremonium. Curr. Genet.* **12,** 337-348.

Skatrud, P. L., Tietz, A. J., Ingolia, T. D., Cantwell, C. A., Fisher, D. L., Chapman, J. L., and Queener, S. W. (1989). Use of recombinant DNA to improve production of cephalosporin C in *C. acremonium. Bio/Technology* **7,** 477-485.

Skatrud, P. L., Hoskins, J., Tobin, M. B., Miller, J. R., Wood, J., Kovacevic, S., and Queener, S. W. (1991). Molecular genetics of β-lactam antibiotic synthesis. *In* "Applied Molecular Genetics of Fungi" (J. F. Peberdy, C. E. Caten, J. E. Ogden, and J. W. Bennett, eds.), pp. 145-159. Cambridge University Press, Cambridge, England.

Smith, A. W., Ramsden, M., Dobson, M. J., Harford, S., and Peberdy, J. F. (1990). Regulation of isopenicillin *N* synthetase (IPNS) gene expression in *Acremonium chrysogenum. Bio/Technology* **8,** 237-240.

Smith, D. J., Bull, J. H., Edwards, J., and Turner, G. (1989). Amplification of the isopenicillin *N* synthetase gene in a strain of *P. chrysogenum* producing high levels of penicillin. *Mol. Gen. Genet.* **216,** 492-497.

Smith, D. J., Burnham, M. K. R., Edwards, J., Earl, A. J., and Turner, G. (1990a). Cloning and heterologous expression of the penicillin biosynthetic gene cluster from *P. chrysogenum. Bio/Technology* **8,** 39-41.

Smith, D. J., Burnham, M. K. R., Bull, J. H., Hodgson, J. E., Ward, J. M., Browne, P., Brown, J., Barton, B., Earl, J. A., and Turner, G. (1990b). Beta-lactam antibiotic biosynthetic genes have been conserved in clusters in prokaryotes and eukaryotes. *EMBO J.* **9,** (3)741-747.

Tobin, M. B., Fleming, M. D., Skatrud, P. L., and Miller, J. R. (1990). The acylCoA:isopenicillin *N* acyltransferase genes (*pen*DE) from *P. chrysogenum* and *A. nidulans:* Molecular characterization and activity of recombinant enzyme in *E. coli. J. Bacteriol.* In press.

Usui, S., and Yu, C-A. (1989). Purification and properties of isopenicillin *N* epimerase from *Streptomyces clavuligerus. Biochim. Biophys. Acta* **999,** 78-85.

van Liempt, H., von Dohren, H., and Kleinkauf, H. (1989). δ-(L-α-aminoadipyl)-L-cysteinyl-D-valine synthetase from *Aspergillus nidulans. J. Biol. Chem.* **264,** 3680-3684.

Veenstra, A. E., van Solingen, P., Huininga-Muurling, H., Koekman, B. P., Groenen, M. A. M., Smaal, E. B., Kattevilder, A., Alvarez, E., Barredo, J. L., and Martin, J. F. (1989). Cloning of penicillin biosynthetic genes. *In* "Genetics and Molecular Biology of

Industrial Microorganisms" (C. L. Hershberger, S. W. Queener, and G. Hegeman, eds.) pp. 262–269. American Society for Microbiology, Washington, D.C.

Wakil, S. J. (1989). Fatty acid synthase; A proficient multifunctional enzyme. *Biochemistry* **28,** 4523–4530.

Weigel, B. J., Burgett, S. G., Chen, V. J., Skatrud, P. L., Frolik, C. A., Queener, S. W., and Ingolia, T. D. (1988). Cloning and expression in *Escherichia coli* of isopenicillin *N* synthetase genes from *Streptomyces lipmanii* and *Aspergillus nidulans*. *J. Bacteriol.* **170,** (9)3817–3826.

Weiner, M. (1986). Series introduction. *In* "Beta-lactam Antibiotics for Clinical Use" (S. F. Queener, J. A. Webber, and S. W. Queener, eds.), pp. iii. Marcel Dekker, New York.

Whiteman, P. A., Abraham, E. P., Baldwin, J. E., Fleming, M. D., Schofield, C. J., Sutherland, J. D., and Willis, A. C. (1990). Acyl coenzyme A:6-aminopenicillanic acid acyltransferase from *P. chrysogenum* and *A. nidulans*. *FEBS Lett.* **262,** (2) 342–344.

Wolfe, S., Demain, A. L., Jensen, S. E., and Westlake, D. W. S. (1984). Enzymatic approach to synthesis of unnatural beta-lactams. *Science* **226,** 1386–1392.

Wyngaarden, M. B. (1988). Large-scale production involving *Cephalosporium acremonium* strain LU4-79-6 *Fed. Reg.* **53,** 43410.

Yelton, M. M., Hamer, J. E., and Timberlake, W. E. (1984). Transformation of *Aspergillus nidulans* by using a *trpC* plasmid. *Proc. Natl. Acad. Sci. U.S.A.* **81,** 1470–1474.

18

Heterologous Gene Expression in Filamentous Fungi

CEES A. M. J. J. VAN DEN HONDEL
PETER J. PUNT
ROBERT F. M. VAN GORCOM
TNO Medical Biology Laboratory
Rijswijk, The Netherlands

MORE GENE MANIPULATIONS
IN FUNGI

I. INTRODUCTION

Filamentous fungi are able to grow on a wide variety of carbon and nitrogen sources and are capable of producing a great diversity of low-molecular-weight primary and secondary metabolites as well as intracellular and extracellular enzymes. In addition, a number of species can carry out (stereo)specific modifications of, for example, aromatic compounds and steroids in so-called bioconversion processes (Barbesgaard, 1977; Bennett, 1985). The chemical diversity of the natural environment (i.e., soil and organic waste) in which these organisms live and have to compete with other (micro)organisms probably underlies their extraordinary metabolic diversity.

By the end of the 19th century, the ability of some filamentous fungi to secrete large amounts of organic acids and enzymes was already recognized. This led to the development of industrial production processes using these organisms. The fermentation industry currently produces a large number of products from filamentous fungi, with an annual production value of several billion dollars.

The availability of genetic manipulation techniques, and the increasing knowledge of how genes are expressed, has provided a good basis for investigating the possibility of improving the current production processes of important products. For this purpose, a number of studies have been performed with several biotechnologically important species (some of which have U.S. Food and Drug Administration GRAS status: *G*enerally *R*ecognized *A*s *S*afe), such as *Aspergillus awamori* and *Trichoderma reesei* (extracellular enzymes); *Aspergillus niger* (organic acids and extracellular enzymes); *Aspergillus oryzae* (extracellular enzymes and fermented food); and *Penicillium chrysogenum* and *Acremonium chrysogenum (Cephalosporium acremonium)* (antibiotics).

Many published reports deal with the expression of heterologous genes in various filamentous fungi, usually providing qualitative data regarding the expression of fungal and nonfungal genes in heterologous fungal hosts. Only a few reports, focusing on high-level production of heterologous (fungal and nonfungal) proteins, describe a more systematic approach to achieving efficient expression of heterologous genes and to evaluating the limiting steps in the process.

In this chapter, we will review the present state of knowledge regarding heterologous gene-expression systems that have been used for efficient production of heterologous proteins. We will first describe current understanding of the organization of transcription control sequences in filamentous fungal genes. Then we will give an overview of the available expression systems, summarize the production yields, and evaluate the steps that limit high levels of production. Finally, we will discuss the prospects of heterologous gene expression for production of foreign proteins.

II. TRANSCRIPTION CONTROL SEQUENCES OF FUNGAL GENES

For efficient expression of genes in heterologous host organisms, it is essential that the gene of interest is flanked by transcription control sequences that are functional in that organism. It is reasonable, therefore, that these sequences should originate either from the host organism itself or from a related species. By the time research on heterologous gene expression in filamentous fungi began, a considerable volume of data already existed regarding gene expression in *Saccharomyces cerevisiae* (Hitzeman *et al.*, 1981; Mellor *et al.*, 1983, 1985; Edens *et al.*, 1984) which, like many other industrially important fungi, belongs to the ascomycetes.

A few *Aspergillus* genes have been isolated by complementation of *S. cerevisiae* mutant strains (see Rambosek and Leach, 1987). Conversely, the successful use of yeast transcription control sequences in filamentous fungi has never been reported. This implies that transcription control sequences from filamentous fungi probably differ fundamentally from the corresponding control sequences of *S. cerevisiae.* Indeed, sequence analysis of 5′ nontranscribed regions of several fungal genes fail to show typical yeast promoter elements. Although TATA- and CAAT-box-like sequences have been indicated in the transcription control (promoter) regions of some fungal genes, in most cases they only distantly resemble the yeast elements. (For a review, see Ballance, 1986; Gurr *et al.*, 1988.)

Thus, it was clear that a detailed knowledge of the structure and function of transcription control sequences in filamentous fungi would be important, both for a better understanding of gene expression and for efficient use of these sequences in heterologous gene expression. Research programs were therefore initiated by several groups, including our own, to address the regulation of gene expression in filamentous fungi. In this section, we will describe several approaches for isolation and characterization of transcription control sequences from filamentous fungi.

A. Isolation of Transcription Control Sequences

The first step necessary for a study of transcription control sequences is the isolation of these sequences. Two strategies have been followed: (1) isolation of random sequences with promoter activity and (2) isolation of transcription control sequences from defined genes.

Several reports have been published dealing with the isolation of transcription control sequences using the first approach. In all cases, a library of random DNA sequences from the host organism was introduced into a "promoter-probe" vector containing a promoterless reporter gene preceded by a unique cloning site. The resulting vectors were subsequently introduced into the host organism and expression of the reporter gene, a dominant selectable marker gene, was monitored by the ability of transformants to grow under selective conditions (Turgeon *et al.,* 1987; Arnau *et al.,* 1988). Subsequently, putative promoter sequences were rescued from transformants to be used for further expression studies. However, since the gene from which the isolated sequences originate is not known, an effect on the expression efficiency by modulation of the promoter regulation is unpredictable. Therefore, any statements regarding promoter efficiencies should include a detailed description of the conditions under which the expression studies were carried out.

Following the second strategy, transcription control sequences from several cloned genes have been isolated and characterized. (For a review, see Gurr *et al.,* 1988.) Subsequently, a number of these sequences have been used for expression studies.

B. Characterization of the Functions of Transcription Control Sequences

Attempts to understand control of transcription in filamentous fungi and analysis of the function of transcription control sequences can be approached in different ways.

One approach, which existed long before the introduction of recombinant DNA techniques, is the isolation and characterization of mutant strains with deregulated expression of genes. [See e.g., Cove (1979) for the nitrate assimilation pathway and Hynes (1978) for the regulation of acetamidase gene expression.] Several genetic loci encoding regulatory genes have been identified and characterized using this approach. For example, in the nitrate reduction pathway, putative regulatory genes whose products mediate nitrate induction *(nirA)* and ammonium repression *(areA)* of the nitrate reductase gene *(niaD)* have been identified and the sites of action have been indicated (Cove, 1979).

With the advent of fungal recombinant DNA technology, approaches involving characterization of cloned transcription control sequences have also become feasible. The simplest of these strategies is the comparison of promoter sequences from different co-regulated genes or of promoter sequences from corresponding genes isolated from different species (see e.g., Gwynne *et al.,* 1987b; Punt *et al.,* 1988; Ward *et al.,* 1988). Putative binding sites for the regulatory protein encoded by the *alcR* gene, which plays a role in the regulation of the synthesis of alcohol dehydrogenase *(alcA)* in *Aspergillus nidulans* (Gwynne *et al.*, 1987b), were identified this way. However, this approach gives no information about whether or not the elements identified represent functional transcription control elements. Therefore, the following more sophisticated methods of analysis have been used.

1. Titration Analysis

In titration analysis, multiple copies of a putative regulator binding site are introduced into the fungal host. Due to the presence of the extra binding sites, the amount of the corresponding regulatory protein available for the original binding site(s) is reduced, which could result in deregulated expression of the gene(s) involved. Functional analyses of the regulatory circuits of the *amdS* system (Kelly and Hynes, 1987; Hynes *et al.*, 1989) and the *qut* system (Hawkins and Roberts, 1989), as well as interspecific functionality of binding sites of the *Aspergillus nidulans amdS* gene in *A. niger* (Kelly and Hynes, 1985), have used this approach.

2. Deletion Analysis

In deletion analysis, deletions or substitutions are made in the transcription control region of the gene of interest, and the promoter mutations are reintroduced into the fungal host after fusion with either the gene itself or a reporter gene from which expression can easily be monitored (e.g., *lacZ;* van Gorcom *et al.*, 1986; van Gorcom and van den Hondel, 1988). Numerous sequence elements involved in fungal transcription control have been identified in this fashion (Hamer and Timberlake, 1987; Davis *et al.*, 1988; Goc and Weglenski, 1988; Soliday *et al.*, 1989; Gómez-Pardo and Penalva, 1990; Punt *et al.*, 1990).

3. Protein–DNA Binding Analysis

In protein–DNA binding analysis, attention is focused on the *in vitro* interaction of putative regulatory proteins with a specific site(s) in the transcription control region. Either purified protein, as in the analysis of the *qa* gene cluster of *Neurospora crassa* (Baum *et al.*, 1987), or nuclear protein extracts are used (Podila *et al.*, 1988; Frederick and Kinsey, 1990). Although the demonstration of *in vitro* protein binding at a specific site(s) gives no information about the *in vivo* function of the binding site(s), in combination with one of the *in vivo* methods described earlier, such analysis provides information about the mechanism of regulation of gene expression.

C. Transcription Control Sequences in Heterologous Gene Expression

For heterologous gene expression, two types of transcription control sequences are used. (1) Control sequences from efficiently expressed fungal housekeeping genes. These genes show a constitutive level of expression. The corresponding transcription control sequences are considered to be strong. (2) Control sequences from genes that are subject to specific metabolic regulation. These genes are expressed efficiently only under specific growth conditions, through regulation at the level of transcription.

Table I gives an overview of different transcription control sequences used for studies on heterologous gene expression. As already indicated, the use of specific fungal transcription control sequences in heterologous gene expression requires a detailed knowledge of the organization and regulation of these sequences. This knowledge is available only for a few well-characterized genes. Nevertheless, very promising results have been obtained with some of these control sequences, as will be described in subsequent sections.

The utility of transcription control sequences for heterologous gene expression would be greatly augmented if such sequences were functional in a variety of fungal species. The successful use of fungal marker genes such as *pyrG, argB, trpC,* and *niaD* in various fungal hosts indicated that transcription control sequences from one fungal host can function in another species. (For a review, see Fincham, 1989; Goosen *et al.,* 1990; Van den Hondel and Punt, 1990.) The results presented in Table II, which illustrate the ability of the transcription control region of the *Aspergillus nidulans gpdA* gene to drive expression of the *Escherichia coli* hygromycin phosphotransferase gene *(hph)* in various filamentous fungi, further demonstrate that transcriptional control sequences can indeed be functional in a wide variety of fungal species.

Vector pAN7–1, containing the *gpdA–hph* fusion (Punt *et al.,* 1987), efficiently confers hygromycin B resistance to a variety of fungal species, including the basidiomycetous species *Schizophyllum commune* (Mooibroek *et al.,* 1990) and *Laccaria laccata* (Barrett *et al.,* 1990). However, from the few cases reported to date, it is clear that not all transcription control sequences function normally in other fungal hosts (Casselton and de la Fuente Herce, 1989; T. L. Smith *et al.,* 1990).

III. EXPRESSION SYSTEMS FOR HETEROLOGOUS GENE EXPRESSION

An expression system consists of two basic components: (1) an expression cassette, which contains all the information necessary for expression

TABLE I.

Fungal Transcription Control Sequences Used for Heterologous Gene Expression in Filamentous Fungi

Promoter[a]	Regulation[b]	Expression[c]	References
adhA (alcohol dehydrogenase; *Aspergillus niger*)	Induction by ethanol	Human tissue plasminogen activator	Upshall *et al.*, 1987
alcA (alcohol dehydrogenase; *A. nidulans*)	Induction by ethanol, threonine, cyclopentanon, ethyl metyl ketone	Avian pathogenic surface antigen	Gwynne, personal communication
		Corticosteroid binding globulin (human)	Gwynne, personal communication
		Endoglucanase *(Cellulomonas fimi)*	Gwynne *et al.*, 1989
		Epidermal growth factor (human)	Gwynne, personal communication
		Glucoamylase *(A. niger)*	Gwynne *et al.*, 1989
		Growth hormone (human)	Gwynne, personal communication
		Interferon $\alpha 2$ (human)	Gwynne *et al.*, 1987a
		Interleukin 6 (human, rat)	Gwynne, personal communication
		Parathyroid hormone (human)	Gwynne, personal communication
		Superoxide dismutase (human)	Gwynne, personal communication
alcA (alcohol dehydrogenase; *A. niger*)	Induction by ethanol	Interferon $\alpha 2$ (human)	Gwynne *et al.*, 1987a
alcC (alcohol dehydrogenase; *A. nidulans*) *(adh3)*	Induction by ethanol	Tissue plasminogen activator (human)	Upshall *et al.*, 1987
amdS (aceetamidase; *A. nidulans*)	Induction by aceetamide, acetate, ω-amino acids	Enterotoxin subunit B *(Escherichia coli)*	Turnbull *et al.*, 1989
amyA/B (α-amylase; *A. niger*)	Induction by starch, maltodextrin	Chymosin (calf)	Korman *et al.*, 1990
amy (taka) [(taka-)amylase; *A. oryzae*]	Induction by starch	α-Amylase *(A. oryzae)*	Ruttkowski, Kahnh, and Gottschalk, personal communication
		Aspartyl protease *(Mucor miehei)*	Christensen *et al.*, 1988
		Chymosin (calf)	Boel *et al.*, 1987
		Lipase *(Thermomyces lanuginosus)*	Boel and Huge-Jensen, 1989

		Triglyceride lipase *(M. miehei)*	Boel *et al.*, 1987; Huge-Jensen *et al.*, 1989
cbh1 (cellobiohydrolase; *Trichoderma reesei*)	Induction by cellulose	Chymosin (calf)	Harkki *et al.*, 1989
glaA (glucoamylase; *A. niger, A. awamori*)	Induction by starch, maltodextrin, maltose	Aspartyl protease *(M. miehei)*	Christensen *et al.*, 1988; Ward, personal communication
		Cellobiohydrolase *(T. reesei)*	Ward, personal communication
		Chymosin (calf)	Cullen *et al.*, 1987; Ward *et al.*, 1989a,b, 1990; van Hartingsveldt *et al.*, 1990a,b
		Glucoamylase *(Humicola grisea)*	Ward, personal communication
		Interferon $\alpha 2$ (human)	Gwynne *et al.*, 1987a
		Lysozyme (hen egg white)	Archer, personal communication
		Preprocathepsin D (human)	Ward, personal communication
		Thaumatin (plant)	Ward, personal communication
		Triglyceride lipase *(M. miehei)*	Boel *et al.*, 1987
gpdA (glyceraldehyde-3-phosphate dehydrogenase; *A. nidulans*)	Constitutive	Aspartase *(E. coli)*	Hunter, Bailey, and Arst, personal communication
		β-Galactosidase *(E. coli)*	Kolar *et al.*, 1988; De Ruiter-Jacobs *et al.*, 1989; Punt *et al.*, 1991
		β-Glucuronidase *(E. coli)*	De Ruiter-Jacobs *et al.*, 1989; Roberts *et al.*, 1989; Punt *et al.*, 1991
		Glucoamylase *(A. niger)*	Punt *et al.*, 1991
		Lysozyme (hen egg white)	Archer *et al.*, 1990
		Phleomycin binding protein *(Streptoalloteichus hindustanus)*	Camels *et al.*, 1991
gpdA (glyceraldehyde-3-phosphate dehydrogenase; *A. niger*)	Constitutive	β-Galactosidase *(E. coli)*	van den Hondel and Punt, 1990
mprA (aspartyl protease; *M. miehei*)	Induction	Aspartyl protease *(M. miehei)*	Gray *et al.*, 1986; Dickinson *et al.*, 1987

(continued)

TABLE I. *(continued)*

Promoter[a]	Regulation[b]	Expression[c]	References
oliC (ATPase subunit 9; *A. nidulans*)	Constitutive	β-Galactosidase *(E. coli)*	Turner, 1990
oliC (ATPase subunit 9; *A. niger*)	Constitutive	Chymosin (calf)	Cullen *et al.*, 1987
pcbC (isopenicillin N synthetase; *P. chrysogenum*)	Secondary metabolism	Expandase *(Streptomyces clavuligerus)*	Cantwell *et al.*, 1990
pe (pectin esterase; *A. niger*)	Induction	Pectin esterase *(A. niger)*	Ruttkowski, Kahnh, and Gottschalk, personal communication
pelA (pectin lyase; *A. niger*)	Induction	Interferon β (human hybrid)	Heim *et al.*, 1990
pg (polygalacturonidase; *A. niger*)	Induction	Polygalacturonidase *(A. niger)*	Ruttkowski, Kahnh, and Gottschalk, personal communication
tpiA (triose-phosphate isomerase; *A. nidulans*)	Constitutive	Human tissue plasminogen activator	Upshall *et al.*, 1987

[a] The fungal gene from which the promoter is used is indicated.
[b] Mode of regulation of the promoter is indicated; in some cases the source of induction is not specified.
[c] The protein produced under control of the indicated promoter is given; see also Tables III, IV, and V.

TABLE II.

Fungal Species Transformed Successfully with the Vector pAN7-1 Containing the Bacterial Hygromycin B Resistance Gene

Transformed Species	Reference
Acremonium chrysogenum	A. W. Smith *et al.*, 1990
Aspergillus awamori	Punt, unpublished observations
Aspergillus ficuum	Mullaney *et al.*, 1988
Aspergillus giganteus	Wnendt *et al.*, 1990
Aspergillus nidulans	Punt *et al.*, 1987
Aspergillus niger	Punt *et al.*, 1987
Claviceps purpurea	Comino *et al.*, 1989
Cryphonectria parasitica	Churchill *et al.*, 1990
Curvularia lunata	Osiewacz and Weber, 1989
Fulvia fulvum	Oliver *et al.*, 1987
Fusarium culmorum	Curragh *et al.*, 1990
Laccaria laccata	Barrett *et al.*, 1990
Leptosphaeria maculans	Farman and Oliver, 1988
Neurospora crassa	Staben *et al.*, 1989
Penicillium roqueforti	Durand *et al.*, 1990
Pseudocercosporella herpetrichoides	Blakemore *et al.*, 1989
Schizophyllum commune	Mooibroek *et al.*, 1990
Septoria nodorum	Cooley *et al.*, 1988
Trichoderma hamatum	Ulhoa *et al.*, 1990
Trichoderma harzianum	Goldman *et al.*, 1990; Ulhoa *et al.*, 1990
Trichoderma viride	Herrera-Estrella *et al.*, 1990
Trichosporon cutaneum	Glumoff *et al.*, 1989

of the protein of interest, for example, transcriptional, translational, and, in the case of a secreted protein, secretion signals; and (2) an appropriate host strain for optimal production of the protein.

To obtain an expression system, a gene-transfer system is needed to introduce the expression cassette into the host of choice. Several sophisticated gene-transfer systems are available for various biotechnologically important filamentous fungi. (For recent reviews see Fincham, 1989; Timberlake and Marshall, 1989; Goosen *et al.*, 1990; van den Hondel and Punt, 1990.)

In this section we will describe the general structure of the expression cassettes that have been designed to facilitate efficient production of extracellular proteins, discuss the strategies that have been developed to obtain strains containing multiple copies of the expression cassette, and describe different host strains that are constructed to facilitate optimal production of heterologous proteins. The results obtained with the expression systems available are given in Section IV.

A. Expression Cassettes Designed for Heterologous Gene Expression

As indicated in Fig. 1, an expression cassette includes the following components: (1) a region comprising all 5′ transcription control sequences; (2) a translation initiation region; (3) in the case of extracellular production, secretion-facilitating sequences (usually a signal sequence); (4) DNA sequences encoding the protein of interest; and (5) a transcription termination and polyadenylation control region.

1. Transcription Control Region

As already mentioned in Section II, two types of transcription control region are generally used in expression cassettes: regions from constitutively expressed genes and regions from highly expressed genes with regulated expression (Table I). The latter type of transcription control region offers the possibility of inducing protein synthesis at a specific stage during fermentation, which can be advantageous when the protein to be produced is unstable or detrimental to the host.

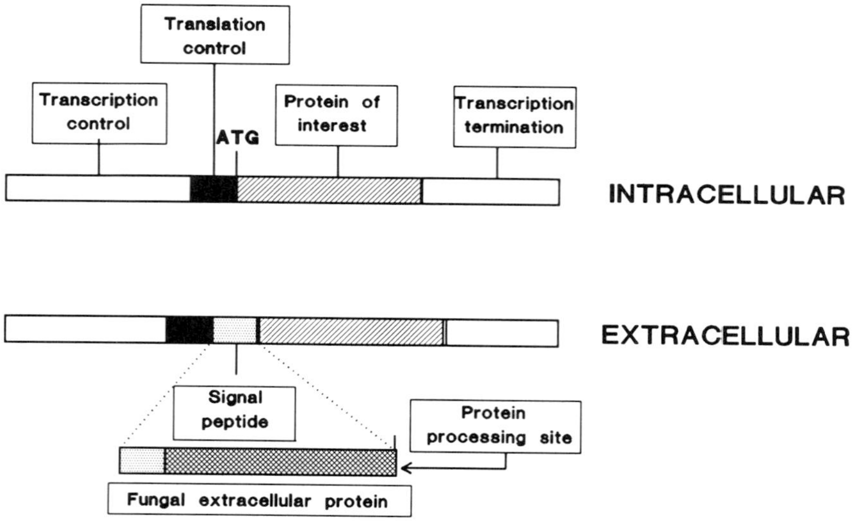

Fig. 1. Schematic representation of expression cassettes designed for intra- or extracellular protein production in filamentous fungi. The cassettes comprise fungal sequences that control both the initiation and termination of transcription as well as the initiation of translation. For secretion of the protein of interest, the cassettes contain either DNA sequences encoding a secretory signal peptide or the complete coding region of an efficiently secreted fungal protein. In the latter case, sequences coding for a protein-processing site have been introduced between the coding region of both proteins in order to obtain the protein of interest in its mature form.

2. Transcription Termination and Polyadenylation Region

At the moment, the mechanism of transcription termination and polyadenylation in filamentous fungi is unknown. However, it is generally assumed that transcription termination and polyadenylation are important in order to achieve optimal levels of mRNA (Davies, 1990). Therefore, all expression cassettes in use contain a transcription termination and polyadenylation region, usually originating from the same gene as the transcription control region.

3. Translation Initiation Region

Little is known about the factors and sequences that are important for efficient translation initiation in filamentous fungi, and there are no systematic studies on this subject. In most expression cassettes, sequences upstream of the ATG have been used that originate from efficiently expressed fungal genes. In some cases, this sequence includes the consensus sequence upstream of the ATG for higher eukaryotic (CCACC*ATG;* Kozak, 1984) or filamentous fungal (${}^{C}_{T}$CA${}^{CC}_{AA}$*ATG;* Gurr *et al.,* 1988) initiation sites (Christensen *et al.,* 1988). The effects of the introduction of the consensus sequences on production levels are not known. In one report, a 2-fold improvement of protein production is reported by changing the sequence directly upstream of the ATG (Gwynne *et al.,* 1987a).

4. Secretion-Facilitating Sequences

A minimum requirement for secretion of proteins is that a sequence encoding a secretory signal peptide is included in the expression cassette. This signal peptide is necessary for the translocation of a polypeptide across the membrane of the endoplasmatic reticulum, the first step in the protein secretion pathway (Blobel and Dobberstein, 1975). In a number of studies, focused on maximal production of heterologous proteins, differences in protein secretion of 2–5-fold were observed when using different signal peptides (Cullen *et al.,* 1987; Harkki *et al.,* 1989; Ward, 1989a; Van Hartingsveldt *et al.,* 1990b).

Recently Ward and co-workers (1989b, 1990) have shown that the secretion of extracellular chymosin can be increased 10–15-fold by the introduction of an additional *N*-glycosylation site in the protein or by an in-frame fusion of the DNA sequence of prochymosin to the complete coding region of the *Aspergillus awamori* glucoamylase gene. The *glaA* fusion approach has also been successfully applied to human interleukin 6 in *A. nidulans* (R. Contreras, personal communication) and, similarly, using the *A. awamori* α-amylase as a "carrier" (Korman *et al.,* 1990).

For convenient construction of various expression cassettes, a great number of expression vectors have been developed that contain handy restriction sites for introducing the coding region of the protein of interest. In most cases, for precise fusion constructions, site-directed mutagenesis or polymerase chain reaction (PCR) methods have had to be employed. The importance of precise fusions is illustrated by Gwynne and coworkers (1987c), who demonstrated a 15-fold increase in the production of human interferon $\alpha 2$, as detected immunologically, after precise fusion of the coding region of mature interferon $\alpha 2$ to the fungal secretion signal.

For obvious reasons, all fungal expression vectors described have been constructed in *Escherichia coli* plasmids. A few representative examples of expression vectors used for the production of intracellular and extracellular proteins in our own laboratory are given in Fig. 2.

B. Amplification of Expression Cassettes

In filamentous fungi, transforming DNA generally integrates into the host genome. This integration occurs by either homologous or nonhomologous integration events. The number of integrated copies varies considerably and depends on both the host strain and the method of selection. In a few systems, predominantly multicopy transformants, carrying tandemly

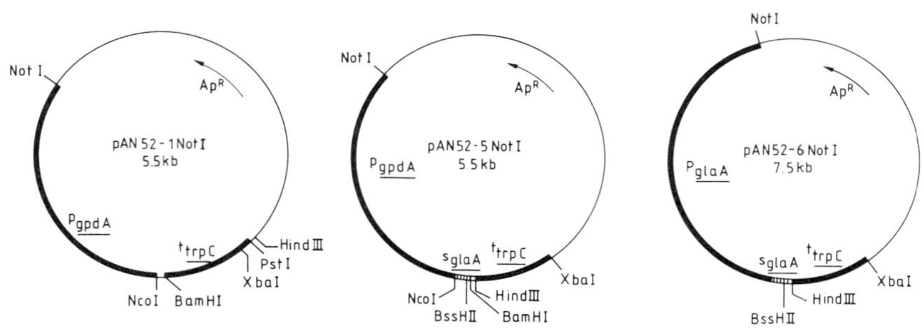

Fig. 2. Schematic representation of three expression vectors constructed and used in our laboratory. pAN52-NotI is a derivative of pAN52-1 (Punt *et al.*, 1987) with a unique *Not*I site and pAN52-5NotI is a derivative of pAN52-4 (Punt *et al.*, 1991) with a unique *Not*I site and a signal sequence identical to the signal sequence of the *A. niger glaA* gene. Vector pAN52-6NotI, a derivative of pAN52-5NotI, contains the transcription and translation control region and the coding sequence of the first 24 amino acids (secretion signal) of the *A. niger glaA* gene (Boel *et al.*, 1984). Only unique restriction sites are indicated. Thick lines represent *Aspergillus* DNA and thin lines pUC18 DNA. p_{gpdA}, transcription and translation control sequences of the *A. nidulans gpdA* gene; p_{glaA}, transcription and translation control sequences of the *A. niger glaA* gene; s_{glaA}, the coding sequence of the first 24 amino acids (secretion signal) of the *A. niger glaA* gene; t_{trpC}, transcription termination region of the *A. nidulans trpC* gene (Mullaney *et al.*, 1985); Ap^R, ampicillin resistance gene.

repeated copies of the vector, are obtained (Kelly and Hynes, 1985; Wernars *et al.*, 1985; Durrens *et al.*, 1986.

Multicopy transformants can also be obtained by co-transformation. Since, in most cases, co-transformation is very efficient (50–80% using most selection markers) and precludes the necessity for often time-consuming vector construction, it represents a convenient method for the introduction of multiple copies of a given expression cassette.

In all industrially important filamentous fungi, with the exception of *Mucor*, in which vector DNA replicates autonomously, (co)transformants that contain more than one integrated copy of the expression cassette can easily be obtained. Strains containing 10 or more copies of the expression cassette can be identified by screening a limited number of transformants.

In general, a higher copy number of an expression cassette results in higher protein production, although in most cases there is no strict correlation between the two. This lack of strict correlation is probably attributable to the different sites of integration of the expression cassette (Van Gorcom *et al.*, 1985; Christensen *et al.*, 1988; Ward *et al.*, 1990).

In the case of the *amdS* selection marker of *Aspergilus nidulans*, multicopy transformants can be detected and selected in a number of fungi by their growth phenotype (Kelly and Hynes, 1985, 1987; Christensen *et al.*, 1988; Harkki *et al.*, 1989). Most fungal species cannot grow readily using acetamide or acrylamide as sole nitrogen source, whereas transformants that contain several copies of the *amdS* gene can. Southern blot analysis of *A. nidulans* and *A. niger* transformants reveals that, in most cases, they contain multiple copies of the *amdS* gene. Multiple copies are probably required for good growth on selective media (Hynes, 1986).

Studies in our laboratory have shown that *A. niger* transformants, selected on acrylamide-supplemented medium, contained a higher copy number (at least 10 copies) than transformants selected on acetamide medium (copy numbers varying from 1 to 10). We recently developed a further improvement of the method to obtain a high frequency of transformants with very high copy numbers (> 100) of an expression cassette. This method requires a cosmid vector which, in addition to the *amdS* selectable marker, contains 4–10 copies of the expression cassette. For example, 4 copies of the *A. niger* glucoamylase *(glaA)* gene were introduced into a cosmid vector. Analysis of six *A. niger* transformants obtained using this *glaA* cosmid, showing rapid growth on acrylamide-containing agar plates, revealed that one transformant contained about 10 copies of the cosmid (i.e., 40 copies of the *glaA* gene) and five transformants contained 20–40 copies of the cosmid (i.e., 80–160 copies of the *glaA* gene; J. Verdoes and C. A. M. J. J. van den Hondel, unpublished observations). Several of these transformants gave 5–10-fold increases in glucoamylase production, although no strict correlation between the copy number and the level of glucoamylase production was found.

A. nidulans transformants containing multiple copies of the transcription control region of *amdS* show a decreased ability to grow on β-lactams and ω-amino acids, such as pyrrolidinone and γ-amino butyric acid (Kelly and Hynes, 1987). This property may be exploited to screen transformants for the presence of multiple copies of the vector and, thus, for multiple copies of the expression cassette. Recently we have found that a similar screening on γ-amino butyric acid is possible for *A. niger,* and that transformants exhibiting impaired growth on plates supplemented with this nitrogen source contain more than 30 copies of the cosmid vector used for transformation. In summary, the *amdS* system used in combination with a multicopy cosmid can yield transformants with very high copy numbers of the expression cassette.

The mitotic stability of transformants containing high copy numbers of an expression cassette has been assessed in a limited number of studies by measuring the protein production or by performing Southern blot analysis after one or several transfers on nonselective medium (Kelly and Hynes, 1985; Wernars *et al.,* 1985; Cullen *et al.,* 1987; Davies, 1990; J. Verdoes and C. A. M. J. J. van den Hondel, unpublished data). In none of these studies was a clear difference observed after several transfers, suggesting that multicopy transformants are stable through a great number of mitotic divisions in most cases.

C. Strains Constructed for Optimal Expression

In addition to an optimal design of the expression vector and a high copy number of the expression cassette in the transformants, host characteristics are likely to play an important role in optimal gene expression and protein secretion. Indications that genetic modification of the host can lead to improvements in production yields have come from recent research by several groups.

One "classical" approach, still in use for strain improvement, is random mutagenesis and screening. In the case of chymosin production by a glucoamylase-production strain of *A. awamori,* several rounds of mutagenesis and screening resulted in a 4-fold increase in the enzyme (Ward, 1989b).

A second approach is the direct elimination of the factors and circumstances limiting protein production. This approach is hampered by the fact that, at present, only a limited knowledge of the process of heterologous gene expression and protein secretion is available. Therefore, only a few obvious limiting factors and circumstances have been revealed using more sophisticated methods.

1. Enhancement of Transcription Factors

In strains containing multiple copies of an expression cassette, the level of factors necessary for an optimal transcription efficiency may be limiting. For most of the transcription control regions used for efficient expression of heterologous genes, it is unknown whether this limitation occurs. However, with the *alcA* expression system such a limitation of the level of a positive-acting regulatory protein (AlcR) has been observed. Therefore, a host strain was generated with a large number of copies of the *alcA* regulatory gene *alcR*. Transformants of this strain, containing multiple copies of an *alcA*-based expression cassette for interferon $\alpha2$, produced significantly higher levels of protein than similar transformants of a wild-type single-copy *alcR* strain (Gwynne *et al.,* 1987c). Similar strategies would be feasible for other systems in which *trans*-acting regulatory proteins are limiting.

2. Elimination of Extracellular Protease Activity

The secretion of proteases into the culture fluid may have a detrimental effect on the efficient production of heterologous proteins in many filamentous fungi. Proteins of interest have been degraded after incubation with the culture fluid of a production strain; for example, complete degradation was observed when human interleukin 6 was incubated with culture fluid of *Aspergillus niger* (I. E. Mattern, personal communication). Protease inhibitors may also be used to monitor the production of extracellular proteins. For example, the production of chymosin by *A. niger* transformants was considerably increased after cultivation in medium containing the protease inhibitors pepstatin and α_2-macroglobulin (van Hartingsveldt *et al.,* 1990b).

To overcome the problem of proteolysis, strains deficient in extracellular protease activity have been generated. Berka and coworkers (1990) constructed an *A. awamori* strain in which a major extracellular protease (aspergillopepsin A, an aspartyl protease) was deleted using a gene-replacement strategy. In our own laboratory, we have used a mutagenesis and selection strategy to obtain *A. niger* Prt$^-$ mutants which exhibit reduced extracellular protease activity (I. E. Mattern and C. A. M. J. J. van den Hondel, unpublished observations). A more detailed genetic and biochemical study of three of these Prt$^-$ mutants indicated that the culture fluid of each of these mutants contained less protease activity. The *prt* mutations were located on different chromosomes. Purification of extracellular proteins from the wild-type and from the three Prt$^-$ mutants by binding to immobilized bacitracin, which selectively absorbs many different proteases (Stepanov and Rudenskaya, 1983), and subsequent desorption (J. M. van Noort, P. van der Berg, J. E. Mattern unpublished data) suggests the presence of several proteases. Some of these protein species are absent from

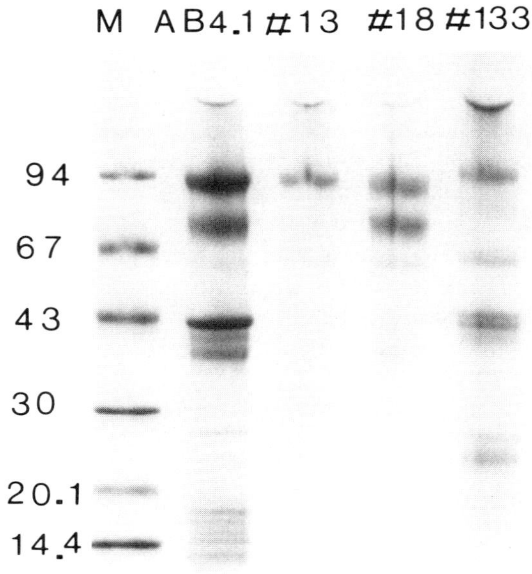

Fig. 3. SDS-PAGE analysis of secreted proteins of *Aspergillus niger* AB4.1 and three Prt⁻ mutants that selectively bind to immobilized bacitracin. Selective binding and elution of extracellular proteins is visualized by SDS-PAGE after bacitracin–sepharose chromatography of medium samples of *Aspergillus niger* AB4.1 (van Hartingsveldt *et al.*, 1987), a *pyrG* mutant of *A. niger* 402; and three Prt⁻ mutant strains derived from *A. niger* AB4.1: #13, #18, and #133 (Mattern and van den Hondel, unpublished data). The molecular weight markers (M) are indicated in kDa.

the mutant strains (Fig. 3). Experiments are currently in progress to verify that all the proteins purified in this way are indeed proteases.

IV. OVEREXPRESSION OF HETEROLOGOUS PROTEINS

In general, the efficient production of heterologous proteins has been studied by (1) introducing into the production host different expression cassettes, often in multiple copies; (2) testing a number of transformants for the amount of protein produced, determining the copy number of the expression cassette, assessing the mRNA level, and characterizing the protein produced; and (3) in some cases, further optimizing the production level by using modified hosts (see Section III,C), optimized media, appropriate large scale fermentation conditions, and classical strain improvement procedures.

In this section we will give an overview of the yields of heterologous proteins that have been produced intra- or extracellularly in fungal hosts. In many cases, only initial production yields are available. Proprietary regulations often preclude the reporting of results from further optimized production processes.

A. Production of Intracellular Proteins

The capacity of various expression signals to direct efficient production of intracellular proteins has been widely assayed using the *Escherichia coli* enzymes β-galactosidase and β-glucuronidase. The activity levels of these enzymes can be determined easily both qualitatively (with plate tests) and quantitatively. In addition, (monoclonal) antibodies are available for Western blot analysis.

Considerable amounts of β-galactosidase and β-glucoronidase are produced by several *Aspergillus* species and *Penicillium chrysogenum* (Table III). In the different fungal species studied, similar yields were found, and were in the order of the production yields of homologous proteins such as the *A. nidulans* GPD protein (Punt *et al.,* 1991). Thus the expression signals used in the different fungal species studied are rather efficient. It is remarkable that, in spite of the high concentration in the fungal mycelium, the enzymes were present in a soluble active form. In other organisms such as *E. coli,* insoluble aggregates are formed when proteins are overexpressed (Primrose, 1986).

Economically important proteins, such as human superoxide dismutase, can also be produced intracellularly, under optimized conditions, at levels of hundreds of milligrams per liter (Table III). Efficient production of the heterologous enzymes involved in antibiotic biosynthesis will also become an important field of research in the future. The expression in *P. chrysogenum* of the expandase gene *(cefE)* of a *Streptomyces* species is a first step in this direction (Cantwell *et al.,* 1990).

B. Production of Extracellular Proteins

1. Heterologous Fungal Proteins

Several of the investigations concerning the efficiency of secretion of heterologous proteins have focused on fungal enzymes. Most of these enzymes, shown in Table IV, are important for the industrial production of

TABLE III.

Intracellular Heterologous Proteins Produced by Filamentous Fungi

Production Organism	Protein	Initial Yield	Reference
Aspergillus nidulans	Aspartase *(Escherichia coli)*	Detectable	Hunter, Bailey, and Arst, Jr., personal communication
	β-Galactosidase *(E. coli)*	13[a]	Punt, unpublished data
	β-Glucuronidase *(E. coli)*	10[a]	Punt, unpublished data
	Superoxide dismutase (human)	≫ 100 mg/l[b]	Gwynne, personal communication
A. niger	β-Galactosidase *(E. coli)*	10[a]	Punt *et al.*, 1991
	β-Glucuronidase	8[a]	Punt *et al.*, 1991
	Penicillin biosynthetic gene cluster *(Penicillium chrysogenum)*	Detectable	D. J. Smith, *et al.*, 1990
A. oryzae	β-Galactosidase *(E. coli)*	1–2[a]	De Ruiter-Jacobs *et al.*, 1989
	β-Glucuronidase *(E. coli)*	5[a]	De Ruiter-Jacobs *et al.*, 1989
Neurospora crassa	Penicillin biosynthetic gene cluster *(P. chrysogenum)*	Detectable	D. J. Smith *et al.*, 1990
P. chrysogenum	Expandase *(Streptomyces clavuligerus)*	Detectable	Cantwell *et al.*, 1990
	β-Galactosidase *(E. coli)*	3[a]	Kolar *et al.*, 1988

[a] The amount of protein produced is indicated as the percentage of total soluble protein.
[b] This yield is obtained after scaling up the production.

foodstuffs, animal feeds, and detergents. A large number of studies has been carried out using *Aspergillus awamori, A. niger,* and *A. oryzae,* for which extensive experience of fermentation and downstream processing has been established already.

The initial level of production of heterologous fungal proteins is in the order of 10–50 mg/liter (Table IV). Under nonoptimized conditions, similar production levels are observed for efficiently secreted homologous proteins. After optimization of the production process, levels of about 3 g/liter can be obtained. This yield has been demonstrated by the groups at Novo Industries and Genencor, in the production of aspartic protease from *Mucor miehei,* by *A. oryzae* and *A. awamori,* respectively. Recently a *Humicola* (or *Thermomyces*) *lanuginosus* lipase, produced by *A. oryzae,* has been brought onto the market, indicating that this protein also is produced efficiently.

In summary, the first results from the studies of expression and production of heterologous fungal proteins are very promising. We expect, there-

TABLE IV.

Extracellular Heterologous Fungal Proteins Produced by Filamentous Fungi

Production Organism	Protein	Yield		Reference
		Initial	Optimized	
Aspergillus awamori	α-Amylase *(A. oryzae)*	1–1.5 g/l	—	Ruttkowski, Kahnh, and Gottschalk, personal communication
	Aspartyl protease *(Mucor miehei)*	50–100 mg/l	~3 g/l	Ward, personal communication
	Cellobiohydrolase *(Trichodenna reesei)*	~50 mg/l	—	Ward, personal communication
	Glucoamylase *(Humicola grisea)*	50–100 μg/l	0.65 g/l	Ward, personal communication
A. nidulans	Aspartyl protease *(M. miehei)*	Detectable	—	Gray *et al.*, 1986
	Glucoamylase *(A. niger)*	—	1.2 g/l	Gwynne *et al.*, 1989
A. niger	Lipase *(M. miehei)*	—	—	Boel and Huge-Jensen, 1989
A. oryzae	Aspartyl protease *(M. miehei)*	50–150 mg/l	~3.3 g/l	Christensen *et al.*, 1988
	Lipase *(H. lanuginosus)*	—	—	Boel and Huge-Jensen, 1989
	Pectine esterase *(A. niger)*	0.8–1.0 U/ml	—	Ruttkowski, Kahnh, and Gottschalk, personal communication
	Polygalacturonase *(A. niger)*	20–30 U/ml	—	Ruttkowski, Kahnh, and Gottschalk, personal communication
	Triglyceride lipase *(M. miehei)*	2 mg/l	—	Huge-Jensen *et al.*, 1989
M. circinelloides	Aspartyl protease *(M. miehei)*	1–12 mg/l	—	Dickinson *et al.*, 1987

fore, that in the near future other interesting fungal enzymes also will be produced successfully in a similar manner.

2. *Heterologous Nonfungal Proteins*

a. Bacterial Proteins. To date, four studies dealing with extracellular production of bacterial proteins by filamentous fungi have been reported: endoglucanase from *Cellulomonas fimi* (Gwynne *et al.*, 1987a); enterotoxin subunit B from *Escherichia coli* (Turnbull *et al.*, 1989); β-glucuronidase from *E. coli* (Punt *et al.*, 1991); and the phleomycin-binding protein from *Streptoalloteichus hindustanus* (Camels *et al.*, 1991). Of these proteins, endoglucanase was secreted into the medium by *Aspergillus nidulans* in the range of 10–15 mg protein per liter, whereas phleomycin binding protein was secreted very efficiently (1.5 g/liter) by *Tolypocladium geodes* (Camels *et al.*, 1991). Enterotoxin subunit B and β-glucuronidase, however, were not detectable in the culture fluid, but were only detectable intracellularly in *A. nidulans* and *A. niger*, respectively. These latter results can be explained either by lack of secretion of these proteins or by complete degradation in the culture fluid. Analysis of these intracellularly located proteins revealed, in both cases, that the signal peptide very probably was removed, indicating that the protein had been translocated into the endoplasmic reticulum. Thus, barring degradation, the absence of enterotoxin subunit β and β-glucuronidase in the culture fluid was probably due to retention of the proteins in one of the compartments involved in the secretion process.

b. Mammalian Proteins. For obvious commercial reasons, many of the studies on heterologous gene expression have concerned mammalian genes. Initially the expression of several of these genes was studied in *Aspergillus nidulans.* Although the production yields were relatively low (about 0.2–2 mg/l; Table V), the proteins produced were biologically active and resembled the authentic mammalian product (e.g., Upshall *et al.*, 1987). Subsequently, Gwynne and co-workers have optimized the expression system in *A. nidulans* using strains that express and secrete *A. niger* glucoamylase under control of the transcription control sequences of *alcA* (Gwynne *et al.*, 1989). Using the *alcA*-based expression cassette and the host strain (see Section III,C), a production level of 1.2 g glucoamylase per liter could be attained (Table IV). Human corticosteroid-binding globulin and human epidermal growth factor have now been produced by *A. nidulans* at levels of 5–10 mg/liter (Table V).

After initial expression studies with chymosin in *A. nidulans,* Ward and co-workers focused their attention on *A. awamori.* Insertion of their *glaA*-based expression cassettes into this fungus resulted in similar levels of secreted product as in *A. nidulans.* Subsequent optimization of production —by fusing the coding region of the chymosin gene to the complete coding regions of efficiently secreted fungal proteins (Korman *et al.*, 1990; Ward *et al.*, 1990); by using improved host strains (Berka *et al.*, 1990; see Section III,C); and by performing several rounds of mutagenesis and screening (Ward, 1989b)—has led to commercially attractive production levels (Ward *et al.*, 1990).

Rather efficient chymosin production levels have been obtained with

TABLE V.

Heterologous Nonfungal Proteins Produced by Filamentous Fungi

Production Organism	Protein	Yield: Initial	Yield: Optimized	Reference
Aspergillus awamori	Chymosin (calf)	10–40 mg/l	>1 g/l	Ward *et al.*, 1989a,b, 1990
	Prepro-cathepsin D	Detectable	—	Ward, personal communication
	Thaumatin (plant)	Detectable	—	Ward, personal communication
A. nidulans	Avian pathogenic surface antigen	0.5–2 mg/l	—	Gwynne, personal communication
	Chymosin (calf)	0.2–2 mg/l	—	Cullen *et al.*, 1987
	Corticosteroid binding globulin (human)	—	5–10 mg/l	Gwynne, personal communication
	Endoglucanase *Cellulomonas fimi)*	20 mg/l	—	Gwynne *et al.*, 1987a
	Enterotoxin subunit B *(Escherichia coli)*	24 μg/g[a]	—	Turnbull *et al.*, 1989
	Epidermal growth factor (human)	1–2 mg/l	5–10 mg/l	Gwynne, personal communication
	Growth hormone (human)	0.5–2 mg/l	—	Gwynne, personal communication
	Interferon-α2 (human)	1 mg/l	—	Gwynne *et al.*, 1987a
	Interleukin 6 (human, rat)	0.5–2 mg/l	—	Gwynne, personal communication
	Parathyroid hormone (human)	0.5–2 mg/l	—	Gwynne, personal communication
	Tissue plasminogen activator (human)	1 mg/l	—	Upshall *et al.*, 1987
A. niger	β-Glucuronidase *(E. coli)*	60 μg/g[a]	—	Punt *et al.*, 1991
	Chymosin (calf)	4–12 mg/l	—	Van Hartingsveldt *et al.*, 1990a,b
	Chymosin (calf)	1–10 mg/l	—	Boel *et al.*, 1987
	Hen egg white lysozyme	1–12 mg/l	—	Archer *et al.*, 1990
	Interferon β (human hybrid)	—	—	Heim *et al.*, 1990
Neurospora crassa	Hen egg white lysozyme	3–10 mg/l	—	Collins, Sheffield, and Connerton, personal communication
Tolypocladium geodes	Phleomycin binding protein *(Streptoalloteichus hindustanicus)*	1.5 g/l	—	Camels *et al.*, 1991
Trichoderma reesei	Chymosin (calf)	0.3–3 mg/l	17–40 mg/l	Harkki *et al.*, 1989

[a] The amount of protein, which was almost exclusively detected intracellularly, has been reported as μg/g wet weight of mycelia.

Trichoderma reesei, using *cbhI*-based expression cassettes (Harkki *et al.,* 1989), and also with *A. niger* (van Hartingsveldt *et al.,* 1990a,b).

c. Nonmammalian Eukaryotic Proteins. Two nonmammalian higher eukaryotic proteins, plant thaumatin and hen eggwhite lysozyme, have been produced by *Aspergillus* and/or *Neurospora* species (Archer *et al.,* 1990; M. E. Collins, P. J. Sheffield, and I. F. Connerton, personal communication; M. Ward, personal communication). In contrast to thaumatin, which could only be produced in detectable amounts, the production level of hen eggwhite lysozyme (1 – 10 mg/liter) was similar to the production level of most of the mammalian proteins (Table V).

C. Evaluation of Overexpression Results

Comparison of recent expression studies on heterologous fungal and nonfungal proteins (Tables III, IV, and V) shows that, in general, intracellular and extracellular production of heterologous fungal proteins is very efficient. However, the extracellular production levels of nonfungal proteins are usually much lower than those of fungal proteins, as a result of problems in transcription, translation, or secretion and degradation. Expression studies with nonfungal model proteins, such as *Escherichia coli* β-galactosidase (Table III), have indicated that efficient production of nonsecreted soluble heterologous proteins is relatively easy to achieve, suggesting that secretion or degradation seems to be the main barrier. Furthermore, the expression cassettes of nonfungal proteins usually are transcribed efficiently, suggesting problems at subsequent steps in expression and secretion.

Unfortunately, a very limited amount of experimental data is available to aid in the identification and solution of the problems encountered with protein secretion and protein degradation. Indications that secretion may be limiting for efficient production of extracellular heterologous proteins were obtained from studies on production of bacterial enterotoxin subunit B in *Aspergillus nidulans* (Turnbull *et al.,* 1989), β-glucuronidase production in *A. niger* (Punt *et al.,* 1991), and chymosin production in *A. niger* (Van Hartingsveldt *et al.,* 1990b) and *A. awamori* (Ward, 1989a).

The influence of different secretion sequences on the efficiency of chymosin synthesis by *A. niger* was studied by van Hartingsveldt and coworkers (1990b). Four expression cassettes, containing different signal sequences or a signal sequence with additional N-terminal sequences from the *A. niger glaA* gene (Fig. 4), were introduced into *A. niger.* Analysis of transformants, which contain one copy of these expression cassettes integrated at the *glaA* locus, revealed that the majority of the chymosin accu-

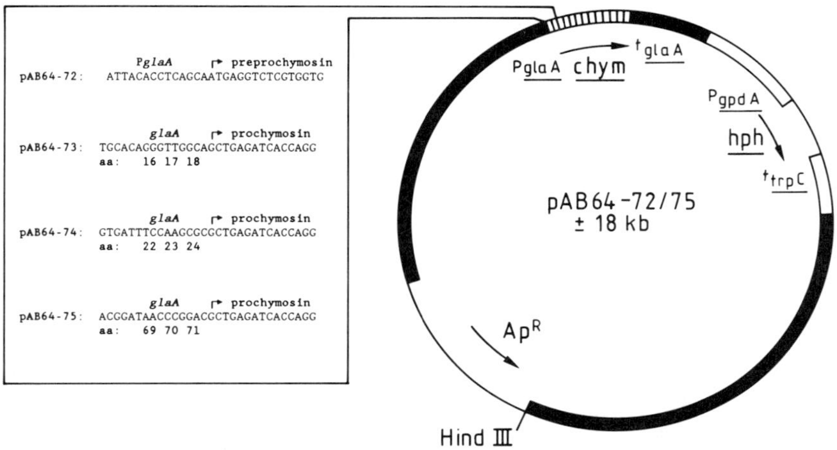

Fig. 4. Schematic representation of four different prochymosin-expression vectors pAB64-72 to -75 (van Hartingsveldt *et al.*, 1990b). Sequences at the glucoamylase-(pre)prochymosin junction are indicated at the lefthand side. Thick lines represent *A. niger* DNA from the *glaA* locus; open boxes, *A. nidulans* DNA; hatched lines, (pre)prochymosin DNA; thin lines, *E. coli* DNA. p_{gpdA}, promoter region of the *gpdA* gene; p_{glaA}, promoter region of the *glaA* gene; t_{trpC}, terminator region of the *trpC* gene; t_{glaA}, terminator region of the *glaA* gene; *hph,* hygromycin resistance gene; Ap^R, ampicillin resistance gene; *chym,* calf (pre)prochymosin. Arrows indicate the direction of transcription.

mulated intracellularly (Fig. 5) only when both signal and propeptide sequences were located upstream of the chymosin coding sequence (pAB64-74). This result suggests that the presence of the propeptide of GlaA protein at the N-terminal end of prochymosin hampered secretion of this protein.

With *A. awamori* transformants containing a few copies of expression cassettes similar to those present in both plasmids pAB64-72 and pAB64-73 (Fig. 4.), most of the chymosin accumulated intracellularly (Ward, 1989a). At present, it is still unclear why chymosin accumulates intracellularly in *A. awamori* while in *A. niger* this accumulation is not observed. However, changing the growth medium for *A. awamori* transformants resulted in improved secretion of chymosin, suggesting that growth conditions play an important role in efficient protein secretion by this organism.

In the case of the bacterial proteins enterotoxin subunit B and β-glucuronidase, essentially all the protein was located intracellularly. Although the signal peptide was probably cleaved off, it is still unclear in which particular compartment(s) of the secretion pathway the proteins are retained.

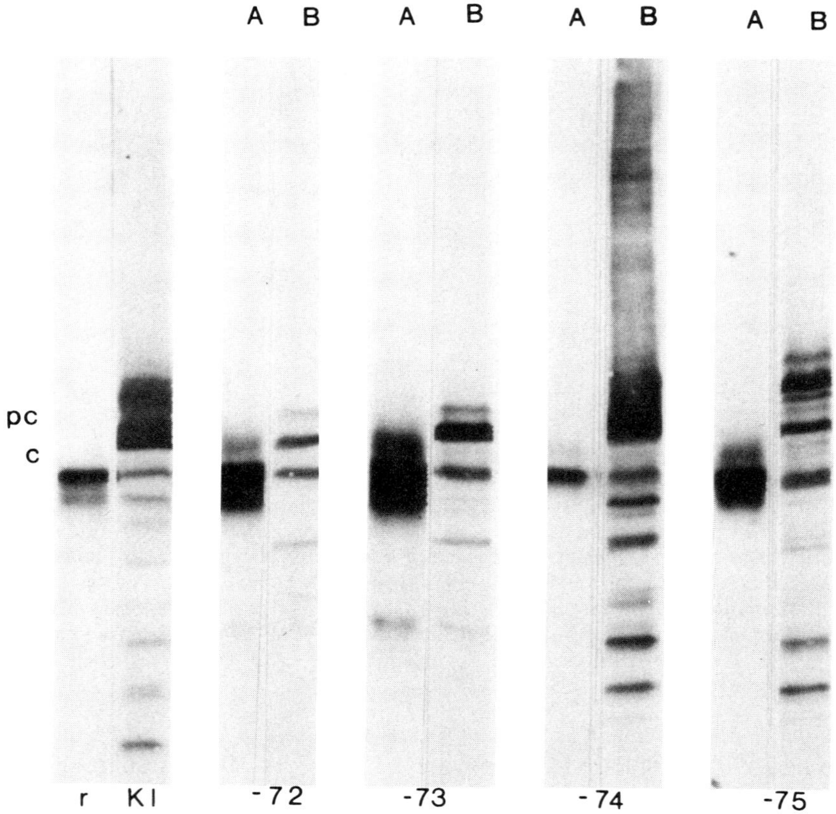

Fig. 5. Western analysis of protein samples from *A. niger* transformants obtained with pAB64-72 to -75. In lanes A, pH 2 treated medium samples of pAB64-72 to -75 transformants are shown and in lanes B, samples of ground mycelial extracts of pAB64-72 to -75 transformants (Van Hartingsveldt *et al.*, 1990b). Lane r contains a sample of a commercial chymosin (c) preparation and lane K1 contains a medium sample of a prochymosin (pc) producing *Kluyveromyces lactis* strain (Van den Berg *et al.*, 1990).

Many filamentous fungi secrete considerable amounts of extracellular proteases into the culture fluid. The recovery of heterologous proteins susceptible to such proteases may be impeded. So, as indicated in Section III,C, considerable effort has been directed toward the construction of protease-deficient strains. Intracellular proteases may also be a problem. For example, Camels and coworkers (1991) recently suggested that impaired production of heterologous proteins could be caused partly by proteolytic cleavage by site-specific proteases of the secretory pathway.

To date, only the problem of extracellular protein degradation has been identified clearly enough so experiments can be performed to circumvent it. Nevertheless, the first nonfungal protein produced by a filamentous fungal host (bovine chymosin, produced by *A. awamori*) has recently been brought onto the market, indicating that optimization of the production process using the more classical strain-improvement strategy can improve production yield to an economically attractive level.

Authentic glycosylation of secreted proteins is also important for the stability of pharmacologically active proteins, and is probably a prerequisite to obtaining permission for their commercial production. Only a few studies have been carried out on the nature of the glycosylation of heterologous proteins in filamentous fungi. In the case of human tissue plasminogen activator (tPA), produced by *A. nidulans,* glycosylation resembles the glycosylation pattern in human cells (Upshall *et al.,* 1987). In addition, overglycosylation of heterologous proteins has been observed in fungi, for example, *Trichoderma reesei* cellobiohydrolase II produced by *A. awamori* (Berka and Barnett, 1989) and *Mucor miehei* aspartic proteinase produced by *A. nidulans* (Gray *et al.,* 1986). In the latter case, this overglycosylation did not affect the activity of the proteinase.

Obviously, further research aimed at a better understanding of protein secretion and glycosylation will be essential for the continuing development of filamentous fungi as production organisms for heterologous proteins.

V. FUTURE PROSPECTS

In this chapter we have described the development and use of various fungal expression systems in the efficient production of heterologous proteins. The results obtained so far are promising. Nevertheless, a considerable fundamental research effort regarding transcription control, mRNA stability and processing, translation efficiency, secretion and glycosylation pathways, and proteolysis is necessary in order to construct fungal strains that produce commercially viable quantities of heterologous proteins. Until this knowledge is available, classical mutation and screening programs still will be important strategies for further improving heterologous protein production.

To date, the most promising results have concerned the efficient production of heterologous fungal enzymes. Stimulated by these results, it is probable that much of the research effort in the near future will concentrate on the production of commercially interesting fungal enzymes using the strategies described by several groups and summarized in this chapter.

Similar strategies also will be used in the production of a second generation of fungal enzymes, the properties of which will have been improved by protein engineering.

The question of whether filamentous fungi will emerge as useful hosts for economically attractive production of mammalian and other nonfungal proteins remains to be answered. However, the group at Genencor has already shown that strains of *Aspergillus awamori* can be "constructed" that produce the levels of biologically active chymosin necessary for commercial production. It will become clear in the next few years if similar results can be achieved for other nonfungal proteins.

ACKNOWLEDGMENTS

We wish to acknowledge all scientists working in the field of heterologous gene expression in filamentous fungi who cummunicated their unpublished data to us. We also thank our colleagues Wim van Hartingsveldt, Jan Verdoes, Hans van Noort, and Ineke Mattern for their contribution to this paper. Furthermore, Peter Pouwels is acknowledged for critical reading of the manuscript and many stimulating discussions. Gerry van Oossanen is acknowledged for typing the manuscript.

REFERENCES

Archer, D. B., Jeenes, D. J., MacKenzie, D. A., Brightwell, G., Lambert, N., Lowe, G., Radford, S. E., and Dobson, C. M. (1990). Hen egg white lysozyme expressed in, and secreted from, *Aspergillus niger* is correctly processed and folded. *Bio/Technology* **8,** 741–745.

Arnau, J., Murillo, F. J., and Torres-Martinez, S. (1988). Expression of TNS-derived kanamycin resistance in the fungus *Phycomyces blakesleeanus. Mol. Gen. Genet.* **212,** 375–377.

Ballance, D. J. (1986). Sequences important for gene expression in filamentous fungi. *Yeast* **2,** 229–236.

Barbesgaard, P. (1977). Industrial enzymes produced by members of the genus *Aspergillus. In* "Genetics and Physiology of *Aspergillus*" (J. E. Smith and J. A. Pateman, eds.), pp 391–404. Academic Press, London.

Barrett, V., Dixon, R. K., and Lemke, P. A. (1990). Genetic transformation of a mycorrhizal fungus. *Appl. Microbiol. Biotechnol.* **33,** 313–316.

Baum, J. A., Geever, R., and Giles, N. H. (1987). Expression of qa-1F activator protein: Identification of upstream binding sites in the *qa* gene cluster and localization of the DNA-binding domain. *Mol. Cell. Biol.* **7,** 1256–1266.

Bennett, J. W. (1985). Molds, manufacturing and molecular genetics. *In* "Molecular Genetics of Filamentous Fungi" (W. E. Timberlake, ed.), pp. 345–366. Liss, New York.

Berka, R. M., and Barnett (1989). The development of gene expression systems for filamentous fungi. *Biotech. Adv.* **7,** 127–154.

Berka, R. M., Ward, M., Wilson, L. J., Hayenga, K. J., Kodama, K. H., Carlomagno, L. P., and Thompson, S. A. (1990). Molecular cloning and deletion of the aspergillopepsin A gene from *Aspergillus awamori. Gene* **86,** 153–162.

Blakemore, E. J. A., Dobson, M. J., Hocart, M. J., Lucas, J. A., and Peberdy, J. F. (1989). Transformation of *Pseudocercosporella herpotrichoides* using two heterologous genes. *Curr. Genet.* **16,** 177–180.

Blobel, G., and Dobberstein, B. (1975). Transfer of proteins across membranes. II. Reconstitution of functional rough microsomes from heterologous components. *J. Cell. Biol.* **67,** 852–862.

Boel, E., and Huge-Jensen, I. B. (1989). Recombinant *Humicola* lipase and process for the production of recombinant *Humicola* lipases. European Patent Application EP 0305216.

Boel, E., Hansen, M. T., Hjort, I., Hoegh, I., and Fiil, N. P. (1984). Two different types of intervening sequences in the glucoamylase gene from *Aspergillus niger. EMBO J.* **3,** 1581–1585.

Boel, E., Christensen, T., and Wöldlike, H. F. (1987). Process for the production of protein products in *Aspergillus oryzae* and a promoter for use in *Aspergillus.* European Patent Application EP 0238023.

Camels, T. P. G., Martin, F., Durand, H., and Tiraby, G. (1991). Proteolytic events in the processing of secreted proteins in fungi. *J. Biotechnol.* **17,** 51–67.

Cantwell, C. A., Beckmann, R. J., Dotzlaf, J. E., Fisher, D. L., Skatrud, P. L., Yeh, W.-K., and Queener, S. W. (1990). Cloning and expression of a hybrid *Streptomyces clavuligerus cefE* gene in *Penicillium chrysogenum. Curr. Genet.* **17,** 213–221.

Casselton, L. A., and de la Fuente Herce, A. (1989). Heterologous gene expression in the basidiomycete fungus *Coprinus cinereus. Curr. Genet.* **16,** 35–40.

Christensen, T., Woeldike, H., Boel, E., Mortensen, S. B., Hjortshoej, K., Thim, L. and Hansen, T. M. (1988). High level expression of recombinant genes in *Aspergillus oryzae. Bio/Technology* **6,** 1419–14.

Churchill, A. C. L., Guffetti, L. M., Hansen, D. R., Van Etten, H. D., and Van Alfen, N. K. (1990). Transformation of the fungal pathogen *Cryphonectria parasitica* with a variety of heterologous plasmids. *Curr. Genet.* **17,** 25–31.

Comino, A., Kolar, M., Schwab, H., and Socic, H. (1989). Heterologous transformation of *Claviceps purpurea. Biotech. Lett.* **11,** 389–392.

Cooley, R. N., Shaw, R. K., Franklin, F. C. H., and Caten, C. E. (1988). Transformation of the phytopathogenic fungus *Septoria nodorum* to hygromycin B resistance. *Curr. Genet.* **13,** 383–390.

Cove, D. J. (1979). Genetic studies of nitrate assimilation in *Aspergillus nidulans. Biol. Rev.* **54,** 291–327.

Cullen, D., Gray, G. L., Wilson, L. J., Hayenga, K. J., Lamsa, M. H., Rey, M. W., Norton, S., and Berka, R. M. (1987). Controlled expression and secretion of bovine chymosin in *Aspergillus nidulans. Bio/Technology* **5,** 369–376.

Curragh, H., Marchant, R., Mooibroek, H., and Wessels, J. G. H. (1990). Genetic transformation of the filamentous fungus *Fusarium culmonum.* (abstract no. 3). British Mycological Society, 8th General Meeting. Nottingham, United Kingdom.

Davies, R. W. (1990). Molecular biology of a high-level recombinant protein production system in *Aspergillus. In* "Molecular Industrial Mycology" (S. A. Leong and R. M. Berka, eds.). Marcel Dekker, New York.

Davis, M. A., Cobbett, C. S., and Hynes, M. J. (1988). An *amdS–lacZ* fusion for studying gene regulation in *Aspergillus. Gene* **63,** 199–212.

De Ruiter-Jacobs, Y. M. J. T., Broekhuijsen, M. P., Campbell, E. I., Unkles, S. E., Kinghorn, J. R., Contreras, R., Pouwels, P. H., and Van den Hondel, C. A. M. J. J. (1989). A gene transfer system based on the homologous *pyrG* gene and efficient expression of bacterial genes in *Aspergillus oryzae. Curr. Genet.* **16,** 159–163.

Dickinson, L., Harboe, M., Van Heeswijck, R., Stroman, P., and Jepsen, L. P. (1987).

Expression of active *Mucor miehei* aspartic protease in *Mucor circinelloides. Carlsberg Res. Commun.* **52,** 243–252.

Durand, N., Reymond, P., and Fevre, M. (1990). Transformations of *Penicillium* to hygromycin B and phleomycin resistance. (abstract no. 47). British Mycological Society, 8th General Meeting. Nottingham, United Kingdom.

Durrens, P., Green, P.M., Arst, H. N., Jr., and Scazzocchio, C. (1986). Heterologous insertion of transforming DNA and generation of new deletions associated with transformation in *Aspergillus nidulans. Mol. Gen. Genet.* **203,** 544–549.

Edens, L., Bom, I., Ledeboer, A. M., Maat, J., Toonen, M. Y., Visser, C., and Verrips, C. T. (1984). Synthesis and processing of plant protein thaumatin in yeast. *Cell* **37,** 629–633.

Farman, M. L., and Oliver, R. P. (1988). The transformation of protoplasts of *Leptosphaeria maculans* to hygromycin B resistance. *Curr. Genet.* **13,** 327–330.

Fincham, J. R. (1989). Transformation in fungi. *Microbiol. Rev.* **53,** 148–170.

Frederick, G. D., and Kinsey, J. A. (1990). Nucleotide sequence and nuclear protein binding of the two regulatory sequences upstream of the *am* (GDH) gene in *Neurospora. Mol. Gen. Genet.* **221,** 148–154.

Glumoff, V., Käppeli, O., Fiechken, A., and Reiser, J. (1989). Genetic transformation of the filamentous yeast *Trichosporon cutaneum* using dominant selection markers. *Gene* **84,** 311–318.

Goc, A., and Weglenski, P. (1988). Regulatory region of the *Aspergillus nidulans argB* gene. *Curr. Genet.* **14,** 425–429.

Goldman, G. H., Van Montagu, M., and Herrera-Estrella, A. (1990). Transformation of *Trichoderma harzianum* by high-voltage electric pulse. *Curr. Genet.* **17,** 169–174.

Gómez-Pardo, E., and Penalva, M. A. (1990). The upstream region of the IPNS gene determines expression during secondary metabolism in *Aspergillus nidulans. Gene* **89,** 109–115.

Goosen, T., Bos, C. J., and van den Broek, H. W. J. (1990). Transformation and gene manipulation in filamentous fungi: An overview. *In* "Handbook of Applied Mycology" (D. K. Arora, K. G. Mukerji, and R. P. Elander, eds.), Fungal Biotechnology, Vol. 4. Marcel Dekker, New York. In Press.

Gray, G. L., Hayenga, K., Cullen, D., Wilson, L. J., and Norton, S. (1986). Primary structure of *Mucor miehei* aspartyl protease: Evidence for a zymogen intermediate. *Gene* **48,** 41–53.

Gurr, S. J., Unkles, S. E., and Kinghorn, J. R. (1988). The structure and organization of nuclear genes of filamentous fungi. *In* "Gene Structure in Eukaryotic Microbes" (J. R. Kinghorn, ed.), Society of General Microbiology Special Publication, Vol. 23, pp. 93–139. IRL Press, Oxford.

Gwynne, D. I., Buxton, F. P., Williams, S. A., Garven, S., and Davies, R. W. (1987a). Genetically engineered secretion of active human interferon and a bacterial endoglucanase from *Aspergillus nidulans. Bio/Technology* **5,** 713–719.

Gwynne, D. I., Buxton, F. P., Sibley, S., Davies, R. W., Lockington, R. A., Scazzocchio, C., and Sealy-Lewis, H. M. (1987b). Comparison of the *cis*-acting control regions of two coordinately controlled genes involved in ethanol utilization in *Aspergillus nidulans. Gene* **51,** 205–216.

Gwynne, D. I., Buxton, F. P., Gleeson, M. A., and Davies, R. W. (1987c). Genetically engineered secretion of foreign proteins from *Aspergillus* species. *In* "Protein Purification: Micro to Macro" (Burgess, ed), pp. 355–365. Liss, New York.

Gwynne, D. I., Sills, A. M., Buch, J. K., Devchand, M., Guo, Z., Skipper, N., Smart, N. J., and, Johnstone, J. A. (1989). Current use and continued development of an expression system derived from ethanol utilization regulon of *Aspergillus nidulans. In* "Proceedings of the EMBO-Alko Workshop on Molecular Biology of Filamentous Fungi" (H. Nevalainen and M. Penttilä, eds.), pp. 129–136. Foundation for Biotechnical and Industrial Fermentation Research, Helsinki.

Hamer, J. E., and Timberlake, W. E. (1987). Functional organization of the *Aspergillus nidulans trpC* promoter. *Mol. Cell. Biol.* **7,** 2352–2359.

Harkki, A., Uusitalo, J., Bailey, M., Penttilä, M., and Knowles, J. K. C. (1989). A novel fungal expression system: Secretion of active calf chymosin from the filamentous fungus *Trichoderma reesei. Bio/Technology* **7,** 596–603.

Hawkins, A. R., and Roberts, C. F. (1989). Molecular interactions between the quinic acid catabolic and shikimate pathways in *Aspergillus nidulans. In* "Proceedings of the EMBO-Alko Workshop on Molecular Biology of Filamentous Fungi" (H. Nevalainen and M. Penttilä, eds.), pp 85–100. Foundation for Biotechnical and Industrial Fermentation Research, Helsinki.

Heim, J., Meyhack, B., and Visser, J. (1990). Novel expression system. European Patent Application EP 0353188.

Herrera-Estrella, A., Goldman, G. H., and Van Montagu, M. (1990) High efficiency transformation system for the biocontrol agents *Trichoderma* spp. *Mol. Microbiol.* **4,** 839–843.

Hitzeman, R. A., Hagie, F. E., Levine, H. L., Goeded, D. V., Ammerer, G., and Hall, B. D. (1981). Expression of a human gene for interferon in yeast. *Nature (London)* **293,** 717–722.

Huge-Jensen, B., Andreasen, F., Christensen, T., Christensen, M., Thim, L., and Boel, E. (1989). *Rhizomucor miehei* triglyceride lipase is processed and secreted from transformed *Aspergillus oryzae. Lipids* **24,** 781–785.

Hynes, M. J. (1978). Multiple independent control mechanisms affecting the acetamidase of *Aspergillus nidulans. Mol. Gen. Genet.* **161,** 59–65.

Hynes, M. J. (1986). Transformation of filamentous fungi. *Exp. Mycol.* **10,** 1–8.

Hynes, M. J., Andrianopoulos, A., Davis, M. A., Van Heeswijck, R., Katz, M. E., Littlejohn, T. G., Richardson, I. B., and Saleeba, J. A. (1989). Multiple circuits regulating the *amdS* gene of *Aspergillus nidulans. In* "Proceedings of the EMBO-Alko Workshop on Molecular Biology of Filamentous Fungi" (H. Nevalainen and M. Penttilä, eds.), pp. 63–71. Foundation for Biotechnical and Industrial Fermentation Research, Helsinki.

Kelly, J. M., and Hynes, M. J. (1985). Transformation of *Aspergillus niger* by the *amdS* gene of *Aspergillus nidulans. EMBO J.* **4,** 475–479.

Kelly, J. M., and Hynes, M. J. (1987). Multiple copies of the *amdS* gene of *Aspergillus nidulans* cause titration of *trans*-acting regulatory proteins. *Curr. Genet.* **12,** 21–31.

Kolar, M., Punt, P. J., Van den Hondel, C. A. M. J. J., and Schwab, H. (1988). Transformation of *Penicillium chrysogenum* using dominant selection markers and expression of an *Escherichia coli lacZ* fusion gene. *Gene* **62,** 127–134.

Korman, D. R., Bayliss, F. T., Barnett, C. C., Carmona, C. L., Kodama, K. H., Royer, T. J., Thompson, S. A., Ward, M., Wilson, L. J., and Berka, R. M. (1990). Cloning, characterization and expression of two α-amylase genes from *Aspergillus niger* var. *awamori. Curr. Genet.* **17,** 203–212.

Kozak, M. (1984). Compilation and analysis of sequences upstream from the translation start site in eukaryotic mRNAs. *Nucl. Acids Res.* **12,** 857–872.

Mellor, J., Dobson, M. J., Roberts, N. A., Tuite, M. F., Emtage, J. S., White, S., Lowe, P. A., Patel, T., Kingsman, A. J., and Kingsman, S. M. (1983). Efficient synthesis of enzymatically active calf chymosin in *Saccharomyces cerevisiae. Gene* **24,** 1–14.

Mellor, J., Dobson, M. J., Roberts, N. A., Kingsman, A. J., and Kingsman, S. M. (1985). Factors affecting heterologous gene expression in *Saccharomyces cerevisiae. Gene* **33,** 215–226.

Mooibroek, H., Kuipers, A. G. J., Sietsma, J. H., Punt, P. J., and Wessels, J. G. H. (1990). Introduction of hygromycin B resistance into *Schizophyllum commune:* Preferential methylation of donor DNA. *Mol. Gen. Genet.* **222,** 41–48.

Mullaney, E. J., Hamer, J. E., Roberti, K. A., Yelton, M. M., and Timberlake, W. E. (1985). Primary structure of the *trpC* gene from *Aspergillus nidulans. Mol. Gen. Genet.* **199,** 37–45.

Mullaney, E. J., Punt, P. J., and Van den Hondel, C. A. M. J. J. (1988). DNA mediated transformation of *Aspergillus ficuum. Appl. Microb. Biotech.* **28,** 451–454.

Oliver, R. P., Roberts, I. N., Harling, R., Kenyon, L., Punt, P. J., Dingemanse, M. A., and Van den Hondel, C. A. M. J. J. (1987). Transformation of *Fulvia fulva,* a fungal pathogen of tomato, to hygromycin B resistance. *Curr. Genet.* **12,** 231–233.

Osiewacz, H. D., and Weber, A. (1989). DNA mediated transformation of the filamentous fungus *Culvularia lunata* using a dominant selectable marker. *Appl. Microb. Biotech* **30,** 375–380.

Podila, G. K., Dickman, M. B., and Kolattukudy, P. E. (1988). Transcriptional activation of a cutinase gene in isolated fungal nuclei by plant cutin monomers. *Science* **242,** 922–925.

Primrose, S. B. (1986). The application of genetically engineered micro-organisms in the production of drugs. *J. Appl. Bacteriol.* **61,** 99–116.

Punt, P. J., Oliver, R. P., Dingemanse, M. A., Pouwels, P. H., and Van den Hondel, C. A. M. J. J. (1987). Transformation of *Aspergillus* based on the hygromycin B resistance marker from *Escherichia coli. Gene* **56,** 117–124.

Punt, P. J., Dingemanse, M. A., Jacobs-Meijsing, B. J. M., Pouwels, P. H., and Van den Hondel, C. A. M. J. J. (1988). Isolation and characterization of the glyceraldehyde-3-phosphate dehydrogenase gene of *Aspergillus nidulans. Gene* **69,** 49–57.

Punt P. J., Dingemanse, M. A., Kuyvenhoven A., Soede, R. D. M., Pouwels, P. H., and Van den Hondel, C. A. M. J. J. (1990). Functional elements in the promoter region of the *Aspergillus nidulans gpdA* gene, encoding glyceraldehyde-3-phosphate dehydrogenase. *Gene.* **93,** 101–109.

Punt, P. J., Zegers, N. D., Busscher, M., Pouwels, P. H., and Van den Hondel, C. A. M. J. J. (1991). Intracellular and extracellular production of proteins under the control of expression signals of the highly expressed *Aspergillus nidulans gpdA* gene. *J. Biotechnol.* **17,** 19–34.

Rambosek, J., and Leach, J. (1987). Recombinant DNA in filamentous fungi: Progress and prospects. *CRC Crit. Rev. Biotechnol.* **6,** 357–393.

Roberts, I. N., Oliver, R. P., Punt, P.J., and Van den Hondel C. A. M. J. J. (1989). Expression of the *Escherichia coli* β-glucuronidase gene in industrial and phytopathogenic filamentous fungi. *Curr. Genet.* **15,** 177–180.

Smith, A. W., Ramsden, M., and Peberdy, J. F. (1990). Transformation of *Acremonium chrysogenum* by vector constructs containing 5′ deletions of the pcbC promoter and the *Aspergillus nidulans gpd* promoter. British Mycological Society, 8th General Meeting. Nottingham, United Kingdom.

Smith, D. J., Burnham, M. K. R., Edwards, J., Earl, A. J., and Turner, G. (1990). Cloning and heterologous expression of the penicillin biosynthetic gene cluster from *Penicillium chrysogenum. Bio/Technology* **8,** 39–41.

Smith, T. L., Gaskell, J., Berka, R. M., Yang, M., Henner, D. J., and Cullen, D. (1990). The promoter of the glucoamylase-encoding gene of *Aspergillus niger* functions in *Ustilago maydis. Gene* **88,** 259–262.

Soliday, C. L., Dickman, M. B., and Kolattukudy, P. E. (1989). Structure of the cutinase gene and deletion of promoter activity in the 5′-flanking region by fungal transformation. *J. Bacteriol.* **171,** 1942–1951.

Staben, C., Jensen, B., Singen, M., Pollock, J., Schechtman, M., and Kinsey, J. (1989). Use of a bacterial hygromycin B resistance gene as a dominant selectable marker in *Neurospora crassa* transformation. *Fungal Genet. Newslett.* **36,** 79–81.

Stepanov, V. M., and Rudenskaya, G. N. (1983). Proteinase affinity chromatography on bacitracin-sepharose. *J. Appl. Biochem.* **5,** 420–428.

Timberlake, W. E., and Marshall, M. A. (1989). Genetic engineering of filamentous fungi. *Science* **244,** 1313–1317.

Turgeon, B. G., Garber, R. C., and Yoder, O. C. (1987). Development of a fungal transformation system based on selection of sequences with promoter activity. *Mol. Cell. Biol.* **7,** 3297–3305.

Turnbull, I. F., Rand, K., Willetts, N. S., and Hynes, M. J. (1989). Expression of the *Escherichia coli* enterotoxin subunit B gene in *Aspergillus nidulans* directed by the *amdS* promoter. *Bio/Technology* **7,** 169–174.

Turner, G. (1990). Expression systems and protein production in filamentous fungi. *In* "Protein Production by Biotechnology" (T. J. R. Harris, ed.). (pp 15–26). Elsevier Applied Science, London.

Ulhoa, C. J., Vainstein, M. H., and Peberdy, J. F. (1990). Transformation of two *Trichoderma* species using hygromycin B and MBC resistance. (abstract no. 3g). British Mycological Society, 8th General Meeting. Nottingham, United Kingdom.

Upshall, A., Kumar, A. A., Bailey, M. C., Parker, M. D., Favreau, M. A., Lewison, K. P., Joseph, M. L., Maraganore, J. M., and Mcknight, G. L. (1987). Secretion of active human tissue plasminogen activator from the filamentous fungus *Aspergillus nidulans. Bio/Technology* **5,** 1301–1304.

Van den Berg, J. A., Van der Laken, K. J., Van Ooyen, A. J. J., Renniers, T. C. H. M., Rietveld, K., Schaap, A., Brake, A. J., Bishop, R. J., Schultz, K., Moyer, D., Richman, M., and Shuster, J. R. (1990). *Kluyveromyces* as a host for heterologous gene expression: expression and secretion of prochymosin. *Bio/Technology* **8,** 135–139.

Van den Hondel, C. A. M. J. J., and Punt, P. J. (1990). Gene transfer systems and vector development for filamentous fungi. British Mycological Society Symposium Volume 18: Applied Molecular Genetics of Fungi, In press.

Van Gorcom, R. F. M., and Van den Hondel, C. A. M. J. J. (1988). Expression analysis vectors for *Aspergillus niger. Nucl. Acids Res.* **16,** 9052.

Van Gorcom, R. F. M., Pouwels, P. H., Goosen, T., Visser, J., Van den Broek, H. W. J., Hamer, J. E., Timberlake, W. E., and Van den Hondel, C. A. M. J. J. (1985). Expression of an *Escherichia coli* β-galactosidase fusion gene in *Aspergillus nidulans. Gene* **40,** 99–106.

Van Gorcom, R. F. M., Punt, P. J., Pouwels, P. H., and Van den Hondel, C. A. M. J. J. (1986). A system for the analysis of expression signals in *Aspergillus. Gene* **48,** 211–217.

Van Hartingsveldt, W., Mattern, I. E., Van Zeijl., C. M. J., Pouwels, P. H., and Van den Hondel, C. A. M. J. J. (1987). Development of a homologous transformation system for *Aspergillus niger* based on the *pyrG* gene. *Mol. Gen. Genet.* **206,** 71–75.

Van Hartingsveldt, W., Van den Hondel, C. A. M. J. J., Veenstra, A. E., and Van den Berg, J. A. (1990a). Gene replacement as a tool for the construction of *Aspergillus* strains. European Patent Application EP 0357127.

Van Hartingsveldt, W., Van Zeijl, C. M. J., Veenstra, A. E., Van den Berg, J. A., Van Gorcom, R. F. M., and Van den Hondel, C. A. M. J. J. (1990b). Heterologous gene expression in *Aspergillus:* Analysis of chymosin production in single-copy transformants of *A. niger. In* "Proceedings of the 6th International Symposium on the Genetics of Industrial Microorganisms, Strasbourg," **1,** 107–116.

Ward, M. (1989a). Production of calf chymosin by *Aspergillus awamori, In* "Genetics and Molecular Biology of Industrial Microorganisms" (C. L. Hershberger, S. W. Queener, and G. Hegeman, eds.), pp. 288–294. American Society for Microbiology, Washington, D.C.

Ward, M. (1989b). Heterologous gene expression in *Aspergillus*. *In* "Proceedings of the EMBO-Alko Workshop on Molecular Biology of Filamentous Fungi" (H. Nevalainen and M. Penttilä, eds.), pp. 119–128. Foundation for Biotechnical and Industrial Fermentation Research, Helsinki.

Ward, M., Wilson, L. J., Carmona, C. L., and Turner, G. (1988). The *oliC3* gene of *Aspergillus niger:* Isolation, sequence and use as a selectable marker for transformation. *Curr. Genet.* **14,** 37–42.

Ward, M., Wilson, L. J., Kodama, K. H., Rey, M. W., and Berka, R. M. (1990). Improved production of chymosin in *Aspergillus* by expression as a glucoamylase-chymosin fusion. Bio/Technology **8,** 435–440.

Wernars, K., Goosen, T., Wennekes, B. M. J., Visser, J., Bos, C. J., Van den Broek, H. W. J., Van Gorcom, R. F. M., Van en Hondel, C. A. M. J. J., and Pouwels, P. H. (1985). Gene amplification in *Aspergillus nidulans* by transformation with vectors containing the *amdS* gene. *Curr. Genet.* **9,** 361–368.

Wnendt, S., Jacobs, M., and Stahl, U. (1990). Transformation of *Aspergillus giganteus* to hygromycin B resistance. *Curr. Genet.* **7,** 21–24.

VI

Appendices

Appendix A

Genetic Analysis with RAPD Markers

JOHN G. K. WILLIAMS
ANNE R. KUBELIK
Central Research and Development Department
E. I. du Pont de Nemours and Company, Inc.
Wilmington, Delaware

J. ANTONI RAFALSKI
SCOTT V. TINGEY
Agricultural Products Department
E. I. du Pont de Nemours and Company, Inc.
Wilmington, Delaware.

MORE GENE MANIPULATIONS
IN FUNGI

I. INTRODUCTION

We have recently described a new class of genetic markers, named *R*andom *A*mplified *P*olymorphic *D*NA (RAPD) markers (Williams *et al.,* 1990). These markers share most of the advantages of restriction fragment length polymorphism (RFLP) markers (Botstein *et al.,* 1980) without many of their disadvantages. In particular, there is no need for species-specific, genomic, or cDNA-clone-derived probes; the assay is nonradioactive; and it can be automated for high throughput applications.

The RAPD assay is based on the use of synthetic 10-mer oligodeoxynucleotides as primers in a DNA-amplification reaction.This amplification protocol differs from more commonly used procedures (Innis *et al.,* 1990) because only a single primer is used, and there is no need to have knowledge of any genomic sequence at the locus being amplified. The single primer is short, increasing the probability that two priming sites occur in the genome close to each other in an inverted orientation. Competition between many initial primer extension products usually results in the synthesis of a few intense DNA bands after 40-45 cycles of amplification.

We have shown that the assay is highly sensitive to nucleotide differences between the primer and the template (Williams *et al.,* 1990), assaying single nucleotide differences. Each DNA band amplified using a 10-mer primer is the result of an interaction between the primer and the genomic DNA template over two priming sites of 10 nucleotides. A typical amplification, showing five major resolvable products, "scans" approximately 100 nucleotides of the template. A single RFLP band revealed by Southern blot analysis using a restriction enzyme that recognizes 6 bases scans only 12 (6 + 6) nucleotides. Nucleotide polymorphisms are therefore easier to identify with RAPD markers than with RFLP markers. This feature makes RAPD markers particularly advantageous for less polymorphic species, for example, soybean (Keim *et al.,* 1989) or tomato.

RAPD markers are functional equivalents of Sequence Tagged Sites (Olson *et al.,* 1989) because knowledge of the primer sequence and amplification product size defines the locus. No exchange of samples is needed to reproduce the results.

Welsh and McClelland (1990) have independently shown that, under different conditions, 20-mer or longer arbitrary sequence primers could amplify a set of DNA products. These products were characteristic of the template DNA genotype and could be used for fingerprinting bacterial species.

The properties of the RAPD assay are described by Williams and associates (1990). Here we present some applications that are of interest to fungal geneticists and discuss possible limitations and extensions of the RAPD assay.

II. MATERIALS AND METHODS

A. Sources of DNA

Neurospora crassa DNA samples were isolated (A. R. Kubelik, J. A. Rafalski, S. V. Tingey, and J. G. K. Williams, unpublished observations) from the strains Oak Ridge FGSC #4488, Mauriceville 1-c*A* FGSC #2225, and 38 progeny (FGSC #4450 to #4487) of a cross between these two parental strains (Metzenberg *et al.,* 1984). All *N. crassa* strains were obtained from the Fungal Genetics Stock Center (Kansas State University). Soybean DNA was isolated (Murray and Thompson, 1980) from the *Glycine soja* accession PI 81762 obtained from Dr. Theodore Hymowitz (University of Illinois).

B. Primer Synthesis

Arbitrary DNA sequences of desired length and GC content were generated by computer. Oligonucleotide primers were machine-synthesized by using phosphoramidate chemistry. After deprotection, the samples were dried under vacuum, dissolved in 200 μl of water, and purified by gel filtration on Sephadex G25 (NAP-5 disposable columns; Pharmacia Inc., Piscataway, New Jersey).

C. Standard Amplification Conditions

Amplification reactions were performed in 25 μl volumes containing 10 m*M* Tris-HCl, pH 8.3; 50 m*M* KCl; 2m*M* $MgCl_2$; 0.001% gelatin; 100 μM each dATP, dCTP, dGTP, and TTP (Pharmacia); 0.2 μM primer; 25 ng genomic DNA; and 0.5 units Amplitaq polymerase (Perkin–Elmer Cetus, Norwalk, Connecticut); overlayed with 1 drop of mineral oil. Amplification was performed in a Perkin–Elmer Cetus DNA Thermal Cycler programmed for 45 cycles of 1 min at 94°, 1 min at 36°, and 2 min at 72°C, using the fastest available transitions between each temperature. Amplification products were analyzed by electrophoresis in 1.4% agarose gels and detected by staining with ethidium bromide.

D. Genetic Mapping Software

A Macintosh version of the MapMaker program (Lander *et al.,* 1987), available from Scott Tingey, was used for the construction of genetic maps.

III. RESULTS AND DISCUSSION

A. RAPD Assay Conditions

The conditions described in Section II generally give satisfactory results, but several factors influence the success of the reaction. Optimization of the amount of genomic DNA and polymerase may be required if weakly-stained bands are seen or a "smear" is observed on the gel. Many methods of genomic DNA preparation are satisfactory. As demonstrated previously (Williams *et al.,* 1990), 10-mer primers with GC contents of 50% or greater work very well. Nonamers can also be used. Longer primers, like the 20-mers of Welsh and McClelland (1990) are usable in the RAPD assay, but we prefer 10-mers for economy of synthesis. Moreover, experiments indicate that, for a 16-mer, the 9 or 10 bases at the 3′ end of the primer influence specificity to a greater extent than do the other bases (J. G. K. Williams, unpublished data).

The exact shape of the temperature profile, which is a property of the thermal cycler, appears to influence the distribution of amplification products. The source and quality of thermostable DNA polymerase is also very important. Preliminary experiments indicate that polymerases from sources other than *Thermus aquaticus* may give radically different amplification products, presumably due to different sensitivities to mismatches between the primer and the template and to their proofreading capacities. To reproduce a RAPD polymorphism, one should precisely follow the conditions under which the polymorphism was originally found.

B. Segregation of RAPD Polymorphisms in Fungi

This example illustrates the use of RAPD markers for genetic mapping in haploid fungi. Genomic DNA samples from the *Neurospora crassa* strains Mauriceville and Oak Ridge were amplified with primer 5′-TGGTCCCTGC. Analysis of the amplification products by agarose gel electrophoresis (Fig. 1, lanes M and O) showed that only one or two DNA segments appeared to be amplified commonly in both genomes, whereas two additional segments (RAPD.1 and RAPD.3) were amplified only from Mauriceville and three additional segments (RAPD.2, RAPD.4 and RAPD.5) only from Oak Ridge. Segregation of the five RAPD markers was examined among 38 haploid progeny of a cross between the two parental strains (Fig. 1, lanes 1–38). The presence of each marker was scored in each of the individuals and correlated with the segregation data for 86 RFLP and phenotypic markers from the same 38 individuals (Metzenberg and Grotelueschen, 1989). Analysis of the data indicates that all five

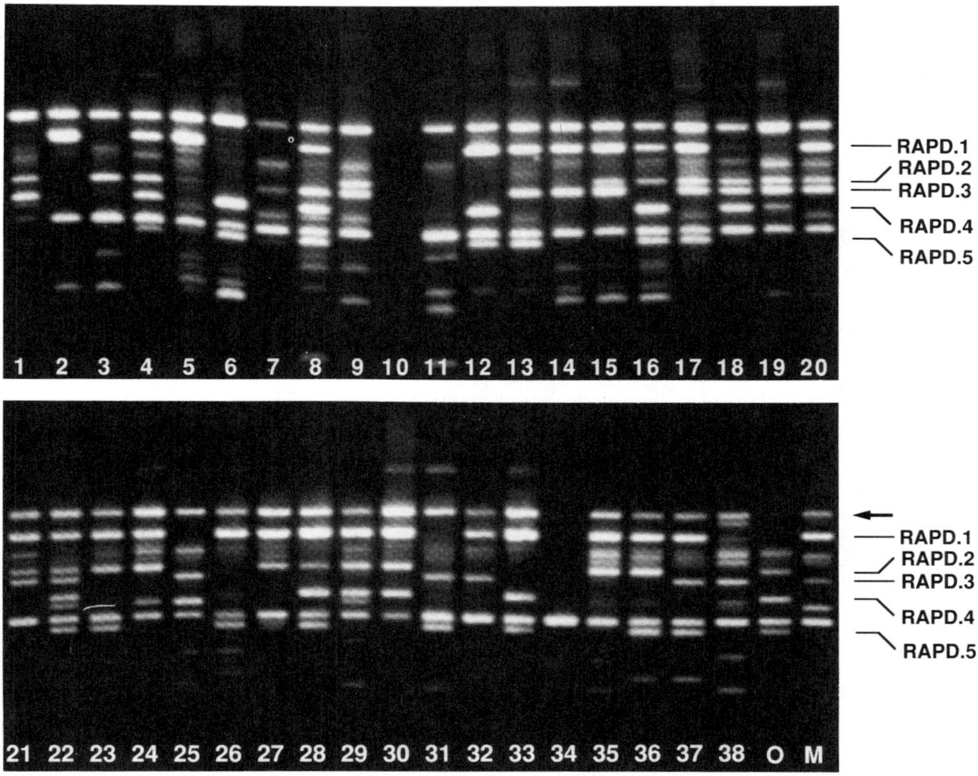

Fig. 1. Segregation of RAPD markers in *N. crassa* progeny. Primer 5′-TGGTCCCTGC was used to amplify DNA from the parental strains Mauriceville (lane M) and Oak Ridge (lane O) and from progeny strains FGSC #4450–4487 (lanes 1–38, respectively) (Metzenberg *et al.*, 1984). Five polymorphic bands (RAPD.1–RAPD.5) are indicated. The presence of each RAPD marker was scored among the progeny and the data were analyzed in the context of an 86-marker RFLP map of *N. crassa* created in the same segregating population (Metzenberg and Grotelueschen, 1989). Multipoint maps and LOD scores were calculated using the MapMaker program (Lander *et al.*, 1987). An apparent polymorphism between the parental strains (arrow) is, in fact, not polymorphic since it amplifies from nearly all the progeny.

markers map with high confidence levels to the *N. crassa* genome at the positions shown in Fig. 2. One DNA segment appeared to be polymorphic; it was amplified from Mauriceville, but only weakly from Oak Ridge (see arrow, Fig. 1, lane M). In spite of this apparent polymorphism between the two parents, this DNA segment was amplified from nearly all the progeny and so it could not be mapped. It appears to be uniparentally inherited.

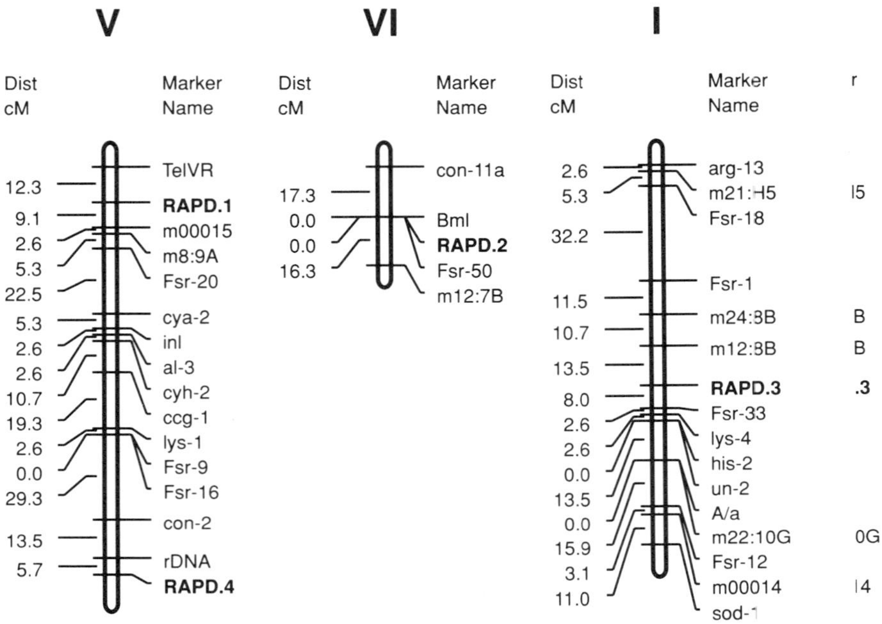

Fig. 2. Linkage groups containing RAPD markers. Roman numerals at the top of each linkage group refer to standard *Neurospora* linkage groups (LG). Linkage groups are depicted with the right arm to the top. RAPDs 1–4 were mapped to the indicated positions from the data in Fig. 1. Calculations from the MapMaker program (Lander *et al.,* 1987) indicate that each of these four markers is at least 10^{13} times more likely to be linked with the indicated marker groups than to be unlinked, and that the indicated map positions of the RAPD markers are at least 500 times more likely than any alternative position. RAPD.5 co-segregated with only one other marker, *cyt-21,* with 0% recombination; this linkage group, consisting of just two markers, is unremarkable and so is not presented in the figure. *cyt-21* has been reported to map to LGI (Kuiper *et al.,* 1988), but did not do so at the significance level (LOD = 4.0) used to calculate this map.

This experiment demonstrates that DNA segments amplified with a single primer map to dispersed and, in general, unlinked genetic loci. Using this approach, Kubelik and co-workers (1991) have constructed a 170-marker RAPD and RFLP map of *N. crassa.*

C. Use of RAPD Markers for the Construction of Genetic Maps in Diploid Organisms

When creating genetic maps in diploid organisms, RAPD markers are dominant and behave as dominant markers in mapping experiments (Williams *et al.,* 1990). Analysis of simulated and experimental data with the

Mapmaker program (Lander *et al.,* 1987) suggests that F_2 populations are not suitable for *de novo* construction of genetic maps consisting only of dominant markers (J. G. K. Williams and J. A. Rafalski, unpublished results). For this purpose, a backcross population or a recombinant inbred population (Burr *et al.,* 1988) is preferable. Doubled haploids, if they can be conveniently obtained without causing rearrangements or other genomic instabilities, are also suitable.

Dominant RAPD markers can be mapped in an F_2 population if they are analyzed with co-dominant RFLP markers. Thus, RAPD markers may be used to enrich a RFLP map created in an F_2 population. Clearly, RAPD markers are particularly well suited for mapping in haploid fungi.

D. Use of Mixed Primers

By using several primers in combination, it should be possible to amplify a wider variety of DNA segments than by using primers individually (Dear and Cook, 1989). To test this idea, we used a set of 7 primers, individually

TABLE I

Pairwise Combinations of Primers[a]

Single primers			Pairwise combinations of primers (New bands/Total bands)					
Primer	Total bands	Primers:	2	3	4	5	6	7
1	6	1	6/8	2/4	2/6	4/7	1/3	1/4
2	4	2		3/5	3/5	3/6	1/1	4/6
3	6	3			1/5	5/10	3/4	1/5
4	6	4				0/0	2/4	0/3
5	5	5					2/4	0/0
6	4	6						2/3
7	6		mean of new bands = 2.2					
mean of total bands = 5.3			mean of total bands = 4.4					

[a] DNA from the soybean line *G. soja* was amplified with seven different primers individually and in all possible 21 pair combinations. The concentration of each primer, when used individually or in combination, was 0.4 μM. Primer sequences (5′–3′) are: 1, CTGATGCTAC; 2, CTGTTGCTAC; 3, CTCTTGCTAC; 4, CTGAAGCTAC; 5, GCAAGTAGCT; 6, TGGTCACTGA; 7, ACGGTACACT. Amplification products were separated in an agarose gel, stained with ethidium bromide, and the number of visible bands were counted. *(Left)* Individual primers. The total numbers of bands obtained with each primer are tabulated. *(Right)* Primers used in pairwise combinations. For each combination, "new" bands are defined as those amplified only by the primer pair (referred to as "double-primer" bands in the text), but not by either primer alone. The total numbers of bands seen with each combination are also tabulated. For example, the entry "6/8" listed for primer combination 1 × 2 indicates that a total of 8 bands were visible, and that 6 of these were "new" bands.

and in all possible 21 pairwise combinations, to amplify DNA segments from the soybean genome. The amplification products were separated by electrophoresis and the number of visible bands were counted for each sample (Table I). An average of 5.3 bands was amplified per single primer, whereas 4.4 bands were amplified per primer pair. This result was unexpected since, in principle, four times as many bands should be amplified by a pair of primers as by an individual primer (Dear and Cook, 1989). When primers are used in pairs, many bands amplified by the individual primers are not seen (Fig. 3). On average, however, 2.2 of the 4.4 bands amplified by paired primers could be identified as being amplified by only one of the two primers in the pair (Table I); this 1 : 1 ratio of single-primer- to double-primer-specific bands is as expected (Dear and Cook, 1989). Under

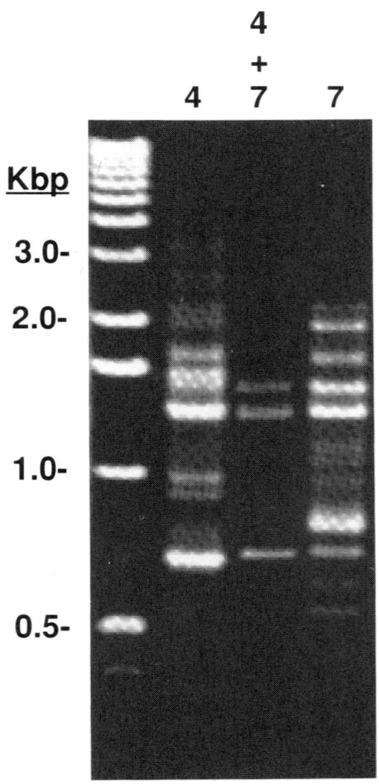

Fig. 3. Addition of a second primer inhibits amplification of some DNA segments. DNA from the soybean cultivar *G. soja* was amplified with primer 4 (lane 4), primer 7 (lane 7), and a combination of both primers (lane 4 + 7). Nucleotide sequences of the primers are given in the legend to Table I.

the conditions of this experiment, the efficiency of obtaining double-primer bands was low compared with that of single-primer bands; 7 reactions using individual primers amplified a total of 37 bands, whereas 21 reactions using paired primers amplified only 46 additional unique bands. This experiment suggests that, although it is possible to generate additional amplification products by using primers in combination, the efficiency of this approach is substantially reduced by an apparent limit on the total number of bands that can be produced by the amplification reaction.

REFERENCES

Botstein, D., White, R. L., Skolnick, M. H., and Davis, R. W. (1980). Construction of a genetic map in man using restriction fragment length polymorphisms. *Am. J. Hum. Genet.* **32,** 314–331.

Burr, B., Burr, F. A., Thompson, K. H., Albertsen, M. C., and Stuber, C. (1988). Gene mapping with recombinant inbreds in maize. *Genetics* **118,** 519–526.

Dear, P. H., and Cook, P. R. (1989). Happy mapping: A proposal for linkage mapping the human genome. *Nucl. Acids Res.* **17,** 6795–6807.

Innis, M. A., Gelfand, D. H., Sninsky, J. J., and White, T. J. (1990). "PCR Protocols." Academic Press, San Diego.

Keim, P., Shoemaker, R. C., and Palmer, R. G. (1989). Restriction fragment length polymorphism diversity in soybean. *Theor. Appl. Genet.* **77,** 786–792.

Kubelik, A. R., Rafalski, J. A., Tingey, S. V., and Williams, J. G. K. In preparation.

Kuiper, M. T. R., Akins, R. A., Holtrop, M., de Vries, H., and Lambowitz, A. M. (1988). Isolation and analysis of the *Neurospora cyt-21* gene. A nuclear gene encoding a mitochondrial ribosomal protein. *J. Biol. Chem.* **263,** 2840–2847.

Lander, E. S., Green, P., Abrahamson, J., Barlow, A., Daly, M. J., Lincoln, S. E., and Newburg, L. (1987). MapMaker: An interactive computer package for constructing primary linkage maps of experimental and natural populations. *Genomics* **1,** 174–181.

Metzenberg, R. L., and Grotelueschen, J. (1989). *Neurospora crassa* restriction polymorphism map. *In* "Genetic Maps" (S. J. O'Brien, ed.), Vol. 3, pp. 3.22–3.29. Cold Spring Harbor Laboratory Press, Cold Spring Harbor, New York.

Metzenberg, R. L., Stevens, J. N., Selker, E. U., and Morzycka-Wroblewska, E. (1984). A method for finding the genetic map position of cloned DNA fragments. *Neurospora Newslett.* **31,** 35–39.

Murray, M. G., and Thompson, W. F. (1980). Rapid isolation of high molecular weight plant DNA. *Nucl. Acids Res.* **8,** 4321–4325.

Olson, M., Hood, L., Cantor, C., and Botstein, D. (1989). A common language for the physical mapping of the human genome. *Science* **245,** 1434–1435.

Welsh, J., and McClelland, M. (1990). Fingerprinting genomes using PCR with arbitrary primers. *Nucl. Acids Res.* **19,** 303–306.

Williams, J. G. K., Kubelik, A. R., Livak, K. J., Rafalski, J. A., and Tingey, S. V. (1990). DNA polymorphisms amplified by arbitrary primers are useful as genetic markers. *Nucl. Acids Res.* **18,** 6531–6535.

Appendix B

Growth Media

J. W. BENNETT
Department of Cell and Molecular Biology
Tulane University
New Orleans, Louisiana

LINDA L. LASURE
Panlabs, Incorporated
Bothell, Washington

MORE GENE MANIPULATIONS
IN FUNGI

While gathering chapters for this book, it became apparent that diverse species have been used for gene manipulations in fungi. We simultaneously became aware that the standard media formulations of several more commonly used fungi, such as *Neurospora crassa* and *Aspergillus nidulans,* are not readily available, or are published in a series of different papers scattered throughout the literature. Since these organisms may be of broad interest, we have assembled information on the "care and feeding" of a number of filamentous fungi.

Media formulations and, in some cases, methods for growing several ascomycetes and fungi imperfecti *(Ascobolus Aspergillus, Cephalosporium, Neurospora, Pencillium,* and *Sordaria)* as well as certain basidiomycetes *(Coprinus, Phanerochaete,* and *Schizophyllum)* are listed here. Finally, media formulations and growth methodology for *Mucor rouxii,* a zygomycete, are also given in this appendix.

I. *ASCOBOLUS* MEDIA

A. Growth Medium (Minimal Medium)

Per liter

Potassium phosphate, monobasic (KH_2PO_4 anhydrous), (50 g/l)	5 ml

Potassium phosphate, dibasic (K_2HPO_4 anhydrous), (60 g/l)	5 ml
Magnesium sulfate ($MgSO_4$), (50 g/l)	5 ml
Urea (100 g/l)	5 ml
Thiamine (10 mg/l)	0.5 ml
Biotin (12 mg/l)	0.5 ml
Oligoelements	0.05 ml
Dextrin	10 g
Agar (for solid medium)	15 g

Sterilize by autoclaving 20 mn at 120°C

B. Crossing Medium

Per liter

Potassium phosphate, monobasic (KH_2PO_4 anhydrous), (50 g/l)	10 ml
Potassium phosphate, dibasic (K_2HPO_4 anhydrous), (60 g/l)	10 ml
Magnesium sulfate ($MgSO_4$), (50 g/l)	10 ml
Thiamine (10 mg/l)	1 ml
Biotin (12 mg/l)	1 ml
Oligoelements	0.1 ml
Dextrin	10 g
Yeast extract	3 g
Cellulose	2.5 g
Agar	15 g

Sterilize by autoclaving 20 mn at 120°C. Pour while mixing continuously.

C. Ascospore Germination Medium

Per liter

Bacto-peptone	15 g
Bacto-agar	15 g

After dissolution in H_2O, add 15 ml of 1.25 N NaOH (final pH: 8.9). Sterilize by autoclaving 20 mn at 120°C.

Oligoelement Solution

Per 0.1 l

Citric acid ($C_6H_8O_7$, $1H_2O$)	5 g
Zinc sulfate ($ZnSO_4$, $7H_2O$)	5 g
Ammonium-iron(II)sulfate ($Fe(NH_4)_2(SO_4)_2$, $6H_2O$)	1 g
Copper sulfate ($CuSO_4$)	0.25 g
Manganese sulfate ($MnSO_4$)	0.05 g

Boric acid ($C_6H_8O_7$, anhydrous)	0.05 g
Sodium molybdate (Na_2MoO_4, $2H_2O$)	0.05 g

Add 1 ml of chloroform to help conservation. Keep in the dark at room temperature.

Biotin

Per 0.1 1

Dissolve 1.2 mg of biotin in 10 ml of $H_20 + 0.05$ ml of 2 N NaOH. Complete to 100 ml. Sterilize by filtration. Keep at 4°C.

Thiamine

Per 0.1 1

Dissolve 1 mg in 100 ml of H_2O. Sterilize by filtration. Keep at 4°C.

II. *ASPERGILLUS NIDULANS* MEDIA FORMULATIONS[1]

A. *Aspergillus nidulans* Complete Medium (CM)

Vitamin Solution [all 0.01% (w/v)]

biotin
pyridoxin
thiamine
riboflavin
p-aminobenzoic acid
nicotinic acid

Store in a brown bottle at 4°C.

Yeast RNA Hydrolysate

Dissolve 20 g yeast nucleic acid (crude) in 75 ml of 2*M* HCl, heat to 100°C for 20 min, cool to room temperature, and titrate to pH 6.2 with 10*M* NaOH. Adjust volume to 100 ml and store at 4°C.

Medium **(use tap water)**

20× nitrate salts	50 ml
1000× trace elements	1 ml
D-glucose	20 g
peptone	2 g

[1] D. Kafer (1977). *Adv. Genet.* **19,** 33–131.

yeast extract	1 g
NZ amine A (Casamino acids)	1 g
adenine (optional except for ad$^-$ strains)	1 g
yeast RNA hydrolysate (optional)	6 ml
vitamin solution	1 ml

Titrate to pH 6.5, adjust volume to 1 liter, autoclave for 15 min at 121°C. Solidify with 15 g agar/liter.

B. *Aspergillus nidulans* Defined Minimal Nitrate Medium

20× Salts

$NaNO_3$	120 g
KCl	10.4 g
$MgSO_4 \cdot 7H_2O$	10.4 g
KH_2PO_4	30.4 g

Add ddH_2O to make 1 liter, store at room temperature. For minimal ammonium medium, substitute 74.5 g NH_4Cl for $NaNO_3$.

Trace Elements

$ZnSO_4 \cdot 7H_2O$	2.2 g
H_3BO_3	1.1 g
$MnCl_2 \cdot 4H_2O$	0.5 g
$FeSO_4 \cdot 7H_2O$	0.5 g
$CoCl_2 \cdot 5H_2O$	0.16 g
$CuSO_4 \cdot 5H_2O$	0.16 g
$(NH_4)_6Mo_7O_{24} \cdot 4H_2O$	0.11 g
Na_4EDTA	5.0 g

Add the solids, in order, to 80 ml of ddH_2O, dissolving each completely before adding the next. Heat the solution to boiling, cool to 60°C, adjust the pH to 6.5–6.8 with KOH pellets. Cool to room temperature and adjust volume to 100 ml with ddH_2O.

Medium

20× salt solution	50 ml
trace elements	1 ml
D-glucose	10 g

Adjust to pH 6.5, add ddH_2O to 1 liter, autoclave 15 min at 121°C. Solidify with 15 g agar/liter.

III. *CEPHALOSPORIUM* AND *PENICILLIUM* MEDIA FORMULATIONS[2,3]

A. Complex Medium

For routine propagation of *Cephalosporium acremonium* and *Penicillium chrysogenum.*

TSB Agar Plates

BBL trypticase soy agar	40 g/liter

TSB Broth

BBL trypticase soy broth	30 g/liter

Both TSB-based media are available from BBL Microbiology Systems, Becton-Dickinson and Co., Cockeysville, Maryland 21030.

B. Minimal Medium

From cultivation of both *P. chrysogenum* and *C. acremonium.*

Czapek-dox Broth — Agar Plates

Czapek-dox broth	35 g/liter

For agar plates, add 10 g/L Difco agar. Czapek-dox broth can be obtained commercially from Difco Laboratories, Detroit, Michigan 48232.

C. Media Used for Production of Penicillin V *(P. chrysogenum)*

Fermentation Medium

whey	7.1%
lactose	0.54%
corn-steep liquor	4.0%

[2] S. W. Queener and R. Schwartz (1979). Penicillins: Biosynthetic and semisynthetic. *In* "Secondary Products of Metabolism" (A. H. Rose, ed.), pp. 35–122. Academic Press, London.

[3] S. W. Queener, S. Wilkerson, D. R. Tunin, J. P. McDermott, J. L. Chapman, C. Nash, C. Platt, and J. Westpheling (1984). Cephalosporin C fermentation: Biochemical and regulatory aspects of sulfur metabolism. *In* "Biotechnology of Industrial Antibiotics" (E. J. VanDamme, ed.), pp. 141–170. Marcel Dekker, New York.

K_2HPO_4 (anhydrous)	0.7%
$MgSO_4 \cdot 7H_2O$	0.3%
$CaCO_3$	1.0%
peanut oil	0.3%
phenoxyacetic acid	0.3%

Adjust pH of medium to 5.4 with sodium carbonate. The amount of phenoxyacetic acid incorporated is increased for high-producing strains.

Vegetative Medium

Growth prior to inoculation into fermentation medium is termed vegetative growth (simply building up cell mass). For *P. chrysogenum,* the fermentation medium is used with the omission of phenoxyacetic acid.

D. Medium Used for Production of Cephalosporin C *(C. acremonium)*

Fermentation Medium

lard oil	10%
beet molasses	3.9%
peanut meal	3.9%
fish meal	3.6%
$CaCO_3$	0.65%
$Ca(OH)_2$	0.1%
DL-methionine or $(NH_4)_2SO_4$	1.0%

DL-Methionine is used for strains early in the lineage of ATCC11550.

Vegetative Medium

corn meal	2.0%
soy flour	1.5%
$CaCO_3$	0.3%
$(NH_4)_2SO_4$	0.1%
methyl oleate	2.0%

With or without DL-methionine at 0.5%. Temperature of incubation is in the range of 25–27°C. Liquid cultures consist of 100 ml broth in a 500 ml Erlenmeyer flask. Inoculated cultures are shaken at 285 rpm (2″ throw).

IV. *COPRINUS CINEREUS* MEDIA FORMULATIONS

A. Complete Media[4]

Bacto yeast extract (Difco)	4 g
Bacto malt extract (Difco)	10 g
dextrose	4 g
Bacto agar (Difco)[5]	15 g
distilled water	1 liter

B. Minimal Media[6]

asparagine	1 g
Bacto agar (Difco)	7.5 g
$MgSO_4 \cdot 7H_2O$	0.06 g
supplements, as required[7]	
distilled water	500 ml final volume (adjust after adding dextrose)
40× NCM salts[8]	12.5 ml
1 mg/ml thiamine	0.5 ml
dextrose[9]	5 g

[4] P. S. Rao and D. J. Niederpruem (1969). Carbohydrate metabolism during morphogenesis of *Coprinus lagopus* (*sensu* Buller). *J. Bacteriol.* **100,** 1222–1228.

[5] Fruiting of *C. cinereus* is inhibited by some types of agar (fruiting on BiTek agar [Difco] is particularly poor). Fruiting on some cheaper grades of agar can be improved by bringing the agar into suspension in distilled water, allowing it to settle, and decanting the supernatant.

[6] D. Moore (1969). Sources of carbon and energy used by *Coprinus lagopus sensu* Buller. *J. Gen. Microbiol.* **58,** 49–56. This represents a simplification of the media suggested by L. Fries (Factors promoting growth of *Coprinus fimetarius* (L.) under high temperature conditions. *Physiologia Plantarum* **6,** 551–563).

[7] Supplements, when required, are added as follows: amino acids, 100 mg/liter; purines and pyrimidines, 50 mg/liter; vitamins, 2 mg/liter.

[8] 40× NCM salts:

NH_4Cl	60 g
$Na_2HPO_4 \cdot 7H_2O$	109.6 g
KH_2PO_4	54 g
Na_2SO_4	11.6 g
Distilled water	1 liter

[9] To avoid caramelizing the sugar, autoclave separately in 50 ml distilled H_2O.

V. *MUCOR ROUXII* MEDIA FORMULATIONS

A. YPG, Yeast Extract–Peptone–Glucose Medium

Difco yeast extract	0.3%
Difco Bacto-peptone	1.0%
glucose, A.R.	2.0%

The ingredients are dissolved in water and the pH adjusted to 4.5–4.8 with dilute sulfuric acid. The medium is sterilized by autoclaving.[10] For solid YPG-agar medium, add 2–3% agar.

Culture Maintenance

The fungus can be maintained by periodic (monthly) transfers on YPG agar slants. For long-term preservation, Perkins'[11] silica gel method is highly recommended; we have maintained sporangiospores of *M. rouxii* alive for nearly 20 years.

Spore Inoculum

Abundant spore suspensions are conveniently harvested from 3–4-day-old solid cultures in 300-ml Roux bottles (50 ml YPG agar, incubated horizontally at 23–28°C). Spores are harvested by adding 10–20 ml sterile distilled water, and gently rubbing the culture mat with a sterile pipet. The spores are washed once with distilled water. Spore suspensions can be stored at 4°C and used within 2 weeks of harvest.

Liquid Cultures

To harvest large amounts of cells, the fungus is cultivated in 2-liter flasks containing 650 ml liquid YPG medium. Cultures are inoculated with a spore suspension (final spore density: $5-10 \times 10^5$ spores/ml) and incubated for 12 hr or longer at 28°C on a reciprocating shaker bath at 100–200 strokes/min. The type of dimorphic development is determined by the atmosphere of incubation.[10,12,13]

[10] S. Bartnicki-Garcia and W. J. Nickerson (1962a). Induction of yeast-like development in *Mucor* by carbon dioxide. *J. Bacteriol.* **84,** 829–840.

[11] D. D. Perkins (1962). Preservation of *Neurospora* stock cultures with anhydrous silica gel. *Can. J. Microbiol.* **8,** 591–594.

[12] S. Bartnicki-Garcia and W. J. Nickerson (1962b). Nutrition, growth and morphogenesis of *Mucor rouxii. J. Bacteriol.* **84,** 841–858.

[13] S. Bartnicki-Garcia (1968). Control of dimorphism in *Mucor* by hexoses: Inhibition of hyphal morphogenesis. *J. Bacteriol.* **96,** 1586–1594.

Mycelial Form

This morphology is typical of aerobic cultures. Culture flasks are fitted with cotton plugs and incubated with agitation, as just described.

Yeast Form

Yeast cells can be grown in pure form by incubation under an anaerobic atmosphere with a high pCO_2. This is accomplished by fitting the flasks with rubber stoppers and glass connectors, and bubbling a stream of N_2 and CO_2 (70 : 30 v/v) through the cultures during the incubation period.[10]

VI. *NEUROSPORA* MEDIA FORMULATIONS

A. Minimal Medium N: 50× Strength Stock[14]

WARNING: Make certain that everything is dissolved before adding the next component. (Conveniently prepared in a large Erlenmeyer flask with a magnetic stirrer.)

Dissolve successively:

Na_3 citrate · 5½H_2O (in 750 ml water)	150 g
(or Na_3 citrate · $2H_2O$ in 775 ml water)	(125 g)
KH_2PO_4, anhydrous	250 g
NH_4NO_3, anhydrous	100 g
$MgSO_4 \cdot 7H_2O$	10 g
$CaCl_2 \cdot 2H_2O$ (dissolve separately in 20 ml water and add the solution slowly)	5 g
biotin stock solution	2.5 ml
trace element solution	5.0 ml

Moderate heating is useful in speeding solution of the citrate and phosphate. Medium pH (single-strength) is about 5.8. No adjustment necessary. Add about 5 ml chloroform as preservative and store the 50× stock solution at room temperature. Single-strength medium is autoclaved after adding sucrose (1 or 1.5%) and agar (1.5%). [See Davis and DeSerres (1970), below (H), for Fries and Nitrate minimal media.]

[14] H. J. Vogel (1964). Distribution of lysine pathways among fungi: Evolutionary implications. *Am. Nature* **98**, 435–446.

B. Synthetic Cross Medium (SC): 2× Strength Stock[15]

Dissolve successively in 3 liters water:

KNO_3	6.0 g
K_2HPO_4	4.2 g
(or $K_2HPO_4 \cdot 3H_2O$)	(5.49 g)
KH_2PO_4, anhydrous	3.0 g
$MgSO_4 \cdot 7H_2O$	3.0 g
NaCl	0.6 g
$CaCl_2 \cdot 2H_2O$ (dissolve separately before adding)	0.6 g
biotin stock solution	0.3 ml
trace element solution	0.6 ml

pH (single-strength) is about 6.5. No adjustment necessary. Add about 2 ml chloroform as preservative and store the 2× stock solution at 5°C. A fine precipitate forms in 2× stock which should be resuspended before sampling. Single-strength medium is autoclaved after adding sucrose (1%) or filter paper (10 × 70 mm strip per 10 ml tube) and agar (1.5 or 2%). Use of filter paper as carbon source allows some strains to be crossed that fail to produce perithecia on SC plus sucrose.

The concentration of amino acid supplements should be minimized because amino nitrogen inhibits crossing. By preference, use amino acid prototrophs as female parents on unsupplemented medium. Supplementation can be avoided by using an auxotroph as fertilizing parent or by combining it with an inactive-mating-type helper strain to form a prototrophic heterokaryon.[16] [See Russo *et al.* (1985) for a modification of Minimal Medium N that can be used for crossing and that can be made up as a 50× stock solution.]

C. Trace Element Solution

In 95 ml water, dissolve successively:

citric acid · $1H_2O$	5.0 g
$ZnSO_4 \cdot 7H_2O$	5.0 g
$Fe(NH_4)_2(SO_4)_2 \cdot 6H_2O$	1.0 g
$CuSO_4 \cdot 5H_2O$	0.25 g

[15] M. Westergaard and H. K. Mitchell (1947). *Neurospora* V. A synthetic medium favoring sexual reproduction. *Am. J. Bot.* **34,** 573–577.

[16] D. D. Perkins (1984). Advantages of using the inactive-mating-type a^{m1} strain as a helper component in heterokaryons. *Neurospora Newsl.* **31,** 41–42.

$MnSO_4 \cdot 1H_2O$	0.05 g
H_3BO_3, anhydrous	0.05 g
$Na_2MoO_4 \cdot 2H_2O$	0.05 g

Total volume is about 100 ml; store at 5°C. Any trace element salt solution may be used that provides roughly the following amounts (mg/liter of IX medium) Zn, 2.0; Fe, 0.2; Cu, 0.1; Mn, 0.02; B, 0.01; and Mo, 0.02.

D. Biotin Stock Solution

Dissolve 5 mg biotin in 100 ml of 50% ethanol.

Conveniently stored at −20°C in 2.5 ml aliquots. Because biotin deficiency has been reported to result in abnormal asci, the concentration has been increased tenfold in some laboratories.

E. Complex Media

There are many variations of so-called "complete" media. The following formula, used for Difco Laboratories "*Neurospora* Culture Agar" (B321) supports growth of almost all auxotrophs (except tyrosine).

Dissolve in 1 liter water:

yeast extract	5 g
protease peptone	5 g
maltose	40 g
agar	15 g

This medium has the disadvantage that conidiation is generally poor. Glycerol Complete Medium (Medium 2)[17] gives good conidiation. This is prepared by supplementing minimal medium with yeast extract, casein hydrolysate (enzymatic), nucleic acid hydrolysate, a vitamin mix, and 0.8% glycerol. (Acid-hydrolyzed casein lacks tryptophan, whereas enzymatic digests such as N–Z–Case retain it.) Good conidiation is also obtained with the complex medium of Horowitz (1947).[18]

[17] E. L. Tatum, R. W. Barratt, N. Fries, and D. Bonner (1950). Biochemical mutant strains of *Neurospora* produced by physical and chemical treatment. *Am. J. Bot.* **37,** 38–46.

[18] N. H. Horowitz (1947). Methionine synthesis in *Neurospora*. The isolation of cystathionine. *J. Biol. Chem.* **171,** 255–264.

Dissolve in 1 liter water:

potassium tartrate ($=C_4H_4K_2O_6$)	5 g
$NaNO_3$	4 g
KH_2PO_4 (monobasic)	1 g
$MgSO_4 \cdot 7H_2O$	0.5 g
NaCl	0.1 g
$CaCl_2$	0.1 g
glycerol	16 ml
hydrolyzed casein	0.25 g
yeast extract	5 g
malt extract	5 g
agar	15 g

This medium should be supplemented with inositol (200 μg/ml) if *inl* strains are to be grown on it. Growth of some auxotrophs such as *trp* and *his* mutant strains are limited without additional supplementation. Growth of certain auxotrophs is inhibited on complete medium. (True for mutants requiring histidine, homoserine or guanine, and for some serine mutants.)

F. Sorbose Medium

Sorbose converts spreading wild-type mycelial growth into restricted colonies. Sorbose in combination with sucrose is toxic, but toxicity is overcome by using small amounts of glucose and fructose in place of sucrose.[19]

To minimal medium add:

sorbose	2%
glucose	0.05%
fructose	0.05%
agar	1.5%

Numerous variations exist. Decreasing sorbose to 1% gives less restricted colonies. 0.1% glucose may be used, omitting fructose. Colonies are more restricted at 34°C than at 25°C.

Caramelization can be avoided by autoclaving the sugars separately from the mineral salts and agar, and combining them after autoclaving. Caramelization can also be minimized by replacing Medium N with Synthetic

[19] H. E. Brockman and F. J. DeSerres (1963). "Sorbose toxicity" in *Neurospora*. *Am. J. Bot.* **50,** 709–714.

Cross Medium, which lacks NH_4 (T. L. Legerton, personal communication).

To obtain colonies at a uniform level in plates, prelayer with about 10 ml as a sterile bottom layer, then add 5 ml as a second layer that contains the conidia or ascospores.

Conidiation occurs when growth of colonies reaches the surface, and the conidia may scatter, resulting in secondary colonies. Conidiation can be delayed by overlaying the inoculated agar with 3–5 ml sterile 0.75% agar medium.

G. Concentration of Growth Factors Required for Optimal Spore Germination and Growth

The following are convenient rule-of-thumb approximations for supplementation of synthetic medium before autoclaving.

Vitamins

10 μg per ml (except inositol, 200 μg per ml). Riboflavin is destroyed if solutions and media are not shielded from light.

Amino Acids

0.2–0.5 mg per ml. Cross inhibitions, due in part to competition for amino acid uptake systems, are common in *Neurospora.* [See Perkins *et al.* (1982).]

Purines and Pyrimidines

0.2–0.5 mg per ml adenine or uracil (or nucleosides), except for guanine mutants, which are inhibited by adenine. (Use 0.2 mg/ml guanosine.)

H. Useful General References for Neurospora Methods

Davis, R. H., and de Serres, F. J. (1970). Genetic and microbiological research techniques for *Neurospora crassa. Methods Enzymol.* **27A,** 79–143. (Includes media formulas.)

Fungal Genetics Newsletter (formerly Neurospora Newsletter). Fungal Genetics Stock Center, Department of Microbiology, University of Kansas Medical Center, Kansas City, Kansas 66103. (A perennial source of valuable information including current bibliographies.)

Perkins, D. D. (1986). Hints and precautions for the care, feeding and breeding of *Neurospora. Fungal Genet. Newsl.* **33,** 35–41.

Perkins, D. D., Radford, A., Newmeyer, D., and Björkman, M. (1982). Chromosomal loci of *Neurospora crassa. Microbiol. Rev.* **46,** 426–570.

Russo, V. E. A., Sommer, T., and Chambers, J. A. A. (1985). A modified Vogel's medium for crossings, mating-type tests and the isolation of female-sterile mutants of *Neurospora crassa. Neurospora Newsl.* **32,** 10–11.

Tatum, E. L., Barratt, R. W., and Cutter, V. M., Jr. (1949). Chemical induction of colonial paramorphs in *Neurospora* and *Syncephalastrum*. *Science* **109,** 509–511.

VII. *PHANEROCHAETE* MEDIA FORMULATIONS

A. Modified Vogel Medium[20]

Vogel Medium (see *Neurospora* Media)
replace biotin with thiamine

add malt extract	3%
add yeast extract	0.25%

B. Minimal Slants[20]

Modified Vogel Medium

glucose	1%
Walseth cellulose	4.5%
agar	1.5%

C. Fruiting Plates[21]

Modified Vogel medium	⅙×
Walseth cellulose	4.5%
agar	1.5%

D. Colony-Forming Medium for Basidiospores[22]

Modified Vogel medium	
sorbose	2%
agar	3%

Adjust pH to 4.8. Heat medium to 50°C. Overlay on spores diluted in 0.1% Tween 80.

[20] M. H. Gold, T. M. Cheng, and M. B. Mayfield (1982). *Appl. Environ. Microbiol.* **44,** 996–1000.

[21] M. Alic and M. H. Gold (1985). *Appl. Environ. Microbiol.* **50,** 27–30.

[22] M. Alic, C. Letzring, and M. H. Gold (1987). *Appl. Environ. Microbiol.* **53,** 1464–1469.

E. Colony-Forming Medium for Conidia[22]

Modified Vogel medium	
sorbose	4%
agar	2%

VIII. *SCHIZOPHYLLUM* MEDIA FORMULATIONS[23]

A. Complete Medium (CYM)

Per liter:

$MgSO_4 \cdot 7H_2O$	0.5 g
KH_2PO_4	0.46 g
K_2HPO_4	1.0 g
Bacto-peptone	2.0 g
Bacto-yeast extract	2.0 g
dextrose	20.0 g

B. Minimal Medium

Per liter:

$MgSO_4 \cdot 7H_2O$	0.5 g
KH_2PO_4	0.46 g
K_2HPO_4	1.0 g
dextrose	20.0 g
thiamine–HCl	120 μg
L-asparagine	2.0 g

IX. *SORDARIA MACROSPORA* MEDIA FORMULATIONS[24]

A. Complete Medium

For 1 liter:

glucose	10 g
yeast extract	3 g
KH_2PO_4	0.5 g

[23] J. R. Raper and R. M. Hoffman (1974). *Schizophyllum commune. In* "Handbook of Genetics" (R. C. King, ed.), pp. 597–626. Plenum, New York.

[24] D. Zickler, G. Leblon, V. Haedens, A. Collard, and P. Thuriaux (1984). Linkage group-chromosome correlations in *Sordaria macrospora:* Chromosome identification by three dimensional reconstruction of their synaptonemal complex. *Curr. Genet.* **8,** 57–67.

K_2HPO_4	0.6 g
$MgSO_4$	0.5 g
mineral concentrate	0.1 ml

For solid medium, 15 g of agar is added.

Mineral Concentrate

Per 100 ml solution:

ascorbic acid · $1H_2O$	5 g
$ZnSO_4 \cdot 7H_2O$	5 g
$Fe(NH_4)_2(SO_4)_2\ 6H_2O$	1 g
$CuSO_4 \cdot 5H_2O$	0.25 g
$MnSO_4 \cdot 1H_2O$	0.05 g
H_3BO_4	0.05 g
$Na_2MoO_4 \cdot 2H_2O$	0.05 g

This solution is stored at room temperature and 1 ml of chloroform is added to prevent contamination.

B. Minimal Medium

For 1 liter:

dextrin	5 g
urea	1 g
KH_2PO_4	0.25 g
K_2HPO_4	0.3 g
$MgSO_4$	0.25 g
biotin	0.1 mg
mineral concentrate	0.1 ml

For solid medium, 15 g of agar is added.

C. Spore Germination Medium

0.44% ammonium acetate either in minimal medium without urea or in a 1.5% solution of Bactopeptone.

D. Restrict-Growth Medium

Restrict-growth medium has the same composition as complete medium except that glucose is replaced by sorbose (10 g/liter). It allows an easy screening for protoplast analysis because of the restricted colonial growth on this medium. Both complete and minimal media allow growth and fruiting of the wild-type strains.

ACKNOWLEDGMENTS

We thank M. Alic, S. Bartnicki-Garcia, C. Novotny, D. Perkins, P. Pukkila, P. Skatrud, W. Timberlake, and D. Zickler for their help in gathering this information.

Appendix C

Genetic Coding Conventions

LINDA L. LASURE
Panlabs, Incorporated
Bothell, Washington

J. W. BENNETT
Department of Cell and Molecular Biology
Tulane University
New Orleans, Louisiana

DNA sequence information is given in the letters A, C, G, and T. RNA sequence information is the same as that for DNA except that U is substituted for T. In addition, the ambiguity codes used are: N = A, C, G, T, or U; R = a purine; and Y = a pyrimidine.

The genetic code is usually portrayed in a 16 block "checkerboard." The codons are given as RNA triplets that represent amino acids (sense) and termination signals (nonsense). Most of the sequence data in this volume are presented as deduced amino acid sequences, in which the notation for different amino acids is given as a single letter of the alphabet. Fig. 1 shows the traditional rendition of the genetic code, with the addition of the single letter alphabetical conventions for amino acids. Fig. 2 is a different format of the genetic code, in which the amino acids are given in alphabetical order, followed by their respective codons. See *J. Biol. Chem.* (1983) **260,** 14–42, for more details on nomenclature and symbolism for amino acids.

MORE GENE MANIPULATIONS
IN FUNGI

First Position	Second Position: U	C	A	G	Third Position
U	UUU] Phe F UUC UUA] Leu L UUG	UCU] Ser S UCC UCA UCG	UAU] Tyr Y UAC UAA *Stop* UAG *Stop*	UGU] Cys C UGC UGA *Stop* UGG Trp W	U C A G
C	CUU] Leu L CUC CUA CUG	CCU] Pro P CCC CCA CCG	CAU] His H CAC CAA] Gln Q CAG	CGU] Arg R CGC CGA CGG	U C A G
A	AUU] Ile I AUC AUA AUG Met M	ACU] Thr T ACC ACA ACG	AAU] Asn N AAC AAA] Lys K AAG	AGU] Ser S AGC AGA] Arg R AGG	U C A G
G	GUU] Val V GUC GUA GUG	GCU] Ala A GCC GCA GCG	GAU] Asp D GAC GAA] Glu E GAG	GGU] Gly G GGC GGA GGG	U C A G

Fig. 1. Genetic code "checkerboard."

A	R	N	D	C	Q	E	G	H	I	L	K	M	F	P	S	T	W	Y	V
Ala	Arg	Asn	Asp	Cys	Gln	Glu	Gly	His	Ile	Leu	Lys	Met	Phe	Pro	Ser	Thr	Trp	Tyr	Val
5′ GCA	CGA	AAC	GAC	UGC	CAA	GAA	GGA	CAC	AUA	CUA	AAA	AUG	UUC	CCA	UCA	ACA	UGG	UAC	GUA 3′
C	C	U	U	U	G	G	C	U	C	C	G		U	C	C	C		U	C
G	G						G		U	G				G	G	G			G
U	U						U			U				U	U	U			U
	or									or					or				
	AGA									UUA					AGC				
	G									G					U				

Termination Signals
UAA (Ochre)
UAG (Amber)
UGA (Opal)

Fig. 2. Alphabetical listing of amino acid representations.

Index

D

I

K

L

Q

R